DESIGN
ENGINEERING

A Manual for Enhanced Creativity

DESIGN ENGINEERING

A Manual for Enhanced Creativity

W. Ernst Eder
Stanislav Hosnedl

CRC Press
Taylor & Francis Group
Boca Raton London New York

CRC Press is an imprint of the
Taylor & Francis Group, an **informa** business

CRC Press
Taylor & Francis Group
6000 Broken Sound Parkway NW, Suite 300
Boca Raton, FL 33487-2742

First issued in paperback 2019

ISBN-13: 978-1-4200-4765-3 (hbk)
ISBN-13: 978-0-367-38887-4 (pbk)

Library of Congress Cataloging-in-Publication Data

Eder, W. E. (Wolfgang Ernst)
    Design engineering : a manual for enhanced creativity / W. Ernst Eder and Stanislav Hosnedl.
      p. cm.
    "A CRC title."
    Includes bibliographical references and index.
    ISBN-13: 978-1-4200-4765-3 (alk. paper)
    ISBN-10: 1-4200-4765-5 (alk. paper)
    1. Engineering design--Handbooks, manuals, etc. I. Hosnedl, Stanislav. II. Title.

TA174.E253 2007
620'.0042--dc22   2007012204

Visit the Taylor & Francis Web site at
http://www.taylorandfrancis.com

and the CRC Press Web site at
http://www.crcpress.com

Dedication

Dedicated to Dr.-Ing. Dr.h.c. Vladimir Hubka, b. March 29, 1924, d. October 29, 2006, whose initiative and pioneering work from the 1960s onward made this book possible. The authors are grateful for the extensive mentoring and lively discussions with Dr. Hubka.

The quiet came and stole away . . .
no darkened urn, no dusty shelf,
no deepest depth of sod interred—
the knowledge left entwined with light—
the written word!
  E.R. Savage, *He who knows the rose: poems,*
  Hatherley Lane Publishing, 2000

The world we have created today, as a result of our thinking thus far, has problems which cannot be solved by thinking the way we thought when we created them. (Albert Einstein)

... an enterprising business is also an enterprising and intelligent organization embedded in a network of other enterprising organizations, in which the quest for profit, while imperative, is not the sufficient condition for success alone. Creativity, employee commitment, investor patience, and professional and trade skills are the other essential parts of the brew. (Will Hutton, "Why the wheels fell off at Rover," *Guardian Weekly*, Vol. 162, No. 13, March 23–29, 2000, p. 12.)

Thoughts without content are empty, perceptions without concepts are blind. (Emmanuel Kant, Introduction to the Transcendental Logic in *The Critique of Pure Reason.*)

Theorie ohne Praxis ist lahm, aber Praxis ohne Theorie ist blind (Theory without practice is lame, but practice without theory is blind). (Banse, G., Grunwald, A., König, W. and Ropohl, G., *Erkennen und Gestalten: Eine Theorie der Technikwissenschaften* [*Discerning and Form-Giving: A Theory of the Engineering Sciences*], Berlin: edition sigma, 2006.)

Study the science of art. Study the art of science. Develop your senses, that everything corresponds to everything else. (Attributed to Leonardo da Vinci [15.4.1452–2.5.1519])

To know how to recognize an opportunity in war, and take it, benefits you more than anything else. Nature creates few men brave, industry and training makes many. Discipline in war counts more than fury. (Niccolò Macchiavelli [1469–1527])

For "war" read "design engineering," for "brave" read "creative," for "fury" read "intuition," yet you cannot do without creativity and intuition. (Comments by W. Ernst Eder.)

The trouble with the world is that the stupid are cocksure and the intelligent full of doubt. (Bertrand Russell, 3rd Earl Russell [1872–1970] *Autobiography* 1969.)

Contents

Part I
Establishing Properties of Designed Technical
Systems 77

Part II
Knowledge Related to Engineering Design
Processes 133

Part III
Knowledge Related to Designed
Technical Systems 267

Part IV
Support for Design Engineering............................. *375*

Contents

xix

List of Figures

Preface

Only if power is exercised through accepted law can we have the luxury of collaborative resolution of conflicts.

W. Ernst Eder
2005

Designing is an activity directed toward an anticipated future goal, and an optimal product delivered in good time and at acceptable cost. Creativity requires open-mindedness: anything that makes designing more efficient, effective, rational, better directed, with a better outcome becomes a welcome addition to the information and heuristic arsenal of designers.

The purpose of this book is to propose and justify a valid, formalized general model of design procedure, especially for innovative design engineering, that is, prescribing a procedure for designing technical systems, presented for use in engineering design practice. Procedures and the model must still be adapted to the actual design situation.

Current information and knowledge about design engineering are presented as a societal and technical process operated mainly by cognitive abilities of human designers. This book contains a survey and map of information and knowledge about systematic, methodical, and intuitive design engineering, and the progressive development of the product (a technical system and/or its operational process) through stages of abstract to concrete modeling.

Methodical designing is the use of established and newly developed methods, both formalized and intuitive, within the engineering design process. Systematic designing is the strategic use of a theory, based on Engineering Design Science, to guide the design process. A combination of formalized and theory-based methods, with a systematic and methodical approach, using systems thinking, and including intuitive working, is recommended.

Attention is focused on the abstract conceptual phases of design engineering, where most of the future cost of a product is committed by explorations and decisions. Yet the more routine phases of embodiment and detail design are important, they are often causes for failures of products—"the devil lies in the detail" [458].

This book is intended for practicing engineering designers in any branch of engineering, especially those involved in innovation projects; planners; managers of design engineering; design teams or project leaders; product planners; architects and industrial designers; teachers, instructors, and students of engineering; researchers into design engineering, to broaden their context, and so forth.

The information is intended for gradual introduction into use within organizations. Engineering designers and others can directly use this book to learn and determine their relevant steps for their particular stage of designing. Progressive successes in small steps should be the goal, leading to an expansion of scope of this information.

Engineering students should use this book during their studies in higher education, in courses or self-study. Instructors should use suitable pedagogical and didactic approaches; their preparation for the courses should include reading more theoretical works by the authors. Designing, its information, relevant knowledge, design theory, and active application including methods should be a prominent feature in all years of engineering study. Instructors should progressively introduce the theoretical and methodical knowledge about design engineering at an appropriate course level and provide, mentor, and supervise appropriate practical project work. The theory of technical system, one aspect of Engineering Design Science, should also be incorporated into all other courses, especially those dealing with the engineering sciences.

For the basic structure of this book, the system of knowledge is described by graphical models. The presented models about design engineering are interconnected, and attempt to reduce the difficulty of entry into systematic and methodical designing for engineering designers.

This book has a short central text, the Introduction, containing brief explanations of the context and scope of design engineering, and the basic systematic models and methods for designing technical systems. The concepts of this Introduction are expanded in Chapters 1 to 12. These present knowledge derived from theory, and experience on which methods are based. They show how and why systematic and methodical knowledge can be adapted to a design situation and how other methods can be integrated. They provide clear definitions and explanations of a consistent terminology, with few synonyms, close to current usage, augmented by a glossary, and indicate additional references and further reading.

The book contains justifications for the strategic and tactical procedures so that engineering designers can understand the ideas and models as paradigms, use them as arguments, and adapt them. The reader should obtain a holistic view of the conception of systematic and methodical designing, start to use the most appropriate models, and obtain a better understanding.

Designing has been a mainly personal activity with few guidelines. In trying to rationalize design engineering by practical application, the recommended procedure for studying and using this book is first to carefully read the Introduction, study its basic models, and try to concretize them for particular cases from your own practice and life:

- A transformation system, its operands, its processes and operations, its technologies, and its operators.
- A technical process, as a particular subset of a transformation process.
- A tangible technical system, as the most significant operator of a transformation process: its life cycle; its properties and their relationships, including its various possible structures; and the tasks of establishing the properties during the design process.

- A design system, including the design process and its operators: a model of a design system; a hierarchical structure of activities, including appropriate design methods; and a procedural model of designing, performing the tasks of establishing the properties in the design process from which a plan and anticipated procedure can be established.
- Engineering design science, its classification, organization, and arrangement.

Then leaf through the chapters to gain an idea of their contents and of the complete models.

Select various design problems for which you have already found solutions at some time in the past. Compare your methods of finding the solutions with the ones recommended here. First choose an appropriate problem for which this book describes the ways of proceeding to a solution. Only after you have gained some orientation, move to other problems. This should give you a first positive experience with the applications and "know-how" of the recommended tactics. Only then should you use the knowledge presented here for more complicated tasks, for example, for establishing the function structure and organ structure for subassemblies or design groups, or for developing "masters" in your specialty.

Working up to speed, while progressively introducing further parts of the procedures for rationalizing the design process, with careful study of the supplements and their graphical models, will allow you to solve your problems with these new tools and use the new procedural strategy. You will still need to concretize the general references in this book for your specialty.

The last step in the application is a complete introduction of the system of design knowledge and methods presented in this book, which requires concretizing the engineering design documents for a specialty and organization. Specific manifestations of design characteristics and properties, or specific masters must be prepared. Consider also the situation with respect to the technical means, especially computers. Many of these also need organizational measures.

Acknowledgments

This book builds on the works of WDK—*Workshop Design-Konstruktion*—an informal international association of scientists, engineers, and educators founded in 1978. A main goal of WDK was rationalizing design work by developing and applying a science about designing—Engineering Design Science—and showing its relationships to the engineering sciences, design processes, and the tangible and process objects being designed. The activities of WDK continue under "The Design Society," an international forum founded in 2000, with several special interest groups.

The first contributions of the authors of this book to solving the problems of rationalizing design engineering were made in the early 1960s. The subsequent developments [278] are characterized by "milestones" represented by the following books (see details in the References):

Mechanical Systems Design, W.E. Eder and W. Gosling, 1965.

Theorie der Maschinensysteme, V. Hubka, 1973; its second edition: *Theorie Technischer Systeme*, V. Hubka, 1984; and its English edition: *Theory of Technical Systems*, V. Hubka and W.E. Eder, 1988.

Theorie der Konstruktionsprozesse (Theory of Design Processes), V. Hubka, 1976.

Konstruktionsunterricht an Technischen Hochschulen (Education for Design Engineering at Technical Universities), V. Hubka, 1978.

WDK 1: Allgemeines Vorgehensmodell des Konstruierens, V. Hubka, 1980; its English edition: *Principles of Engineering Design*, V. Hubka, 1982; and the second English edition: *Engineering Design*, V. Hubka and W.E. Eder, 1992.

Practical Studies in Systematic Design, V. Hubka, M.M. Andreasen and W.E. Eder, 1988.

Einführung in die Konstruktionswissenschaft, V. Hubka and W.E. Eder, 1992; and its English edition: *Design Science*, V. Hubka and W.E. Eder, 1996.

WDK 21—ED—Engineering Design Education—Ausbildung der Konstrukteure—Reading, W.E. Eder, V. Hubka, A. Melezinek and S. Hosnedl, 1992.

WDK 24—EDC—Engineering Design and Creativity—Proceedings of the Workshop EDC, W.E. Eder (editor), held at Pilsen, Czech Republic, November 1995.

This path was motivated by the literature and intensive discussions. In particular, the WDK conferences—International Conference on Engineering Design (ICED), since 1981—and associated workshops, especially nine of AEDS-SIG in Pilsen, were fruitful in developing the theory, and in obtaining and incorporating new insights.

We acknowledge the contributions of many persons, participants at the ICED, colleagues, and friends too numerous to mention. Thanks are due to several people for proofreading and their comments. The remaining errors are the responsibility of the authors and the editor.

W. Ernst Eder, author and editor
Kingston, Ontario, Canada

About the Authors

W. Ernst Eder, professor emeritus, Ing. (Austria), M.Sc. (Wales), Dr.h.c. (University of West Bohemia) was born in Austria in 1930. He was educated in both England and Austria. He graduated from the primary Advanced Technical College in Austria (TGM—Technologisches Gewerbe-Museum, Vienna) in 1951, and was awarded the state-registered title of "Ingenieur" in 1957. Ten years of industrial experience in both countries consisted of employment in design offices of organizations producing alpine forestry equipment, power transformers and switchgear, and steel processing plant.

His first academic appointment was at the University College of Swansea in 1961, where he started his involvement in teaching engineering design. In 1968 he was admitted to the degree of M.Sc. in Engineering from the University of Wales. Further appointments took him to the University of Calgary (1968–1977), Loughborough University of Technology (1977–1981), and the Royal Military College of Canada, Kingston, Ontario (1981–2000). Even though he retired from full-time employment, he is still heavily involved in design teaching at RMC.

Ernst Eder has attained an international reputation in systematic design. He has published more than 130 papers on design methodology and on engineering education, and has coauthored or edited 13 technical books on the subject of theory for design engineering and the design process, including the *Proceedings for the International Conference on Engineering Design*, ICED 87 Boston. He was a founder member of the Design Research Society (UK) in 1966, and of The Design Society in 2000. He was on the editorial board of three journals, and still acts on one.

His other interests include correct use of and possible improvements in the International System of Units (SI), and knowledge engineering, including applications of microcomputers. His extracurricular activities embrace Scottish Country Dancing (including editing eight books of dances, several of which he devised), golf, bridge, walking, photography, and others.

Ernst joined the American Society for Engineering Education (ASEE) in 1968, has since then been a member of the Educational Research and Methods (ERM) Division, and has been active in the Design in Engineering Education Division (DEED) for several years. He received the 1990 ASEE Fred Merryfield Design Award. In 1997, the Ernst Eder Young Faculty Award was instituted by DEED to be given for the best paper from young faculty presented at the ASEE Annual Conference. He is now a life member of ASEE.

In November 2005 he was admitted to the degree of Doctor *honoris causa* at the University of West Bohemia in Pilsen, Czech Republic.

Stanislav Hosnedl, professor, Ing., C.Sc., Doc. was born in 1942 in Czechoslovakia. He graduated (Ing.) in design engineering at the University of West Bohemia in Pilsen, Czech Republic, in 1964 in the field of machine tool design. He was employed in the engineering design and research departments in SKODA Concern Enterprises in Pilsen until 1990. He received his first doctorate degree (C.Sc.) at the University of West Bohemia in 1984 with the thesis "Complex Function Analysis of Driving Mechanisms of Machine Tools Using Computers." Since 1990 he has been a lecturer in the Department of Machine Design at the Faculty of Mechanical Engineering of the University of West Bohemia in Pilsen. From 1990 until 2000 he was the head of the Department of Machine Design with a staff of 30. He received his second doctorate (Doc.) here in 1992 with the thesis "Computer Integrated Design Using Feature Based Machine Elements and Parts." From 2001 to fall 2006 he was appointed Vice Dean of the Faculty of Mechanical Engineering.

His teaching, research, and industrial activities concern design engineering of machine elements, theory of technical systems, methodics of knowledge-integrated design engineering, and implementation of knowledge of engineering design science into CAD systems, machine elements, and manufacturing and transporting machines. He teaches both undergraduate and postgraduate students including those from other Czech and foreign universities, and has organized several courses for practicing engineers in major industries. He has supervised several completed Ph.D. research projects, and is currently supervising students involved in the above topics.

Dr. Hosnedl is the author or main coauthor of two books; has two patents on inventions, more than 110 papers including more than 40 in foreign publications, more than 100 research reports; and is the main author of a large software package used in industrial practice and education since 1980. He is also coauthor of more than 30 realized engineering design projects of heavy machine tools utilized in industrial companies all over the world. He has led or participated in leading more than 20 research projects and over 11 teaching-related projects. He was the author and manager of "Courses for Practicing Engineering Designers" approved by the Czech Ministry of Education, and chief manager of the state research plan (1999–2004). He has been chairman, cochairman, or member of more than 35 international and 10 national conferences and workshops on design engineering.

He has been a scientific committee member of numerous national and international conferences. Dr. Hosnedl was cochairman of the steering committee of the "National Office of the European Union programme TEMPUS" (Prague, 1994–1996), member of the steering committee of the "National Society for Machine Tools" (Prague, 1994–2000), and advisory board member of the "International Society WDK" (Zurich secretariat, 1990–2000). He is a member of "UNESCO International Centre for Engineering Education—UICEE" (Melbourne secretariat, since 1999), a founding member of the successor to the WDK Society, "The Design Society," a worldwide community (Glasgow secretariat, 2000) and a member of its advisory board (since 2000), an editorial board member of the *Journal of Engineering Design* (London, since 2000), an advisory board member of "Design Education Special

Interest Group—DESIG" of the Design Society (Glasgow secretariat, since 2003), a chairman of the "Applied Engineering Design Science Special Interest Group—AEDS-SIG" of the Design Society (Pilsen secretariat, since 2003), and a member of the Czech Association of Mechanical Engineers (Prague secretariat, since 1992).

List of Symbols

Symbols may be combined in useful ways to express the concepts, for example, FuDtPr = functionally determined properties. Figure-specific symbols are used in several figures; see legends.

A	active and reactive—for environment
Ac	action
AI	artificial intelligence
Ass	assisting
Au	auxiliary
B	building
C	constructional
CAD	computer-aided design
CAE	computer-aided engineering
CAM	computer-aided manufacture
CIM	computer-integrated manufacture
Cn	connecting
COTS	commercial off-the-shelf products
CPM	critical path method
CPU	central processor unit
Des	design
Desr	designer
DFMA	design for manufacture and assembly
DfX	design for X
Di	distribution
Dim	definitive (dimensional)—for layout
Dt	determined
E	energy
Ec	electrical
EDS	engineering design science
Ef	effect
Eff	effector
El	elemental
Em	element
Eng	engineering
Env	environment
Est	esteem

Ev	evoked
Expt	experiment
F	factor
Fb	feedback
FD	design-internal factors of the design situation
FE	environment factors of the design situation
FEM	finite element method
FMECA	failure mode, effects and criticality analysis
FO	organization factors of the design situation
FT	task factors of the design situation
Fu	function
G	general
H	design operation hierarchy
HKB	herstell Kosten Berechnung (manufacturing cost calculation)
Hu	human
I	information
IDEF	standard for integration definition for function modeling
InvMach	invention machine; see TIPS
IPD	integrated product development
L	living things, animal, vegetable, and so forth
LC	life cycle
Lf	life
Liq	liquidation
M	material
Md	model
Me	means
ME	machine element
Mfg	manufacturing
Mgt	management
Mn	main
Mo	mode—for example, of action
MS	machine system
No	quantity
O	originality
Od	operand
OEM	original equipment manufacturers
Op	operation
Opp	operational
Org	organ
Orgsm	organism
Ot	operator
P	process, or design step, stage
Pa	partial
Pc	principle
PDM	product data model/management
PERT	program evaluation review technique

Pl	planning
Pr	property
Prd	production
Prel	preliminary—for layout
Prep	preparation
Pro	propelling
Pu	purpose
QFD	quality function deployment
RAM	random access memory
RC	regulating and controlling
Rec	receptor
Rl	relationship
(s)	as subject
S	system
Sci	science
SI	le systeme international d'unites
Sit	situation
Simul	simulation
Str	structure
Tg	technology
TIPS	theory of inventive problem solving
TP	technical process
TrB	trans-boundary
TQM	total quality management
TRIZ	see TIPS
Trf	transformation
TS	technical system
TS-"sort"	sort of technical system
TTS	theory of technical systems
VA/VE	value analysis/value engineering
W	working—for means
Wt	weighting factor
Σ	sum, collection or aggregate, for example, of operators, effects, operands, properties, and so forth

▭	process
◇	decision process
▱ △	property (including function)
▢ ⬭ ○	operator (e.g., technical system)

$\longrightarrow$ effect (output of operator, input to operand of process)

$\longrightarrow$ input to/output from process
               input to/ouput from operator

               operator that delivers effect(s) to an operand

Introduction

I.1 INTRODUCTION

The highest human achievement is "creating something" with potential benefit for mankind [146,152]. Examples are artistic works, aesthetic expression, goods and services, artifacts and processes from **designing** and/or engineering, scientific **knowledge** from research, and so forth. Design engineering is solving technical problems, finding suitable and preferably optimal solutions for the given task.

NOTE: Additional comments are given in NOTES to clarify some aspects. References are numbered in square brackets [], and listed in alphabetical order of first author at the end of this book. Entries in the glossary are marked by **bold** at their first appearance.

The subject of this book is designing products: technical processes and **technical systems**, with a substantial contribution from engineering. Designing must account for organizational, economic, cultural, societal, climatic and other factors, and consider **hazards** and **dangers**, and their **risks**.

Most **organizations**, even "not-for-profit" groups, need to cover their fixed and variable costs. Organizations generally offer goods and services to global or local potential **customers**, to generate income and surplus funds. Goods and services try to satisfy the needs expressed or anticipated by an interest **group** [388,389], **stakeholders**, legislators, and so forth.

Goods and services comprise the products of an organization, including artifacts, and energy. A "product" is defined as a "result of any process" [10], classified as: (1) *hardware*: a tangible, material object, with countable quantity; (2) *software*: information and an intangible object (an insurance policy, a law, a computer program, and so forth); (3) *service*: the intangible result of an activity performed at the interface between a supplier and a customer; this includes provision of electricity, water, fuel, transportation, garbage removal, policing, wholesale and retail, advertising, delivery of information, providing ambience, and so forth; and (4) *processed material*: a solid, liquid, or gaseous (bulk) material that can be measured in units of volume, mass, energy, and so forth, for example, plastic pellets, fuel, grease, coolant liquid. The classification used in this book is (A) living, (B) inanimate material, (C) energy, and (D) information (L, M, E, I), and "I" includes signals, commands, and so forth. Engineering products are also differentiated from others.

Goods are manufactured, that is, the result of manufacturing is a product. Services include the results of transformations of tangible objects, energy and information, in their (internal) structures, (external) forms, location (in space), and time—usage, maintenance, servicing, upgrading, renewal of goods, and providing suitable energy and information. Goods and services must generally be thought out, planned, laid out, designed, before they can be available. They are purchased by customers for their own use, or for use by other people. Information generated also has "customers" or "stakeholders" inside and **outside** that organization (see Chapter 3).

A need exists for workable definitions of several terms, connected into coherent expressions.

NOTE: When trying to build a scientific discipline, or to explain phenomena or ideas, the **means** of expression must have a clearly delimited and compelling interpretation suitable for the field. Such clearly defined **terms** are classified as *termini technici* (singular: *terminus technicus*). This step is pertinent for this book (see Section 12.2). Several definitions are included in the Glossary. Some of the basic **sciences**, for example, cybernetics and **systems theory**, supply important inputs for Engineering Design Science (EDS), yet an agreed terminology is lacking. The aim is toward **precision**, comprehensive understanding, and up-to-date completeness as far as possible. Various terms are accompanied by symbols or abbreviations for formalization, and shorthand notation, and are listed.

A science about the subject of design engineering (or engineering design, expressions often used interchangeably) is developed in stages, beginning as follows:

Science investigates existing phenomena to obtain knowledge (see Sections 12.3 to 12.5). A science is a system of knowledge, a structure that defines the elements and relationships, and a suitable arrangement of the information about a phenomenon. Knowledge (and science) is a subset of information. Science is used as accumulated systematized knowledge, especially when it relates to the physical world, and theory denotes the general principles drawn from any body of facts (as in science) [2,13].

Each science has an agreed boundary, preferably isolated so that the system is not influenced by interactions with other regions. The science contains ordering characteristics, organization, categories, systematization, codification, records of structures of knowledge, theories and hypotheses about the region and its **behaviors**, including mathematical expression where possible, preferably published and peer reviewed. Practical application is not a prior condition.

A science usually does not contain details and concrete expressions about practical applications. On the basis of the science, a theoretically useful way of classifying information from and for the practice can be shown, including information about a **manifestation** of the phenomenon that may be abstracted into the science. For instance, zoology states the taxonomy and the properties for recognizing that an animal should be given a particular (Latin) species name. That animal is not part of the science, it is a manifestation of the phenomenon that is abstracted into the science.

NOTE: Properties of a car include: power, torque (or tire force), and fuel consumption. Manifestations of fuel consumption can be: (1) mass or volume of fuel used and distance traveled by the car, with values of x 1/100 km or y mi/gal; or (2) brake specific fuel consumption (BSFC) of the motor, with values of u g/kWh or v lbm/Hp.h. Characteristics (curves on a graph) for the three properties can show: (a) BSFC in relation to power and rotational speed; (b) BSFC in relation to brake mean effective pressure and rotational speed; and (c) BSFC in relation to output torque at the clutch and rotational speed.

Research and formulation of theories is related to scientific activities, and may be classified into fundamental or applied. Fundamental research sometimes acts only for obtaining knowledge, with little view to application. Scientists in fundamental research are involved in discovering, producing, and expanding the forefront of the boundaries of their science. They are not normally involved in existing information, nor in relationships, except for teaching future researchers. Nevertheless, these individual areas incidentally require a general awareness of history and the humanities. Scientists in applied research are also involved in discovering, producing, and expanding knowledge, but aim mainly toward information for a specific application.

Research for human activities, generating knowledge and plausible scientific theories, follows four parallel paths (see Section 12.4): (1) the classical experimental, *empirical* way of independent observing, for example, by protocol studies, experiments, and so forth: describing, abstracting, modeling, and formulating hypotheses and theories—observations can only capture a small proportion of thinking, usually over short time spans; (2) *participative observation*, the observer is a member of the design **team** and takes part in the observed process, for example, [248]—observations may be biased by the observer's participation in the process; (3) the reconstructive, *detective* way of tracing past events and results by looking for clues in various places [438]—reconstructions can never fully capture the original events, human memory is limited, records of information about events are stored in many separate chunks at different locations in the brain and need to be reconstituted for recall (see also Sections I.8.8 and 11.1.1); and (4) the speculative, reflective, *philosophical* way of hypotheses, theories, modeling, and testing. In designing as a subject for research, the empirical ways include elements of self-observation, and impartial observation of experimental subjects. None of these paths can be self-sufficient, they must be coordinated to attain internal consistency and plausibility.

Art allows free expression with intent to produce items that appeal to the senses. Art, the results of artistry, is related to craft, the result from craftsmanship. The emphasis is less on utility, and more on aesthetic and sensual appeal. Arts may be separated into representational and performing arts. The activity of art appears in some form in most human activities.

Designing in contrast involves planning and executing (or having executed) an envisaged task, including writing, graphical work, products, and so forth. Art plays a role in engineering [2,13]. The scope and approach between sciences and engineering show distinct differences.

NOTE: An artificial system can be a process system or a tangible object system. These two terms need to be correctly understood:

1. A *process* (P) is a change, procedure, or course of events taking place over a period of time, in which an object transforms, or is transformed, from one state to a preferably more desirable different state, generally called a TrfP. The smallest convenient steps in a process are called *operations*. Examples are crushing seeds to extract oil, transporting a load from one place to another, building a bridge, recording information, and so forth (see Chapter 5).
2. An *object system* is a tangible, real, material entity.

3. A *technical process* (TP) is that part of a TrfP performed mainly by or with the help of outputs (effects) delivered by a technical system.
4. A TS is an object system with a substantial engineering content, which is capable of solving or eliminating a given or recognized problem, that is, providing *effects* (at a particular time) to operate a process. A TS consists of constructional parts and their relationships (see Chapters 6 and 7). Examples are a car manufacturing facility (industrial plant), a car (machine), a motor (engine) in that car (machine or assembly group, or module), a connecting rod assembly in that motor (assembly group or module), a threaded stud in the connecting rod assembly (constructional part), a building, and so forth.

Designing should be distinguished from a design process. Designing implies that humans are the only operators performing the process. Yet designers use tools, knowledge of various kinds, and external representations, and are subject to management, and their environments. Design process is more suitable for these transformation processes. Design engineering and engineering design process should be used where the product is a technical system, TS(s), and technical process, TP(s). The addition of "(s)" signifies that the TP and TS is the subject of the design process. Human cognitive abilities and skills (latent and developing) are essential, a computer alone cannot design. Some artificial intelligence (AI) techniques can almost complete designing for some TS (e.g., VLSI electronic chips).

Designing in engineering has the purpose of creating future *operating* artifacts (TS), and the **operational processes** (TP) for which they can be used, to satisfy the needs of customers, stakeholders, and users. These artifacts may be able to actively operate, or to be operated as a tool by a human being. This purpose is accomplished by designing suitable technical means (TP and TS), and producing the information needed to **realize** and **implement** a product. Designing something useful with a substantial engineering content, usually within market constraints, distinguishes engineering from scientific or artistic activity. Therefore *design engineering*, combining art, craft and science, is the activity and subject of this book.

Design engineering explores alternative solution proposals, and delivers proposals for appearance and presence, and manufacturing specifications for a designed product. Architecture, styling or industrial design are not specifically included as customers for the systematic design processes (DesP) presented in this book, except where they influence design engineering. A substantial difference exists between artistic designing and design engineering, yet both have much in common.

DesP can emphasize the *artistic elements*, external appearance, ergonomics, marketing, customer appeal and satisfaction, and other (mainly external) properties of the artifact. This includes color, line, shape, form, pattern, texture, proportion, juxtaposition, and so forth. For industrial products, this is the scope of *industrial design* [204,337,541,542]—in the English interpretation—artistic design. Louridas [395] describes such designers as bricoleurs [85] or tinkerers who collage divergent ideas to form a complex product. The task given to or chosen by the designers is usually specified in rough terms. Designing consists of conceptualizing possible future artifacts, then rendering or physical modeling to provide a "final" presentation, for management approval. The artifact can be made as a single item, or in quantity. Economic assessments are common, technical analysis is often absent. Designing is intuitive, with emphasis on "creativity" and judgment, and is used in architecture, typographic design, fine art, and so forth.

NOTE: "Intuitive" is used in this book in a wider sense than for computer–human interfaces, where it implies recognition of an icon, as distinct from keyboarded computer commands.

In contrast, design engineering emphasizes the *internal functioning*, actions, operability, functionality and life cycle, that is, a TS, and its operational process (TP). The DesP should anticipate intended and unintended usage of the TP and TS, its manufacturing processes, environment, and so forth. Designing usually proceeds from a given design brief, which may be questioned and adapted, by developing a design specification to obtain a full understanding of the problems, and to obtain criteria for selecting among possible alternative proposals—*clarifying the problem* [25,26]. When searching for candidate solutions and investigating their behavior, several abstract structural elements are available, that is, transformation processes, technologies, **functions**, **organs**, constructional parts, and others (see Sections I.7.1, I.9.1.5, I.9.1.7, I.12.7, and I.12.9, and Chapter 6). The elements from TrfP operations to organs can be used for *conceptualizing*. The hardware components in configuration and parametrization are used for *embodiment* in sketch layouts and dimensional layouts, and for *detailing* in detail and assembly drawings, parts lists, and so forth, or their computer equivalents. The elements and structures are always present, but need not be used. The engineering DesP can thus range from purely intuitive to **systematic** and **methodical**, and prototypes and test rigs may be used to verify parts or complete TP and/or TS.

In designing, many choices are open; their range of validity and appropriateness depends on the circumstances, and the person who is choosing. This is a "nondeterministic" process, the verb to **establish** is used to describe the process of generating a preferred solution, and deferring those solutions that are not considered appropriate or **optimal**. "Establishing" shows that there is a determining connection between one concept and another, but acceptable alternatives exist for selection. "Determining" implies an analytical frame of mind and a single valid result, especially for realized values to be measured. This topic is expanded below for "causality" and "finality."

Designing has been claimed as "not rational, but creative and intuitive." The 1960s revealed that design engineering can and should be rational, and much can be explained and taught, for example, [140,281]. A scientific analysis of design engineering (distinct from the scientific analysis of "designs") led to proposals for rationalizing design engineering. Rationalizing implies change. In design engineering the intuitive and the rational must cooperate (see also Section 12.1.9).

Design engineering is a complex activity, a complex progression in the development of a "best" solution to a given problem [365]. It is creative, but contains procedural and routine aspects. It must be methodical and systematic, include creative thinking, clarifying thinking, critical thinking, and many other forms, and must use existing information, including but not restricted to scientific knowledge. The information for designing lies predominantly in the collection of existing areas of knowledge and knowing. Design engineering involves some skills and abilities, and their phenomenological realization, craft. Design engineering is applied to the progress toward defining an object or process that fulfills a purpose.

New inventions and science spin-off developments must be formed within the existing information bases—many engineering developments occurred before the sciences had been formulated. For design engineering the information includes the engineering sciences, culture, societal organization, economics, market development, aesthetics, and other areas, at macro and micro levels—and general awareness is often sufficient. The relationships among the areas of information must be clear and explicit, with sufficient integration. For instance, in a thermodynamic process (chemical to mechanical energy conversion), pressure must be contained (strength, mechanics of materials), cooling is needed (heat transfer), energy must be supplied and extracted (fluid dynamics, mechanics, machine elements), and so forth. The islands of information need to be integrated into a cohesive whole. "What designers did *not* know appears as consequential in its own way as what they *did* know. . . . they didn't even know what they didn't know" [556, p. 45], and ". . . difference between science and engineering . . . may have epistemological implications. In science what you don't know about is unlikely to hurt you . . . . In engineering, however, bridges fall and airplanes crash, and what you don't know about can hurt you" [556, p. 269, Note 55] (see also Figure 12.6). "You" includes designers, users, customers, stakeholders, and society.

Engineering education and continuing learning during practice (see also [163]) should aim to achieve *competency* of engineers, technologists, technicians, and so forth, in analyzing and (more importantly) in synthesizing (designing) technical systems. This requires knowing internalized information of objects and DesP, and awareness of where to find recorded and experiential available information. Competency includes [161,456,458] the following:

1. **Heuristic**- *and practice-related competency*—ability to use experience and precedents [71], design principles [314], heuristics [365], information and values (e.g., of technical data) as initial assumptions and guidelines, and so forth.
2. *Branch- and subject-related competency*—knowledge of a TS-"sort" within which designing is expected (completed during employment); typical examples of TS-"sorts" should be included in education (i.e., in addition to conventional and newer machine elements, see Section 7.5), and should also show the engineering sciences, pragmatic information, knowledge and data [98,556], and examples of realized systems.
3. *Methods-related competency*—knowledge of and ability to use methods, following the methodical instructions under controlled conditions, and eventually learning them well enough to use them intuitively—for diagnostics, analysis, experimentation, information searching, representing (in sketches and computer models), creativity [153], innovative thinking, and systematic synthesizing [307,308,318].
4. *Systems-related competency*—ability to see beyond the immediate task, analytically/reductionistically and synthetically/holistically, to take account of the complex situation and its implications, for example, life cycle engineering [62,160,186,237,244,582], or economics.
5. *Personal and social competency*—including team work, people skills, transdisciplinary cooperation, obtaining and using advice, managing

subordinates, micro- and macroeconomics, social and environmental awareness, cultural aspects, and so forth [162]; and the associated leadership and management skills.

6. *Socioeconomic competency*—including awareness of costs, prices, returns on investment, micro- and macroeconomics, politics, entrepreneurial and business skills, and so forth.

These competencies are related to creativity (see Section 11.1.7).

Personal development concerns confidence, leadership, assertiveness, emotions, autonomy, morality, aesthetic sensibility, integrity, purpose, motivation [165], inter-personal relationships, and so forth. Time is needed for engineers to accumulate and integrate an information system (IS), including "tricks of the trade," "know-how," and "know-what" (heuristic information) about products and design methods and approaches—about 10 years to become competent for a particular TS-"sort".

"Design methods and theory can constrain a problem enough to make it comfortable to mess with. These are valuable ways to HELP solve design problems, they are not "musts," only guidelines; but beware, they can also be used as crutches to "explain" procrastination. Useful advice is to first try to solve it in QUICK AND DIRTY ways, especially for graphical work (sketches) and calculations (using very simple models), and refine later if needed" [482] to achieve safety, economy, **functionality**, and others.

Engineers need to be aware of the functioning of an organization within an economic system. *Products* must be marketable at an economic rate of return, ethically and morally acceptable, aesthetic for customers and users, ergonomic for users and maintainers, and so forth (see Chapter 3). Engineers must also protect the intellectual property of an organization (see Section 11.3.2).

If a product is intended to be visually attractive and user-friendly, its **form** (especially its external shape) is important—a task for industrial designers, architects, and similar professionals. If a product should work and fulfill a purpose (e.g., mechanical), its *function* is important—a task for engineering designers within or across the conventional engineering disciplines. If a product is to be made, its design for *manufacturability* is important—a task that involves production engineering. Other aspects of the life of a tangible product require involvement of different specialists, for example, for disposal and liquidation at the end of its life.

The *degree of novelty* in design engineering (see also Section 6.11.2) ranges between

1. *Novel designing*—likely in constructional groups, machine elements (class II complexity) for design engineering, or complete machines (class III complexity) for industrial design—"radical **technology**" [98], "radical design" [556].

2. *Redesigning* (including "reverse engineering")—for changes of functions, variants in size and performance, constructional and manufacturing alterations, **modular** adaptations, configuration tasks, or direct adoption of an existing system—"normal technology" [98], "normal design" [556].

The majority of design problems (about 95%) are tasks of redesigning. Previous experience, tacit internalized knowing and recorded information of

Objectives, design conditions	Design engineering	Artistic—architectural— industrial design
The object to be designed, or the existing (designed) object	Transformation Process and Technical System; primary: functioning, performing a task	Tangible Product; primary: appearance, functionality
Representation and analysis of the object as designed, and its "captured design intent"	Preparing for TS(s) manufacture, assembly, distribution, and so forth AI, CAD/CAM/CIM	Rendering for presentation and display, product range decisions
Design process (for the object), methodology, generating the "design intent"	Theories of designing, Engineering Design Science, formal design methodologies	Intuitive, collaborative, interactive designing
Properties of the object as output of designing (see Figure I.15)	Internal properties, to generate external properties	External properties to achieve customer satisfaction
Design phenomenology	Empirical, experimental and implementation studies	Protocol studies
Responsibilities	Professional, ethics, reliability, safety, public, legal liability, enterprise, stakeholders	Organization, stakeholders (Architecture adds organizational and contract responsibility)
Location	Design/drawing office	Studio

Literature   *(see also Figure 7.10)*
Eder, W.E. (2004a) 'Integration of Theories to Assist Practice', in *Proc. IDMME 2004 5th International Conference on Integrated Design and Manufacturing in Mechanical Engineering,* Bath, UK, 5–7, April 2004

FIGURE I.1   Characteristics of designing.

existing products and of previous DesP, is the start for many innovations—"dirty blackboards" are extensively used [48]. The published systematic models of DesP (e.g., [304,305,314,315,370,457]) attempt to lay out a complete design process for novel products, from which designers choose the portions they employ. Differences in designing can be characterized as in Figures I.1 and 7.10.

NOTE: This listing shows a contrast of extremes, rather than an assessment of all aspects of designing. Architects are responsible for the external appearance and internal arrangement of spaces. They are usually also responsible for the management of large-scale contracts, including coordinating with civil, structural, mechanical, electrical, and others. Architects are credited for successful projects, but any liability for damage or loss of life and property caused by engineering work will be charged to engineers.

For new or revised products, *designing*, thinking out, needs smaller stages of progress, in smaller sections (parts, assembly groups). They often need to be recursively subdivided into smaller "windows" [438], "form-giving zones" (see Chapter 2—operating instruction OI2.12, Sections 4.2 and 4.5.1, and Figure 2.16), to recombine selected alternative solutions.

Designing is a cognitive–conceptual processing of information, that also contains routine work, and can be supported by prescribed methods. Typical activities include (1) *analysis,* using **causality** as a premiss, and mathematical models—for example, the engineering sciences; (b) *synthesis*, using **finality** as the aim, including creativity, to find and select among candidate solutions for a TP(s) and TS(s); (c) *management* to formulate, direct, and control activities toward the goals; (d) *decision making*, and formulating the criteria for decisions; (e) *problem solving* as a detail procedure within designing; and others (see also the NOTE in Section I.11.1).

Consequently, designers usually work in *teams*, and must have adequate people skills and competencies [456,458]—related to working methods that engineers can apply. Design tasks are usually too large for one person, and the range of required information is too broad, including potential users, manufacturers, marketers, economists, and others. Specialists in these areas at times act as designers, team members, and design consultants. Designers have various levels of ability and competence—design engineering combines the work of design engineers (registered professional engineers), technologists, technicians (e.g., draftspersons), analysts, consultants, and so forth, collectively known as *engineering designers*. One aim of this book is to deliver appropriate information to enable engineering designers to perform their work more efficiently and effectively. This book therefore emphasizes design engineering, applied to a tangible object system (TS), the operational process for which the object system is used (TP), and its other life cycle processes—TP(s) and TS(s) are tools for a human purpose. *Design engineering* is used as a *terminus technicus*. The TS being designed is equally important to the processes of designing, and equally important is the context (situation) within which designing takes place. The triad of "subject–theory–method" is used as a guideline (see Figure I.2).

NOTE: As formulated in cybernetics [351], "both theory and method emerge from the phenomenon of the subject." A close relationship should exist between a *subject* (its nature as a concept or product), a basic *theory* (formal or informal, recorded or in a human mind), and a recommended *method*—the triad "subject–theory–method." The theory should describe and provide a foundation for explaining and predicting "the behavior of the (natural or artificial, process or tangible) object," as subject. The theory should be as complete and logically consistent as possible, and refer to actual and existing phenomena. The (design) method can then be derived from the theory, and take account of available experience. One aim of this book is to separate the considerations of theory from considerations of method.

In design engineering, the TP(s) and TS(s) are the subject of the theory and the method. The *theory* should answer the questions of why, when, where, how (with what means), who (for whom and by whom), with sufficient precision. The theory should support the utilized *methods*, that is, how (procedure), to what (object), for the operating subject (the process or tangible object) or the subject being operated, and for planning, designing, manufacturing, marketing, distributing, operating, liquidating, and so forth. the subject. The method should also be sufficiently well adapted to the subject, its "what" (existence), and "for what" (its anticipated and actual purpose)—see Figure 8.7. The phenomena of subject, theory, and method are of equal status. Using the convention suggested by Koen [365], underscoring the second letter of a word indicates its heuristic nature: "a m̲ethod is a p̲rescription for a̲nticipated f̲uture a̲ction, for which it is h̲euristically imperative that y̲ou a̲dapt it f̲lexibly to y̲our c̲urrent (e̲ver c̲hanging) s̲ituation"—and nearly all words in this book should have the second letters underscored.

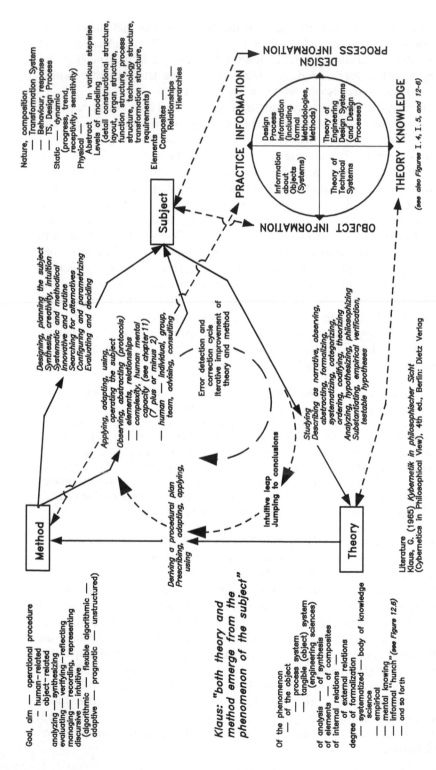

FIGURE I.2   Relationships among theory, subject, and method.

Methods are heuristic, ". . . a plausible aid or direction . . . is in the final analysis unjustified, incapable of justification, and potentially fallible" [365, p. 24]. "The *engineering method* is the use of heuristics to cause the best change in a poorly understood situation within the available resources" [365, p. 59].

The *subject* and its context needs to be explained in general terms, then interpretations and conjectures can be introduced, together with consistent terminology, the *theory*. Finally, an application of the presented information can be shown, especially the *methods* to enhance the procedures and results of design engineering. A theory can range from an incomplete mental image of circumstances and their behavior (a hunch, intuition), to a verified, comprehensive, published (recorded), accepted, formalized and codified **statement** about the phenomenon.

NOTE: In the last 30 to 40 years, many ideas and proposals have been made for improving and rationalizing design work and attaining optimal TP and TS. During this period the importance of designing has been recognized as the source of the properties of a system. The ideas and proposals cover: (1) the operation and performance (and related properties) of TP and TS, and their development during design engineering; (2) the properties and working methods of engineering designers, including cognitive abilities, knowing, experience, open-mindedness, creativity, reflection, personality, and so forth; (3) the social aspects of cooperation, *team work*, awareness, and willingness to cooperate with customers and stakeholders, internal and **external to** an organization; (4) *management* of the organization, the range of products, and the processes of design engineering; and (5) the *societal* contexts of designing, legal, economic, environmental, and other factors. Points (1) and (2) are typified by such works as Pahl [457] for TS, and their systematic development using formalized methods, and [110,111,335] with respect to the human designers and their mental processes. Design (and some "industry best practice") methods have been introduced to design engineering, with support from organization managements. In human thinking processes, point (2), **synergy** occurs by bringing together various thoughts—mental association takes place to bring about new insights that are not just the sum of the individual thoughts.

Various classes of properties are listed in ISO 9000:2000 [9, p. 26/3.5.1/Note 3]: physical, sensual, behavioral, temporal, ergonomic, and functional. Only the physical, temporal, and functional apply to TrfP and to TS. All may apply to the human system (HuS).

I.2 OUTLINE OF CONTENTS AND STRUCTURE

This book contains the most comprehensive survey to date of the **state of the art** in design engineering. Conventional prejudices need to be overcome, and some of the problems that surround designing and designers anticipated. A new region of knowledge is introduced, with a potential to improve design engineering as a process, and to improve the TS, the "designs." Information needs to be structured to be useful for engineering designers. These concepts need to be introduced into engineering practice (see Section 11.5). The optimal way of applying such information is to gradually combine the traditional ways (i.e., intuitive and routine procedures) with the concepts and methods presented in this book, and other known methods. This book, *structured* as a manual, consists of a general overview (this Introduction), and an encyclopedia of topics with more detail in the chapters. The theories and their adaptations for application in engineering practice are explained, and demonstrated on some worked examples (see Chapter 1). The applications are particularly useful for conceptualizing, with appreciation for layout and detailing, a major obligation for professional engineers. The book also shows the context of organizing and managing for design engineering, and its societal role.

NOTE: The Introduction and the chapters are based on References [147,153,287,290,299,304, 305,314,315]. Advances of the theory have been incorporated, especially in clarifying some definitions and making them more precise. This book is focused on *design engineering*, designing of technical systems (TS) and technical processes (TP), but must consider many other aspects, for example, the manufacturing processes for the TS(s), the processes that take place internal to and across the boundaries of the TS(s), marketing, economics, societal interactions, and so forth. Theoretical knowledge about design engineering is surveyed in Section I.11, and about the TP(s) and TS(s) in Section I.9.

I.3 DESIGNING

Designing is a process of *formulating a description* for an anticipated process system and/or an object system that is intended to transform an existing situation into a future situation to satisfy needs.

Typical *process systems* may be transportation processes, travel services, catering, maintenance and repair, and so forth. The process system usually is performed by engaging an *object system*, but the process must often be designed before, or simultaneously with, the object system. For instance, manufacturing processes need appropriate manufacturing machinery (TS) to perform them—the TS must be, or must have been, designed to be suitable. Typical object systems may be furniture, aircraft, plant pots, syringes, packaging, manufacturing machinery, and so forth. External appearance, light-weight construction, or internal functioning may be important aspects. Designing is human-initiated. The human as an individual or as a societal group recognizes or reacts to *deficiencies* in an existing situation, usually formulated as *requirements*. The human then defines the envisaged *goals* for a preferably improved future situation for the stakeholders, then establishes the likely *means* (TP and/or TS) to overcome the deficiency, and directs the activities toward the goals.

Designing therefore involves *anticipating* a future situation, its complexity, contexts, and consequences. The properties of the TP(s) and TS(s) need to be established, so that they probably overcome the deficiency—in the opinion of the designers, substantiated by arguments, modeling, and simulation. The results of designing cannot be fully predicted. Designers must therefore frequently *reflect* [167,506,507] on their results, use positive and constructive questioning and criticism, search for possible alternatives, and for positive and negative consequences. They must then apply corrective actions to improve the results—an example of adaptive open-loop *feedback* control. The resulting description of the future systems should be brought into a **state** as ready as possible for realizing, that is, for manufacture of the TS(s), and implementation of the TP(s). Outside the scope of designing itself are the steps of planning, realizing (manufacturing and implementing), distributing, using the TP(s) and TS(s), and final disposal, but the implications of these steps must be carefully considered.

Industry and other organizations try to achieve *continuous improvement*, relating to a product or process, and to the organization. Yet a product or process can and should only be *upgraded* at distinct time intervals, to preserve some continuity (compare Sections I.9.1.7 and I.11.9).

I.4   WHY MANUAL FOR DESIGN ENGINEERING?

This book presents information about design engineering, a process operated mainly by human designers. It is a best approach to a complete survey and map of current engineering design knowledge. The questions, intermediate goals, and methods to find suitable *means* to fulfill the requirements of the engineering design task are discussed. The book covers the progressive development of the product through various stages of modeling, from or through abstraction to concrete. Useful aims and questions to ask are shown, for each step and substep, model, and method that can be used and recommended. Flexible application of methods, adapted to the problem, is possible, necessary, and encouraged by showing the changes that are likely to be useful. This book provides a comprehensive and coordinated view, on a coherent theoretical basis, of the current state of knowledge about design engineering.

Even if there is some similarity in designing for other products, TP(s) and TS(s) have a special place, their substantial engineering content. They have a family relationship that follows stricter laws, with a more scientific basis. The design process can be more systematic, and theories, models and methods can be more applicable. Human imagination, opportunism, idiosyncrasy, and intuition are significant in solving problems (see Section I.11.1). Yet many opportunities for innovation and optimization are lost if the engineering designers do not adopt newer methods and theory-based models. There is increasingly a need to be first on the market. Maintaining the "status quo" in design methods can lead to a decline in design capabilities.

Much research has been published about humans as designers, for example [85,110,111], and about proposed design methodologies, for example [370,457,498]. Many observations have occurred, including participative [248] and reconstructive [438] investigations. Possible applications of computers in design engineering have been investigated, especially the use of AI and information technologies (IT). Philosophical, reflective, theoretical, empirical, experimental, and observational research have gathered information about designing, and especially about design engineering, since about 1930. Publications—books, papers, presentations, and so forth—include the new knowledge about design engineering. Most of this work is scattered. The current patchy distribution of that knowledge has led to the main weaknesses—a lack of unified solutions, varied terminology, different forms of presentation, and selection of nonuniform addressees for knowledge. These have deterred entry into this knowledge region.

Any science tries to find an arrangement of knowledge about a subject that is systematic, comprehensive, and logical, even if it cannot achieve the full rigor of mathematics [240] (see Section 12.3)—and even mathematics is necessarily

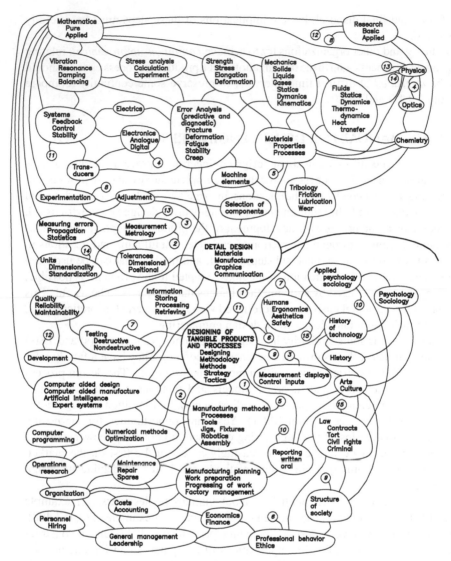

Examples of islands of object information

FIGURE I.3   Transition from scattered information to categorized knowledge.

incomplete, starting from unprovable axioms [365]. The subject, the process of design engineering and the products being designed, leads to a proposal for an EDS. Knowledge about design engineering can be categorized, codified, abstracted, systematized, and structured (see Figure I.3). The resulting form of this knowledge includes surveys, taxonomies, models, and other forms.

NOTE: The left side of Figure I.3 shows "islands" (elements) of information that influence design engineering, and some of the relationships among the islands. Engineering designers

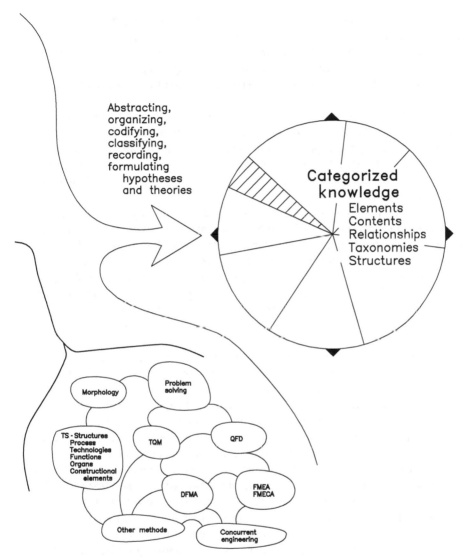

Abstracting,
organizing,
codifying,
classifying,
recording,
formulating
hypotheses
and theories

Categorized
knowledge
Elements
Contents
Relationships
Taxonomies
Structures

Problem
solving

Morphology

TS - Structures
Process
Technologies
Functions
Organs
Constructional
elements

TQM

QFD

DFMA

FMEA
FMECA

Other methods

Concurrent
engineering

Examples of islands of design process Information

FIGURE I.3 Continued.

need to understand the islands and their context (see quotation attributed to Leonardo da Vinci), some in detail, others as awareness. Current engineering education concentrates on the upper quarter of "object information."

I.4.1 CONCEPT

A combined systematic approach to designing is recommended, using appropriate methods and intuitive working. The model of an ideal systematic and methodical procedure is based on EDS [315], and supporting literature [147,153,287,290,299,304, 305,314]. *Systematic* design engineering is the heuristic-strategic use of a theory to guide the design process. *Methodical* design engineering is the heuristic use of newly developed and established methods in engineering design, including theory-based and industry-best practice, strategic and tactical, and formalized and intuitive methods.

The start of the procedural model, Figure 4.1, is an origin of a design problem, either product planning for an organization, or assignment of a problem from management to a design section. The proposed solution (preferably in manufacturable or implementable detail) is the **output** of a TrfP: design engineering. This ideal procedure needs to be adapted to the actual problem and situation (see Chapter 4). Applications are shown in Chapter 1.

A premature entry into a design problem is usual in many organizations; in most cases the problem statement "as given" is accepted without critical review—the problem is "understood" without real clarity or definition. For example, the stated requirement is for a milling machine, lawn mower, and electrical transformer. A more abstract formulation would be of transformation of raw material to a usable part, cutting grass, or changing the voltage of an electricity supply. Yet all design work should start by clarifying the design problem, especially with respect to customers' real needs and wishes, economics, environmental, and life cycle impacts, and so forth [189].

Once the problem is clarified, the majority of design problems are entered at a more concrete stage than "conceptualizing" (see Chapters 2 and 4), typically with requirements for functioning. The "problem as given," and therefore the design task deals with an existing product "line," a TS-"sort," for which conceptualizing (and the more abstract forms of modeling) can be limited, or is not used—the organ structure for a particular TS-"sort" remains constant, that is, the relevant models of (e.g.) transformation processes and main functions exist and remain unaltered.

The complete "top-down" procedure, and generating models of more abstract structures of a TP(s) and TS(s) can be important if a radically new solution should be found. The question "Why wash clothes?" may lead to the concept of disposable clothing as a radical solution for a specialized problem situation—with environmental consequences. Stepping back into the abstract, "bottom-up," can be valuable for organs—smaller sectors of a system where innovation shows promise, mainly for (recursively separated) design groups, subassemblies, or mechanisms.

One problem concerns industries—design engineering is often regarded as an accounting "overhead." Parallel to the efforts to scientifically investigate design engineering, many industrial organizations discovered the core importance of design engineering for their survival and financial health. In the English-speaking regions the emphasis was on "industrial design," correcting the appearance and ergonomics of TS. A systematic and theory-based coverage of all necessary properties, using consistent models of the concepts and methods, is still to be accepted.

Change is not easy. Unless advantages are seen, and changes are actively demanded, engineering designers have little incentive to change. Introducing the presented theoretical, methodical, heuristic, and experience information into an organization can and should be done in small steps.

I.4.2 ADVANTAGES

The system of EDS presented in this book is significant. The concept of the system is based on the triad "subject–theory–method," that is, a *theory* about a *subject* allows a *method* to be defined and heuristically applied, for using or for designing the subject (see Section I.1 and Figure I.2). The system is focused on design engineering of technical processes (TP = TS-operational process) and TS, and includes design engineering information about TP and TS, and engineering DesP. The representation of the topics is a set of generally valid and mutually interconnected graphical *models* with describing comments—see the figures in this Introduction and in the chapters. The aim of this system is a "methodical procedure," a recommended systematic and methodical procedural model, and a transparent system (map) of knowledge about and for design engineering (see Figures I.4, I.5, and 12.7, and Section I.11.6). The system is based on theory and engineering practice, supported by the industry experience of the authors, and mirrors past and current levels of design engineering. The system is open, and compatible with items of theoretical knowledge, practical design engineering, IPD and related concepts, CAD, CA, AI, and IT technologies, and their developments, and so forth.

This can significantly support the development of a system of design knowledge, including systematic concretization to any level of specialized TS-"sorts" and their developments. It supports flexible, systematic, and stepwise implementation in design

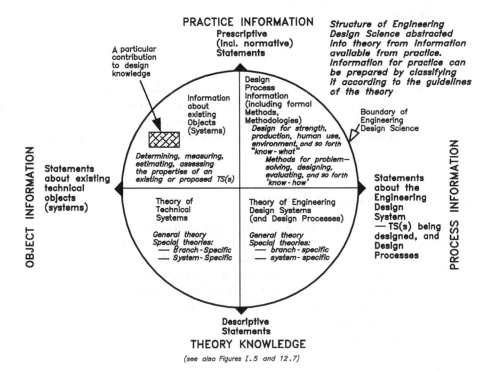

FIGURE I.4   Model (map) of engineering design science—survey.

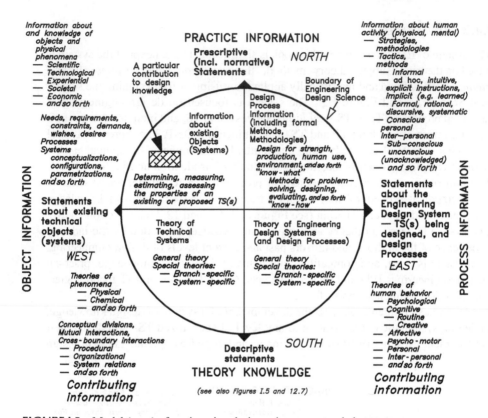

FIGURE I.5   Model (map) of engineering design science—extended survey.

engineering, research, teaching, and practice. It potentially enhances the creativity of engineers and students, by encouraging use of systematic thinking, improved "top-down" and "bottom-up" methodical approaches, and intuitive thinking—a coordinated, integrated, and flexible use of these modes is needed for all engineering creative activities.

I.5 HYPOTHESIS, THEORY, SCIENCE

Normally, the main aim of science is to study what exists, and to explain it in a generally agreed way (see Sections I.1 and 12.3). Scientists claim to proceed from observation to formulate the explanations, and isolate the phenomenon to be studied. The information obtained is abstracted and generalized. The natural and sociological sciences developed mainly from studying the phenomena from nature, societies, and humans, by observing, recognizing, perceiving, and understanding. Only after the knowledge was tested, systematized, generalized into laws and theories, and verified by experiments and by practical applications, did the sciences become effective instruments of human society and, secondarily, a source for higher levels of abstraction.

NOTE: "... sciences have grown out of the practical concerns of daily living: ... mechanics out of problems raised by architectural and military arts (*authors' interjection: 'arts' as activities of humans that result in artifacts*), ... economics out of problems of household and political management, and so on. To be sure, there have been other stimuli ..." [433, p. 8]. Kant's "Copernican Revolution" states that our minds imprint laws on nature. "Nature, to be commanded, must be obeyed," Francis Bacon, "aphorisms."

In most cases, scientists make *conjectures* about possible explanations, based on hunch, insight and feel. Experiments try to verify those conjectures, and refine them into *hypothesis*. Further refinement and reformulation based on observations lead to an agreement on the explanation, an "accepted truth," which is as complete and coherent as possible at that time—a theory.

Each science deals with a small range of phenomena—and over time the range of each science tends to narrow as further specialized sciences arise. Many sciences are regarded as "pure," with little regard for possible application, others are adapted to applications, for example, engineering sciences.

NOTE: Aims, attitudes, knowledge, and procedures of science, differ from engineering. "Science is supposed to advance by erecting hypothesis and testing them by seeking to falsify them. But it does not. ... the environmental determinists of the 1960s looked always for supporting evidence ..." [484, pp. 79–80]. "The fuel on which science runs is ignorance. Science ... must be fed logs from the forest of ignorance that surrounds us. In the process, the clearing we call knowledge expands, but the more it expands, the longer its perimeter and the more ignorance comes into view" and "The forest is more interesting than the clearing" [484, pp. 271–272].

In contrast, engineering intends to create what does not yet exist, in a form that is likely to work. Engineering, and especially designing, needs designers to be aware of the whole range of existing information and its complex interactions. Engineering designers need to take into account and accommodate all possible influences of scientific, technical, economic, societal, political, and other areas to achieve a successful and optimal designed system (see Figure I.3). What is beyond the "clearing" can hardly be used to design technical systems [165].

NOTE: In designing, a solution proposal remains a hypothesis until it is definitively accepted as an optimal solution with sufficient confidence on the basis of an evaluation (see Chapter 9). Proven experiences are valuable in helping engineering designers. Any gaps emerging through reorganization need to be closed, bring an enrichment of information and new ideas, and inspire designers. Preparing documentation is important for introducing systematic and methodical designing. Time is needed, but processing available material can achieve values that are important for design engineering.

I.6 SYSTEM

A system within a defined boundary consists of elements, and their mutual relationships, within an environment (see also Section 12.2.1). This boundary may be generally recognized, or can be defined as convenient or arbitrary for a specific purpose. Elements may be of various kinds, tangible, abstract, or conceptual. Relationships exist among elements internal to the system, and across its boundary to elements in the environment. Some parts of the environment are directly or indirectly active or reactive, other parts are remote and have little or no influence. The combination of elements and their relationships define a *structure* for the system. One set of relationships is the *arrangement* of elements relative to each other.

Any system has primary, assisting and secondary **inputs**, and primary and secondary **outputs**—while in the system, these are sometimes called throughputs. Inputs are subjected to a change of state, a *transformation* or processing to produce the outputs. Transformations usually proceed in discrete steps or continuously, and may be natural or artificial, or a combination.

Systems form a hierarchy, from simple one element systems, to compound systems, to global systems—systems of higher complexity consist of lower systems and their relationships. The behavior of a higher system is an aggregate of behaviors of lower systems, including synergies—that is, the higher system exhibits behaviors that arise through interactions of lower systems.

I.6.1 SPECIALIZED SYSTEMS FOR DESIGN ENGINEERING AND THEIR MODELS

An artificial (human-made) system can be a *process system* or an *object* (*tangible, real*) *system* (see NOTE in Section I.1). These two types are always interconnected, as shown in Figure I.6 and Chapter 5.

A clear separation is preferred, following the triad "subject–theory–method" (see Figure I.2). (1) One aspect is an actual *transformation system*, including its theory, its TrfP, Figure I.6, and typically its five operators, especially its executing operator, the TS as it exists in its final designed or its realized state—the "west" hemisphere in Figures I.4, I.5, and so forth. (2) The other aspect is *design engineering* as a system, including its theory, its processes and methods, and object-related heuristics to guide designing and developing the TrfP and TS(s), the "east" hemisphere.

Figure I.6 shows a general model of a transformation system (TrfS), with its TrfP, its technology (Tg), and its five operators. The TrfP can take place if (and only if) (1) all operators are in a state of being *operational*, they should be able to operate or be operated, if appropriate inputs are delivered to the operator, for example, the TS is able to run, it may be stationary or idling (see Section I.9.1.3); (2) an operand in state Od1 is available; and (3) both are brought together in a suitable way, that is, with an appropriate technology.

Collectively, input, throughput, and output for a TrfP is called the *operand*, which can consist of materials, energy, information, and biological matter (including living things, especially humans)—M, E, I, L. The TrfP and its operand are **topologically** "external to" the operators—that is, aspects of topology are important for this book.

NOTE: See also the NOTE in Section 5.2 concerning M, E, I, L.

An artificial transformation needs a technology (Tg), which is driven by *effects* (Ef) delivered by one or more of the *operators*. The technology (Tg) (definite article),

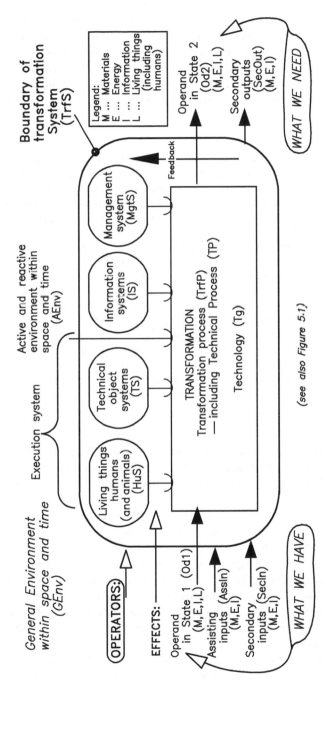

FIGURE I.6 General model of the transformation system.

describes the nature of the direct interaction between (1) a "**main** effect" as (active and reactive) output of an operator at its "main effector" with (2) an "operand," and causes the operand to be transformed. Transforming the operand is only active when a set of effects is exerted onto the operand. The operators comprise a HuS, a TS, an environment system (EnvS), an information system (IS), and a management system (MgtS). The execution system comprises HuS, TS, and AEnv. The operators interact with one another to deliver the effects, see Figure 5.1, part 3. Inputs and outputs of operators consist of materials, energy, and information (M, E, I).

NOTE: The singular, "operand," "technology," and so forth, should be understood as "a set of operands" and "a set of technologies," and so forth. The total transformation Od1 to Od2 consists of an aggregate of transformations in all operations and their synergies. Each operation in the transformation has its technology. The total technology consists of an aggregate of technologies in all operations and their synergies. The total transformation is thus a function of the total technology; see Figure 5.1, part 3.

For example: (1) A venturi is an operational TS without moving mechanical parts, it is "capable of guiding a fluid (the operand if it is present) to increase its velocity and then reduce it, and consequently to react and reduce its effective pressure and then increase it, at a mass flow rate," whether *moving* fluid is present or not. (2) A water jet as an operational TS is capable of cutting a stone (material as operand OdA) by the effect of kinetic energy and contact with a material surface (the technology TgA), if the stone is present. The water jet in this operational view is considered internal to the TSA. (a) If we now "zoom in" to a more detailed view, that water jet fulfills a TS-internal function of the TSA "form a high-speed water jet"—whether the stone is present or not. Using the function of TSA as the TrfP of TSB, the input water to the process (if present) is now Od1B, and the TSB exerts its effects to compress and deliver the water in a high-speed jet (Od2B), using the technology TgB of "sucking, transporting, pressurizing, shape forming."

The concept of the transformation system (TrfS), is not well known, most engineering industries and appropriate academic disciplines are concerned only with TS. In contrast, chemical engineering deals almost exclusively with the TrfP and TP, and is much less concerned about the TS.

Operators IS and MgtS usually act indirectly, through operators HuS and TS (and AEnv), their direct effects to the operand occurs by signals/commands to the execution system to perform their tasks, to deliver their effects. IS can act directly on an information process, and MgtS can act directly on a management process. Material and energy, as processed by a TS (or other operator), usually acts directly as a physical effect to change the operand. Information can act physically and logically to change the operand, a differentiation may be useful. The HuS can also act indirectly through the TS (e.g., "hammer;" see NOTE below), the effects delivered by the HuS are active (or reactive) at the operator-operand interface. Nevertheless, the HuS and TS (and AEnv) can still be operational and operating (or being operated), even if the operand is not present. It may be useful for a particular TrfS to differentiate "direct" and "indirect effects."

NOTE: A hammer is a TS with no moving internal parts. If held by another operator, by a human in his hand, the hammer can be swung (operated) to hit a piece of metal (an operand). The energy and control of the hammer face are supplied by the human. On impact, the hammer is decelerated (as a shock—the technology), the energy is partly converted to heat, and partly into strain energy internal to the operand metal—elastic and plastic deformation. The center of gravity of the hammer is not at the hard impact face, thus a shock wave travels through the hammer toward the handle, and is felt by the hand, as a secondary output of the TS. Physical contact of the operator "hammer" with the operand is necessary, and M, E, I are exchanged. Both energy and mass are conserved.

Inputs to the transformation system include the operand that is to be transformed in the TrfP from its initial state (Od1), main inputs to the operators, assisting inputs to the process and to the operators (AssIn), and secondary inputs to the process and to the operators (SecIn), mostly disturbances. Outputs from the transformation system include the operand in its ending state (Od2), and secondary outputs (SecOut) from the process and the operators. Some of the secondary outputs can be beneficial, some can be reused for other purposes, and some are disturbances, pollutants, and other negative influences acting on the operators, the active and reactive environment, and the general environment. Feedback usually exists from outputs (as measurements, comparisons with set points) to inputs to adjust the outputs closer to the desired states (see also Section I.12.6). That significant part of the TrfP that mainly or only needs the effects exerted and delivered by a TS is called a TP.

I.7 PRODUCTS, PROCESSES, TECHNICAL SYSTEMS—RESULTS FROM DESIGN ENGINEERING AND MANUFACTURING

Outputs of an organization process are "products," operand in state Od2, intended to provide the operating revenue for an organization (see Figure I.7), and include processed natural "produce." Among *organization products*, consideration is limited to those with a substantial engineering content, that is, TS, which perform their technical role by driving a useful TP. Nonengineering artifacts and processes are excluded, although many of the considerations will also apply to them.

Descriptions of classes of products offered in Section 6.11.10 are incomplete, not unique, boundaries are fluid and overlap. They provide a rough scale to differentiate TP and TS from other products [164]. These classifications refer to the TS-operational process, operand, technical process, technology, effects delivered by the TS, and complexity of the system. Products may appear in more than one classification. A product from one organization may be an input for another. Product classes include artistic works, consumer products, consumer durables, bulk or continuous engineering products, industry products, industrial equipment products, special purpose equipment, industrial plant, configuration products, infrastructure products, intangible products, software products, and so forth. Products may be for own interest and pleasure, purchase, consumption, application in domestic situations, industry (assembly into an organization's own product—OEM, COTS), the organization's

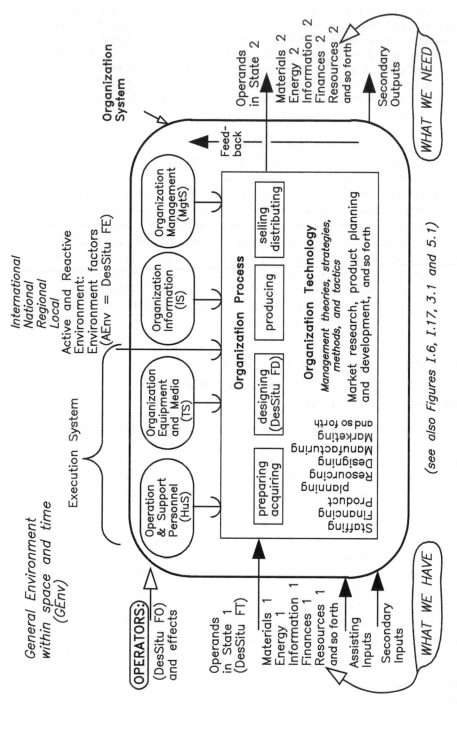

FIGURE I.7 Model of the organization system.

own manufacturing operation—plant. The theories, methods, and models in this book apply to the range of products from *consumer durables* with a substantial technical content, to *industrial plant*. They can probably be adapted to other products.

I.7.1 IMPLICATIONS AND CONTEXT

Industrial designers tend to be the primary designers for consumer products and durables, engineering designers tend to be primary for technical systems. Both kinds of designers cooperate in design teams, which should include manufacturing, sales, and other experts.

Industrial designers and engineering designers may be employed in an organization-wide process of integrated product development (IPD), working in three parallel streams: (1) marketing and sales, (2) designing, and (3) preparing for manufacture. Products from IPD are generally made in larger quantities, intended for a consumer market, and do not necessarily display an engineering content; see Section 11.4. Those IPD-products that are TS need design engineering in addition to industrial design. Some engineering products do not need industrial design. IPD and design engineering overlap, but the terms and procedures are not coincident (see Figure 7.10).

A major difference between design engineering and industrial design is the interpretation of the phase of "conceptualizing." *Industrial designers* tend to solve the problems of appearance, desirability, attractiveness, and usability. Novelty and innovation may be a strong consideration. Their conceptualizing consist mainly of preliminary sketches of external possibilities—a direct entry into hardware (the constructional structure) and its representation. The sketches are progressively refined, and eventually *rendered* (drawn and colored, and modeled by computer or intangible materials) into visually assessable presentation material, full artistic views of the proposed artifact. Considerations of the necessary engineering take place, but often at a rudimentary level. Industrial design and IPD usually work *outside inwards*, defining the envelope, thus constraining the internal actions. Presenting the results to higher management is an important part of the skills of industrial designers. Similar considerations apply to architecture. Technical problems are passed on to design engineering, the engineering designers are expected to follow the decisions of industrial design— the TP/TS solutions remain within the limitations imposed by the chosen appearance solution.

In contrast, *engineering designers* tend to solve the problems of making something work, including manufacturability and other life cycle related properties. They work from critical zones for capability of functioning, for example, form-giving zones, *inside outwards*, defining the internal operational means first, constraining the outside. Novelty may be a consideration, but reliability (control of risks), operational safety, and achievability of functioning is usually primary.

Engineering designers can conceptualize in several abstract structures of the TP(s)/TS(s), including transformation operations, technologies, functions, organs, configuration, and parametrization (see Sections I.12.6, I.12.7, I.9.1.5, I.9.1.7, and I.12.9, and Chapter 6). External appearance of the final artifact should be considered, but tends to be a result of the internal considerations of design engineering.

Nevertheless, the external form of some designed TS(s) must dictate the space available for internal arrangements, for example, aircraft or ships. Constraints for designers may come from ergonomics, and law and societal conformance. Presentation of the results to higher management is not usual, design engineering thus tends to be less visible.

A close, knowledgeable and mutually sympathetic cooperation between industrial and engineering designers is beneficial for the product and the organization. However, this book is intended to support the abilities, skills, experience, competence, creativity, and ingenuity of engineering designers by presenting relevant theory and methods.

Designers cannot grasp the whole diversity of a TP(s)/TS(s) at the same time. They can direct attention to any part of the Trf(s) or TS(s), and to any part of their mental model [435], and then dive into detail. Human working memory is strictly limited [101,416–418,420] (see Section 11.1.1). Designers need to externalize their thoughts in suitable sketches and models, and interact with them to expand the scope of their working memory. This book shows forms of sketches and models with which engineering designers can record their thoughts, and transfer between levels of abstraction.

NOTE: A model is a mental or physical representation (including its information content) of an implemented or proposed TP, a realized or proposed TS, or an idea or hypothesis [304], by suitable means, for example, sketches. Relationships between a model and the original are generally the laws of similarity. Modeling techniques have been expanded by application of computers.

Suitable similarity between the model and the original is of interest, that is, which properties are to be expressed in the model, and what purpose is to be served. A prototype of a TP/TS permits determination of most of the properties relevant to the final system. A model only permits determination of certain properties, such as behavior, structure or form, which lend their name to the appropriate model. A model always has a definite purpose, for example, determining the measures of properties, or for checking and verifying, communicating, or instructing. Aspects can be summarized by a "model of models" [469], which demonstrates four dimensions: (1) *Context* ranges in a spectrum of variety from *abstract to concrete*, from *conceptual to material*, and from *general to specific*. (2) *Function* and purpose can be one or a combination of: (a) *describing*, to explain some aspect of the model and the reality, in a theory or in a narrative; (b) *predicting*, to foresee some aspect of behavior, and to quantify it; (c) *exploring*, to investigate behavior under possible changes of circumstance; (d) *planning and designing*, to propose new or novel applications or devices; or (e) *prescribing*, as a normative or heuristic instruction. (3) *Medium* can be one or a combination of: (a) *verbal*; (b) *mathematical/symbolic*; or (c) *imagal/graphical*. (4) *Mode* of usage can be: (a) *iconic*, a reality in a two- or three-dimensional representation that is usually recognizable—drawings, space models of machines or workshops, photographs, forms of mathematical equations, and realistic verbal descriptions, where the similarity between reality and model are noticeable; (b) *similitic* (used as signs), or analog, a limited number of properties of the model are similar to the real system—static and dynamic properties of a reality can be imitated or simulated by this use, which includes graphs and diagrams, and models that rely on laws of similarity, for example, for fluid flow, electrical or magnetic fields, thermal conduction, computer simulation programs, and others; or (c) *metaphoric*, a mathematical, verbal or graphical symbol represents a context, for example, technical system by the metaphor "TS."

The relationship, especially in context and function, between a mental model and the modeled reality is complex, and depends on usage of the model. Mental models are formed by *abstractive documentation*, by apperceiving and abstracting from a physical reality, a 1:1 mapping, and their properties are elaborated by generalizing and theorizing. When a physical reality is being operated, the mental model is applied to the reality in a 1:1 situation of *empirical/observational documentation*, including operating, experimenting, controlling, simulating, and so forth. If a physical reality does not exist at the time, an *empirical/speculative documentation* can be developed in a $1:n^m$ situation by planning, designing, establishing, realizing, **concretizing**, predicting, and so forth. Each of these operations can take place in several stages, with alternatives available at each stage. The resulting combinations may reach large numbers—combinatorial complexity. Of the generated alternatives, a smaller number may be viable and usable, and can be selected by explorative sampling of the combinations, and thus learning about their potentials.

When designers dive into detail, they also recall relevant general and professional information, for example, mental models of the surrounding constructional structure. Nevertheless, designers comprehend the total problem through a restricted "window" [439], as a conceptual or constructional design zone, including form-giving zone. The boundaries of that window are determined by the design task, the knowing and the organizational position of the individual.

For the purposes of a design process, engineering designers can and should draw an arbitrary suitable boundary around the TrfP(s), and the TS(s), that is of immediate interest at that time. These boundaries can and will change as design engineering progresses, zooming the window in and out, and abstracting and concretizing changes. Systems are hierarchical, but we can only consider one level at a time. Case study 1.3 "Smoke Gas Filter," illustrates this by showing the change of emphasis and attention to 1.3.1 "Rapper." This case example shows that an assisting input (mechanical shock to shake particles off the plate electrode) can be solved by using a "function" of the larger problem, and treating it as the TrfP of the subproblem (subsystem) "rapper"—now the shock is the main effect of the TS(s).

NOTE: The change of window is shown on a water jet cutting system; see NOTE in Section I.6.1.

Assisting inputs, secondary inputs, and so forth, also need effects to suitably transform them, and these are problems at the next lower level of complexity and detail, a zoom-in window.

The examples of products in Section 6.11.10 show that the traditional divisions among engineering disciplines (e.g., mechanical, electrical, computer, etc.) are not particularly useful. Is a speed-controlled electric motor (e.g., audio tape drive) a product of electrical engineering (electrical and magnetizing sections), or of mechanical engineering (shafts, bearings and housings, magnetized rotors, and stators), or of electronics (controller circuits)? The word "or" is obviously misplaced, that motor is a product of all of these. Many products are made under the direction of one (group of) discipline(s), and used under another, for example, machines for mining, road construction, or robotic assembly of printed circuit boards.

Engineering designers need a broad range of general and specialized knowing and contexts, available recorded information, and good support in the organization, to cover all properties of the TS(s) they intend to design, compare Figures I.3, I.5, I.8, 6.8, 6.9, and 6.10.

The useful life of a product is usually limited. Even after normal "life ended," some artifacts find reuse (see Section 6.11.9). A different type of organization may extend or revive the life (e.g., railways run by volunteers as tourist attractions); an item may be used as a display or monument.

NOTE: Materials (etc.) that are no longer useful or usable can be *recycled* into raw material, or sent to disposal. Constructional parts (or subassemblies) can be upgraded, altered, brought back into a usable state by *remanufacturing* or *refurbishing*, or upgraded to a newer version by *reengineering*.

I.8 INFORMATION, KNOWLEDGE, DATA

Information, knowledge, data, and others are relative concepts, the terms are imprecise and ambiguous, with unclear boundaries. They are dynamic, their contents change with time. The terms depend on the interpretations of individuals, and their state of knowing and awareness.

In this book we use *information* as a general term for all instances of this phenomenon (see Section 12.1). This use of information as the primary term fits with ISs, information technology, knowledge engineering, and knowledge-based systems.

Information can be defined as a statement of meaning assigned to a static or dynamic phenomenon or thought, ISO 9000:2000 [10], article 3.7.1 states information is meaningful data. Information is carried by data and by objects (M, E, I, L). Knowledge can be defined as information representing a meaning assigned on the basis of theoretical and practical context to a static or dynamic phenomenon or thought. Data can be defined as information assigned on the basis of conventions, that is, without implied context. It need not have information content, and can then be defined as nonprocessed/natural, or artificial/processed expressions for revealed or potential information perceivable by human senses, or by technical means, measurement.

I.8.1 INFORMATION, GENERAL

The *contents* of information can concern tangible, process, and cognitive or conceptual objects. Process objects can be subdivided according to their intention: (1) for a useful application, manufacturing, distributing, operating, or disposing of a tangible or process object or (2) for designing a tangible or process object [164]. Cognitive objects include thoughts, ideas, intuitions, feelings, associations, **apperceptions**, and so forth about tangible or process objects. Various processes form the relationships among these constituents. Typical *subjects* for information are everyday life, biology, sociology, physics, mathematics, agriculture, history, arts, geography, and so forth, and designing and design engineering. The *constituents* of information include data, observations, evidence, rules, theories, knowledge,

(A) **Actual properties of an existing TS can be completely arranged into the classes shown in this table.**

(B) **Properties of a TS(s) to be designed must preferably fulfill all requirements that arise from each process in the TS life-cycle, and from the operators of each of these processes, in an optimal way** *(see Figure 6.15)*.

	Class	Symbol	Description		
EXTERNAL PROPERTIES	Pr1	PuPr	Purpose properties		
	Pr1A	FuPr	Function properties — behavior — effects properties	*Purpose*	*With respect to:*
	Pr1B	FuDtPr	Functionally determined properties — parameters, properties conditional on functioning (operating)	*of TS(s)* *Operational* *Process*	LC6
	Pr1C	OppPr	Operational properties		*Particular*
	Pr2	MfgPr	Manufacturing properties, planning and preparation — realization properties, manufacture, assembly, adjustment, packaging, etc.	LC4	*phases* *of the TS(s)* *life-cycle*
	Pr3	DiPr	Distribution properties, maintenance and service organization, warranty, consulting	LC5	*(see Figures* *1.13 and*
	Pr4	LiqPr	Liquidation properties	LC7	*6.11)*
	Pr5	HuFPr	Human factors properties — ergonomics, esthetics, psychology, cultural acceptability		
	Pr5A		In manufacturing, LC4		
	Pr5B		In distribution, LC5		
	Pr5C		In operation, LC6		
	Pr5D		In liquidation, LC7		
	Pr6	TSFPr	Properties of factors of other TS (in their operational process)		
	Pr6A		In manufacturing, LC4		
	Pr6B		In distribution, LC5		
	Pr6C		In operation, LC6	*Factors of*	
	Pr6D		In liquidation, LC7	*particular*	
	Pr7	EnvFPr	Environment factors properties	*operators of*	
	Pr7A		Social, cultural, geographic, political and other societal factors	*each TS(s)* *lifecycle*	
	Pr7B		Materials, energy and information — TP/TS inputs — effects of and on environment — TS-material — effects of and on environment — TP/TS secondary outputs and TS disposal	*phase* *(see Figures* *1.6 and* *5.1)*	
	Pr8	ISFPr	Information system factors properties — Including law and societal conformity, cultural, political, and economic considerations, information availability, and so forth	*Relationships:* With respect to a transformation process within	
	Pr8A		Scientific information	the transformation	
	Pr8B		Technological information	system, designing	
	Pr8C		Societal information	is not useful	
	Pr8D		Legal information	unless this purpose	
	Pr8E		Cultural information	is fulfilled.	
	Pr8F		Other information		
	Pr9	MgtFPr	Management factors properties	Design engineering delivers the quality	
	Pr9A		Management planning — product range	of the TS(s) as	
	Pr9B		Management of design process	designed.	
	Pr9C		Design documentation — design report, version control	Manufacturing gives the quality of con-	
	Pr9D		Situation — management climate, personnel relationships, and so forth	formance, quality control, quality	
	Pr9E		Quality system — quality of design, quality control, quality assurance	assurance for purchased parts.	
	Pr9F		Information properties — licensing, intellectual property, and so forth	Management should	
	Pr9G		Economic properties — costs, pricing, returns, financing, and so forth	be concerned with life cycle	
	Pr9H		Time properties — delivery, planning, process durations, repair, maintenance, and so forth	assessment and engineering.	
	Pr9J		Tangible resources —availability, accessibility, and so forth		
	Pr9K		Organization — goals, personnel, and so forth	Quality management	
	Pr9L		Supply chain properties — availability, delivery time, reputation, reliability, and so forth	system - ISO 9000:2000	
INTERNAL PROPERTIES	Pr9M		Other management aspects		
		DesPr	Engineering design properties	*Cause of*	*Designing*
	Pr10		Engineering design characteristics	*all TS(s)*	*(see Figures*
	Pr11		General engineering design properties	*external*	*1.16 and*
	Pr12		Elemental engineering design properties	*properties*	*2.1)*

(see Figures 6.8, 6.9 and 6.10 for additional information)

FIGURE I.8 Classes of properties of technical systems.

experience, and so forth. Information with respect to its *credibility* ranges over fact, observation, guideline, belief, myth, prejudice, hearsay, to deliberate misinformation. Typical *forms* of information include data, observations, experiences, explanations, heuristics, generalizations, rules, hypotheses, knowledge, and theories, and their verification or proof. Information may be more or less structured, from an informal collection to a formalized and verified system. For other classes see Section I.8.4.

Information can act as *operator* of a TrfP (i.e., an information process); see Section I.9.1. Information as a system then mainly provides guidance to the other operators.

Information can also act as *operand* of a TrfP—it can be processed (see Section I.9.1). The changes can influence the content and structure (internal) of information, its form (verbal, graphical, symbolic), location (internalized, externalized), and time (see Figure I.9).

	Transformation process (TrfP and technical process (TP) *(see Figure I.6)*	TRANSFORMATION			
		Of structure	Of form	Of space coordinate	Of time coordinate
		Processing	Manufacturing	Transporting	Strong
OPERAND	Material M	M convert (process)	M transform (form, shape)	M transport	M store
OPERAND	For example	Iron ore —> steel	Scantling —> workpiece	Workpiece in storage bin — wp on assembly line	Workpiece in storage bin
OPERAND	Energy E	E convert	E transform	E transport	E store
OPERAND	For example	Hydraulic —> electrical	50000 V —> 220 V	Power station —> consumer	In accumulator (battery, spring)
OPERAND	Information I	I convert	I transform (translate)	I transport	I store
OPERAND	For example	Graphical —> digital	German text —> English	News sender —> receiver	Speech on magnetic tape
OPERAND	Human (animal) Hu (L)	Hu convert	Hu transform	Hu transport	Hu store
OPERAND	For example	Sick —> healthy	Natural hand —> prosthesis	In London —> in Toronto	At home (in bed)
	Typical verbs of operand transformation or of TS-internal function	Convert, rectify, oscillate, process	Enlarge increase, reduce connect, separate join, divide collect, diffuse combine, dissipate connect, interrupt couple, disconnect emit, absorb manufacture produce	Conduct, isolate guide, release	Store, retrieve hold, release keep, reject

(see also Figures 7.11 and 9.9)

FIGURE I.9 Basic classification of transformation processes (TrfP) and technical processes (TP).

Processing of information can be performed by humans, and by a computer (or any other machine), for example, for purposes of controlling another process. The computer programs are then usually referred to as *information technology*. Some information technology is referred to as *knowledge-based* systems, an accepted inconsistency, and includes an implied accepted true belief (see Section I.8.5), for developing and using meaning, understanding, and opinions.

NOTE: For *transformation* operations, the formulations should normally contain a verb (or verb phrase) and a noun (or noun phrase) that specify what should be done to the operand being changed.

I.8.2 INFORMATION, RECORDED

If information can be expressed and formulated into words (e.g., a thought, theory, hunch, or imagination), images and symbols, it can be transferred to a tangible medium—even feelings can be expressed; otherwise communication is unlikely. Information can be *recorded* in available repositories, stored in an accessible form, unordered to ordered and classified, for example, scientific. It can be retrieved, if a suitable method for classifying and searching is available. Classification is usually by hierarchies of classes, but relationships among items of information in different branches tend to be lost. Classification systems can be based on a "flowchart" or multisubject matrix. Records cover the forms of information listed in Section I.8.1.

I.8.3 INFORMATION, GENERATING AND USING

The processes of generating and using information can be shown in their relationships (see Figure I.10). Each process (in rectangular boxes) consists of a sequencing of messages appropriately formulated from an initial result (in an elliptical box) to achieve another result. The upper section shows developments of knowledge and experience, often starting from observed or conjectured data. The lower section indicates application of information and internalized, tacit knowledge (knowing) to achieve other results. These various forms of information can be recognized and brought into meaningful relationships (see Figure I.11). The information, as design process and object information, can then be categorized and codified as in Figure I.3.

I.8.4 INFORMATION, TRANSMISSION, COMMUNICATION

A model of *communication* to transmit messages is shown in Figure I.12. This model shows that information (as operand) can be transmitted from one person or record to another. The first subprocesses formulate the information suitable for the transmission media and conditions. The subprocess "transmit as message" may be performed by a TS—a communications device. A sequence of symbols encodes the message, but the symbols remain meaningless to the transmitting TS. The last processes in Figure I.12 receive and interpret the received message, either as executable commands, or as information to be understood by humans. A message can be characterized by several dimensions; see Figure I.12 and Section I.8.1: contents, subjects, constituents, credibility, forms. Similar schemes may be found elsewhere, for example [553].

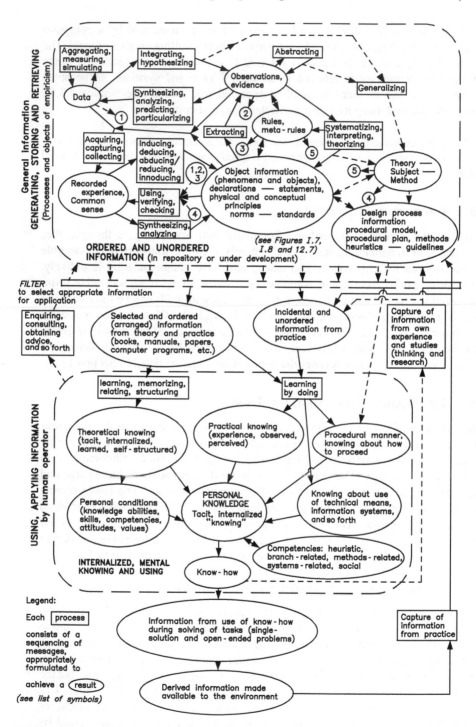

FIGURE I.10 Characterization of information—processes and relationships.

Levels of content and generality	Types of Information	Description	Kinds of information according to location within an information system	
			External — generally available "Knowledge as recorded object"	Internalized — in the human mind "Knowing as mental process"
	General information (observations, evidence, anecdotes, opinions, instructions, etc.)	Generally available facts and opinions, including statements and narratives, given interpretation for a static or dynamic occurrence on the basis of a convention, some relationships — Usable advice for working, incl. enterprise product catalogs and application instructions	Recorded, factual, raw information, partly processed with or without relationships (context), unclassified, not verified, some is known to be false Can be recorded into an information System, including: — Masters — Catalogs — Proforma (forms) — Standards, regulations, laws	Tacit (quiet and unexpressed), internalized knowledge (a) Object knowledge related to the real or imagined existing world — declarative knowledge (b) Combined from "nature and nurture," inborn (genetic) and learned (by experience and formal instruction) (c) Structured idiosyncratically by each individual person, both structured and unstructured tacit knowledge, (d) Working and interpreting (establishing meaning) on processing it into usable "proceduralized" knowledge,
	Scientific Knowledge	Agreed according to an accepted disciplinary matrix, paradigm — [Kuhn] — Hypothesis, formal model, theory — Physical sciences, social siences	Form: — Hard copy form — Microfilm/negative form — Electronic computer records — In transit — communication, signal, command, instruction	(e) Partly recallable with more or less difficulty, synthesized on recall from many related fragments, (f) Partly explainable, mostly intuitive and emotion-based (1) Cognitive (related to rational thinking), (2) Affective (related to feelings, emotions), (3) Psycho-motor (related to movement)
	Experience Knowledge	Rules and Principles — perceptions, observations, interpretations, opinions — Repository of observations for abstracting into scientific form,	System of knowledge → science: codified, classified, categorized, abstracted, related, integrated	A) Verbal (in words, expressions) B) Graphical (in images) C) Symbolic (in semantic meanings, including numbers and algorithms)
	General Knowledge	Real, conjectured, and philosophical — Including opinion, hearsay, anecdote, belief, even at times prejudice — partly processed		— Mental process (related to changes as they happen or are caused by human actions) — Learning and experience transforms "knowing" into applicable "proceduralized" knowledge — Related to skills and abilities — Human capabilities as initially learned (abilities),
	Data	Continuous or discrete evidence, normally numerical data and experiments, from observations and experiments, distinctiveness of information content, measures of properties — unprocessed		— As developed by more extensive practice (skills) — And as combined (competencies), — But limited by mental and physical human resources (long-term and working memory)
	Object Information — Heuristics	Stepwise processing of object information during designing, usually iterative and recursive, steps may not be easily identified in actual procedures	Published and otherwise available: — Procedural advice, heuristics, prescriptions, techniques, — methods, methodologies	Abstracting — concretizing Analysing — synthesizing Dividing — combining proposing and testing hypotheses Imagining Thinking
	Design Process Information — Methods	Stepwise progress through a process, usually iterative and recursive, steps may not be easily identified in actual procedures	— Advice, prescriptions, techniques for application of procedures and their prescriptions	

Object information — Descriptive, Prescriptive
Sub-classes — structuring and abstracting
Descriptive
Prescriptive
Normative
and Normative
Process

Kinds of information according to scope

Conceptual picture: "greenhouse" — propagating the week
"botanical garden" — displaying the strong
"supermarket" — providing choice

Research model: "hard science" — experimental, empirical
"soft science" — social-participatory, observational
"reflective" — theorizing from available information
"speculative" — generalizing from gut feelings

Literature:
Kuhn, T.S. (1970) The Structure of Scientific Revolutions (2. ed). Chicago: University of Chicago Press
Kuhn, T.S. (1977) The Essential Tension: Selected Studies in Scientific Tradition and Change, Chicago: University of Chicago Press

FIGURE I.11 Characterization of information—definitions and perceptions.

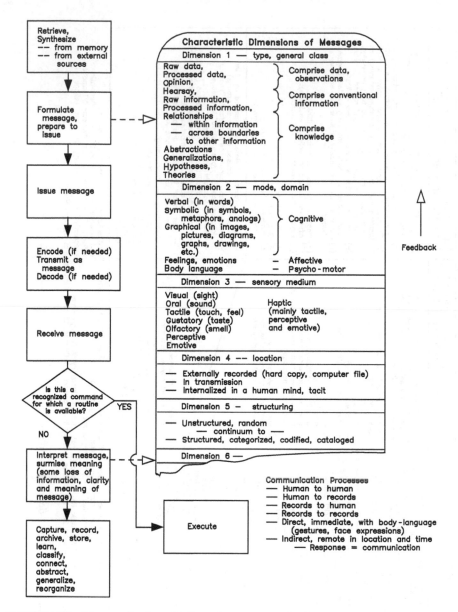

FIGURE I.12   Characterization of information—communication.

I.8.5 KNOWLEDGE, GENERAL

Empiricism claims that experience is the source of all knowledge. This refers to the heuristics [365] and knowledge of practice, and knowledge that has been abstracted and codified into hypotheses and theories of the sciences. Relationships among elements of information reveal the structure. Processing often tries to generalize and summarize information—*abstracting*—and bring it into relationships with other information items—*codifying* (relating, hypothesizing, theorizing, etc.) and *structuring*. Codified knowledge thus represents the "accepted or warranted true belief" (by an individual and a group) on the basis of evidence (or lack of it) and description (narrative and theory) of phenomena, at that particular time. Humans (and computers as TS programmed by humans) rely on currently accepted *true belief*, *truth* as agreed abstractions, and as recognized useful understandings based on other experience.

Information is processed into knowledge by deduction, induction, reduction/ abduction, or innoduction [174]; see Section 12.1. Scientific knowledge is not the only sort [152], much of informally structured information is of little interest for science, yet is necessary for technological (engineering) application. This includes information gained directly from experience, which may not have been formalized and recorded, or may not even be capable of being formalized. Examples are [98,133,188,342,499,521,556], standards, codes of practice, regional laws, and so forth.

Processing usually progresses from full contents of information, through informal and general structuring—a *typology*—to scientific categorization (a *taxonomy*). Processing has various aims: (1) holistic, comprehending an overall picture vs. reductionist or atomistic, isolating the elements; (2) synthesizing, placing together in possible arrangements vs. analyzing—the arrangement of information for these two purposes should be different, unless selective search by computer is used; analysis can use the arrangement of the traditional engineering sciences, synthesis usually needs an arrangement according to achieved output effects; (3) system (functioning) vs. detail (components, constructional parts); (4) among phenomena vs. among other information elements; and (5) in a progression via hypotheses, axioms, theorems, and corollaries, to theories. Both directions of each are usually necessary for full understanding.

With accumulation of further evidence (some of which will not fit the accepted interpretations), what is accepted as currently valid knowledge will need to be revised, and new proposals made to overcome the deficiencies [376,377]. This results in a change of the disciplinary matrix, a paradigm shift, a scientific revolution, replacement by a new theory, a change that may take substantial time. Established proponents will usually resist change. Examples of such change are Newton's laws of motion, relativity, quantum theory, chaos theory, and so forth.

I.8.6 KNOWLEDGE, RECORDED

Codified information held in tangible repositories can be called recorded knowledge, suitably cataloged. This conforms to the usage of "knowledge-based systems," which often rely on AI, that is, an interpretation of processed, stored, codified information.

I.8.7 DATA

Data usually refers to concrete properties (see also Section I.8). Figure I.11 shows that data may be continuous, consisting of observations and evidence, or discrete, measured values of properties of a TS or of other naturally occurring or artificial phenomena. Data is represented by agreed primitives, the symbols consist of alphanumerics, iconics, and others. Symbols have agreed meanings, for example, numerical symbols, units of measurement (e.g., "number of items" as a unit); text data; and symbolic data. Data is regarded as context free, but this is doubtful.

I.8.8 INFORMATION, INTERNALIZED, TACIT

When using available information, the cognitive processes of a human lead to various forms of *knowing* [116,117,175,224,232], which is internalized, tacit, mentally structured, incorporated into a person's acquired experience, and thus not accessible to other persons. It is individual and idiosyncratic, but all have much in common.

The human mind can act consciously (in working or long-term memory), subconsciously (intuitively), or unconsciously (instinctively) (see also Section 11.1). Mental actions have three domains: *cognitive* [68], thought; *affective* [371], feelings; and *psycho-motor*, physical actions and their control—see dimension 2 in Figure I.12.

Tacit knowledge can concern tangible or process objects, usage of objects, problem solving and designing [164], managing, and so forth, and mental constructs resulting from association of ideas. Storage within the brain is thought to occur in chunks of information of varying size and connectivity, and in several parts of the brain according to the mind's senses and abilities.

Internalized information is stored initially as *declarative* object or process information. When it has been learned well enough, a person is not conscious of use, it becomes *procedural* object or process information, know-how and knowing (see Figure I.11), and its use is "intuitive."

From this internalized information, the human mind (consciously or subconsciously, unaided or reminded by sketching in graphical, verbal, and symbolic media) produces mental constructs. Some of these result from *reasoning*, in a "forward" (causality) or "backward" (finality) direction; a combination of the two is usually effective. Procedural object or process information reappears as subconsciously applied methods (compare [351]) evidenced in skills and competencies.

I.8.9 KNOWLEDGE, INTERNALIZED, TACIT, UNDERSTANDING

Information is a major source of experience and understanding, which evolve from interpreting, extracting meaning and values, and recognizing relationships and patterns in the available information. Research can process this through hypotheses, and refine it into a theory, as part of the codifying process of information. "It seems that a large amount of knowledge has to be taken into account in a highly integrated way for *understanding* to take place" [270, p. 569].

Mental processing of information to develop understanding and knowledge can result in both tacit and recorded knowledge, and occasionally also wisdom. Different arrangements of information are needed for research (extending the scope of knowledge), for archival collection (searching according to disciplinary categories), and for designing (searching to achieve effects); see [193,194,498] and Chapter 9. The internalized knowledge of an organization's employees can be captured; see Chapter 10 [195].

I.8.10 Intelligence

Intelligence manifests itself as appropriate behavior revealing knowing and under-standing. It arises from internalized processing of information into knowing, that is, developing know-how, by abstraction and association. How much of intelligence is nature vs. nurture is still in question.

Intelligence may be classified as crystallized (declarative, knowing) and fluid (proceduralized), compare classes of object and design process information (see Figures I.4, I.5, and 12.7).

Artificial intelligence attempts to mimic human information processing in small areas of interest, using computers for, for example, heuristic programming, genetic algorithms, computer-extracted information, computer-generated advice, and computer-control for machines.

I.8.11 Summary

The subject of information is complex, Section I.8 has tried to bring some order into this region (see Section 12.2). The *quality of information* that meets the needs of engineering designers must be addressed (see Chapter 9).

I.9 TECHNICAL OBJECT AND PROCESS SYSTEM

Summarizing from Section I.1, in a transformation system (TrfS) someone (HuS) and something (TS), in an environment (AEnv), with information (IS), and management (MgtS), does something (TrfP and TP) to something (Od1) to produce a different state (Od2) to satisfy someone and something. Preferably, the TS that is used should be designed and manufactured to be optimal for its TP in the given circumstances.

The goals for this section are to indicate the theoretical support for the prescriptive knowledge related to engineering DesP (see Section I.12.7) and transformation systems (TrfS)—especially TP and TS (see Section I.11.7).

I.9.1 Transformation Systems

As outlined in Section I.10.1, existence and life consists of and depends on *processes* of change—TrfP (see Figures I.6 and 5.1). Many processes are *natural*, others have natural elements (e.g., agricultural activities), still others are artificial. Boundaries between natural and artificial are poorly defined. Artificial processes, not necessarily continuous or linear, are triggered by needs or wishes, and are not available from nature without actions and means provided by humans. They are driven by effects delivered by operators, and together with them form a system. For example, if a person feels ill, a medical doctor makes a diagnosis, assisted by tests and measurements helped by devices—machines (mechanical, electrical, electronic, etc.). Then a therapy (partly performed by machines, partly by medicines produced by machines) is recommended (prescribed).

The *quality of life* depends on the quality of TrfP, and on their output products. These are intended to satisfy human needs, considering restrictions and constraints, and depend critically on the operators of transformations, especially TS. Therefore the quality of life is a function of the results of transformations, and thus of the applied TrfS at the time and in the space (see Figure I.6). A survey of products is included in Section 6.11.10.

NOTE: Preferably the applied transformation system (TrfS), should be the best (optimal) system for the situation as it exists at that time. This also implies that a degree of optimality should exist in each of the elements of the TrfS—not necessarily that each element should be optimal in itself.

I.9.1.1 Transformation (Trf), Transformation Process (TrfP), Transformation System (TrfS)

Each transformation may be regarded as a system—a TrfS (see Figure I.6), consisting minimally of a TrfP, and operators (Op).

The TrfP shown in the rectangular box, with its *operations* (see Figure 5.4), is driven by the applied *technology* (Tg, definite article), the cause of the transformation. This TrfP accepts an input, the *operand* in state Od1—in analogy with mathematics, those variables that are to be changed by a function to generate the desired output values. This operand is transformed (reactively) to an output state Od2, which is more suitable for the stakeholders—achieving state Od2 is the intended *purpose*. Each transformation consists of one or more partial transformations, that at their smallest (convenient) limit are called *operations*. Each operation can take place only if a suitable technology (Tg) is applied (see Section I.6.1 and Figure 5.1C).

The operand may be a simple discrete or continuous object, or a complex object consisting of several to many items. The operand can consist of inanimate material, animate material (humans, other animals, or plants), energy, or information (M, L, E, I)—these are *interdependent* aspects, but one may be more important than others for the purpose. The transformation may involve one or more properties of the operand (e.g., temperature, pressure, structures, space, time, etc.). Special names are often used for transformations that change the (internal) structure, (external) form, location (in space), or time dimension (see Figure I.9).

The TrfP also accepts assisting inputs (AssIn) to help perform the transformation, and secondary inputs (SecIn), many of which are probably undesirable—disturbances. The operand in state Od2 is always accompanied by secondary outputs (SecOut), many of which are undesirable, but some may be adapted for other purposes.

Typically five *operators* (Op), shown as the rounded boxes, Figure I.6, are active or reactive to directly or indirectly implement the technology: humans (and animals), technical systems, the active and reactive environment (within the general environment), information systems, and management systems. These accept primary, assisting, and secondary inputs (M, E, I), and deliver their active and reactive *effects* (Ef), the round-based arrows, and secondary outputs. Normally, the states, manifestations and values of the outputs are assessed and measured (evaluated), compared to a desired or required target/limit state. A *feedback* (Fb), the upward arrow, connects assessments of results with inputs, and can influence these states by applying changes to the input of the operand and the operators, in an attempt to correct errors in the results.

Investigating the purpose of the TP(s) and TS(s) should include: (1) usage analysis—which helps to establish the application, including (for the TrfP and the TS) user instructions, maintenance instructions, feedback reports from users, maintainers/repairers, customers, stakeholders, and so forth; (2) structuring of processes; (3) anticipating the needs of users and other persons or organizations, social and environmental effects, and so forth.

The operand should be in a suitable state Od1 to accept the change into the intended state Od2. This investigation should therefore include: (1) the general structure of TrfP, which can be used as a checklist for design engineering. This involves identifying the operations and relationships, arrangement, inputs and outputs (operand, assisting, and secondary), and so forth of TrfP—preparing, executing and finishing operations, main and assisting operations (i.e., propelling, **regulating** and **controlling**, connecting and supporting, auxiliary operations, etc.), alternative operations and relationships, and possible failure mechanisms of the operation and the technology (see Chapter 5); (2) identifying factors that influence the results (factors of the design situation); (3) recognizing possible assisting and secondary inputs (including disturbances) and outputs (usable and polluting), and ways of avoiding damaging influences from them; (4) setting up a representation of the TP, and identifying the technology, the operators, and possible failure mechanisms of the technology and the operators; (5) describing general properties, magnitudes that can be used to describe the TrfP (e.g., for evaluations and decision making).

I.9.1.2 Technical Process

A TP is an artificial TrfP that results (primarily or exclusively) from operating a TS (see Chapter 6), via an appropriate technology (see Chapter 5): *primarily*, if other operators supply effects directly to the operand, or (indirectly) act by operating the TS (e.g., an electric hand-held drilling machine); *exclusively*, if the TS acts alone. A TP is a special case of a TrfP. The statements about TrfP are equally valid for the TP. The technical process is driven by effects delivered topologically external to the TS.

NOTE: Process normally refers to the TrfP (see ISO 9000:2000, article 3.4.1 [10]), that is, not only to manufacturing processes. Technical system normally refers to the tangible (object) system—unless additional qualifying words are used. A TP can be extracted from a TrfP by focusing on those operations that involve an existing or assumed operational TS—but does not preclude adding other operations. The TS-operational process delivers the effects for the real or potential TP.

The main operator of many artificial transformations (TrfP or TP; see Figure I.6) is usually a TS, and is often helped and operated by humans (Hu)—these, together with active and reactive parts of the environment, are the execution system. When the TS is active or reactive in its TS-operational process it exerts *effects* (Ef) on the operand to operate the TP.

Effects consist of material, energy, and information (M, E, I), with one of the three as primary. Effects may be exerted directly or indirectly by an operator: *directly* when the effect acts from an operator (HuS, TS, AEnv, IS, MgtS) by means of a

technology (Tg) onto the operand (Od)—IS in the case of an information or manufacturing process, MgtS in the case of a management process; *indirectly* when the
actions of an operator act through another operator to produce the direct effect—for
example, HuS → TS → Od or IS → HuS → TS → Od.

NOTE: When a TrfP generates a product (i.e., the operand in state Od2 is a commecializable
commodity), the TrfP is regarded as a manufacturing (or production) process. Manufacturing
processes include IS in their execution system (operators) to deliver the information about what
is to be made, and how—for example, engineering drawings and manufacturing preparation
documents of constructional parts (see Figures I.13 and 6.14 and the NOTE in Section 6.6)—
and have output in state Od2 consisting of *products* from artificial transformations. The quality
of a TP (assessed mainly by the quality of the output operand Od2) depends critically on the
quality of the TS and of the human operator systems.

I.9.2 TECHNICAL SYSTEMS

Technical system are man-made, tangible material objects that perform a useful task.
TS comply with the laws of nature, especially for the ways in which they work, act, and
function.

A TS has the (internal) *capability* of processing its (primary, assisting, and
secondary) inputs into (primary and secondary) outputs, that is, across its boundaries. These inputs and outputs may consist of materials, energy, and information
(M, E, I), that is, not living things or humans. The outputs include the *effects* that can
be delivered via a technology to transform the operand in the TP—this is the purpose
of a TS. This capability is a *potential*, which is realized as an effect (action or reaction)
only when TS is operational and capable of being operated, with an operand present.
For example, a portable fire extinguisher has its potential and readiness to "deliver
extinguishing fluid when needed" (its TP)—when it is triggered by a suitable input
action from a human.

All societies, cultures, and civilizations depend on the sorts of TS that they have
available. Developments in societies, cultures, and civilizations can only take place
if the TS is simultaneously developed. Developments in TS must become "ripe" by
accumulation of information and needs, a continual and mutual interaction between
society and technology.

Various classification criteria for TS can be identified, including the traditional disciplines of the engineering sciences, branches of engineering, the *novelty* of the future
TS(s); stages of market development; sequence of demand to design engineering
and manufacturing; scale of production of the TS(s), size relationships, complexity; manufacturing location, standardization; market; life ended, and so forth (see
Section 6.11). For instance, regarding complexity, four typical hierarchical levels are
defined (see Figure 6.5): level IV, plant; level III, machines (including electronics);
level II, assembly groups (subassemblies); and level I, constructional parts.

I.9.2.1 TS-Internal Processes

Each TS is capable of performing its internal and cross boundary actions or reactions
to deliver its effects. The *behavior* of a TS, the sequencing of states through which

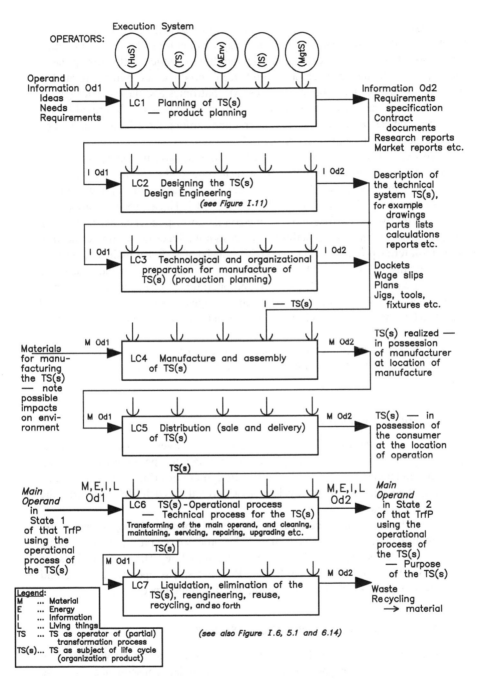

FIGURE I.13   General model of life cycle of TS as a sequencing of TrfS.

the TS passes in response to its inputs, results from these actions or reactions. Each such action results from a *mode of action* (way of operating) based on an action principle—mechanical, electrical, electronic, building, chemical, other discipline or engineering branch (industry sector)—"high tech" products are hybrids related to mechanical, computer, and other disciplines. Mechatronics and nanotechnology are the result of trends such as miniaturization. Action principles are described by the engineering sciences. Relationships and interactions among the effects described by engineering sciences must also be considered.

We are dealing with two separate parts of the transformation *system*, TrfS, Figure I.6, and their link represented by a technology: (1) *TS-internal processes and functions*: the TS, a tangible object, acts as operator, one of the "object" parts of the "product" defined in ISO 9000:2000 [9,10]. These processes are exclusively described by the TS-functions. A one-to-one correspondence exists between TS-internal processes and TS-functions—for design engineering we do not need to consider TS-internal processes. (2) *Effects* (Ef): consist of M, E, I as main outputs of an operator (not only TS), delivered by a technology, received by an operand, to achieve the operation within the TP/TrfP. (3) TrfP and operations, TrfP: the "process" part of the product; see ISO 9000:2000 [9,10]—the changes experienced by an operand from Od1 to Od2 have only an indirect influence on the descriptions of TS-functions.

NOTE: For TS-internal functions, the formulations should normally contain a verb (or verb phrase) and a noun (or noun phrase) that specify what should be done to the object being changed.

I.9.2.2 TS-Life Cycle, Phases

Any TS life cycle consists of different life phases, it "lives" through these TrfP. A typical sequence of TrfP includes phases of origination, operation and disposal, and their dependencies (see Figure I.13). For design engineering, the requirements on the TS(s), the *subject* of the life cycle, can be derived for each life-phase from the TrfP and its operators.

In most of the life cycle processes, the TS(s) is the operand. During manufacture, the "information about the TS(s) as designed" becomes an operator of that life cycle process, the drawings are the information for manufacture. The realized TS(s) is operator in its TS-operational process, the TP. The TS(s) is operator and operand during tests or experiments for development, and in maintenance.

I.9.2.3 TS-Properties, Classes, Relations, "Scales"

Every TS carries its *internal* and its *external properties*, which exist whether they have been deliberately designed, occur as an unintended consequence, or arise "incidentally" from the designed structures. The complete system of classes of TS-properties that has been found useful for design engineering is shown in Figures I.8 and 6.8 to 6.10. Their relationships are complex, many properties influence several property classes. For instance, the external property of "stiffness" of an existing machine tool is caused by the stiffness of all individual contributing constructional

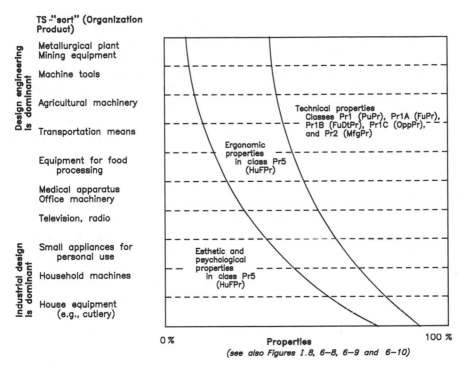

FIGURE I.14 Weighting of selected classes of properties for some "sorts" of technical systems.

parts, the stiffness of all organs connecting the parts, and the arrangement (configuration) of the constructional parts. Properties of TS(s) to be designed must fulfill all requirements from each process in the TS life cycle (i.e., also their operators)—the classes of external properties (and at times some of the internal properties) provide a guideline for establishing a design specification—clarifying the problem.

Within these classes, the *importance* (weighting) of properties is different for different sorts of TS (TS-"sorts"; see also Sections I.1, I.4.1, and I.11.3); see Figure I.14. Industry products with predominant engineering content are typically the upper examples. The lower examples are consumer products with dominant industrial design, aesthetics, ergonomics, and customer psychology.

The TS-properties change among the different *states* of existence for each TS, for example, various life cycle phases of a TS(s). The states of properties exist (see Section I.12 and Figure I.8) and change under various operating states, the "duty cycle" of the TS: (1) at rest, no operation; (2) during start-up; (3) during normal operation—idling, full-power and part-load, overload, and so forth, for self-acting operation (automatic), or running and ready to be operated by another operator, for example, human or another TS; (4) during shutdown, ending an operational state and returning to "at rest" conditions; (5) in fault conditions—(a) internal faults—overload, safe trip-out, breakage or equivalent and (b) external faults—damage, wrecking, and so forth; (6) at "life ended."

The same property can be compared among different future and existing TS by *evaluating* them. Some properties are *quantitative*, their values can be measured on *scales* defined from acknowledged units of measurement, for example, power, speed, and so forth. Scales of measurement are absolute, with a zero point defined from a natural limit (e.g., the Kelvin scale), or relative, with an arbitrary zero point (e.g., the Celsius scale). States of other properties are *qualitative*, such as appearance or other manifestations, and can only be *estimated* (assessed). Ranking may be possible and useful, and a value assigned. It may be useful to collect manifestations into *sets*, and assign numerical values where needed (see Section 2.4.3).

I.9.2.4 TS-Quality

TS-quality is related to the capability of a future or existing TS to fulfill the assumed or given requirements within its life cycle. Quality is a relative measure, a degree of excellence, measured or perceived relative to a requirement for a TS(s), or relative to an envisioned or perceived ideal.

Quality may depend on a physical or conceptual viewpoint, whether the TS(s) is in its state as designed, as manufactured, or in some parts of its operational process, and so forth.

The value of TS-quality can be characterized (measured) by forming a ratio between (1) an aggregate of the states of properties for the achieved (or existing) states of the considered TS(s) and (2) a similar aggregate of the reference values of these properties for the assumed, established, or existing states of a reference ideal (or of a competitor, etc.) TS. For this purpose, values need to be assigned to manifestations of qualitative properties. Evaluations can be with or without weightings. The states or values of properties may be given different weightings to give emphasis to the important and desired ones. Different weightings for properties also apply to the various TS-"sorts" (see Figure I.14). The total value or partial value of TS-quality can be obtained, if all or only selected properties are evaluated. In most cases only the partial user value of the TS-quality is considered, and is assessed from viewpoints of customers or stakeholders. At times, this user value is considered as the total value of TS-quality (see also Section 6.7).

The optimal quality of a TS(s), as the goal of design engineering, the expected output of a design process, depends on the situation in that part of the life cycle under consideration, that is, the design situation, the manufacturing situation, the operating situation, and so forth. It is consequently different for each task. The condition for obtaining optimal quality of a TS(s) is the existence of several alternatives of TS-structures from which the most promising may be chosen—this must be the principle of an engineering design strategy, a procedural model, a procedural plan, and the tactics of procedure, including design methods. An absolute optimum does not exist, the situations are complex and depend on human opinions, but an optimal TS can usually be established.

NOTE: ISO 9000:2000 [9,10], article 3.1.1 differentiates between "inherent" properties, and "assigned" properties. Under market conditions, price is a monetary value "assigned" to a product by a seller as a target, and potential buyers assess that offering price together with other available information to decide whether to buy. The actual exchange is then agreed between a seller and a buyer [388,389], the monetary value of the product, now an "inherent" quantity.

I.9.2.5 TS-Structures

TS-structures are elemental design properties of a TS. They are always present (see Figures I.8, 6.8 to 6.10, and Section I.9.2.3). The *TS-structures* consist of elements, their relationships within and across the (chosen) TS-boundary, and may be formed from different viewpoints and levels of abstraction.

The main (cross boundary) task of a TS is to exert *effects* onto the operand of the TP with the aim of transforming it from state Od1 to state Od2. This main TS-ability—its *operating behavior*—is caused by appropriate TS-structures, which should preferably be optimal for the operating situation. The relevant relationship is (1) the TS-structure determines the properties and resulting short-term and long-term *behavior* of the TS during its operational process (TP) and all other life-cycle processes, within statistical tolerance, but not necessarily fully predictable and (2) the *same* properties and resulting behavior can be obtained from *different* TS-structures—this permits finding alternative principles, modes of action, structures, and so forth.

TS-quality depends on TS-properties, and therefore on TS-structures. For design engineering, the recognized structures should be helpful in achieving an optimal TS(s) by exploring, establishing, describing, and evaluating the proposed TS at various levels of abstraction. TS-structures are needed for TS-models and their representations, and are established step by step in systematic and methodical design engineering.

The most concrete form of TS-element for a future or existing TS (compare Chapter 6) is the TS-constructional part, or component, usually the manufacturable hardware (solid, liquid, or gas); see Sections I.7.1, I.12.6, I.12.7, I.9.1.5, I.9.1.7, and I.12.9, and Chapter 5.

A more abstract TS-element is a *TS-organ*, a function-carrier, which is formed by the *action locations* on each constructional part where two or more constructional parts interact with each other, or a constructional part interacts with an operand. An organ is a connection in principle without regard to the supporting materials of the constructional parts. An *organ* fulfills one or both of two basic functions: "connecting, carrying, supporting" or "acting or reacting." For design engineering, the more important is the *acting* or *reacting organ* (of various types, usually simply referred to as an organ), which is necessary during the operation of a TS. These organs provide the means for functioning, the means to realize the (internal and cross boundary) functions of a TS, for example, the interacting teeth of two gear wheels, the internal surface of a valve contacting the operand fluid (see Section 1.2). Other organs are *evoked* by the need to assemble and manufacture the constructional parts, they are passive during TS operation, they do not contribute to motions or other main functions. *Elemental* organs are simple contacts, for example, between a shaft and a hole. *Partial* organs are parts of complete organs, for example, action locations on a constructional part. Organs may be combined into organ groups, complex organs, and organisms. Connection organs or *organ connectors* are the analogs of constructional parts.

An action location (site or locality) is a point, line, surface, volume, and so forth on a constructional part, at which an action may take place. It is one part of an acting or reacting organ, a partial organ, a location where a TS-internal effect is to be transferred from one constructional part to another, or the output of a TS where it exerts its effects on an operand (or another TS). An action or reaction process is a process that is capable of resulting in an action, especially a TS-internal process. An action or reaction medium is any medium (solid, fluid, etc.) capable of transmitting an action. It is not necessarily an "active medium," it may at times be inactive, reactive, or countering the desired actions—for example, air resistance. An action chain is a chain of elements capable of performing an action or reaction, in a function structure, organ structure and constructional structure, and includes the effector action location where the technology acts to change the operand.

Still more abstract is the *TS-function*, which describes the capabilities of an organ to perform TS-internal and TS-cross boundary tasks. These tasks may consist of receiving TS-inputs, processing these inputs and intermediate states into intermediate states and outputs, and delivering TS-outputs. The most important are the *TS-main functions*, which deliver the main *effects* that are the main purpose of a TS and form a (possibly branched) chain of capabilities. The TS-main functions are always accompanied by *TS-assisting functions*.

The TS-internal assisting functions can be (1) *TS-auxiliary* functions, for accepting, preparing, delivering, and eliminating material; (2) *TS-propelling* functions, for accepting, preparing, delivering, and eliminating energy; (3) *TS-regulating* and controlling functions, for accepting, preparing and delivering signals, detecting and processing measurements, feedback, making information available; (4) *TS-connecting* and supporting functions, to provide internal connections within the TS. The *TS-cross boundary* functions may be: (5) *TS-receptors* to transfer *TS-inputs* (M, E, I) from outside to inside the TS-boundary; (6) *TS-effectors* to transfer *TS-outputs* (M, E, I), especially effects, from inside to outside the TS-boundary; and (7) *TS-connectors* to support the TS relative to the fixed system (Earth).

The *TS-function structure* (the structure of TS-functions) is defined by its elements (functions) and their relationships. The function structure gives the engineering designer a means to evaluate the operational states and failure possibilities of a (existing or future) TS, especially in the early stages of a novel engineering design process (DesP).

Similar considerations apply to the TS-structures consisting of organs, the TS-organ structure, and to the structures consisting of constructional parts, the TS-constructional structure. Examples are shown in Chapter 1, and Figures 2.15 and 2.18.

I.9.2.6 TS-Inputs and TS-Outputs

Technical systems in general only operate, that is, perform their TS-operational process, during times when suitable *main inputs* (M, E, I) are supplied to the TS. Then the TS delivers the desired *main outputs* (M, E, I), that is, the effects that can cause the transformation of the operand in the TrfP. Various *assisting inputs* may also be needed for the TS to operate correctly or adequately.

At all times, *secondary inputs* can influence the TS, mostly as disturbances. They can cause malfunction and deterioration of the operating or nonoperating TS. At all times, but especially when the TS performs its TP, *secondary outputs* (including displays to indicate the TS state) can occur. These mainly influence the environment of the TS, including its human operator.

I.9.2.7 TS-Development in Time

Progress is described by changes that usually enhance the capabilities, weight, cost, performance, appearance, and other properties of a TS-"sort." With elapsed time, successive TS are usually developed to use less material, increase functionality, and so forth—the "state of the art" changes with time. This also depends on tendencies, fashions, and so forth. Laws (regularities) of development can be recognized (after the events), but can hardly be predicted.

These changes may result from improvements in the TP, in the Tg, and in the TS itself, in its functions, organs, and constructional parts. Usually these are small stepwise changes, but over a sufficiently long time they may be regarded as a gradual evolution. If a new process, technology, or partial TS becomes available, a larger stepwise change may occur, and be recognized as a "technical breakthrough" or leap. Rapid changes in the TS-"sort" (and in the state of the art) that occur during short time periods result in "dynamic concepts" [477]—and little change will be experienced during longer periods—"static concepts." The result is a surge–stagnate sequence [331] of developments.

I.9.2.8 TS-Taxonomy

In addition to the classifications of TS outlined in Section I.9.1 and Chapter 6, TS can be divided into classes according to various criteria. For design engineering, one criterion is the *degree of abstraction*, in analogy to biology: phylum, class, family, genus, species. Typically, an increased number of different TS is included as the classes become more abstract, whilst the number of established properties decreases—a phylum defines the TP that the TS have in common, a class designates a common Tg and main TS-functions, a family exists with a common function structure and organ structure, a genus establishes a common arrangement, and the species has its constructional structure fully defined (see Figure 6.17).

I.9.2.9 Theoretical Knowledge about TS-"Sorts"

In addition to the general TS, and the related generalized theories, many different specialized TS exist at various levels of abstraction—TS-"sorts" (see Section 6.11.10). For design engineering, a specialized IS of TS branch knowledge can be set up for each of these TS-"sorts"; see Chapter 7. Much of the information will normally be common among several specialized TS, for example, the applicable engineering sciences will be the same for different TS.

I.10 DESIGN ENGINEERING, ENGINEERING DESIGN PROCESS

Design engineering is the process of designing of technical (object) systems and process systems. Design engineering applies particularly to designing the internal and cross boundary capability for action of a TS. The main task and goal of design engineering is to establish a full description of an optimal TS for the given conditions and requirements, which often include the full description of an optimal TP. Design engineering should produce this description in the shortest possible time, and with highest efficiency and effectiveness, and at an acceptable cost, "right first time."

This full description should be sufficiently complete for manufacture of a TS(s) or implementation of a TP(s). It should anticipate any foreseeable circumstances that may arise during the remaining life of the designed system, including manufacture, distribution, operation (its duty cycle; see Section I.9.2.3), and disposal, during normal operation, fault or error conditions, and anticipated misuse.

NOTE: An *optimal* TP and TS should also be sufficiently insensitive to changes in its manufacturing and operational processes, *robust*, that variations of properties within tolerance limits do not bring the overall performance outside the allowed operating range [344,530,531].

Optimality can best be achieved if there is a possibility of generating and exploring several alternative solution proposals, and selecting among them in a reliable way. Searching for alternatives involves engineering creativity—a learnable human ability.

Achieving an optimal TS(s) needs a suitable technology of an engineering design process, a way of performing and proceeding during design work. Only methodical and systematic designing can heuristically hope, but not guarantee, to reliably achieve an optimal TS(s).

Designers therefore need broad information and need to know about TS in general, about operational processes for these TS(s), about other life cycle processes, and about engineering DesP (see Figure I.3). They also need information about societal, cultural, economic, and other conditions, compare Chapter 3. During design engineering, there is always a possibility and necessity of questioning, conferring and receiving information from consultants, specialists, and colleagues.

I.10.1 Design Engineering—Needs of Society and Their Satisfaction by Technical Systems

Society and individuals have many needs and wishes. These are manifested in the essential tasks of organizing their lives, establishing a livelihood, avoiding hunger, obtaining security, trying to improve the quality of life, and so forth. Needs and wishes also arise as a result of establishing a society, interrelationships, infrastructure, trade, communication, social and political administration, housing, industry, agriculture, recreation, transportation, power, community services, laws, culture, community, and so forth (see Section 11.1.4). Technologies (definite article) also enter the humanities (book printing and libraries, theater equipment), fine arts (acrylic paints, brushes, sculpture tools), and so forth.

A need implies that a suitable product is unavailable. Something must be (artificially) transformed from an existing state into a different state to satisfy the need. Any such changes consequently demand that *technology* (no article) must be applied. The applicable state of technology depends on the state of the society, which in turn depends on the available level of technology, one cannot *develop* without the other. As technology is modified, so the society must change, and vice versa—some gains will be made, but something must also be lost.

Development also implies that values are created—science, engineering, and technology cannot be value free [26,471,472,494]. Values are interdependent with society and products in a complex way (see Figure I.15A), resulting in needs, requirements, wishes, desires, visions, dreams—different levels of demand for satisfaction—but also restrictions and constraints. Some of these relationships are supporting, others may lead to conflicting requirements or results. Added *functionality* always costs something, may conflict with safety and efficiency, or have ergonomic, aesthetic, economic, ecological, and other consequences. Evaluating these relationships for specific cases and situations is always needed. Figure I.15B shows design engineering, as a central activity, in at least three axes of influence on life and society.

Historically, societies, cultures, and technologies first progressed by developing and *using tools*. Energy was initially delivered and control performed by the human, and energy was later provided by domesticated animals.

An essential element of this development, and in all progress, is that failures and errors occur, and are observed and overcome [466,467]. Attempts to explain these failures and errors requires *learning*, an increase in information and knowledge—individuals *learn* from the situation, discover new ways of operating, transfer this information into experience—and in the process "forget" or "unlearn" the obsolete information.

The second stage of progress, *mechanization*, was accomplished by using tools to develop machines, including operational machines (water pumps, ground tilling devices, printing presses, weapons, and defense equipment) and transporting machines (e.g., boats, the wheel).

The third stage, *powered mechanization*, developed energy delivering or converting machines. Humans, with limited force capability and power output, were replaced by *mechanical prime movers* (first technical-scientific revolution)—water wheels, turbines, combustion engines.

In the fourth stage, *automation*, control devices for automatic use of machines were developed. Humans are slow and unreliable, with limited mental capabilities (compare Figure 6.1), and were replaced in *regulating and controlling* functions by machines (second technical-scientific revolution), initially as mechanical automation—mechanical governors, sequence regulators, analog electronics, fluidics, and so forth.

Recently, with *computerization*, electronic machines were developed for information processing, regulating, and controlling. Routine decisions can be algorithmized, and some of the decision-making functions transferred to these machines. Integrated structures, *mechatronics* and *robotics*, combine computers and their programs into mechanisms, and monitor them by sensors (transducers).

(A) **Values in technical activities**   *Effects of properties of TS(s) on other systems*

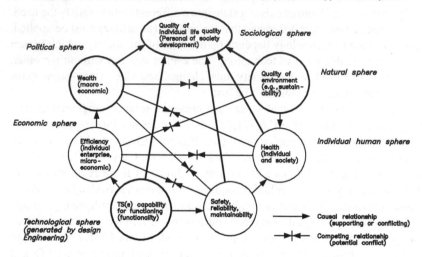

(B) **Centrality of design engineering in context**

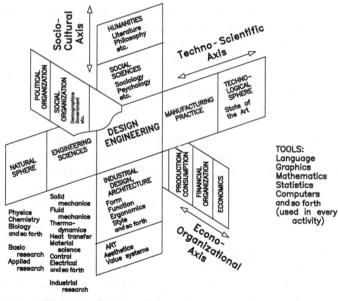

Literature
Part A
Ropohl, G. (1990) 'Die Wertproblematik in der Technik' (Value Problems in Technology), in
   Roozenburg, N. and Eekels, J., *WDK 17 — EVAD — Evaluation and Decision in Design*
   — Reading, Zürich: Heurista, pp. 162 — 182
   — VDI (2000) VDI Richtlinie 3780: *Technikbewertung — Begriffe und Grundlage* (Technology
   Evaluation — Terminology and Fundamentals), Düsseldorf: VDI-Verlag
Part B
Dixon, J.R. (1966) *Design Engineering*, New York: McGraw-Hill
Roseman, M.A. & Gero, J.S. (1998) "Purpose and Function in Design: from the Socio-Cultural to
   the Techno-Physical", *Design Studies* Vol. 19 No. 2, p. 161—186
Hundal, M.S. (1997) *Systematic Mechanical Designing: a Cost and Management Perspective*,
   New York: ASME Press

FIGURE I.15   Role of design engineering in context of technology and society.

The computers can even "learn" the desired response by genetic algorithms and AI, usually without ability to explain the response.

Craftsmen usually performed the earlier developments in slow evolutions. Only with the introduction of "division of labor" in the industrial revolution were the processes of designing and manufacturing separated, and formal scientific knowledge developed and applied.

Different cultures and sections of society will be at different stages of development in a nonlinear progression, and development stages may overlap. The best available technology (definite article) in a specific field at any one time is called the *state of the art*, and different levels may be found in such fields as education, health care, sports, entertainment, manufacture, transportation, and so forth. Knowing, know-how and know-what available to a person is termed the *personal state of the art* [365].

I.11 ENGINEERING DESIGN SYSTEM—STRUCTURE

Engineering designers are the most important factor in design engineering. Other factors play significant roles. All factors are interrelated, have relationships to factors outside designing, and constitute a system. The model of the engineering design system shown in Figure I.16 is derived from the model of a general transformation system.

The TrfP of design engineering comprises: (1) The initial operand—information in state Od1, ready to be processed, that is, the given requirements for the TP(s) and TS(s) as input to the design process, the nature of the task at the start of designing, explicitly stated, generally implied, or normative/obligatory [10, article 3.1.2], including wishes, desires, dreams, restrictions, and constraints. This operand may be a design or requirements specification, a design brief, a request for proposals, and so forth from a customer or marketing. (2) The processed operand of designing—information in state Od2, that is, a full description of an anticipated TP(s) and TS(s) to optimally fulfill the given requirements, as the goal of the design process, closely approximated by the output information. (3) An engineering design process, a transformation of information, from (1) state Od1 "list of requirements," to (2) state Od2 "description of a TP(s) and TS(s)" (see Section I.12). This includes procedures employed by engineering designers, the *design technology*, containing formalized and informal methods, techniques and intuitions (see Section I.12.2).

This transformation is realized or influenced by typically five *operators* of this process: (1) *Engineering designers*, the most important operator, with various levels of physical and cognitive skills and abilities, tacit knowing, motivation, and responsibility—for the overall project or only a small section. A model of the ideal designer indicates the necessary knowing (see Chapter 2). This also involves characterizing the engineering design process by *criteria for evaluation* of design engineering and of the designer, and deriving the educational requirements for engineering designers (see Section I.11.2). (2) *Working means*, the technical assisting means available to designers—tools, equipment, and so forth, including computers (see Section I.11.3 and Chapter 10). (3) *Working conditions*, the active and reactive environment in which the engineering design process takes place, the working climate, TS-"sorts," and market; see Section I.11.6. (4) *ISs*—information about and for

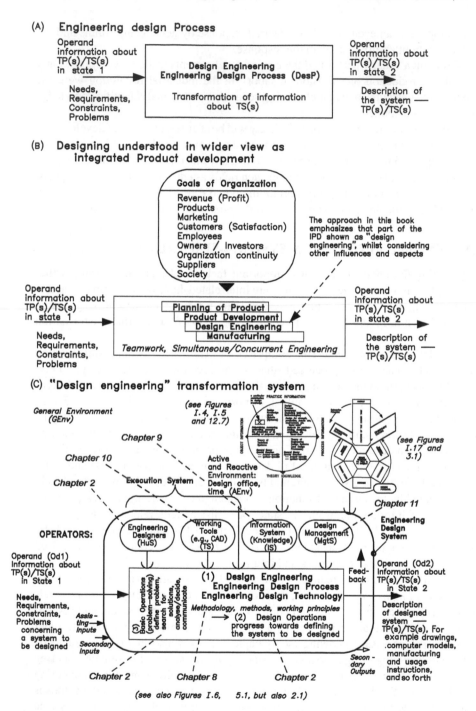

FIGURE I.16 Models of the engineering design process and engineering design system.

design engineering, that is, objects and phenomena, technical information, computers, programs/applications (e.g., CAD, engineering science analysis, finite element methods, AI, IT), internet and database records, recorded, collected, and codified into an IS (see Sections I.11.4 and 9.1, and compare Figure I.3). (5) *Design management*— management of the design process and its tasks, organization; leadership system; see Sections I.11.5 and 11.3.1 (i.e., distinct from the management of the product range and the organization; see Section 11.3.2).

The *situation* at the current stage of designing influences all the operators, the design process itself, the formulation of EDS, and various assisting and secondary inputs (see Chapter 3). Some secondary outputs occur as a result of the design process and its operators.

I.11.1 Engineering Designer (HuS)

Engineering designers are the most important operators of engineering DesP. Psychology (see Section 11.1) delivers insights into "designerly thinking" [109,110], rational/logical and intuitive. Knowledge about designing minds needs to be incorporated in expositions of design engineering [438, p.12]; the contribution of the individual is essential. Every idea or transformation of the TrfP(s) and TS(s) being designed is channeled through somebody's individual cognitive "machinery," and is subject to the restrictions and laws of the human mind. Design engineering has to be managed to obtain an optimal solution. The level of engineering design creativity needs to be improved to answer the growing needs of customers and the challenges presented, for example, decreasing natural resources, or increasing complexity of new technologies. Knowledge and experience about engineering thinking is needed to eliminate costly and harmful errors. Consequently, the mental processes of engineers, and the cognitive system, need to be set inside considerations in design engineering.

Designers pick up the relevant information by apperception from the available assortment of external representations, directing attention to interesting spots [435]. They encode the information into mental representations, based on mental models that they have learned. These representations are revised by producing and interacting with (transient) written and graphical notations. Contemplating these embodiments of thoughts, and comparing them to other relevant representations of information, designers refine their thoughts to mental representations better fitting the situation— and externalize them again. Consequently, the forms of external representation need to be considered, which is one purpose of this book.

This externalized material indicates the progress of the engineering design process, and constitutes the only means for supporting and organizing the internalized core processes of design engineering, the thought processes of individuals. Externalized representations in models of various kinds, as presented in this book (see Section I.7.1), provide visual feedback—their form and information content is obviously of vital importance for supporting the individual mind and for team communication.

NOTE: With respect to *mental capabilities*, see Section 11.1, the literature uses words such as thinking, learning, intellect, inspire, association, ideas, intuition, instinct, serendipity, idiosyncrasy, reflection [167,506,507], skill, ability, flair, talent, spontaneity, judgment, feel, flexibility, conscious, subconscious, preconscious, unconscious, and creativity (see Section 11.1.7).

With respect to *information*, words include experiences, expert knowledge, heuristics (see Section I.1), internalized, recorded, and codified.

With respect to *procedural aspects*, words are problem solving (see Chapter 2), iterative, recursive, decomposition/composition [236], opportunistic, algorithmic, innovative, inventive, evolutionary, team work, specialization, efficiency [10, article 3.2.15], and effectiveness [10, article 3.2.14].

All these concepts seem to be essential to designing and design engineering, but as individual statements none of them captures the essence and complexity of design engineering. Given the same problem, no two persons or teams will produce identical proposals, concepts, or embodiments.

I.11.2 WORKING MEANS (TS)—TECHNICAL SUPPORT FOR DESIGN ENGINEERING

All technical tools and equipment are included here: writing and drawing implements, sketch pads, pencils, tracing paper, drafting machines, calculation equipment, computer hardware, communications hardware, desks, filing cabinets, copying and printing equipment, and so forth (see Chapter 10). Older tools such as pencil and paper are still potent, especially in the early stages of designing. The quality of these items has some effect on the quality of designing.

I.11.3 ACTIVE AND REACTIVE DESIGN ENVIRONMENT—INFLUENCE ON DESIGNING

Engineering designers work within a physical, mental, and social environment that influences their performance. The physical environment includes the space, display facilities, heating/ventilating, and other building factors of the spaces in which designing takes place. The mental and social environment is partly due to management actions, for example, security of the designer in the organization, or time-limited work under contract for a certain design order or contract.

I.11.4 INFORMATION SYSTEM—SUPPORT FOR DESIGN ENGINEERING

This includes recorded and tacit information, within and outside the organization, including literature, standards, codes of practice, information about TS-"sorts" (Section I.7 and Chapter 6), research results, memories, and so forth (see Chapter 9). A special place is occupied by computer software; see Chapter 10 and [9].

I.11.5 MANAGEMENT SYSTEM—SUPPORT FOR DESIGN ENGINEERING

Because of the complexity of the design process (Figure I.16), the management operator is important for the effectiveness of design engineering. Formulating the goals for the organization and for design engineering is critical; see Section I.11.6 and Chapter 3.

Two main duties emerge. One is to manage, plan and rationalize the organization, including the range of organization products, and to define the tasks set for designers (see Section 11.3.2). This includes IPD, but part of this planning may be included in the design process (see Figures I.16 and 2.1). The other duty is to manage the process and progress of engineering design itself on specific projects, including managing the design personnel and the documentation that results (see Section 11.3.1). This also involves managing the interface between the two duties.

Compared with other management tasks, managing the engineering design process reveals special issues. Some tasks of the MgtS are critical to long-term survival of the organization, for example, (a) managing information, (b) coordinating design engineering with manufacturing (concurrent or simultaneous engineering), and marketing, and (c) ensuring timely delivery of products with appropriate properties.

All measures to rationalize the design process should be coordinated by the MgtS, and contained explicitly in the management goals for the design teams. Among these tasks of management is *version control*, keeping track of changes to the design documentation, and ensuring that the current version (e.g., drawings) is used for manufacture and life cycle processes of the TS(s).

I.11.6 ENGINEERING DESIGN SITUATION

During design work, the state of the TP(s) and TS(s) at a specific time, and the state of the operators of design engineering, are collectively termed the design situation. Figure I.17 and Chapter 3 shows the influence of various factors, which change as designing progresses.

The classes of *internal factors* describe the design system at a particular stage of design engineering, and the *design potential* available to the organization. They include the *operators*:

FD1 *Humans*, individual designers and teams, other specialists, for example, from manufacturing, cost estimating, purchasing, sales, customer service, and so forth (see Section I.11.1).
FD2 *Technical systems*, working means, including computer aids (see Section I.11.2).
FD3 The *active and reactive environment*, organization factors that directly influence design engineering (see Section I.11.3).
FD4 *Information systems*, of systems and of DesP (see Section I.11.4).
FD5 *Leadership/management*, including goals and objectives (see Section I.11.5).

The classes of *external factors of the situation* are provided by the general environment

FT Factors of the design task, the current state of establishing the TP(s) and TS(s) during design engineering; problems and consequences that arise; planning, selecting methods; and thinking about future products, assisted by classifications of TS (see Chapter 6).

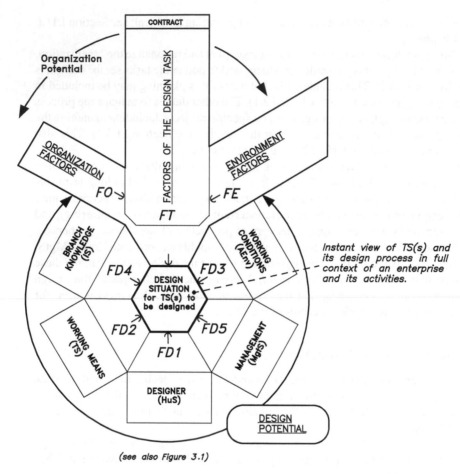

FIGURE I.17   Design situation—general model.

FO   Organization factors, administration, policies—manufacturing, selling, servicing, and so forth.

FE   Environment factors: human society, local and world economy, cultural, social, and political factors; and nature, local and world climate, environment impacts, pollution, and so forth.

The external factors include the organization potential, with contributions of cooperation, management, acquisition, human, financial, know-how, marketing and technology, and these contribute to the innovation potential of an organization [86].

The factors of the design situation are hierarchical: the design system, in the organization system (see Figure I.7), in a national economic and cultural system, in the world economy, and subject to the laws of nature. The connections and mapping for orientation must be understood.

The information for the process of design engineering is collected, an arrangement of concepts is shown in the model of Figures I.4, I.5, and 12.7 (see Section 12.4). This information is divided into two basic classes, from the morphology in Section 12.1.6: (1) information about (existing, "as is" state) TP(s) and TS(s), the operand of designing, the "west" hemisphere (see Chapters 5 to 7); and (2) information about DesP, the transformation of information from needs to full descriptions of proposed TP(s) and TS(s), the "east" hemisphere, outlined in Chapters 2 to 4; and information about prescriptions and heuristics for future ("as should be" state) TP(s) and TS(s). Each of these is further divided into two sections: (1) descriptive information, theory, the "south" hemisphere and (2) prescriptive (including normative) information from and for practice, the "north" hemisphere.

Figure I.4 shows these contents as four sectors bounded by two axes. "Normative" statements are also "prescriptive," but with obligatory application. The two "theory" quadrants (south) show a distinction between the general constituents of the science, and the specialized sciences derived from them for different TP and TS-"sorts" (see Chapter 7). The science-based information is placed inside the boundary of EDS, other information needed for designing is shown in Figure I.5 surrounding the boundary (see also Figure I.3).

I.11.7 Prescriptive and Methodical Information Related to Designed TP(s) and TS(s)

Normally, designing aims to establish a TP(s)/TS(s) that shows some improvement over previous systems, TP and TS are subject to development in time (see Section I.9.1.7).

For engineering designers to start their design work, and to achieve the necessary values of the properties, they must have available information about quantitative and pseudo-quantitative relationships of individual properties. For instance, the strength of a TS (and of its constructional parts) is usually a basic requirement for its ability to function, its durability and reliability—an expression of *finality*. Strength in turn depends on the structure and form, and the dimensions (sizes) of the TS (and of its constructional parts)—an expression of *causality*. C. Bach (1906) solved the questions of strength by recommending calculation methods for loading conditions, applied stress, allowable maximum stress, and guidelines for factors of safety, to establish the necessary dimensions, forms (shapes), and surface qualities. These are useful as first estimates; see quote from [482] in Section I.1. Newer analytical methods provide greater **accuracy** of prediction, for example, finite element methods, but at an increased cost of design engineering.

The validity of these calculations, and the assumptions made to enable them (e.g., that a calculation method is "conservative"), is really only shown when the TS is operating (see Figure I.18). Acceptance tests after manufacture of a TS can confirm some of the calculations and assumptions. Corrections in the design properties (classes Pr10, Pr11, and Pr12) can best be made if a recognized "error quantity" can be fed back from a more positive determination of the achieved values of other properties, *iteration*.

The feedback loop from (experimentally) determining a property to establishing the desired configuration and value of the property (designing) is illustrated in Figure I.18, on examples of technical and economic properties as a time function

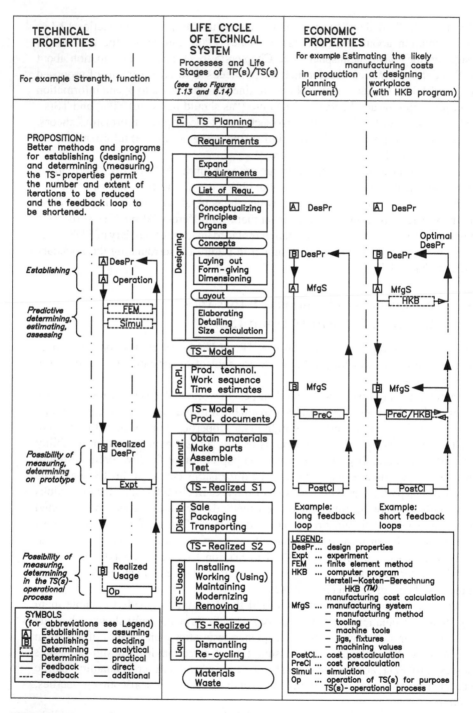

FIGURE I.18 Interval between establishing and possible measuring (determining) the properties of technical systems.

of the TS-life cycle. This feedback loop can be shortened by simulating the conditions expected in the TS-operational process, and therefore in the TP. For instance, manufacturing costs can be assessed in the design office (see Section 10.2). If a phenomenon (e.g., a failure mode) is unknown or ignored, a capability to function may suddenly and unexpectedly be lost, catastrophic failure may occur [434,466,467, 488,572].

NOTE: Diagnosing and researching failures is generally a way in which new information for engineering can be found. Experience is also gained by tracing fault occurrences detected in TP/TS (and in the other TrfP-operators), faults described by jargon as gremlins, bugs, viruses, snafus, and so forth.

Many other properties show a similar time interval between "establishing" in the engineering design office and "determining" (measuring, experiencing) in operation. The "control loop" of feedback that results from this interval should be as short as possible. Any changes in the description of the TS(s) at more concrete engineering design stages tend to be more costly, the "time to market" is decreasing, liability for failures is being applied more strictly, and so forth.

I.11.8 Design for X (DfX)

The information necessary to design a TP(s) and TS(s), to achieve the required states, values and manifestations of TS-properties by design engineering, is partly tacit internalized by the engineering designers, and partly available in the literature and other records, including standards and codes of practice. This information should be collected and arranged for use during design engineering. Classification according to TS-properties, functions, and output quantities (effects) has proved best, and is known as "Design for X," where "X" is a particular property or class of TS-properties (see Section 9.1.5).

Specialized theoretical knowledge can and must exist related to each specialized class of TS-"sorts" (see Section I.12.8). Related methodical information of "Design for X" (DfX) can be developed (see Chapter 4), based on the structure and content of generalized information.

I.11.9 Developments (Changes) of Properties with Time of Existing TS

Development has two major aspects. One is concerned with a particular TP/TS, and its progress from idea to realization and maturity for marketing at a certain time. In this view, development takes place as a part of the process of design engineering and manufacturing, and is intended to raise the product toward an acceptable technical level, and closer to the state of the art.

The second aspect concerns the developments of the technical level and the state of the art for a succession of TP/TS over time (see Section I.9.2.7 and Figure 6.16).

The succession of members of *development in time* of a TS is often divided into "generations," signifying a major observable change. At any one time various

(A)   **Relationships in developments in Time of technical systems**

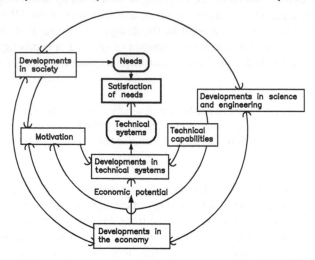

(B) **Transportation speeds**

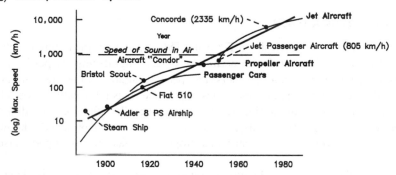

FIGURE I.19   Developments in time of technical systems.

measures of the parameters exist for different uses and conditions. Further measures are expected, related to development and manufacture of products. Considering individual parameters reveals aspects of the progress of developments in TP/TS over a time period (see Figure I.19A).

Investigating the technical level, typical values and states of properties of a TP/TS-"sort," development curves are obtained that indicate the technical levels existing at various times and under various conditions. Figure I.19B, demonstrates the general relationships, and the development of speed in individual sorts of transport systems. Each individual TS-"sort" shows a development curve for speed that has a characteristic shape, often an asymptotic approach to an upper limit, which is expressed by a law of nature, for example, the maximum attainable speed from thrust generated by aircraft propellers, or limits imposed by properties of materials such as strength. A special case of these limits is conditioned by the environment of the TS, for example, the speed limit set by law, or road surface. The development curve then need not approach the limit asymptotically, it may even intersect the limit, and the limit may also be subject to a trend in time.

I.12 ENGINEERING DESIGN PROCESS (DesP)

Design engineering aims to propose a TrfP(s)/TP(s) and a TS(s) that is anticipated to fulfill a given duty. *Requirements* should be specified, and are best classified according to the classes of TS-properties, Figures I.8 and 6.8–6.10. The TP(s) and TS(s) can then be *established*, by searching the solution space, considering alternative candidate proposals, evaluating, and deciding/selecting. A TP(s) or TS(s) as proposed can be *evaluated* with respect to fulfillment of the requirements. Any deficiencies can be corrected by iteration.

The model of an engineering design process is shown in Figures I.16 and 2.1, and is a concrete form of the TrfP, Figure I.6. If the demands and requirements of industry are included, the aims of design engineering are (1) a *description* of an optimal TP(s)/TS(s); (2) in the shortest possible *time*; and (3) with optimal *efficiency* and *effectiveness* (see Section I.10).

Procedures that can be heuristically used in design engineering range from trial and error, in stages to a full methodical and systematic procedure (see Figure I.20).

Efficient and effective is an alternation and mixture of intuitive and systematic procedures, flexibly adapted to the demands of the design situation, including complexity, originality, and other aspects. Yet only a planned and conscious, iterative and recursive, and systematic and methodical engineering designing procedure can effectively approach an optimal solution.

The optimal TP(s) and TS(s), designed and implemented or manufactured for the given requirements, should perform its tasks with appropriate efficiency and effectiveness, including life cycle costs, impact on the environment, and so forth. Rational and effective design engineering needs information about the design process, influencing factors must be analyzed; see next section.

I.12.1 STRUCTURE OF THE ENGINEERING DESIGN PROCESS

As with any other TrfP (see Figure I.6), the engineering design process can be analyzed (see Figure I.16). It can be hierarchically divided into identifiable *partial processes*, that is, phases, stages, steps, and so forth, which consist of a sequencing of *operations* (see Figures 2.1E and F). The main *working operations* transform the operand "information" from the given requirements (state Od1) to a full description of the TP(s) and TS(s) (state Od2). The working operations are accompanied by assisting operations—management operations (analogous to regulating and controlling) and *auxiliary operations* (other information processing operations).

Engineering Design Situations, Figure I.17 (see Section I.11.6 and Chapter 3), occur in actual design problems. Routine tasks are repeats of a situation for subsequent problems, and methods should be adapted to them.

The engineering designer anticipates and establishes the future TP(s)/TS(s), including finding any consequences, for example, as in life cycle engineering [62,160,186,237,244,582]; see also Sections I.1, I.7.1, and I.10. The designer also

		Technology of engineering design Processes (DesP)	General characteristics of Procedure	Comments
Information Support (IS)	A	Trial and error, cut and try	A series of trials (attempts) should find the solution for individual steps, and lessons from the result. Appraisal (evaluation) of the solution for fitness is then necessary	TS(s): accidental quality, possibly no result / DesP: time intensive / Prep: none
	B	Intuitive DesP (traditional)	The experienced designer spontaneously finds a fairly useful (but not optimal) solution. This procedure is mostly connected with creativity, inspiration, feeling	TS(s): usable, not optimal / DesP: very quick / Prep: long experience
	C	Methodical pragmatic and intuitive DesP	Combination of types "B" and "C", in which some steps are accomplished intuitively and by leaps, others methodically and according to a pragmatic or theory-based plan	TS(s): usable to optimal / DesP: quick / Prep: experience, schooling
	D	Methodical and systematic DesP	Procedure of a designer working systematically, using methods and a strategic plan, considering all aspects of the object to be designed and boundary conditions for the problem	TS(s): good quality possible / DesP: could be slow / Prep: schooling, training
Computer Support (TS)	E	Partially computer supported DesP	A computer is used to support some operations of the design process, for example calculation, drawing, storage. Better suited for methodical design work	TS(s): fairly good quality / DesP: possibly quicker / Prep.: schooling, esp. in branch
	F	Integrally computer supported DesP	A computer is intensively applied, and the majority of steps are supported by computer. The designer remains the key operator of the DesP	TS(s): fairly optimal quality / DesP: quicker / Prep.: long, in branch and organizational
	G	Automated DesP	Automated total or partial process of designing, in which only the data for the case are entered, and the solution is output with no further human intervention	TS(s): optimal quality / DesP: quick / Prep.: long, in branch and organizational

Legends:
Prep.   ...   Preparation of designer for an engineering design process
DesP   ...   Engineering design process
TS(s)   ...   Technical system to be designed

FIGURE I.20   Characterization of engineering design processes according to technology (strategy) of designing.

predicts and synthesizes the structures and behavior of the TP(s)/TS(s). Design engineering is a major stage in realizing a system (an organization product) (see Figure I.13).

Design engineering establishes all *properties* of the future system (see Chapter 6). Properties are important, about 80% of initial cost of a realized TP(s)/TS(s), and about 80% of running costs, are established by planning and design engineering, but expenditure for designing is less than 20% of initial cost; see Figure 10.5. Yet "the devil lies in the detail" [457].

The classification of properties shown in Figure I.8 is explained in Figures 6.8 to 6.10. The properties of a TS(s) are an aggregate of the properties of the parts, and important properties are generated by relationships (interactions) among the parts—synergy occurs.

Requirements for TP(s) and TS(s) are continually expanding. The aim should be shortest design time, with few iterations and recursions, and few subsequent alterations to the design documentation due to final implementation of the TS-operational processes—"right first time."

Contradictions and conflicts among the requirements often occur, especially in establishing the constructional structure of the TS(s). Compromises must be reached before an optimal constructional structure is achieved. The precise interdependence of some requirements and anticipated external properties with the proposed constructional structure is not known in advance. Several searches, iterations, recursions, and experiments may be needed. The limited mental working capacity of human designers also leads to the need for iterative and recursive working (see Chapters 2 and 11). Such reasoning processes constitute a *hermeneutical* circle [221] or spiral, and an *epistemic-ontological* process of striving toward constructive understanding of the desired transformations, functions, organs, and constructional parts, in order to create a product as an embodiment of the requirements of a customer [436].

The mixture of properties achieved by design engineering, that is, the quality of a future TP(s)/TS(s), is also a function of the recorded and codified *information* available to the designer (see Section 12.1). The designers' internalized (tacit) information, knowing (declarative and procedural information [270, pp. 363–365]), and thinking influences the result. Appropriate education and experience are needed to achieve the competencies (see Section I.1)—theoretical and practical, including experience of designing (see Section 11.6).

I.12.2 Technology and Operations in an Engineering Design Process

Design engineering involves *predicting* and *synthesizing* a TP(s)/TS(s), which needs compromise and avoidance of conflicts or contradictions in the relationships among individual parts. A new entity is created, with optimal properties for the envisioned task, for which generating alternatives and variants of goals and means, principles and embodiments, is important. Designing involves analyzing, evaluating, selecting, and deciding, searching for information, verifying, checking, reflecting [167,506,507], and so forth. Making the results of designing useful involves representing (e.g., graphically) and communicating.

The technology of design engineering describes possible ways of proceeding, including methods and heuristics, to perform the transformation of information from state Od1 to state Od2.

The technology is applied to operations, which exist at several hierarchical levels; see Figure I.21. Each operation at one level includes several or all of the operations at the next lower level. The top of this figure shows the engineering design system (see Figure I.16). Level 1 of Figure I.21 indicates the conventional stages of management of design engineering, macro-operations. Level 2 shows typical engineering design operations (see Chapter 4 and Figures I.22 and 4.1). The basic operations are shown at level 3 of Figure I.21, and constitute the cycle of problem solving with its

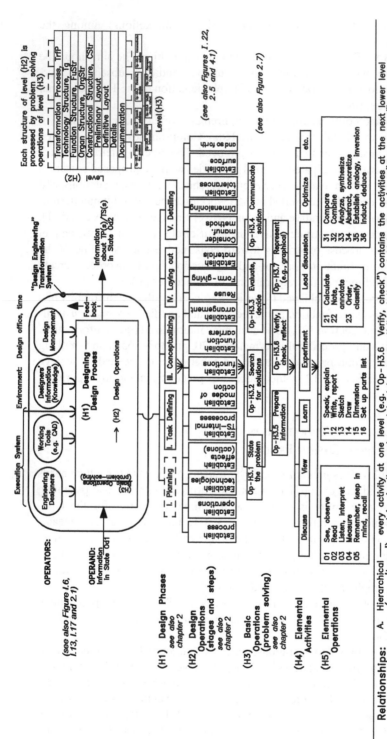

FIGURE I.21 Survey of hierarchy of activities in the design process.

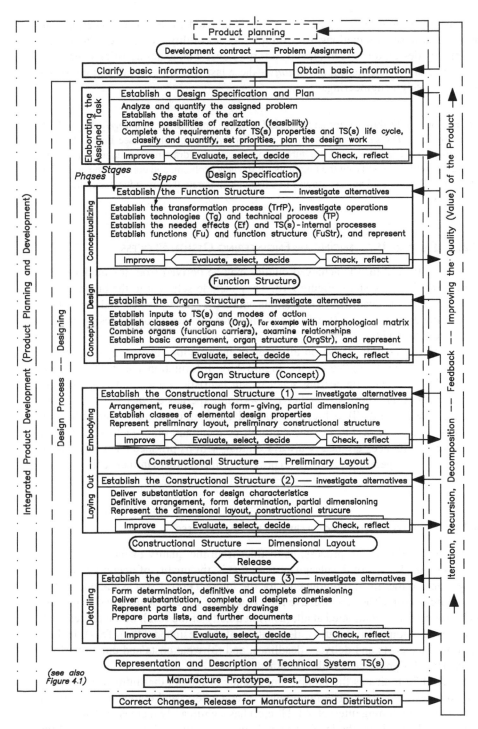

FIGURE I.22 General procedural model of the engineering design process.

auxiliary operations (see Figure 2.7). This is adapted to design engineering, but can be used in other situations, for example, for feasibility studies (see Chapter 2), preliminary design for tendering, and so forth. Level 4 indicates assisting operations and level 5 shows **elemental** *operations*—these last two are general, and no longer specific to design engineering.

NOTE: Figure I.22 shows steps and models for design engineering of a *novel* TP(s)/TS(s); see Figure 2.10. The same scheme can be used for *redesign* problems (see Figure 2.11) by abstracting from a constructional structure (layout or assembly drawing) to recognize its organs and their functions, modifying the functions for the new problem, and concretizing according to the "novel" scheme. The case studies in Chapter 1 illustrate the procedures and the use of the models.

I.12.3 Factors Influencing an Engineering Design Process

Influencing factors for the achieved quality may be derived from the model of the engineering design system, Figure I.16. They include self-cost to the organization, time to market, and deadlines of operations within the phases, stages, and steps of the engineering design process (DesP). They influence the quality of the design process and of the future TP(s)/TS(s).

Design engineering draws on all areas of information, in varying proportions depending on the TS(s)-"sort" (see Figure I.3). Information about TS is available in the pure and engineering sciences, in heuristic values and processes, in agreed conventions (norms, standards, laws), in experience [98,133,188,342,499, 521,556], and so forth. To achieve an optimal TS(s), information on the utilization and economic value of TP/TS, and cultural, political and other factors need to be considered. Influences are nearly always mutual, actions produce reactions, design engineering and operating TP(s)/TS(s) provides information to influence the existing information.

I.12.4 Quality of the Engineering Design Process

The quality of the design process, as distinct from the quality of the TP(s)/TS(s), depends on the factors of the design system, that is, the quality of operators, of technology, and of the given requirements for the future TP(s)/TS(s) as the operand of design engineering. They are decisive for: (1) whether the engineering design process can be successfully completed at all, (2) what quality of the result of the process will be achieved—output information, description of the designed TP(s)/TS(s), including committed costs and delivery deadlines, and (3) what costs and duration of the engineering design process will be attained.

Figure I.23 indicates the relative importance of factors that influence the quality of the design process. The importance of computers, and computer-aided (CA) technologies (including internet, AI, IT, etc.) has significantly increased. This results in growing demands on the level of knowledge, scientific and heuristic, needed for design engineering.

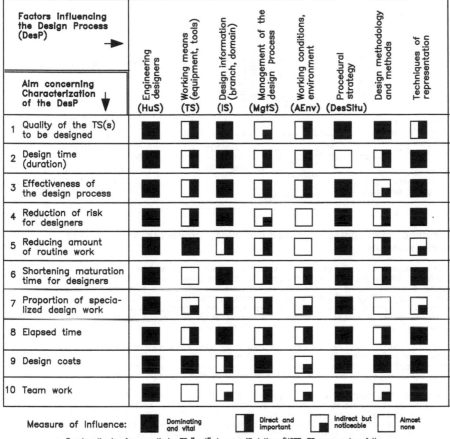

FIGURE I.23   Influence of some factors of the engineering design system for a particular design situation.

I.12.5 STRATEGIES AND TACTICS FOR DESIGN ENGINEERING

Strategies look for broad answers to the question: "How can a procedure for a complex task be generated?" Tactics deal with a single or small groups of steps.

Designers use methods for performing activities and operations that are repeated, for example, solving problems. Many methods that are learned and internalized, often by "trial and error," form the basis of "intuitive" actions, and are likely to be used from the subconscious.

Formalized methods are prescriptions for activities (phases, stages, steps, operations) that are intended to be heuristically useful for a purpose, characterized in Figure I.24 (see Chapter 8). Application is voluntary, unless a (normative/obligatory) law enforces usage. They are developed and published based on formal or informal theory, and pragmatic and heuristic utility. Methods may be strategic, tactical, or both. Strategic methods tend to be more comprehensive, and more difficult to apply.

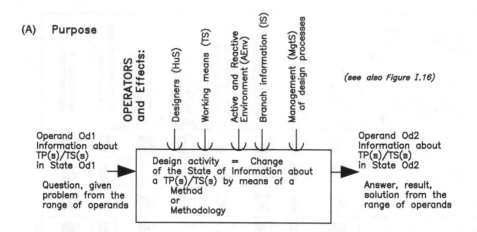

(A)   Purpose

(see also Figure I.16)

(B)   Method, methodology, procedural plan, procedural manner and working principle

The term design method is defined to mean a system of methodical rules and instructions.   These are intended to guide   and   establish the way of proceeding for executing a certain design activity and to regulate the interaction with the available technical means.   They are valid under the assumptions that a particular "normal" designer, with "normal" technical information (tacit and recorded), is acting within certain "normal" environment conditions.   A model which reflects all these elements is represented in this figure as process model of a design activity.

Each method can usually be sub-divided into more elemental methods, or combined into more comprehensive methods.   A coordinated set of design methods is termed a design methodology.   This term can cover different sets of design methods, for example for a region of activities or branch specialization (product family).

Each element of a method can be regarded as a working principle.   It usually delivers a universally valid instruction for appropriate action (conduct) in certain situations. Examples of such a working principles are: the guideline of providing constant wall thickness when establishing the form of a casting, or the endeavor to achieve minimal manufacturing costs in all design activities.   As different levels of a hierarchy, a system of working principles plus further references can represent a working method, and a set of methods forms a methodology.

The existence of a method permits the development of a plan of action (procedural plan), which establishes the rules for behavior in a design activity for a particular (concrete) case.   A method can deliver a starting point for a series of plans of action which will change according to an altered task definition (referring to branch and kind of design activity), design situation, deviations from "normal" branch information, and the technical and organizational conditions of the task.

The individual way (of a certain designer) in which a design task is performed is called a procedural manner or designer's mode of action.   It can be derived from a method or an action plan, and depends on how far a certain designer deviates from the "normal". The extent of the knowledge and ability of a designer, tiredness and similar factors influence the utilized procedural manner.

Each design process can fundamentally be structured with the help of a <u>general</u> procedural model into more or less complex partial processes, phases, down to design operations and steps (see Figures   I.21,   I.22,   2.5,   2.7 and   4.1   The elements of procedure that emerge are also processes, within which the state of information is changed.   Each of these elements is therefore directed toward a (more or less) precisely formulated goal which is evident from the procedural model.   In order that these processes can proceed methodically and according to plan, directed toward the goal and under the given boundary conditions, corresponding rules of behavior and methodical advice must be available.   These are contained in the methods or in the working principles, which can regulate the work as reference points.

FIGURE I.24   General study of engineering design methods.

(C) Characteristics of methods

Each method carries a number of features which characterize that method from different points of view. Some of the features can also be used as organizing criteria, which help in generating classifications. These features of methods can be arranged crudely into two classes:
— Features which characterize the method as a tool, this is with respect to its application;
— Features which characterize the method as information.

The features of a method as a tool can be obtained with the help of the following key-words or questions:
— which goal (purpose) should be served by the method? For which areas of the designer's activity is it applicable (see figure I.21)?
— what is its breadth of application for a product family? Is it usable only for one field (branch, domain) or for several?
— is it only usable for certain conditions, or must certain conditions be fulfilled? What resources are necessary?
— for whom (type of operator, designer) is the method intended? For an individual person or for a team? What preconditions must the person fulfill?
— from what source did the method come? In which branch, area of knowledge, discipline, branch or domain did it originate? Is it based on pragmatic experience, or on a theory?
— how does the method work (behave)? On which phenomena it is based? What is the 'mechanism' of the method?
— which time demands does application of the method make?

A method can be characterized as information by means of questions about general properties of information:
— are the contents correct?
— Is its effect verifiable? (Sources, experiences)
— Are the contents complete, not only excerpts? Are further details available?
— Is the format of the description unequivocal, clearly understandable?
— How old is the information about this method? Is its status still current?
— Is the method of unrestricted application?

We term the system of design methods and working principles as design tactics (see Chapter 8); in contrast to design strategy (e.g., a procedural model, see Chapter 4) which has the task of providing the basic structure. Design tactics and strategy are the fundamental areas of design methodology.

(D) Application of design methods in practice

A design method or methodology should usually be learned on a few simple examples, before a real design task is attempted. Familiarity with the procedures of a method is usually a requirement to achieve a measure of success with that method. Learning a method during a critical task is often a recipe for failure.

FIGURE I.24 Continued.

Advantages are usually only realized after several applications, and then achieve a return on the investment of effort [505].

Design engineering is a highly individual activity involving recorded and tacit knowledge (see the NOTE in Section I.11.1). Designing is also a societal problem, of teamwork and team building, exchanging information, cooperating, mutually supporting, and so forth; compare Section 11.2, helped by many tools and methods, and a wide range of information.

A useful classification of DesP according to novelty of the TS(s) includes novel and innovative design engineering, or redesigning (see Sections I.1 and 6.11.2). Most design problems are redesigning, often as "reverse engineering" (see NOTE in Section I.12.2).

Engineering designers need a suitable path to support creativity, and toward a solution of their tasks, including intuitive and systematic activities. They need a planned process for achieving their goals, an appropriate manner of conduct in their typical situation, and guidance toward effective work. The presented methods should help designers to answer the question "how?," in what way, with what procedure, how applied, depending on the existing situation. Several methods—and a methodology— need to be considered.

The problems of introducing new knowledge into engineering practice also remains to be solved (see Sections I.13.2 and 11.5).

I.12.6 SYSTEMATIC STRATEGY FOR DESIGN ENGINEERING

The chosen strategy is presented in Figures I.22, I.24, 2.1, and 4.1, and Chapters 2 to 4. The strategy is generally valid for design engineering of any TP(s) and TS(s) (see Chapters 5 to 7). It offers possibilities of variation, search for alternative principles and means, at several procedural places and levels of abstraction (see Chapter 6), and delivers favorable conditions for optimal solutions. Each subproblem or subsystem can be treated in a procedure according to the model.

The strategy allows a transparent design process, structured into clear design operations, which enable comprehensive documentation of the developing system and DesP, and shows effective feedback loops. The knowledge and laws of the theory can be applied to prescribe a procedure and method; see "subject–theory–method" in Section I.1. Nevertheless, iterative and recursive working is essential within and between each of the stages.

Some simplified graphical models are included in this Introduction, with more complete models presented in the chapters. The simpler models are complex, and complete enough that they will need repeated and careful study—they present important elements, their relationships within the model and interrelationships to other models, and relevance for application. In engineering design applications, the nature of the modeled system can be better and recognized quickly, especially with additional explanations incorporated into the models. Two basic models form the basis for our procedural model:

1. The model of the TrfS (see Figures I.6 and 5.1) describes the artificial transformation of an operand. If an operand (M, E, I, L) in a TrfP is treated during an adequate time period by a suitable technology, its state changes (from state Od1 to state Od2). Delivering necessary effects (M, E, I), is the task of the operators: HuS, TS, AEnv, IS, and MgtS (see Sections I.6.1 and 5.2).
2. The model of the TS, in its operational process (TP), is an operator of the TrfP. If the inputs (M, E, I) to a TS are processed at a point in time internal to the technical system in a suitable way, the technical system delivers *effects* (M, E, I) as outputs. A causal chain exists, from TS-inputs via TS-internal processes, effects and technology, to changes experienced by the operand.

Different structures of the TS are possible (see Chapter 6), and are constituents of the "elemental design properties" in Figures I.8 and 6.8 to 6.10. Structures may be recognized at different degrees of abstraction, using different elements: functions, organs, and constructional parts; see Sections I.7.1, I.12.7, I.9.1.5, I.9.1.7, and I.12.9, and Chapter 6. The elements of each structure, and their relationships within the TS and across its boundary describe the nature of the TS(s) at different levels of abstraction.

The engineering design system (Figure I.16) is a transformation system (see also Section I.11). The engineering design process in this system transforms the information about requirements (input to design engineering) into the information about a TP(s)/TS(s) that is suitable to fulfill the requirements. This design transformation is accomplished by the effects delivered by the operators of the engineering design process—designer, technical means, design environment, IS, and management system—following the general model of the TrfP, Figure I.6. The technology of design engineering is the design procedure described in this book. The applied methods that act as instructions and recommendations in the procedural model are predominantly derived from the area of logic, philosophy, and cybernetics.

The model allows effective management of design projects and complete documentation. The procedures, methods, and models can be used in parts or as a whole, in various levels of abstraction, combined with intuitive thinking, and many other published methods.

I.12.7 TACTICAL ENGINEERING DESIGN METHODS AND WORKING PRINCIPLES

Tactical methods are those used heuristically in smaller steps of a strategy. They are supported by tactical working principles, generalized guidelines that record prior experience (see Section 8.2). These are identical with the strategic procedural principles mentioned in Section I.12.8, but are applied to finding means, TS(s) and TrfP on any level of complexity (see Section I.9.1).

I.12.8 STRATEGIC "PROCEDURAL PRINCIPLES" FOR "SYSTEMATIC/STRATEGIC DESIGN ENGINEERING PROCEDURE"

The prime condition for reaching an optimal solution for an engineering design problem, to anticipate a TP(s)/TS(s), is the existence of *alternative* solution possibilities. These can be found in the more *abstract* levels of modeling for the operational processes (TP), and for the TS(s) in the function structure, organ structure, and constructional structure (see Sections I.7.1, I.12.6, I.12.7, I.9.1.5, I.9.1.7, and I.12.9, and Chapter 6). These structures are used in several design phases, level H1 of Figure I.21: clarifying the task, conceptualizing, laying out (embodying), detailing (elaborating), and at several intermediate levels of abstraction (see Sections I.1 and I.7.1, and Chapters 2 to 4).

Design engineering involves a series of steps: defining (intermediate) goals; finding plausible alternative means to solve the goals; evaluating the alternatives; selecting and refining the most promising among them; and in the end establishing them well enough to allow manufacture of the TS(s), and implementation of the TrfP, especially of the TP. This constitutes *optimizing* in concepts and principles, and when necessary using mathematical formulations and methods. Good records of considered alternatives should be kept, to allow backtracking if a chosen alternative becomes less than optimal in a subsequent stage.

The TS(s) can imply choices among possible strategic principles for design engineering, depending on the design situation—top-down, progressing from a holistic view toward the detail, or bottom–up, from detail toward overall definition (see also Section I.4.1).

It is usually not possible to solve the requirements directly, that is, to give an immediate answer to the question of engineering DesP: "*With what means* can the problem be solved?" Finding (establishing) the optimal means, a TS(s), and its operational process, a TP(s), demands different directions related to strategic procedural principles and tactical working principles—used individually or in combinations (see Chapter 8):

- From the *abstract* to the *concrete* (concretization, sometimes helped by abstraction)
- From the *rough* (approximate) to the *refined* (complete, accurate, and precise)
- From the *preliminary/provisional* to the *definitive*
- From the *general* to the *special* or specific
- From the *complicated* to the *simple* (but usually with a need to reintegrate—recursion)
- From low to high probability
- From possible to optimal

At times, working in the opposite direction is useful to open further possibilities or to retrace steps, characteristic of *iteration* for designing. The effectiveness of iteration, and so forth (see Chapter 2) depends on: (1) theoretical knowledge, information, conceptual understanding and awareness, generalized and specialized, about objects, processes and phenomena (see Section I.8); (2) existing experiences, that is, practical object and design process information, and related skills and abilities; (3) applied methods (i.e., methodical information and skills in applying methods); and (4) the quality of the starting point of the approximation, for example, the given design requirements in the design situation, or the initial starting solution for iterative refinement.

I.12.9 Recommended Systematic Strategic Procedure
for Design Engineering

The recommended procedure for systematic design engineering follows from the structures and properties of TP(s)/TS(s), and is shown in Figures I.22 and 4.1. Once a suitable operational process is identified for a future TP(s)/TS(s), the list of requirements can be derived. On that basis, the TS-cross boundary and TS-internal functions and their relationships can be formulated as a function structure. Organs that can implement these TS-functions can then be established, and by exploring the possible arrangements of organs, an organ structure. These organs can then be concretized and embodied into constructional parts and their relationships, a constructional structure. At each of these stages, additional (evoked) functions, organs and constructional parts may be recognized, which are needed to complete the requirements—a "function–means" hierarchy (see Figure 5.5). This is the basis for the procedural model in Figures I.22 and 4.1 (see also Sections I.12.1 and I.12.2).

I.12.10 TACTICAL METHODS AND WORKING PRINCIPLES FOR DESIGN ENGINEERING OPERATIONS

Questions need to be asked about the TP(s) and TS(s), and their engineering design process.

The operations of design engineering must be adapted to particular design situations (see Section I.11.6 and Chapter 3). The process of design engineering should be adapted from the procedural model (Figures I.22 and 4.1) by establishing a *procedural plan* for the design situation, the design problem, its TS-"sort," and its highest (useful) level of abstraction (see Section 7.4). The procedural plan can then be flexibly followed by the individual designer or design team, using their own *procedural manners*. During this procedure, the design situation will change, and the procedural plan and manner will need to be reviewed and adapted.

Tactical procedures are mostly related to problem solving. These basic operations (Figures I.21 and 2.7) can and should be frequently used with iterations and recursions. Reflection [167,506,507] is essential to learn from the results of design engineering.

I.13 ENGINEERING DESIGN SCIENCE

Engineering Design Science has drawn on and evolved with the information of various other disciplines. In addition to the general theories, we need specialized EDSs, Theories of Technical Systems, and Theories of DesP (see Chapter 7). Each of these concretize the general information for a particular phylum, class, family, genus, or species of TS.

Some consideration of human psychology for designing is needed (see Section 11.1). The human is an operator of the TrfP, and an operator, repairer, and maintainer of the TS, thus consideration of *ergonomics* is essential, relevant information is available, for example [584].

TrfP and TS are usually involved in an *organization system* (Figure I.7), which forms a part of human society, and enters considerations of products, markets, revenues, profit and profitability, design management, and so forth into the surroundings of design engineering.

The existing information about design engineering, as the process and human activity to create a TP(s)/TS(s), can be isolated, abstracted, and generalized into a science about design engineering—EDS [315]; see Figures I.4 and 12.7—a multi-dimensional region. Two important axes range between poles on the vertical axis in Figure I.4 of (1) *descriptive* theory and (2) *prescriptive* information; and on the horizontal axis of (a) existing TP/TS and (b) the design system (see Section I.11.6).

Outside the boundaries of EDS, other regions of information influence designing (see Figure I.5).

EDS (see Section 12.4) is a theory, intending to answer the questions of engineering designers regarding their designing tasks and procedures, preferably in forms that they can adapt as methods to their own design engineering problems. Adapting is difficult and passes through stages, from theory to generalized methods, next into adapted methods that are directly applicable in a given design situation, and finally introducing the methods into engineering practice. EDS should therefore support the mode of action, the procedural manner of designers, as hierarchical activity elements and relationships.

The four regions of design information shown in Figures I.4 and I.5 should be based as intensively as possible on scientific study, that is, logic and research. Information from practice should also be collected and codified (see Section 12.1.9). Theory should be prepared for use in practice, for example, by developing prescriptive information and methods. This information should be built into a logically unified system where the interrelationships among units are apparent.

The information introduced and explained in this book is presented mostly in *prescriptive statements*, which belong in and around the "northeast" quadrant of Figures I.4 and I.5, advice about design procedures and about considerations for the TP(s)/TS(s). The form of this information is different from the theoretical knowledge of *descriptive statements* (see Chapters 2 to 6 and 12) situated in the lower two quadrants. Prescriptive information should be easily accessible, by being arranged into suitable classifications, and cross-referenced.

I.13.1 RELATED SYSTEMS

Several other systems of prescriptive information for design engineering have been proposed, including design methodologies developed in continental Europe, for example [295,370,457,498]. They are similar to the system presented here, but are based on pragmatic considerations and personal experience, and are generally not as complete.

Comparisons were made to several more recent theories and methodologies [172], with the conclusion that they are conditionally compatible with the concepts presented in this book. References for earlier developments included [21,283, 287,295,299,304,314,315,370,457,498]; for procedural models [20,21,24,33,34,43, 67,457]; for pseudo-theories [39,40,421,523,528]; for set-theoretic proposals [236,393,394,500,543,588–591,593]; for AI applications [77,226,257,258,518]; concerning situatedness [173,174,227,348,375,492]; and concerning constructional structures [112,551,552,574–576].

Several management tools have been proposed. "Continuous improvement" (TQM) aims to continually search for faults, omissions, and ways of making the procedures and the products of an organization better. "Benchmarking" aims to identify the advantageous factors of a competitor organization to improve one's own procedures and products. "Concurrent engineering" or "simultaneous engineering" aims to design the product and its manufacturing and implementation processes at the same time, it is applicable mainly in the constructional structure.

Relevant aspects of these systems have been incorporated into EDS [315], and into the descriptive and prescriptive information presented in this book.

1.13.2 ROLE OF ENGINEERING DESIGN SCIENCE

Introducing EDS and its related methods into engineering practice is the subject for Section 11.5. Procedure and steps that can be followed are outlined, gradual introduction in small steps provides some successes, which in turn encourages further steps.

The role of EDS in education, pedagogical (curriculum, theories, strategies, etc.), and didactic aspects (presentation, supervision, coaching, projects, etc.), is the subject for Section 11.6.

Part I

Establishing Properties of Designed Technical Systems

The process of establishing all necessary design properties is described in Chapters 2 to 4. The main procedure follows the general procedural model, Figure 4.1, using the captions from that figure to guide the process. It is augmented by the basic operations of problem solving, Figure 2.7. This procedure must be adapted to the problem, as demonstrated by the case examples included in this part.

PROCEDURAL NOTE: Notes such as this add comments to the engineering design process.

1 Engineering Design Processes: Case Examples

A few case examples should illustrate applications of the concepts introduced in the introduction and chapters of this book. The information presented is mainly situated outside the "map" (Figures I.4, I.5, and 12.7), but is closely related to the "north–east" quadrant. Steps in each case study are according to the Procedural Model, Figure 4.1, and use the models and explanations of the book.

The first case, the tea machine, presents a novel design problem where industrial designers and engineering designers collaborated. The second case, the water valve, illustrates an engineering redesign process. The third and fourth cases, the smoke filter and one of its subsystems, show a novel engineering design problem where a subproblem is identified and treated as a novel technical system (TS).

The cases demonstrate that an engineering designer can idiosyncratically interpret the strict models to suit the problem and develop information in consultation with a sponsor. Opinions will vary about whether any one requirement should be stated in the class of properties as shown, or would be more appropriate in a different class.

These cases have been edited to show the final results after iterations from later steps of the design process, recursions, revisions, trial starts, and corrections. Such trials would have made the cases too long. The cycle at the end of each major section is therefore marked "reviewed."

PROCEDURAL NOTE: The models presented in the Introduction and the chapters are adapted to the problem, and the needs and preferences of the engineering designer—theoretical accuracy, correctness, completeness, and consistency are not necessary factors in using these models. We do not wish to make the case examples too rigid, for example, the alternatives of PERMIT, ENABLE, ALLOW are small changes in meaning, but indicate that these are tools of the designer, not normative/compulsory forms of statement.

The case studies illustrate applications of the relevant models for conceptualizing a new or revised TS, allowing the reader to assess the technical feasibility of the resulting proposals.

The stages and steps of the engineering design process leading up to the final detail and assembly drawings, or their computer equivalents, do not ever need be completed. At an incomplete state, progressing to the next step will usually allow the engineering designer to learn more about the previous step, and add to the results of that step— try out, learn, go back to revise, progressively develop the output information for each step using iteration and recursion. For instance, the evoked functions in the morphological matrix of each case could only be added after an investigation of some of the candidate organ structures, or even a preliminary layout. Consistency of formulations among the steps is desirable (e.g., in the transition from the function structure to the morphological matrix), but not essential. "Active" and "reactive" functions may be solved in the same row of a morphological matrix. If manufacture is intended, the final documentation must be as complete and correct as possible.

1.1 TEA MACHINE

A small enterprise wishes to enter the market with an automatic device, to prepare and deliver a hot tea beverage closely approaching traditional tea. The peculiarities of tea drinkers should be accommodated as far as possible.

Tea leaf, the raw material made from the tea plant (*Thea sinensis*), is either fermented or not, rolled, and broken or cut into small pieces. The resulting raw tea may be packaged, sold as loose tea leaf in a container, or sealed into tea bags. The tea beverage is usually made by bringing boiling water together with the tea leaf (loose, in a perforated metal "egg" or "spoon," or in a porous fabric bag), allowing the mixture to stand with possibly some mild agitation for 30 s to 5 min infusion time, and then separating the spent tea leaves from the beverage. The quality of the beverage depends on the chemical constituents of the tea leaf, especially its caffeine and aromatic oils (quick extraction) and tannins (extracted later in the water-contact period, causes browning and "strength" of tea).

NOTE: Various words are used in different regions for preparing tea, for example, mash, brew, steep, and wet. Milk, lemon, or an alcoholic liquid may be added.

(P1) Establish a list of requirements—investigate alternatives

P1.1 Establish rough factors for all life phases (purpose, operand, technology, operators, inputs, outputs of each life stage)

Tea as a beverage should be brewed from water and tea leaf with minimum intervention from human hand.

Final containment for the beverage should be a serving container—traditionally a tea pot.

P1.2 Analyze life phases, establish requirements on the technical process and technical system

Long maintenance-free life, and easy cleaning were prime demands.

P1.3 Analyze environment of the individual life cycle processes, especially TS(s)-operational process (TP) and its users

Usually indoors, no special requirements were recognized.

P1.4 Establish importance (priority level) of individual requirements, processes, and operators (fixed requirements—wishes)

Household or organizational customers, appearance and ease of use was most important.

P1.5 Quantify and tolerance the requirements where possible

One cup minimum, ten cups maximum for each usage. Problems may be
    a. Dispensing tea leaf and water in correct proportions.
    b. Establishing the right conditions for extracting the active chemical constituents.
    c. Regulating the extraction time, interrupting the contact between water and tea leaf.

P1.6 Allocate the requirements to life phases, operators, and classes of properties

An easily recognized human interface is important.

P1.7 Establish requirements for a supply chain, and environmental concerns

During distribution of the TS(s), packaging to protect against outdoor environment.

P1.8 Reviewed

(P1) Output   **Design specification** (see below)

Requirements are listed only under the most relevant TP or TS-property (see Figures I.15, 6.8–6.10) as judged by the engineering designer, and, not repeated in any other relevant property class. Indication of priority— F fixed requirement, must be fulfilled; S strong wish; W wish; N not considered.

Pr1 Purpose
F   Should prepare tea (infusion from prepared leaf material) comparable with that from conventional ways, self-acting and unsupervised once the device has been started.

Pr1A Function properties
F   Should separate the beverage from the tea leaf material after the infusion period.
F   Should allow serving into separate drinking cups.
S   Should keep the tea beverage sufficiently hot.
S   Should avoid or prevent chemical interaction of beverage and containing materials/structures, no leaching of chemical elements from container to liquid.

S   Should not need to remove tea leaf immediately after the cycle ends.

N   Any milk, sugar, lemon, alcohol, and so forth should be added later.

Pr1B Functionally determined properties

F   Quantity variable, minimum one cup, maximum ten cups (240 to 2400 ml).

F   Strength of the beverage should be variable from "weak" to "strong"— infusion time between 30 s and 5 min, use 1 to 5 tea bags, 5 to 25 ml loose leaf.

F   Water should be boiling (98°C minimum) on contact with the tea leaf material.

S   Infusion container should be preheated to about 50°C before boiling water is introduced.

W   Serving container should be preheated to about 50°C before tea beverage is introduced.

W   Normal environment is a kitchen, an office space, and so forth.

F   Use drinking water, 10 to 20°C, limited contents of calcium compounds (hardness).

Pr1C Operational properties

F   Free-standing device, not attached to a work-top or table.

F   Device not connected to a water supply.

F   Working from domestic electric supply (e.g., 110 V AC, 60 Hz, 15 A maximum for North America, or 220 V AC, 50 Hz for Europe, provide electrical connector plug for destination country).

F   Electrical safety to relevant standards, test by Underwriters Laboratory (UL) and Canadian Standards Association (CSA), apply control symbol on rating plate.

F   Life expectancy minimum 5 years.

S   Easy maintenance, especially with respect to hard water.

S   Easy repair for an expert.

W   Minimum space requirements for use of the device—TS-operational process.

Pr2 Manufacturing properties

F   Within limitations of own manufacturing facilities, plus OEM parts.

F   Small series, about 200 as initial quantity.

W   Use OEM-available constructional parts where possible, modular construction.

Pr3 Distribution properties

S   Easy transportation, packaging for safety, security of TS(s).

W   Minimum requirement for packaging.

Pr4 Liquidation properties

S   Materials recyclable, marked with material code.

W   Easy disassembly.

Pr5 **Human system factors**—for all life cycle phases

F Particularly important: Customer appeal, contemporary styling, should fit most decoration tastes—involvement of an industrial designer is essential (preferably a professional), from TS(s)-conceptualizing onwards.

F Easy and clear operation, clear controls, ergonomic interfaces to human.

F Safe operation, no contact of operator with hot device parts.

S Easy cleaning, internal and external.

Pr6 **Technical system factors**—for all life cycle phases

S Minimum requirements on other TS (as operators) during life cycle of device.

S Avoid special tools for manufacture, assembly, testing, and maintenance of TS(s).

Pr7 **Environment factors**—for all life cycle phases

F No significant environmental impact for materials used in device.

Γ High energy efficiency, low energy consumption.

S Conform to ISO 14000:1995.

Pr8 **Information system factors**—for all life cycle phases

S Minimum requirements for information, during TS(s) life cycle, avoid special additional information (including training), provide clear user instructions.

S Conform to ISO 9000:2000.

Pr9 **Management and economic factors**—for all life cycle phases

F Delivery of finished devices in 8 months from contract.

F Price to customers less than $50.00 (2005).

Pr10–Pr12 **Design Properties** (if any prespecified)
None.

(P2) Establish a plan for the design work—investigate alternatives

P2.1 Analyze and categorize the technical process and technical system from viewpoints that influence design work and planning

Involvement of an industrial designer was considered essential, early contact was made.

P2.2 Select overall strategy, partial strategies and operations for important partial systems

Full systematic procedure as in Figures 4.1 and 2.10.

P2.3 Establish degree of difficulty of the design work

Design difficulties could arise in transferring boiling water to and from tea leaf. It could also be difficult to heat the serving container (tea pot) before accepting the tea beverage.

P2.4 Establish tasks of operators and individual staff members

Needs designers with good communication capabilities for team work.

P2.5 Plan the anticipated design work, under time pressure

Six weeks as target, intermediate target dates need to be set for several stages.

P2.6 Estimate the design costs

Not available.

P2.7 Establish further goals, for example, masters, forms, catalogs, and so forth.

Precedents were available in coffee machines, but were not directly applicable for tea.

P2.8 Reviewed

(P2) Output   **Design specification and plan** (see above).

(P3a) Establish the Transformation Process—investigate alternatives

P3.1 Analyze the TS(s)-operational process—TrfP, TP

TS(s)-operational process should accomplish the main transformation; see Figure 1.1.

P3.1.1 Establish input to the operational process—TP operand state Od1

Use water from domestic faucet, tea leaf loose or prepackaged.

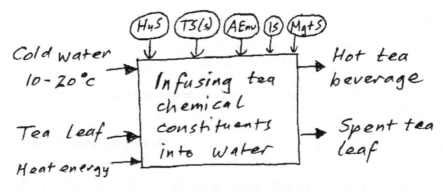

FIGURE 1.1   Tea machine—transformation process—black box.

P3.1.2 Establish operators of the operational process—assumed

As first assumption, human to measure and fill water, measure and supply tea leaf, set infusion time; TS(s) to accept electric power.

P3.1.3 Establish technological principle, technology—operations, sequencing

Heat water: source of energy: electric current phenomenon:
  a. Radiation
  b. Conduction (and convection)—preferred
conversion of electric power to heat by electrical resistance.
  a. Metal resistor—preferred
  b. Semi-conductor resistance
  c. Use water as resistance—danger
Immerse tea leaf
  a. Tea leaf into and out of water
  b. Water into and out of container that holds tea leaf—preferred
      i. With tea leaf loose—difficult cleaning may catch tea leaf in strainer
      ii. With tea leaf in container—porous bag, metal "egg" or "spoon"
Move liquids
  a. Use steam pressure—caution—needs pressure-limiting relief, needs vacuum relief for cooling
  b. Mechanical pump
  c. Gravity
Measure the states of properties: (1) water and tea leaf quantities, (2) water temperature, (3) steam temperature and pressure, (4) temperature of device parts, (5) time, (6) electrical conductivity, and (7) color of beverage.
Color is usually coordinated with "strength" and taste of the tea beverage for a given variety of tea leaf, including herbal and fruit teas.

P3.1.4 Establish necessary effects acting on the operand—technology in individual operations

Contain water and steam, release water and steam, transfer heat, limit maximum over-pressure, relieve vacuum, and so forth.

(P3a) Output   **Transformation process** (see Figure 1.2).

Also marked on the TrfP is an indication whether the main operator of an operation should be Hu, TS(s), AEnv, or a combination. Various alternatives were suggested.

PROCEDURAL NOTE: Compare with Figures 5.1, 5.3, and 5.4 to check whether any important elements may be missing.

(P3b) Establish the TS-Function Structure

P3.2 Work out the complete function structure of the TS(s)

The TS-function structure should deliver the necessary effects (see Figure 1.2).

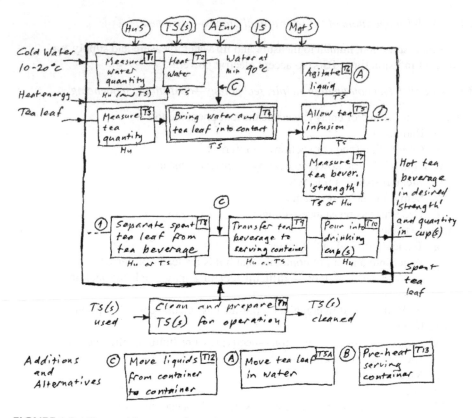

FIGURE 1.2   Tea machine—transformation process—structure.

P3.2.1 Establish distribution of effects between humans and technical systems (and environment)

See P3.1.2.

P3.2.2 Establish TS(s)-main functions of individual technical systems

One TS(s), or possibly two—for example, machine, and separate serving container.

P3.2.3 Allocate and distribute the main function among individual TS

P3.2.4 Establishing additional functions from an analysis of the relationships "human-technical system" and "environment-technical system"

Allow accepting tea leaf and cold water, accept electrical energy, accept settings for operation, and see P3.1.4.

P3.2.5 Establish the assisting functions from analysis of the transformation functions: auxiliary, propelling, regulating and controlling, and so forth

Needs measuring of states of operand and TS(s).

P3.2.6 Establish possibly evoked functions from important properties

Needs some automatic control of TS(s).

P3.2.7 Establish TS(s)-inputs, check of TS(s)-output regarding environment.

Electric power conduction, by grounded conductor system (three-wire power cord).

P3.3 Represent TS-function structures

See Figure 1.3.

P3.4 Assess and evaluate the variants of the function structure

For the machine, preferable alternatives are A, C1, C2, and C3, otherwise use functions in the main structure.

P3.5 Reviewed

(P3b) Output   **Function structure** (see Figure 1.3).

Also marked on the FuStr is an indication of an order of importance for the solvable functions. Various alternatives are suggested, for example, changing a function from active to reactive, or adding a possible function to complete the TS(s)-internal structure. The designer added symbols to show which functions were responsible for the effects exerted on the operand, and the operations in Figure 1.2 that are caused by these TS-effects.

PROCEDURAL NOTE: Compare with Figure 6.4 to check whether any important elements may be missing.

(P4) Establish the TS-organ structure—investigate alternatives

P4.1 Enter TS-functions from function structure into first column of morphological matrix

P4.2 Find action principles, possible modes of action

P4.3 Establish organs or organisms as means (function carriers)

See Figure 1.4.

PROCEDURAL NOTE: Compare with Figure 4.4 to check whether any important elements may be missing. In this case study, the engineering designer allowed some inconsistency in the formulation of the functions between Figures 1.3 and 1.4.

If we assume that there are no conflicts among the partial solutions in this morphological matrix, 1,285,956,000 combinations are "possible"—this number is too large for convenient use, and results from combinatorial complexity. Because of actual conflicts, the "potentially feasible" number of combinations will be much smaller.

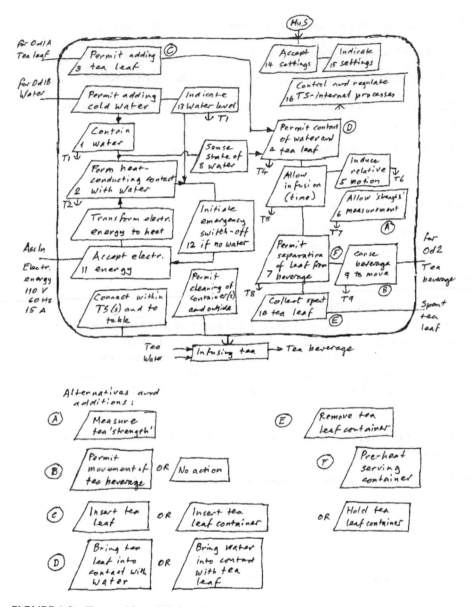

FIGURE 1.3   Tea machine—TS-function structure.

P4.4 Combine individual organs to a unity (a proposed organ structure)

P4.5 Evaluate organ structure proposals and select the best

At this point, the industrial designer contributed an exhaustive exploration of topology in principle (see Figure 1.5), as possible solutions to function 1. Columns 1 to 5 represent three separate containers, columns 6 and 8 represent two containers (one performing double duty), and column 7 represents one container performing all duties. After eliminating those that are identical or not feasible, he used an iconic representation (see Figure 1.6). It is not clear whether these are intended as top views or front views.

The engineering designer explored the combinations of functions 1 (variants a, b, and c), 2, 3, and 4 from the morphological matrix (Figure 1.4; see Figure 1.7), using one or two containers. She decided that arrangement **B-1** was the only promising one, the others were likely to be too complicated. Next she investigated the selected arrangement **B-1**, adding combinations of organs for functions 7 and 9; see Figure 1.8. The problems arising from a device with two containers, **B-1-I**, were considered to be too difficult. The engineering designer chose an arrangement with three containers, variants **B-1-II** and **B-1-III**. These were considered almost equivalents, except that **B-1-II** showed more restrictions on constructional arrangements, the serving container could only be placed below the brewing container. Variant **B-1-III** could be constructionally rearranged in many of the ways investigated in Figures 1.5 and 1.6, columns 1 to 5.

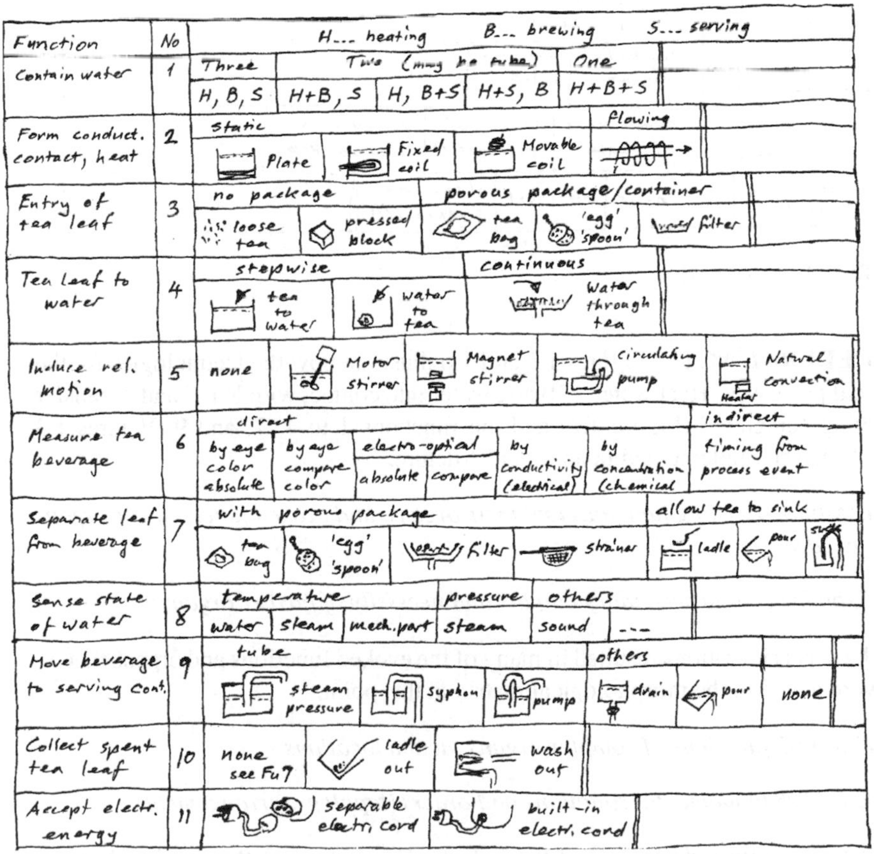

Function	No	H... heating B... brewing S... serving					
Contain water	1	Three: H, B, S	Two (mug to tube): H+B, S	H, B+S	H+S, B	One: H+B+S	
Form conduct. contact, heat	2	Static: Plate	Fixed coil	Movable coil	Flowing		
Entry of tea leaf	3	no package: loose tea, pressed block		porous package/container: tea bag, 'egg' 'spoon', filter			
Tea leaf to water	4	stepwise: tea to water, water to tea		continuous: Water through tea			
Induce rel. motion	5	none	Motor stirrer	Magnet stirrer	circulating pump	Natural convection	
Measure tea beverage	6	direct: by eye color absolute, by aye compare color, electro-optical absolute, compare, by conductivity (electrical), by concentration (chemical)					indirect: timing from process event
Separate leaf from beverage	7	with porous package: tea bag, 'egg' 'spoon', filter, strainer				allow tea to sink: ladle, pour, suck	
Sense state of water	8	temperature: water, steam, mech. part		pressure: steam	others: sound, ---		
Move beverage to serving cont.	9	tube: steam pressure, syphon, pump			others: drain, pour, none		
Collect spent tea leaf	10	none see Fu.7	ladle out	wash out			
Accept electr. energy	11	separable electri. cord		built-in electri. cord			

FIGURE 1.4 Tea machine—morphological matrix.

Function	No						
Initiate emerg. switch-off	12	eject cord plug from TS(s) body	TS(s)-internal switch with RESET button	none			
Indicate water level	13	graduated sight glass	separate measuring jug	flow measure			
Accept settings	14	analog		digital			
		rotary knob	sliding knob	key buttons			
Indicate settings	15	analog		digital			
		pointer on knob	scale on knob	LED display			
Control TS(s)-internal processes	16	mechanical		electrical	electronic	digital-electronic	
		thermal	bi-metal				
EVOKED Fu							
to 1		see partial solutions to Fu 9					
to 9 steam limit max. p.		relief valve		relief plug		control	
		weight	spring	pop-up	pop-out	pressure sensor + actuator valve	
to 9 steam avoid vacuum		none	open valves				
to 9 stop or start flow		use boiling pressure	actuated valve				
to 12 detect water shortage		temperature of heater	weight of container(s)	pressure on container base			
to alternative (F) pre-heat		none	electrical	steam from container(s)			

FIGURE 1.4 Continued.

PROCEDURAL NOTE: Figures 1.5 and 1.6 illustrate how the technologies for the technical processes "Heat water," "Bring water into contact with leaf," and "Transfer to serving container" (Figure 1.2) and the functions 1 to 4, 7, and 9 (Figures 1.3 and 1.4) can be implemented in an organ structure.

P4.6 Establish assisting functions and their organs, and evoked functions for organs where needed

P4.7 Examine evoked secondary outputs and necessity of further organs

These considerations resulted in many of the evoked functions and organs that were added in the morphological matrix at this point.

P4.8 Establish functions of complex organs, action locations

P4.9 Establish fundamental situations and orders (topology, arrangement)

The engineering designer and the industrial designer together explored the constructional arrangements in detail (see Figure 1.9) with respect to general

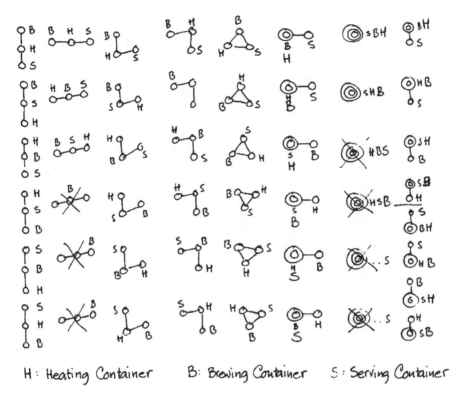

H: Heating Container   B: Brewing Container   S: Serving Container

FIGURE 1.5 Tea machine—function 1—topology in principle. (Reprinted from Hubka, V., Andreasen, M.M. and Eder, W.E., *Practical Studies in Systematic Design*, London: Butterworths, 1988. With permission from Elsevier.)

considerations. Another joint investigation produced proposals for liquid transfer, function 9 (see Figure 1.10).

P4.10 Represent final technical system proposals as organ structures with enough detail for evaluation and selection

The engineering designer produced more comprehensive organ structures (see Figure 1.11). She recognized that there was little to restrict the constructional arrangement, the tubing connecting the containers could allow many of the arrangements in Figures 1.9 and 1.11. She decided that an organ structure with three valves, variant (b), would best fulfill the requirements of the design specification. This decision evoked another function that could have been entered (with partial solutions) into the morphological matrix:

To 9   Keep steam pressure within limits—interrupt heating |actuate valves| combine interruption and valve actuation.

Using interruption of heating was considered less desirable, the heat retention in the materials would cause thermal "inertia" that could make control difficult. Continuous steam pressure (or repeated pressurizing) implied that

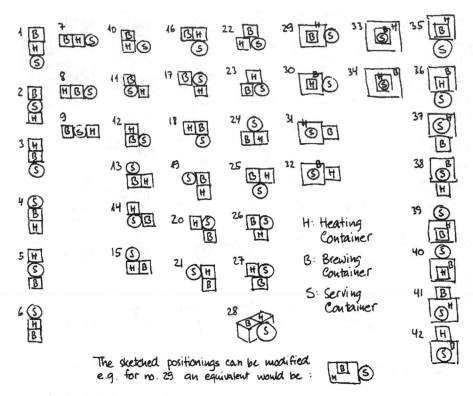

FIGURE 1.6 Tea machine—function 1—topology in iconic representation. (Reprinted from Hubka, V., Andreasen, M.M. and Eder, W.E., *Practical Studies in Systematic Design*, London: Butterworths, 1988. With permission from Elsevier.)

probably an extra cup of water would be needed in the heating container—two cups to produce one cup of tea, eleven cups of water for ten cups of tea.

P4.11 Reviewed

PROCEDURAL NOTE: When such an evoked function is recognized, it is usually useful to enter it into the developing morphological matrix; see comments to P4.6 and P4.7 above. Such entry can assist if review and revision is needed. It can also be useful to enter the evoked function(s) into the function structure, and even the relevant (technical) processes into the transformation process structure. It may then be possible to recognize combinations of functions (and processes) that would simplify the constructional structure. The resulting documentation can also help when radical innovations are intended. In this case, a transfer of the FuStr (and TrfP) to a graphics computer application can help the processes of updating the diagrams; see case study 1.3.

PROCEDURAL NOTE: Compare with Figure 6.6 to check whether any important elements may be missing.

(P4) Output   **Organ structure** (see Figure 1.11), variant (b).

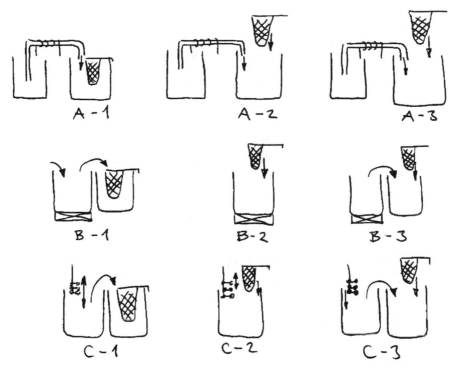

FIGURE 1.7　Tea machine—functions 1 to 4. (Reprinted from Hubka, V., Andreasen, M.M. and Eder, W.E., *Practical Studies in Systematic Design*, London: Butterworths, 1988. With permission from Elsevier.)

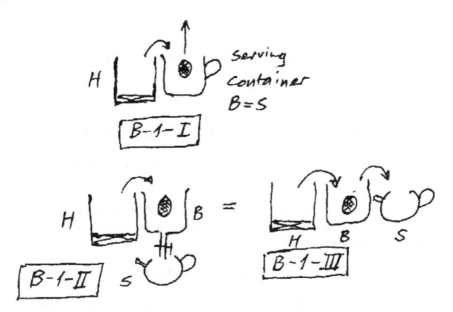

FIGURE 1.8　Tea machine—added functions 7 and 9—draining vs. transferring.

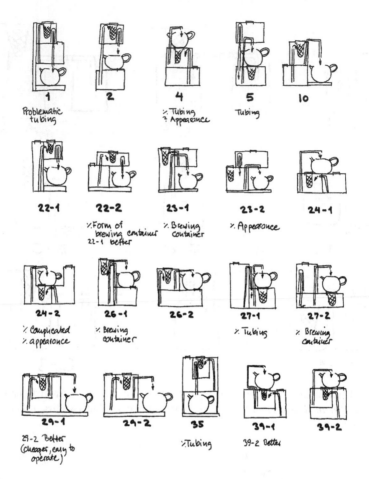

FIGURE 1.9 Tea machine—general arrangements. (Reprinted from Hubka, V., Andreasen, M.M. and Eder, W.E., *Practical Studies in Systematic Design*, London: Butterworths, 1988. With permission from Elsevier.)

(P5a) Establish the constructional structure (1)—investigate alternatives

P5.1 Establish and Analyze the Requirements for the Constructional Structure and Constructional Parts

This review, and a more detailed design specification, delivered preliminary considerations about materials and manufacturing methods for containers and all other constructional parts.

P5.2 Establish a Conception of the Constructional Structure

At this point the industrial designer contributed a set of preliminary layouts, as constructional arrangements that realize the chosen organ structure, with

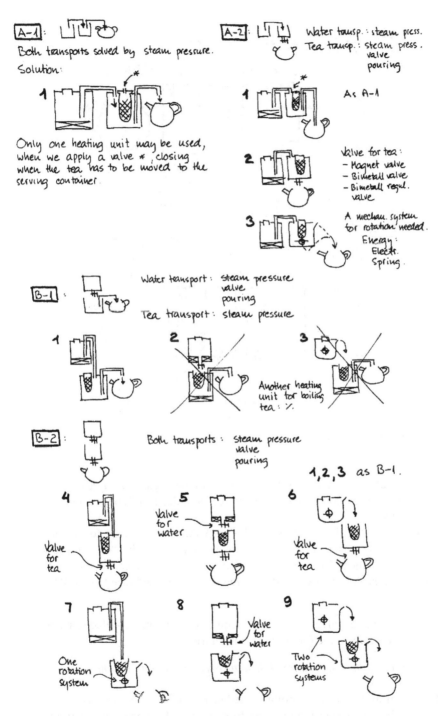

FIGURE 1.10 Tea machine—liquid transfer. (Reprinted from Hubka, V., Andreasen, M.M. and Eder, W.E., *Practical Studies in Systematic Design*, London: Butterworths, 1988. With permission from Elsevier.)

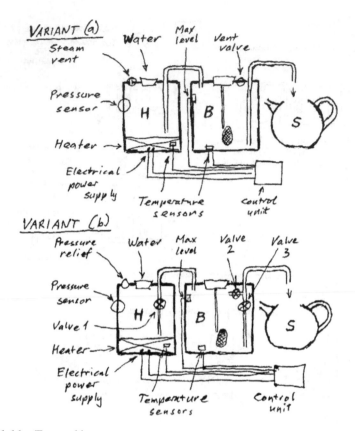

FIGURE 1.11   Tea machine—organ structures.

a rough evaluation according to five criteria (see Figure 1.12). The engineering designer explored functional surfaces and details of construction (see Figure 1.13).

The holder for the tea bag should keep the bag as close to the floor of the brewing container as possible to accommodate the "one cup minimum" requirement. Two, three, or more tea bags may need to be held, for up to ten cups. A further evoked function was recognized (see note to P4.10 above).

To 16   Actuate valves in correct sequence and timing—electromagnet valves| one rotary valve for all three tube entries

P5.3 Establish Rough Constructional Structure as Preliminary Layout

The industrial design at this point looked promising (see Figure 1.14).

The subsequent stages and steps from Figure 4.1 appear to be routine design work for both design engineering and industrial design. We have chosen not to complete these stages and steps. Nevertheless, the importance

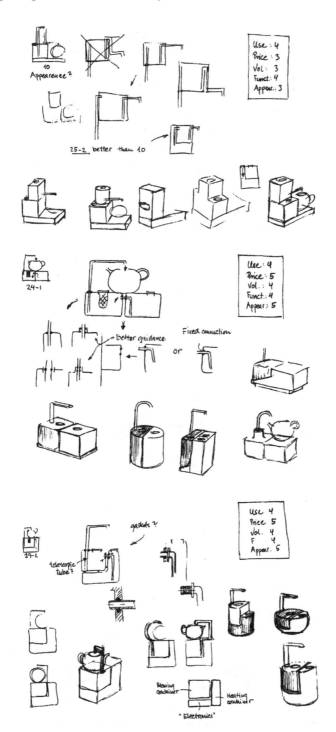

FIGURE 1.12   Tea machine—preliminary layouts—appearance. (Reprinted from Hubka, V., Andreasen, M.M. and Eder, W.E., *Practical Studies in Systematic Design*, London: Butterworths, 1988. With permission from Elsevier.)

of these subsequent steps must be emphasized, as many fault conditions may unintentionally be introduced in the embodiment and detail phases.

(P5a) Output   **Preliminary layouts**

(P5b) Output   **Constructional structure**

(P6) Output   **Manufacturing documentation**

NOTE: This case study was originally performed as a student exercise in an industrial design department. It was edited by Prof. Eskild Tjalve [540,541], Institute for Machine Design, Technical University of Denmark. We express our thanks to him for allowing us to use his drawings and considerations. The case study was published in [305], and was completely revised by the authors for this book to illustrate a full use of the procedural model, Figure 4.1, in the project.

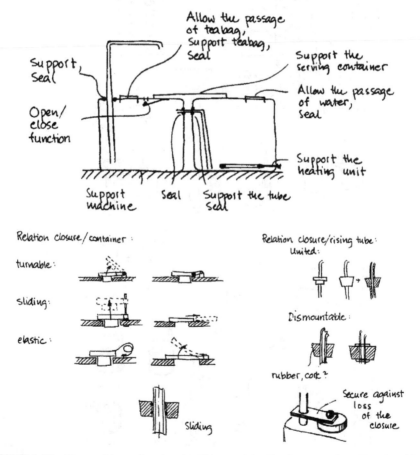

FIGURE 1.13   Tea machine—functional surfaces and details of construction. (Reprinted from Hubka, V., Andreasen, M.M. and Eder, W.E., *Practical Studies in Systematic Design*, London: Butterworths, 1988. With permission from Elsevier.)

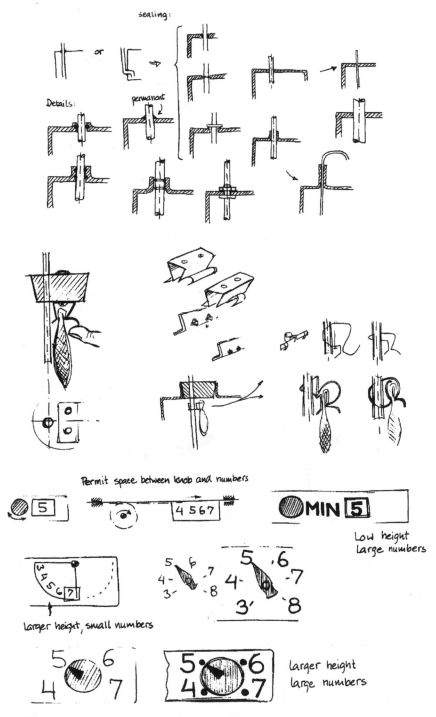

FIGURE 1.13 Continued.

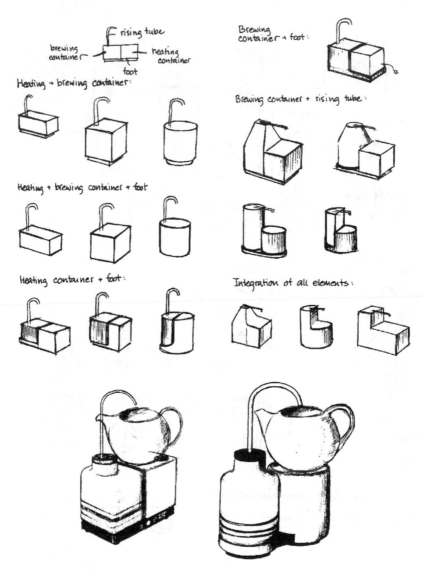

FIGURE 1.14 Tea machine—rendered representation of proposals. (Reprinted from Hubka, V., Andreasen, M.M. and Eder, W.E., *Practical Studies in Systematic Design*, London: Butterworths, 1988. With permission from Elsevier.)

1.2 WATER VALVE REDESIGN

Dissatisfied with available water valves, a small organization issued a contract to investigate a redesign. Water valves in pipe diameters 20 mm ($\frac{3}{4}$ in.) to 50 mm (2 in.) are usually produced with rising spindles, and are consequently tall—the handwheel is situated far from the pipe, and its position varies with valve opening (see Figure 1.15).

The contract stated preference for a nonrising spindle, with the handwheel as close to the body as possible. The new valve mechanism should fit into the existing valve body. The handwheel as a purchased OEM part, and its connection to the spindle should remain unaltered.

(P1) Establish a list of requirements—investigate alternatives

P1.1 Establish rough factors for all life phases (technology, operators, inputs, outputs of each life stage)

This is intended as a replacement for an existing valve mechanism.

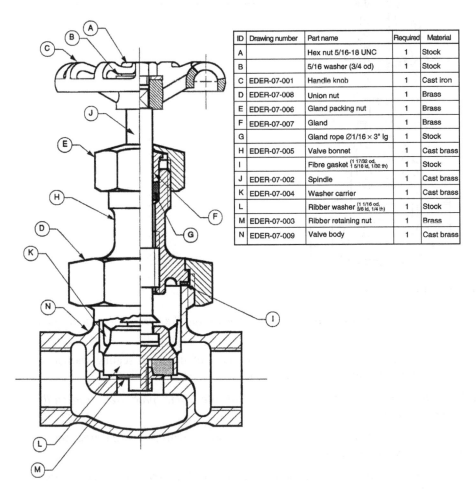

ID	Drawing number	Part name	Required	Material
A		Hex nut 5/16-18 UNC	1	Stock
B		5/16 washer (3/4 od)	1	Stock
C	EDER-07-001	Handle knob	1	Cast iron
D	EDER-07-008	Union nut	1	Brass
E	EDER-07-006	Gland packing nut	1	Brass
F	EDER-07-007	Gland	1	Brass
G		Gland rope Ø1/16 × 3" lg	1	Stock
H	EDER-07-005	Valve bonnet	1	Cast brass
I		Fibre gasket {1 17/32 od, 5/16 id, 1/32 th}	1	Stock
J	EDER-07-002	Spindle	1	Cast brass
K	EDER-07-004	Washer carrier	1	Cast brass
L		Ribber washer 1 1/16 od, 3/8 id, 1/4 th	1	Stock
M	EDER-07-003	Ribber retaining nut	1	Brass
N	EDER-07-009	Valve body	1	Cast brass

FIGURE 1.15 Water valve—general assembly. (Reproduced by permission of The Royal Military College of Canada, from Mechanical Engineering drawing EDER-07-010.)

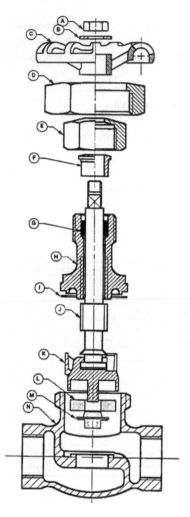

ID	Drawing number	Part name	Required	Material
A		Hex nut 5/16-18 UNC	1	Stock
B		5/16 washer (3/4 od)	1	Stock
C	EDER-07-001	Handle knob	1	Cast iron
D	EDER-07-008	Union nut	1	Brass
E	EDER-07-006	Gland packing nut	1	Brass
F	EDER-07-007	Gland	1	Brass
G		Gland rope Ø1/16 × 3" lg	1	Stock
H	EDER-07-005	Valve bonnet	1	Cast brass
I		Fibre gasket { 1 17/32 od, 5/16 id, 1/32 th}	1	Stock
J	EDER-07-002	Spindle	1	Cast brass
K	EDER-07-004	Washer carrier	1	Cast brass
L		Ribber washer { 1 1/16 od, 3/8 id, 1/4 th}	1	Stock
M	EDER-07-003	Ribber retaining nut	1	Brass
N	EDER-07-009	Valve body	1	Cast brass

FIGURE 1.15   Continued.

P1.2 Analyze life phases, establish requirements on the technical process and technical system

Life and repair of the mechanism, especially of the main sealing washer arrangement, should not be jeopardized.

P1.3 Analyze environment of the individual life cycle processes, especially TS(s)-operational process (TP) and its users

The valve should be usable indoors and outdoors (with frost protection) for cold or hot water services.

P1.4 Establish importance (priority level) of individual requirements, processes, and operators (fixed requirements—wishes)

Easy assembly into an existing pipeline was considered essential.

P1.5 Quantify and tolerance the requirements where possible

Rated pressure 1 MPa (150 p.s.i.), permitted temperature range 2 to 95°C (36 to 203°F).

P1.6 Allocate the requirements to life phases, operators, and classes of properties

Otherwise no special requirements.

P1.7 Establish requirements for a supply chain, and environmental concerns

Not considered at present.

P1.8 Reviewed

(P1) Output   **Design specification** (see below)

Requirements are listed only under the most relevant TS-property (see Figures I.15 and 6.8–6.10) as judged by the engineering designer, and, not repeated in any other relevant property class, with an indication of priority— F fixed requirement, must be fulfilled; S strong wish; W wish; N not considered.

Pr1 **Purpose**
F   Must be able to close and seal against the maximum expected water pressure, 1 MPa (150 p.s.i.), equivalent to 100 m water column height.

Pr1A **Function properties**

Pr1B **Functionally determined properties**
F   Overall height of the valve should be as small as possible, and constant, independent of valve opening position.
S   Flow resistance (pressure drop) through the valve should be minimized for all flow rates.
W   The valve should indicate its opening position.
S   Wear on the main sealing washer should be minimized, with no relative rotation to the seat whilst axial sealing force is applied.

Pr1C **Operational properties**
F   Preferably no increase in operating torque for opening or closing compared to the existing valve (Figure 1.15).
F   No decrease in life expectancy compared to the existing valve (Figure 1.15).

Pr2 **Manufacturing properties**
F   Within limits of available manufacturing facilities.

Pr3 **Distribution properties**
F   Packaged for avoiding damage during transport, no special appearance needs.

Pr4 **Liquidation properties**
F   All materials must be capable of and marked for recycling.

Pr5 **Human system factors**—for all life cycle phases
F   Easy to operate, clockwise rotation of the handwheel (looking towards the valve body) should close the valve.

Pr6 **Technical system factors**—for all life cycle phases
None.

Pr7 **Environment factors**—for all life cycle phases
F   Must be able to withstand outdoor conditions for service above freezing point ($0°C$), or be protected from freezing, temperature range 2 to $95°C$ (36 to $203°F$).

Pr8 **Information system factors**—for all life cycle phases
F   Must conform to national piping standards.

Pr9 **Management and economic factors**—for all life cycle phases
None.

Pr10, Pr11, Pr12 **Design properties** (if any prespecified)
F   The revised mechanism should fit into the existing valve body.

(P2) Establish a plan for the design work—investigate alternatives

P2.1 Analyze and categorize the technical process and technical system from viewpoints that influence design work and planning

P2.2 Select overall strategy, partial strategies, and operations for important partial systems

Modified systematic procedure from Figures 4.1 as shown in Figure 2.11.

P2.3 Establish degree of difficulty of the design work

None anticipated.

P2.4 Establish tasks of operators and individual staff members

P2.5 Plan the anticipated design work, under time pressure

Four weeks as target, intermediate target dates need to be set for several stages.

P2.6 Estimate the design costs

Not available.

P2.7 Establish further goals, for example, masters, forms, catalogs, and so forth.

Precedents were available in the literature.

P2.8 Reviewed

(P2) Output   **Design specification and plan**

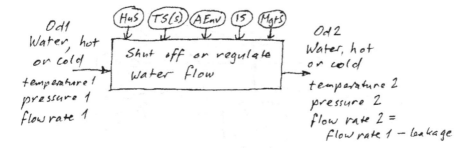

FIGURE 1.16   Water valve—transformation process.

(P3a) Establish the transformation process—investigate alternatives

P3.1 Analyze the TS(s)-operational process—TrfP, TP

The TS(s)-operational process should deliver the effects to accomplish the main transformation (see Figure 1.16).

PROCEDURAL NOTE: This figure is only included here for interest, and as a check on the TS-function structure, which will result from analysis of the existing valve.

PROCEDURAL NOTE: Compare with Figures 5.1, 5.3, and 5.4 to check whether any important elements may be missing.

P3.1.1 Establish input to the operational process—TP operand state Od1

P3.1.2 Establish operators of the operational process—assumed

P3.1.3 Establish technological principle, technology—operations, sequencing

P3.1.4 Establish necessary effects acting on the operand—technology in individual operations

Not completed for this project.

(P3a) Output   **Transformation process**

(PRev5) Establish the existing TS-organ structure by reverse engineering

At this point the procedural model is modified (see Figure 2.11). The revised design process starts from the assembly drawing (Figure 1.15), and could be termed "reverse engineering." The full assembly drawing was first transformed into a "skeleton" representation, ignoring all material wall-thicknesses (see Figure 1.17).

(PRev5) Output   **Existing organ structure** (see Figure 1.17)

(PRev4) Establish the existing TS-function structure by reverse engineering

The designer now looked for significant contacts between surfaces on neighboring constructional parts, and to the active environment. Initially,

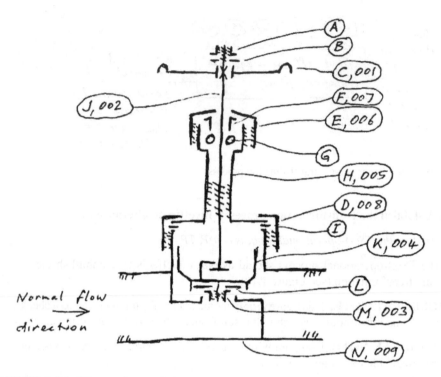

FIGURE 1.17 Water valve—existing organ structure. Part ID and drawing number are according to Figure 1.15.

the designer looked for elemental organs in one region, the seal to the valve stem.

Constructional parts	Surface	Organ designation
H + E	Screw threads external/internal	ElOrg1
E + F	Conical rings top/inside	ElOrg2
F + J	Cylindrical guidance external/intcrnal	ElOrg3
F + H	Cylindrical guidance external/internal	ElOrg4
F + G	Annular end face to rope	ElOrg5
G + H	Cylindrical pressure contact	ElOrg6
G + J	Cylindrical pressure contact	ElOrg7
G + H	Annular conical end face	ElOrg8

Acting or reacting together, these elemental organs form an *organ group* that seals the water pressure to the active and reactive environment for minimum leakage. It also allows resealing if wear causes leakage, by tightening item E, without disassembling any parts. Consequently, the designer listed all function groups detected in the existing valve, with their expected functions

Constructional parts	Organ group	Function
Hu hand, C	Receptor	Accept torque and rotary motion to open/close
C, A, B, J	Handwheel mounting	Transmit torque and motion to spindle
J, E, F, G, H	Spindle seal	Seal water pressure from active environment
		Allow seal adjustment to reseal
H, J	Motion screw	Transform torque to axial force
		Transform rotary motion to axial motion
H, D, I, N	Bonnet connection	Connect and seal bonnet mechanism to body
K, N	Carrier guide	Guide carrier to keep washer about parallel to seat
J, K	Carrier connection	Transmit axial force to carrier
		Allow carrier to rotate independent of spindle
K, L, M	Washer holder	Keep washer in place on carrier
L, N	Washer seal	Allow closing and regulating of water passage
N, H, J	Bore surface effector	Accept and react to water pressure
		Guide water flow
N, pipe (2x)	Pipe connection	Support TS(s), connect to fixed system
		Seal connection against water pressure

PROCEDURAL NOTE: As designers gain more experience, they need not perform these steps in such detail, but will learn to recognize the organ groups and their functions from the assembly drawing.

(PRev4) Output   Existing function structure

The designer could now draw the function structure of the existing water valve. At this point, she decides to revise the functions for the new design specification and follow the remaining steps of the procedural model (Figure 4.1).

(P3b) Establish the revised TS-function structure

P3.2 Work out the complete function structure of the TS(s)

P3.2.1 Establish distribution of effects between humans and technical systems (and environment)

P3.2.2 Establish TS(s)-main functions of individual technical systems

Several functions of the existing structure, and their solutions, should remain unchanged.

Constructional parts	Organ group	Function
Hu hand, C	Receptor	Accept torque and rotary motion to open/close
C, A, B, J	Handwheel mounting	Transmit torque and motion to spindle
H, D, I, N	Bonnet connection	Connect and seal bonnet mechanism to body
K, L, M	Washer holder	Keep washer in place on carrier
L, N	Washer seal	Allow closing and regulating of water passage
N, H, J	Bore surface effector	Accept and react to water pressure
		Guide water flow
N, pipe (2x)	Pipe connection	Support TS(s), connect to fixed system
		Seal connection against water pressure

Revised functions were needed according to the design specification.

Function

Constrain axial motion of spindle to handwheel
React axial force through spindle and bonnet to body
Transform torque to axial force
Transform rotary motion to axial motion away from handwheel for clockwise rotation (implies left-hand screw thread or similar)
Guide spindle for rotation
Constrain rotational motion of axially moving part
Indicate valve opening position
Seal water pressure to active environment for motion inducing parts
Allow seal adjustment to reseal
Guide carrier to keep washer about parallel to seat

Transmit axial force to carrier
Allow carrier to rotate independent of moving part

These functions were represented in a function structure (Figure 1.18).

PROCEDURAL NOTE: Compare with Figure 6.4 to check whether any important elements may be missing.

P3.2.3 Allocate and distribute the main function among individual TS

P3.2.4 Establish additional functions from an analysis of the relationships "human-technical system" and "environment-technical system"

P3.2.5 Establish the assisting functions from analysis of the transformation functions: auxiliary, propelling, regulating and controlling, and so forth

P3.2.6 Establish possibly evoked functions from important properties

P3.2.7 Establish TS(s)-inputs, check of TS(s)-output regarding environment

P3.3 Represent TS-function structures

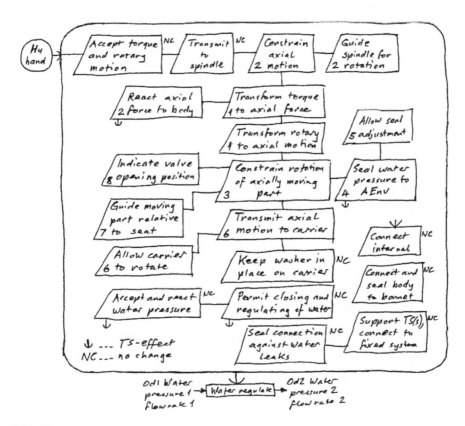

FIGURE 1.18 Water valve—function structure.

P3.4 Assess and evaluate the variants of the function structure

Not used separately.

P3.5 Reviewed

(P3b) Output   **Revised function structure** (see Figure 1.18)

NOTE: Part of the purpose of developing the abstraction of this function structure is to disconnect the designer's conscious mind from the constructional parts of the existing valve, and allow a process similar to incubation to expand the potential solution field.

(P4) Establish the revised TS-organ structure—investigate alternatives

P4.1 Enter TS-functions from function structure into first column of morphological matrix

P4.2 Find action principles, possible modes of action

P4.3 Establish organs or organisms as means (function carriers)

A morphological matrix developed by the designer is shown in Figure 1.19.

PROCEDURAL NOTE: Compare with Figure 4.4 to check whether any important elements may be missing. In this case study, the engineering designer allowed some inconsistency in formulating the functions in Figure 1.19 compared to Figure 1.17.

If we assume that there are no conflicts among the partial solutions in this morphological matrix, 25,920 combinations are "possible"—this number is too large for convenient use, and results from combinatorial complexity. Because of actual conflicts, the "potentially feasible" number of combinations will be much smaller—for example, function 5, partial solutions (a) and (b) can only be combined with function 4, partial solutions (1) and (2).

P4.4 Combine individual organs to a unity (a proposed organ structure)

The designer explored several proposals for organ structures (see Figure 1.20).

PROCEDURAL NOTE: Compare with Figure 6.6 to check whether any important elements may be missing.

P4.5 Evaluate organ structure proposals and select the best

The designer made the following informal evaluations:
1. Guiding the washer carrier and moving part in the body is preferable to guiding in the bonnet, results in reduced overall height.
2. Guiding the washer carrier on the spindle threads does not seem good.
3. Washer carrier should be able to rotate for each closing, even if only a few degrees, otherwise "wire-drawing" could damage the seat.
4. Clockwise rotation for closing requires left-hand threads on or in spindle and moving part. Are taps and dies available?

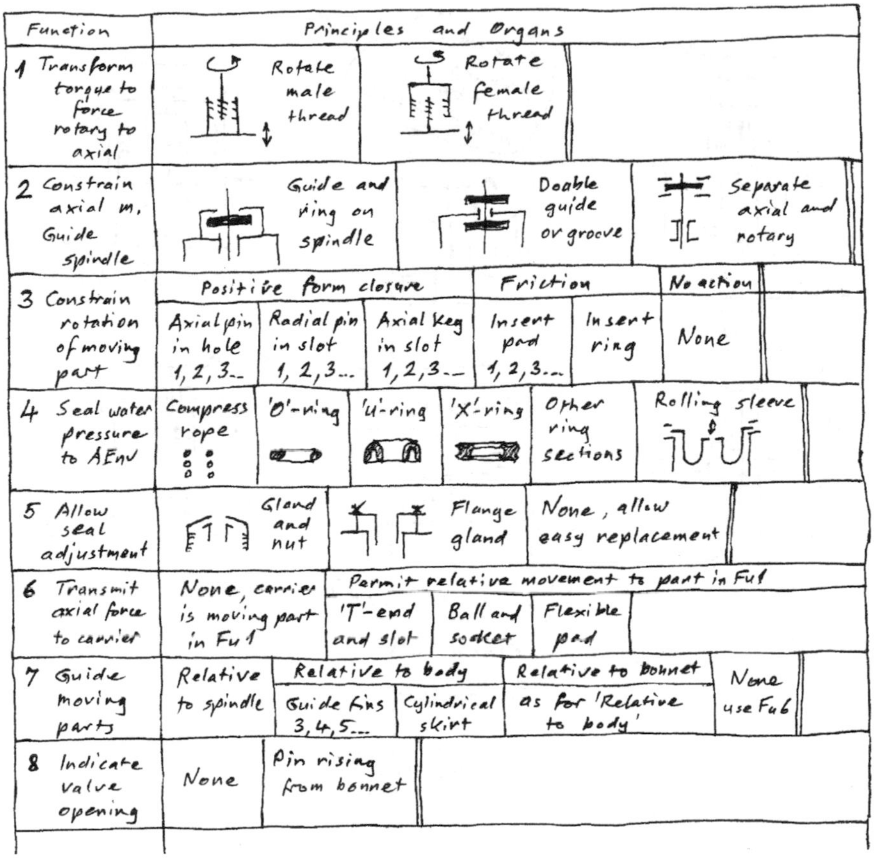

Function	Principles and Organs					
1 Transform torque to force rotary to axial	Rotate male thread	Rotate female thread				
2 Constrain axial m, Guide spindle	Guide and ring on spindle	Double guide or groove	Separate axial and rotary			
3 Constrain rotation of moving part	Positive form closure			Friction		No action
	Axial pin in hole 1, 2, 3..	Radial pin in slot 1, 2, 3...	Axial key in slot 1, 2, 3..	Insert pad 1, 2, 3...	Insert ring	None
4 Seal water pressure to AEnv	Compress rope	'O'-ring	'u'-ring	'X'-ring	Other ring sections	Rolling sleeve
5 Allow seal adjustment	Gland and nut	Flange gland	None, allow easy replacement			
6 Transmit axial force to carrier	None, carrier is moving part in Ful	Permit relative movement to part in Ful				
		'T'-end and slot	Ball and socket	Flexible pad		
7 Guide moving parts	Relative to spindle	Relative to body		Relative to bonnet as for 'Relative to body'	None use Ful	
		Guide fins 3, 4, 5...	Cylindrical skirt			
8 Indicate valve opening	None	Pin rising from bonnet				

FIGURE 1.19 Water valve—morphological matrix.

5. A rotating female thread provides one shoulder on the spindle and only an added ring to give a second shoulder, but needs a slotted or split plate to fit into the groove, which would allow dirt to enter the valve bonnet. An alternative could be an extra nut on the spindle.
6. A top seal increases the height of the valve.
7. A skirt on the moving part that holds a seal ring in a groove reduces the height of the valve.
8. A rope seal allows resealing by tightening, other rings are difficult to reseal, they would need replacement.
9. A leakage seal in a skirt of the moving part performs multiple duties, it seals, and it constrains rotation.
10. A leakage seal in a skirt of the moving part creates an enclosed space in the bonnet that could become a pressure chamber if the seal leaks. This space must be vented to the outside.
11. An external thread on the moving part (or washer carrier) could be difficult to cut, especially if that part is to be guided in the body.

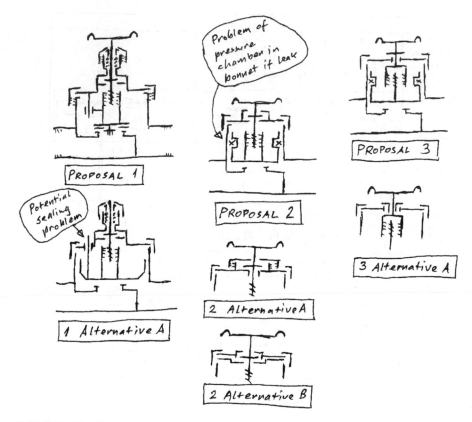

FIGURE 1.20   Water valve—organ structures.

12. A ring on the spindle needs an additional constructional part to trap
    it for axial motion constraint.

Based on these considerations, the designer chose Proposal 2, Alternative B.

P4.6 Establish assisting functions and their organs, and evoked functions for organs where needed

P4.7 Examine evoked secondary outputs and necessity of further organs

P4.8 Establish functions of complex organs, action locations

P4.9 Establish fundamental situations and orders (topology, arrangement)

P4.10 Represent final technical system proposals as organ structures with enough detail for evaluation and selection

Not used for this project.

P4.11 Reviewed

(P4) Output   **Revised organ structure**

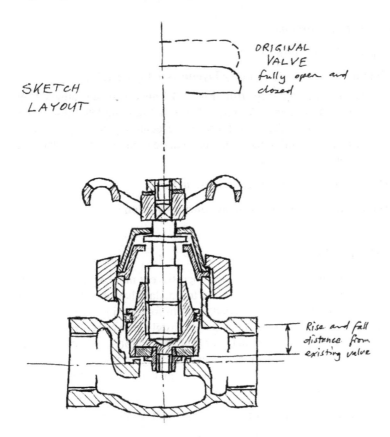

SKETCH
LAYOUT

ORIGINAL
VALVE
fully open and
closed

Rise and fall
distance from
existing valve

FIGURE 1.21   Water valve—preliminary layout.

(P5a) Establish the revised constructional structure (1)—investigate alternatives

P5.1 Establish and analyze the requirements for the constructional structure and constructional parts

Not completed for this project.

P5.2 Establish a conception of the constructional structure

P5.3 Establish rough constructional structure as preliminary layout

The designer then produced a sketch layout close to scale (see Figure 1.21).

P5.3.1 Define constructional design groups

P5.3.2 Establish rough form-giving of each constructional design group

P5.3.3 Elaborate the rough constructional structure—integrate into the developing total constructional structure (synthesis)

P5.3.4 Represent preliminary layout(s)

P5.3.5 Choose optimal variants

Not completed for this project.

(P5a) Output **Revised preliminary layouts** (see Figure 1.21)

The subsequent stages and steps from Figure 4.1 appear to be routine design work for design engineering. We have chosen not to complete these stages and steps. Nevertheless, the importance of these subsequent steps must be emphasized, as many fault conditions may unintentionally be introduced in the embodiment and detail phases.

(P5b) Output **Revised constructional structure**

(P6) Output **Revised manufacturing documentation**

1.3 SMOKE GAS FILTER

Environmental requirements are becoming progressively more severe, and retrofitting of equipment to bring operating plant up to standard happens more frequently. A thermal electric generating station needed a retrofit for cleaning the exhaust smoke from the boilers, a smoke gas filter. The contract was given to a local enterprise with some experience. The enlightened management of the enterprise asked the designers to provide full documentation of their engineering design processes, so that information would not be lost for future contracts. The design team decided to use computer applications to document all steps and models.

(P1) Establish a list of requirements—investigate alternatives

P1.1 Establish rough factors for all life phases (technology, operators, inputs, outputs of each life stage)

Transportation of the completed smoke gas filter from the organization to the plant must be possible; rail and road transport profiles, and access to the plant must be investigated.

P1.2 Analyze life phases, establish requirements on the technical process and technical system

Acceptable state of cleanliness of the exhaust gases should be established from the current laws and standards.

P1.3 Analyze environment of the individual life cycle processes, especially TS(s)-operational process (TP) and its users

The filtration system must be weather-proof. All collected materials must be safely available for extraction and disposal.

P1.4 Establish importance (priority level) of individual requirements, processes, and operators (fixed requirements—wishes)

Several general descriptions and data were found, see references in the following steps.

P1.5 Quantify and tolerance the requirements where possible

This power station was used as a "peak-lopping" facility, it ran part-load for long periods.

P1.6 Allocate the requirements to life phases, operators, and classes of properties

P1.7 Establish requirements for a supply chain, and environmental concerns

P1.8 Reviewed

(P1) Output   **Design specification** (see below)

Requirements are listed only under the most relevant TS-property (see Figures I.15 and 6.8 to 6.10) as judged by the engineering designer, and, not repeated in any other relevant property class, with an indication of priority—F fixed requirement, must be fulfilled; S strong wish; W wish; N not considered.

Pr1 **Purpose**
F   Must accept maximum 75 m^3/s of smoke gas at input temperatures 125°C to 400°C.
F   Must reduce content of particulate solids with a minimum efficiency of 95%. Particulates consist mainly of fly ash, but may also contain other contaminants.

Pr1A **Function properties**
F   Must be capable of running effectively at part-load down to $\frac{1}{6}$ power output, expected minimum smoke rate of 12.5 m^3/s.

Pr1B **Functionally determined properties**
F   Must reduce content of harmful chemical agents with a minimum efficiency of 90%. These chemicals include SO_2 (or H_2SO_3), SO_3 (or H_2SO_4), CO, NO_x, and so forth. This requirement may be deferred to a later date, when designing the particulate filter has progressed sufficiently.

Pr1C **Operational properties**
S   Flow resistance (pressure drop) through the filter should be minimized for all flow rates.
F   Electrical safety to relevant standards, test by UL or CSA, apply control symbol on rating plate.
F   Life minimum 5 years.
S   Suitable for easy maintenance.

Pr2 **Manufacturing properties**
F   Must accommodate available manufacturing facilities.

Pr3 **Distribution properties**
F   Must be transportable by road and rail to power station site.

Pr4 **Liquidation properties**

Pr5 **Human system factors**—for all life cycle phases
F   Access for inspection of gas channels and other parts must be provided.
F   Safety is essential, human exposure to smoke gases and high voltage must be eliminated.

Pr6 **Technical system factors**—for all life cycle phases
S   Minimum requirements on other TS (as operators) during life cycle of device.
S   Avoid special tools for manufacture, assembly, testing, and maintenance of TS(s).

Pr7 **Environment factors**—for all life cycle phases
F   No significant environmental impact for materials used in device.
F   High energy efficiency, low energy consumption.
F   Withstand wind load of 350 km/h wind (e.g., Texas), or 150 km/h wind plus 1 m transient snow load (e.g., Buffalo, NY).
S   Conform to ISO 14000:1995.

Pr8 **Information system factors**—for all life cycle phases
S   Minimum requirements for information, during TS(s) life cycle, avoid special additional information (including training), provide clear user instructions.

Pr9 **Management and economic factors**—for all life cycle phases
F   Delivery of finished devices in 15 months from contract.
S   Conform to ISO 9000:2000.

Pr10, Pr11, Pr12 **Design properties** (if any prespecified)
None.

(P2) Establish a plan for the design work—investigate alternatives

P2.1 Analyze and categorize the technical process and technical system from viewpoints that influence design work and planning

P2.2 Select overall strategy, partial strategies and operations for important partial systems

P2.3 Establish degree of difficulty of the design work

P2.4 Establish tasks of operators and individual staff members

None.

P2.5 Plan the anticipated design work, under time pressure

Two months as target for design completion, delivery of technical system within 15 months, intermediate target dates need to be set for several stages.

P2.6 Estimate the design costs

Not available.

P2.7 Establish further goals, for example, masters, forms, catalogs, and so forth.

Precedents were available; see references below.

P2.8 Reviewed

(P2) Output   **Design specification and plan**

(P3a) Establish the transformation process—investigate alternatives

P3.1 Analyze the TS(s)-operational process—TrfP, TP

TS(s)-operational process should accomplish main transformation (see Figure 1.22A).

P3.1.1 Establish input to the operational process—TP operand state Od1

P3.1.2 Establish operators of the operational process—assumed

see Figure 1.22A.

P3.1.3 Establish technological principle, technology—operations, sequencing

The main available technologies are shown in Figure 1.22B. Probably the most favorable is electrostatic precipitation; see for example [76,99,390].

P3.1.4 Establish necessary effects acting on the operand—technology in individual operations

This technology needs a direct-current corona, and collector plates that need frequent cleaning (see Figure 1.22C).

(P3a) Output   **Transformation process**

PROCEDURAL NOTE: Compare with Figures 5.1, 5.3, and 5.4 to check whether any important elements may be missing.

(P3b) Establish the TS-function structure

P3.2 Work out the complete function structure of the TS(s)

Some preliminary estimates were made using information and data from [76,99,390].

With a flow rate of $Q = 75$ m^3/s, at a recommended maximum flow velocity of 2 m/s, the required flow cross-section will be 37.5 m^2, approximately 6×6 m.

The flow velocity from the boiler was given as 5 m/s, diffusion to the lower velocity was needed.

At turn-down, a flow rate of 12.5 m^3/s, and a recommended minimum flow velocity of 1 m/s, the flow cross section should be 12.5 m^2. The total cross-section can be divided into three separate sections.

(A) Transformation process — black box

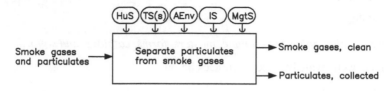

(B) Technologies

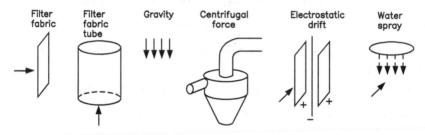

(C) Transformation process — main flow

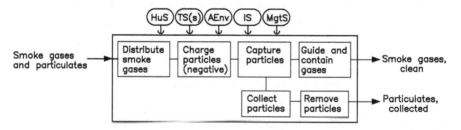

FIGURE 1.22   Smoke filter—transformation process.

If the collector plates are assumed 6 m high, the width of flow channels should be 6.25 m for maximum flow rate, or 2.08 m for turn-down rate. If this width is divided into assumed 3 banks of 9 channels, each is 260 mm wide. For the reduced flow rate, banks of 9 channels could be shut off from the flow.

With a particle drift velocity assumed at $V = 0.11$ m/s, particles should take about 1 s for the maximum drift distance of 115 mm.

Using the simple Deutsch–Andersen equation: $\eta = 1 - \exp(-VA/Q)$ and assuming a collector plate width (in the flow direction) of 8 m, therefore total active collector plate area of $6 \times 8 \times 27 \times 2 = 2592$ m^2, the expected efficiency should be $\eta = 97.76\%$ and the necessary collection current (assumed 200 A/m^2) should be 518 mA.

The corona wires can be rated to deliver 0.5 mA/m, the total length of corona wire should then be 1036 m. Using an active length of 6 m, 174 wires would be needed, just over 6 per channel. If 7 per channel are used, they can be placed within the first 3 m at 0.5 m spacing, allowing sufficient channel length to precipitate the charged particles.

Using a corona voltage of 60 kV, the direct-current power needed would be 31 kW, delivering a power ratio of 6.91 $W/(m^3/min)$.

The flash distance (60 kV to ground) should be larger than the half-channel width, assume 200 mm for safety. Creep distances over solid insulator should be checked, probably 600 mm.

One "rapper" (cleaning device) may be used for cleaning the wires of each electrostatic section, and one rapper per collector plate would be recommended.

P3.2.1 Establish distribution of effects between humans and technical systems (and environment)

P3.2.2 Establish TS(s)-main functions of individual technical systems

P3.2.3 Allocate and distribute the main function among individual TS

P3.2.4 Establish additional functions from an analysis of the relationships "human-technical system" and "environment-technical system"

P3.2.5 Establish the assisting functions from analysis of the transformation functions: auxiliary, propelling, regulating and controlling, and so forth.

P3.2.6 Establish possibly evoked functions from important properties

P3.2.7 Establish TS(s)-inputs, check of TS(s)-output regarding environment

P3.3 Represent TS-function structures

See Figure 1.23.

P3.4 Assess and evaluate the variants of the function structure

The proposed variant was approved.

P3.5 Reviewed

(P3b) Output   **Function structure**

PROCEDURAL NOTE: Compare with Figure 6.4 to check whether any important elements may be missing.

(P4) Establish the TS-organ structure—investigate alternatives

P4.1 Enter TS-functions from function structure into first column of morphological matrix

P4.2 Find action principles, possible modes of action

P4.3 Establish organs or organisms as means (function carriers)

The morphological matrix is shown in Figure 1.24.

PROCEDURAL NOTE: Compare with Figure 4.4 to check whether any important elements may be missing.

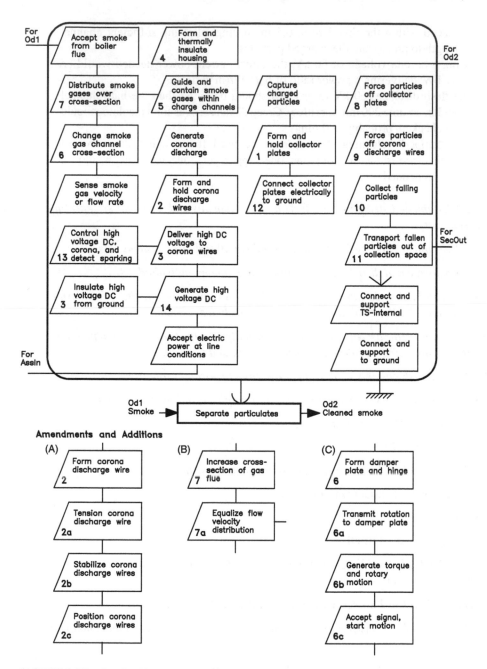

FIGURE 1.23   Smoke filter—function structure.

Function	No	Principles, Partial Organs			
Form and hold collector plates	1	Flat plate	Wavy plate	Shielded plate	Pocket plate
Form corona discharge wires	2	Round wire	Square wire (twisted)	Strip (punch-sheared)	
Tension corona discharge wires	2a	Weight suspended	Frame mounting		
			Spring tensioned	Screw (turn-buckle)	Cross-bar supports
Stabilize corona discharge wires	2b	Stabilizing bar and loops	Stabilize frame at lower end		
Position corona discharge wires	2c	Support bar for corona wire transverse to bus-bar	Top bar of frame of function 2a		
Deliver high DC voltage to corona wires	3	Bus-bar in air space, solid support insulators, clearance holes for high voltage DC leads down to support bar for corona wire	Transformer-style ceramic bushing to lead high voltage DC to bus-bar in smoke space, distribute to support bars for corona wires		
Form and thermally insulate housing	4	Outer shell to contain all working parts, prevent escape of smoke gases, support to ground, allow access (doors and stairs) for inspection, reduce temperature drop of smoke gases, etc.			
Guide and contain smoke gases within charge channels	5	Baffles (a) At top of collector plates between corona wires (b) At base of collector plates shaped to allow particles to fall down into particle collecting space (c) At housing sides to deflect smoke gases from side to end charging channel			
Change smoke gas channel cross-section	6	Dampers hinged at channel ends		Dampers hinged at outer walls	
Distribute smoke gases over cross-section	7	Long flared tube (about 5°)		Short increase in cross-section with splitter plates	
	7a	None	Perforated plate across flow direction	Flow straightener tubes as for wind tunnels	
Force particles off capture plates	8	Subproject 1.3.1, see below			
Force particles off corona discharge wires	9	Subproject 1.3.2 not completed for this example			
Collect falling particles	10	Subproject 1.3.3 not completed for this example			
Transport fallen particles out of collector space	11	Subproject 1.3.4 not completed for this example			
Connect collector plates electrically to ground	12	Subproject 1.3.5 not completed for this example			
Control high DC voltage, corona, and sparking	13	Sub project 1.3.6 not completed for this example; may need computer-control hardware and software			
Generate high voltage DC	14	Sub project 1.3.7 not completed for this example			

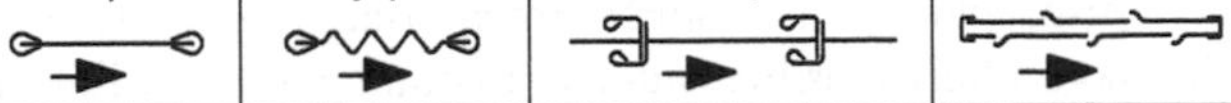

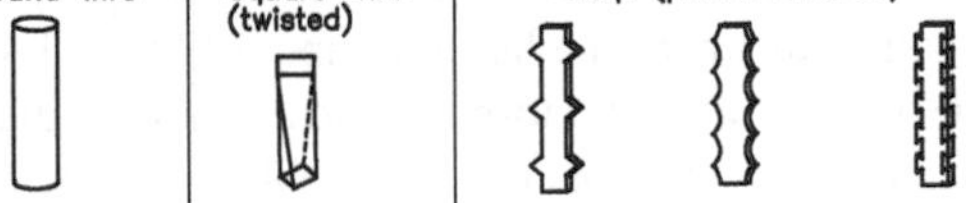

FIGURE 1.24 Smoke filter—morphological matrix.

If we assume that there are no conflicts among the partial solutions in this morphological matrix, 7680 combinations are "possible"—this number is too large for convenient use, and results from combinatorial complexity. Because of actual conflicts, the "potentially feasible" number of combinations will be much smaller for the main investigation. Each of the subprojects for functions 8 to 14 will have its own morphological matrix, again with a large number of "possible" combinations.

During development of the morphological matrix, the designers recognized that some functions were too concise to be solved. Function 2 was expanded as shown in the morphological matrix, and this was transferred as amendment (A) to the function structure (Figure 1.23). Function 7 was also expanded and transferred as amendment (B). Function 6 was anticipated to need expansion, shown as amendment (C).

The project could at this point be recursively separated into smaller, relatively self-contained portions. The functions 8 to 14 were considered suitable for separate treatment, for example, by a smaller team. Function 8 "Force particles off collector plates" is illustrated in Section 1.3.1. Coordination of teams must be a high priority, decisions by one team usually affect the problems faced by other teams.

PROCEDURAL NOTE: The procedure of separation shown in Section 1.3.1 demonstrates the recursive repeat of the full procedure shown as an inset in Figure 2.10. If desired, the structure models from this recursion could be combined with the models for the main TS(s), but may make the models too complicated for easy review and understanding.

P4.4 Combine individual organs to a unity (a proposed organ structure)

P4.5 Evaluate organ structure proposals and select the best

P4.6 Establish assisting functions and their organs, and evoked functions for organs where needed

P4.7 Examine evoked secondary outputs and necessity of further organs

P4.8 Establish functions of complex organs, action locations

P4.9 Establish fundamental situations and orders (topology, arrangement)

Combined treatment.

P4.10 Represent final technical system proposals as organ structures with enough detail for evaluation and selection

The designer attempted several combinations of organs to solve groups of functions (see Figure 1.25). The results for functions 2 and 3 revealed that 12 arrangements could be identified, depending on which header was combined with which corona wire support and tensioning. An evaluation by weighted rating was performed (see Figure 1.26).

P4.11 Reviewed

(P4) Output   **Organ structure**

(A)   Functions 3, 13, and 14: Generate and control high voltage DC

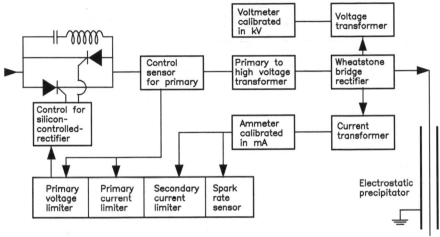

Literature:
Ogelsby, S. Jr. and Nichols, G.B. (1978) *Electrostatic Precipitation*, New York: Dekker

(B)   Functions 2 and 3: High voltage DC to corona wires
        Variant 1

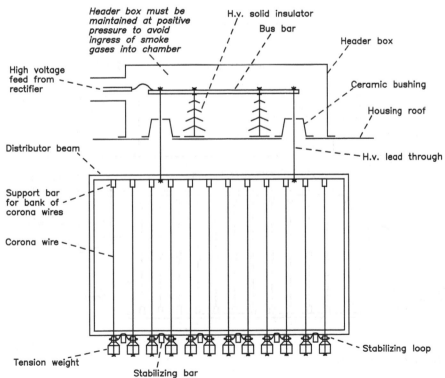

FIGURE 1.25   Smoke filter—organ structures.

(C)   Functions 2 and 3: High voltage DC to corona wires
         Variants 2a and 2b

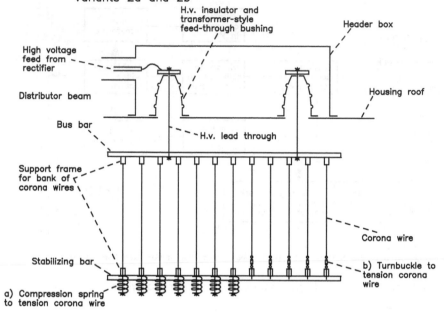

(D)   Functions 2 and 3: High voltage DC to corona wires
         Variant 3

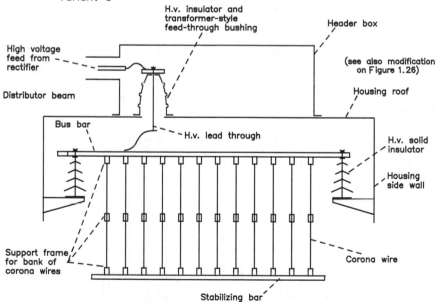

FIGURE 1.25   Continued.

Legend:
Combinations of solution proposals from Figure 12.25:
A ::: Bus bar in header, support insulators in header, variation 1 upper
B ::: Bus bar in housing, multiple feed-through bushings as support insulators, variation 2 upper
C ::: Bus bar in housing, support insulators in housing, variation 3 upper
1 ::: Corona wires tensioned by weights, variation 1 lower
2 ::: Corona wires tensioned by compression springs, variation 2a lower
3 ::: Corona wires tensioned by turn-buckles, variation 2b lower
4 ::: Corona wires supported along length by frame members, variation 3 lower

Selection criteria:
a ... Accuracy of positioning corona wires between collector plates
b ... Consistency and sensitivity of tensioning of corona wires
c ... Possible problems of contamination, e.g., of insulators
d ... Additional facilities needed, e.g., air ballast for sealing
e ... Housing space requirements
f ... Overall complexity of installation
g ... Potential reliability
Results:
W ... Weighting of criteria: 1 lowest priority, 10 highest priority
R ... Rating of solutions: 1 worst, 5 best
% ... Percentage of ideal (total weight x best rating = 265)

Header box
Housing roof
H.v. solid insulator
Housing side wall

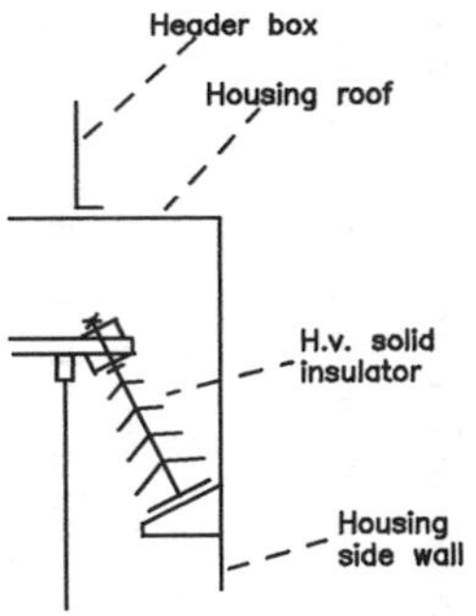

NOTE: Variant C could be improved (e.g., by one point for criterion 'c') by setting the solid insulators at an angle, the base of the insulator close to the housing side wall, the tip of the insulator closer to the corona wire support, allowing sufficient flash distance between the corona wire support and the housing side wall, see diagram above.

Criterion	W	A-1 R	A-1 WxR	A-2 R	A-2 WxR	A-3 R	A-3 WxR	A-4 R	A-4 WxR	B-1 R	B-1 WxR	B-2 R	B-2 WxR	B-3 R	B-3 WxR	B-4 R	B-4 WxR	C-1 R	C-1 WxR	C-2 R	C-2 WxR	C-3 R	C-3 WxR	C-4 R	C-4 WxR
a	10	4	40	5	50	5	50	5	50	4	40	5	50	5	50	5	50	4	40	5	50	5	50	5	50
b	10	5	50	5	50	4	40	5	50	5	50	5	50	4	40	5	50	5	50	5	50	4	40	5	50
c	8	4	32	4	32	4	32	4	32	5	40	5	40	5	40	5	40	2	16	2	16	2	16	2	16
d	7	4	28	4	28	4	28	4	28	5	35	5	35	5	35	5	35	5	35	5	35	5	35	5	35
e	7	4	28	4	28	4	28	4	28	5	35	5	35	5	35	5	35	3	21	3	21	3	21	3	21
f	6	2	12	3	18	3	18	4	24	3	18	3	18	3	18	4	24	4	24	4	24	4	24	5	30
g	5	4	20	4	20	4	20	4	20	5	25	5	25	5	25	5	25	3	15	3	15	3	15	4	20
TOTAL	53		210		226		216		232		243		253		243		259		201		211		201		222
%			74.2		85.3		81.3		87.5		91.7		95.5		91.7		97.7		75.8		79.6		75.8		83.8
Rank			11		6		8		5		3=		2		3=		1		10=		9		10=		7

Solution proposal combination

FIGURE 1.26 Smoke filter—weighted rating of organ structures.

PROCEDURAL NOTE: Compare with Figure 6.6 to check whether any important elements may be missing.

(P5a) Establish the constructional structure (1)—investigate alternatives

P5.1 Establish and analyze the requirements for the constructional structure and constructional parts

Further work towards a preliminary layout needs more detailed information, for example, see [447], and is beyond the scope of this case example.

The subsequent stages and steps from Figure 4.1 appear to be routine design work for design engineering. We have chosen not to complete these stages and steps. Nevertheless, the importance of these subsequent steps must be emphasized, as many fault conditions may unintentionally be introduced in the embodiment and detail phases.

(P5a) Output   **Preliminary layouts**

(P5b) Output   **Constructional structure**

(P6) Output   **Manufacturing documentation**

1.3.1 RAPPER FOR ELECTROSTATIC SMOKE GAS FILTER

(P1) Establish a list of requirements—investigate alternatives

P1.1 Establish rough factors for all life phases (technology, operators, inputs, outputs of each life stage)

P1.2 Analyze life phases, establish requirements on the technical process and technical system

P1.3 Analyze environment of the individual life cycle processes, especially TS(s)-operational process (TP) and its users

P1.4 Establish importance (priority level) of individual requirements, processes and operators (fixed requirements—wishes)

P1.5 Quantify and tolerance the requirements where possible

P1.6 Allocate the requirements to life phases, operators, and classes of properties

P1.7 Establish requirements for a supply chain, and environmental concerns

P1.8 Reviewed

(P1) Output   **Design specification** (see below)

Requirements are listed only under the most relevant TS-property (see Figures I.15 and 6.8–6.10) as judged by the engineering designer, and, not repeated in any other relevant property class, with an indication of priority—F: fixed requirement, must be fulfilled; S: strong wish; W: wish; N: not considered.

Pr1 **Purpose**

F   Must, at set or variable time intervals, provide vibration to collector plates and wire discharge electrodes to dislodge the accumulated particle material—this part of the case study treats only the function 8 in Section 1.3, Figure 1.23: "Force particles off collector plates."

Pr1A **Function properties**

Pr1B **Functionally determined properties**

Pr1C **Operational properties**

Pr2 **Manufacturing properties**

Pr3 **Distribution properties**

Pr4 **Liquidation properties**

Pr5 **Human system factors**—for all life cycle phases

Pr6 **Technical system factors**—for all life cycle phases

Pr7 **Environment factors**—for all life cycle phases

Pr8 **Information system factors**—for all life cycle phases

Pr9 **Management and economic factors**—for all life cycle phases

Pr10, Pr11, Pr12 **Design properties** (if any prespecified)
None.

Other requirements are "inherited" from "parent" problem, Section 1.3.

(P2) Establish a plan for the design work—investigate alternatives

P2.1 Analyze and categorize the technical process and technical system from viewpoints that influence design work and planning

P2.2 Select overall strategy, partial strategies, and operations for important partial systems

P2.3 Establish degree of difficulty of the design work

P2.4 Establish tasks of operators and individual staff members

P2.5 Plan the anticipated design work, under time pressure

P2.6 Estimate the design costs

P2.7 Establish further goals, for example, masters, forms, catalogs, and so forth.

P2.8 Reviewed

These items are taken over from the parent problem, Section 1.3.

(P2) Output   **Design specification and plan**

(P3a) Establish the transformation process—investigate alternatives

P3.1 Analyze the TS(s)-operational process—TrfP, TP

P3.1.1 Establish input to the operational process—TP operand state Od1

P3.1.2 Establish operators of the operational process—assumed

P3.1.3 Establish technological principle, technology—operations, sequencing

P3.1.4 Establish necessary effects acting on the operand—technology in individual operations

These items are taken over from "parent" problem (Section 1.3). The resulting black box transformation process is shown in Figure 1.27A.

(P3a) Output   **Transformation process**

PROCEDURAL NOTE: The black box representation of this transformation process does not need to be reformulated, it can be accepted as identical to function 8 in the "parent" problem, Figure 1.23. Compare with Figures 5.1, 5.3, and 5.4 to check whether any important elements may be missing.

(P3b) Establish the TS-function structure

P3.2 Work out the complete function structure of the TS(s)

P3.2.1 Establish distribution of effects between humans and technical systems (and environment)

P3.2.2 Establish TS(s)-main functions of individual technical systems

P3.2.3 Allocate and distribute the main function among individual TS

P3.2.4 Establishing additional functions from an analysis of the relationships "human-technical system" and "environment-technical system"

P3.2.5 Establish the assisting functions from analysis of the transformation functions: auxiliary, propelling, regulating and controlling, and so forth.

P3.2.6 Establish possibly evoked functions from important properties

P3.2.7 Establish TS(s)-inputs, check of TS(s)-output regarding environment

P3.3 Represent TS-function structures

P3.4 Assess and evaluate the variants of the function structure

The resulting function structure is shown in Figure 1.27B.

PROCEDURAL NOTE: Compare with Figure 6.4 to check whether any important elements of the function structure may be missing.

P3.5 Reviewed

(P3b) Output   **Function structure**

(A) Transformation process — black box

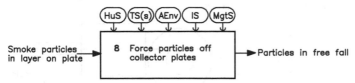

(B) Function structure

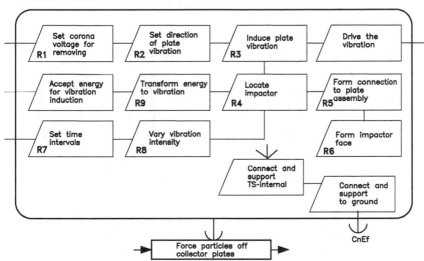

(C) Morphological matrix

Function	No	Principles, partial organs					
Set corona volt for removing	R1	On	OFF				
Set direction of plate vibration	R2	Shear	(particle to plate)		Tension		
		Up/down	Side to side		Normal to plate		
Induce plate vibration	R3	Impact					Burst of forced cycles
		Solenoid pull-in	Solenoid spring return	Geared motor gravity			Vibrator
Locate impactor	R4	Hinged on plate assembly	Hinged on motor shaft	On solenoid rod end			None
Form connection to plate assembly	R5	Anvil on plate assembly					Bolt connection
Form impactor face	R6	Fixed hammer face	Mobile ring mass				None
Set time intervals	R7	Fixed	Variable motor speed	Variable timing controller			
Vary vibration intensity	R8	Fixed	Variable motor lifter position				

FIGURE 1.27 Rapper for electrostatic smoke filter.

(D)   Organ   structures

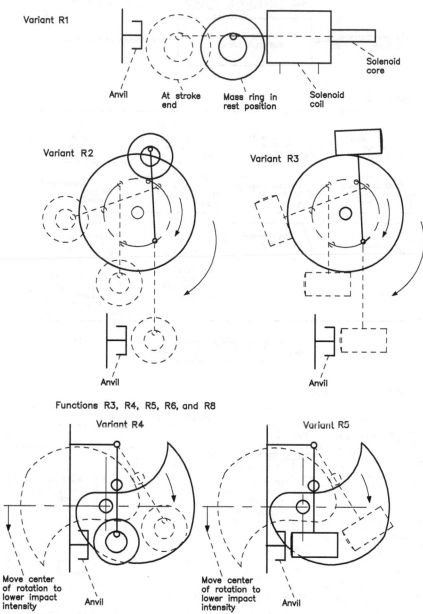

Functions R3, R4, R5, and R6

Variant R1

Solenoid core

Anvil   At stroke   Mass ring in   Solenoid
             end   rest position   coil

Variant R2   Variant R3

Anvil   Anvil

Functions R3, R4, R5, R6, and R8

Variant R4   Variant R5

Move center   Move center
of rotation to   of rotation to
lower impact   Anvil   lower impact   Anvil
intensity   intensity

FIGURE 1.27   Continued.

(E)   Evoked functions
            For variant R2

Form drive disk
Fix drive disk to shaft — evoked organ
Form drive pin
Fix drive pin to drive disk — evoked organ
Form hinge bearing
Fix hinge bearing to hammer plates — evoked organ
Form hinge bolt
Connect hammer plates and hinge bearing — evoked organ
Form hammer pin
Connect hammer pin to hammer plates — evoked organ
Form mass ring
Retain mass ring over hammer pin — evoked organ

(F)   Preliminary layout
            For Variant R2

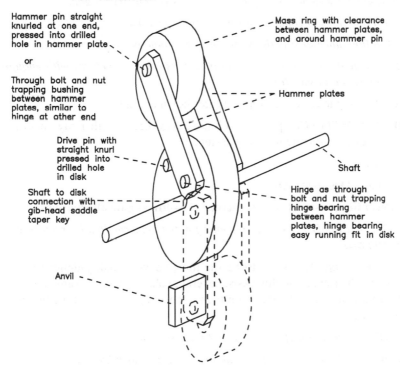

FIGURE 1.27   Continued.

(P4) Establish the TS-organ structure—investigate alternatives

P4.1 Enter TS-functions from function structure into first column of morphological matrix

P4.2 Find action principles, possible modes of action

P4.3 Establish organs or organisms as means (function carriers)

The resulting morphological matrix is shown in Figure 1.27C.

PROCEDURAL NOTE: Compare with Figure 4.4 to check whether any important elements of the morphological matrix may be missing.

If we assume that there are no conflicts among the partial solutions in this morphological matrix, 649 combinations are "possible"—this number is too large for convenient use, and results from combinatorial complexity. Because of actual conflicts, the "potentially feasible" number of combinations will be much smaller.

P4.4 Combine individual organs to a unity (a proposed organ structure)

P4.5 Evaluate organ structure proposals and select the best

P4.6 Establish assisting functions and their organs, and evoked functions for organs where needed

P4.7 Examine evoked secondary outputs and necessity of further organs

P4.8 Establish functions of complex organs, action locations

P4.9 Establish fundamental situations and orders (topology, arrangement)

P4.10 Represent final technical system proposals as organ structures with enough detail for evaluation and selection

The resulting organ structure proposals are shown in Figure 1.27D.

PROCEDURAL NOTE: Compare with Figure 6.6 to check whether any important elements of the organ structure may be missing.

P4.11 Reviewed

(P4) Output   **Organ structure**

(P5a) Establish the constructional structure (1)—investigate alternatives

P5.1 Establish and analyze the requirements for the constructional structure and constructional parts

Variants "R2" and "R3" seem to be the most conventional, judging from the available literature. On this basis, several evoked functions were recognized for variant "R2," together with their connecting organs; see the list in Figure 1.27E.

P5.2 Establish a Conception of the Constructional Structure

P5.3 Establish Rough Constructional Structure as Preliminary Layout

A preliminary layout was developed (see Figure 1.27F).

The subsequent stages and steps from Figure 4.1 appear to be routine design work for design engineering. We have chosen not to complete these stages and steps. Nevertheless, the importance of these subsequent steps must be emphasized, as many fault conditions may unintentionally be introduced in the embodiment and detail phases.

(P5a) Output   **Preliminary layouts**

(P5b) Output   **Constructional structure**

(P6) Output   **Manufacturing documentation**

Part II

Knowledge Related to Engineering Design Processes

Typical life phases of TP/TS are shown in Figures I.13 and 6.14, which includes design engineering and its "engineering design system" (see Figure I.16). The life cycle also includes as sixth process the TS-operational process, the TP(s), the prime purpose of the TS(s).

NOTE: Designing is often regarded as creative, which implies that there is little scope, need for, or benefit from systematic or methodical approaches. A common perception is that design engineering consists of preparing engineering drawings—layouts, details, assemblies, and their parts lists. Both of these misrepresent the essential content of designing and design engineering. Human designers with cognitive abilities are the essential participants of designing, especially in their interactions with reality as natural and artificial objects, and with graphical and other representations (see Figure I.10). Design engineering must include clarifying the problem, conceptualizing, laying out, and detailing (see level H1 of Figure I.21), and a combination of systematic, methodical, intuitive, cognitive, and creative activities.

The general trend of development in societies and civilizations is toward improvements to the overall effectiveness of processes, *rationalization*, "to make more efficient and effective by scientifically reducing or eliminating waste of labor, time, or materials" [2]. This should also be true of design engineering. Part II consists of Chapters 2 to 4, related to the processes of designing and redesigning of TP(s) and TS(s), design engineering, and are coordinated with the chapters of Part III, and the case examples in Part I.

2 Engineering Design System (DesS) and Engineering Design Processes (DesP)

Design engineering, including novel designing and redesigning, is performed on a proposed future technical process, TP(s), and technical system, TS(s), as operand (Od), the subject of the design process. The operand, operators, and relationships among them and to the outside, form the Engineering Design System, see Section I.11. The information presented in this chapter is situated in and around the "east" hemisphere of Figures I.4, I.5, and 12.7.

NOTE: Since the 1960s, many ideas and proposals have been made for improving and rationalizing design engineering and attaining optimal TP and TS. These ideas and proposals cover (1) technical aspects of operation and performance (and all related properties) of TS(s), considering the TP(s), and their development by design engineering; (2) the properties and working methods of the *engineering designer*, including recorded information, tacit and internalized information, cognitive abilities, knowledge about and use of formalized methods, experience, open-mindedness, willingness to obtain opinions and suggestions, creativity, psychology (see Section 11.1), and so forth; (3) social aspects of cooperation, team work (see Section 11.2), awareness and willingness to work together with all customers and stakeholders, internal and external to an organization; (4) *management* of the organization (see Section 11.3), of the range of products, and of the processes of designing in general and design engineering in particular; and (5) the societal contexts of designing, legal, economic, environmental, and so forth.

Many developments have taken place for point (1), typified by such works as [457] with respect to the designed systems and formalized methods, and [110,111,335] with respect to the human designers and their mental processes. Design methods (and some "industry best practice" methods) have been introduced into engineering design practice, by courses in universities and colleges, by commercial vendors (e.g., QFD, TRIZ, and so forth) and by active support from organization managements.

As regards point (2), *synergy* occurs in human thinking processes—mental association brings about new insights that are more than the sum of individual thoughts, but which must be brought out of the mind for interaction and implementation. Design engineering starts from given requirements, and proceeds to a complete description of a TP(s) and TS(s), and is complicated and demanding. The requirements for TP(s)/TS(s) are becoming more extensive and abundant. Increasing competition and globalization has made it necessary to realize the optimal balance among the required and achieved state of properties (including cost, price, delivery time, and so forth) of the TS in the shortest time to market, see Chapter 3. This implies the shortest design time with few iterations and recursions, and few alterations from design documentation of the TS(s) to its realization (and TP implementation)—"right first time."

Some dilemmas concern conflicts, see Section I.12.1, and the limits of working (short-term) mental capacity of human designers, see Section 11.1.

2.1 GENERAL MODEL OF THE ENGINEERING DESIGN SYSTEM AND ENGINEERING DESIGN PROCESS

The general model of the engineering design system, which includes the engineering design process (DesP), can be derived directly from the model of the transformation system (TrfS) (Figures I.6 and 5.1) by concretizing each of the system elements. The corresponding elements are transformation process (TrfP) = engineering design process, HuS (operator) = engineering designer, and Tg = engineering design methodology and methods. The result is shown in Figure 2.1; see also Figure I.16.

As for other types of process (see Chapter 5), the quality of results of a DesP are influenced by all factors of the engineering design system, Figure 2.1.

The model defines *design engineering* as a process of transforming information from a customer's or sponsor's statement of requirements, a given design brief (Od1), to a full description of the proposed future TS(s) and its TP(s) (Od2), that is anticipated to fulfill the requirements, including, manufacturing documentation, and implementation instructions.

The model shows a *basic structure* for the DesP, including assisting processes relating mainly to information, and management processes related to regulating and controlling. Auxiliary, propelling, and connecting and supporting processes in the general process structure, Figures 5.1 and 5.4, are usually absent.

The model shows the operators (execution system) that exert effects onto the operand (information) of the DesP (1) *engineering designers*, their professional profiles, personal characteristics, tacit/internalized knowing and experience—the most important factor in designing; (2) *technical means*, working means and their use, for example, tools, equipment, computers and applications (programs), and so forth; (3) *working conditions* and design situation, see Chapter 3, the active and reactive environment (AEnv)—conditions in time and space in which design engineering takes place. In addition to the execution system, the model shows other factors that influence the "designing" transformations and their timing and economics; (4) the technology of design engineering, working methods used in designing (general and engineering *design methods*), including techniques of representation; (5) the state of *information* (available information system, IS), particularly general and technical information (object information), information about working methods, methods and techniques of representation (design process information), and specialist design information (domain knowledge—recorded, and internalized tacit), see Chapter 9; and (6) *management* and control of the DesP and of the design personnel, quality of management, teamwork and team building, conflict resolution, time management, psychological and sociological factors, goal setting, staffing, financing, and so forth, see Sections 11.2 and 11.3. The *technology of design engineering*,

(A) Definition — design engineering (selected from among many available definitions)

Operand (information) in state Od1:

Needs, requirements, constraints placed on the TP(s)/TS(s)

→

| Design Engineering
Engineering Design Process
(DesP) |

Operand (information) in state Od2:

Description of the TP(s)/TS(s), full information for possible manufacture

→

Design process — definition: Design engineering means transforming the given problem statement into a full description of a (realizable) technical system and its operational process. The direct content of the design process consists of thinking out (conceptualizing) and describing the structures of a technical system and its operational process (see Figures 6.3 and 5.4).

Consequences
(1) The region of action of design engineering comprises the full range from the simplest constructional part to the highest level of complexity of technical systems (see Figure 6.5).
(2) Design engineering also includes and influences a set of various "foreign" operations (e.g., management of design, see also Figure 2.1D, and 2.1E).
(3) The given problem statement and the description of the TS(s) and TP are items of information, the design process therefore has the general characteristics of an information-transforming process.

(B) Related Terms

Design engineering, according to the above definition, must be regarded as a collective term for other words in normal use, for example

Designing implies	When
Developing	New and unknown TS(s) are to be designed
Adapting	Prior examples of related TS(s) are available
Project engineering	Complex TS(s) are to be designed, for example plant
Planning	TS(s) as constructional systems are to be designed
Organizing	Humans are present as operands (to be managed)
Conceptualizing	Organ structures (concepts) are the object of action
Laying out	Constructional structures are being designed (as layouts)
Form-giving	Form (e.g., shape) of constructional parts is the aim of designing
Designing (architectural, industrial, artistic)	External form of a product (appearance and human interaction) is being designed

(C) Engineering design process — model (Engineering design system)

This model results when all elements of a general transformation (compare Figures I·6 and 5.1) are correlated into the system of design engineering.

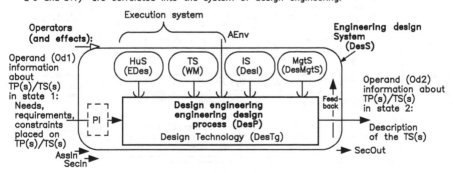

(see also Figures I.6 and I·11)

FIGURE 2.1 Engineering design system—definition, model, terminology of design engineering.

Element 1
    Operand (of designing) ⟶ the TS(s) to be designed (and possibly manufactured)
       and/of the TP(s) to be implemented

Element 2
    Technology (of designing) ⟶ design methodology,
              design methods, strategies, tactics, principles

Element 3
HuS (Desr)	... engineering (and other) designers		
TS (WMe)	... working means of designers	Additional elements:	
IS (DesI)	... design information	Fb	... feedback
AEnv	... design office, time	PI	... planning (of product)
MgtS	... design management		

(D)   Scope of the engineering design process

broader view — Integrated product development model
    ⟶ defining needs + product planning (PI) + industrial design +
    design engineering (in the narrower view) + developing
    (experimental) the realized TS(s)
Narrower view
    ⟶ design engineering as main task

(E)   Internal structure of the engineering design process

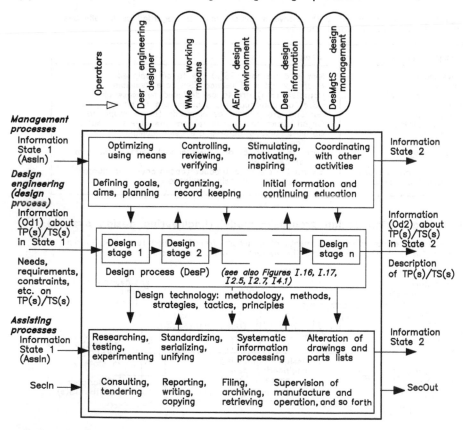

FIGURE 2.1   Continued.

design methodology, is the region of knowledge about strategy and methods in design engineering (see Chapter 8), and indicates useful combinations of individual methods.

2.1.1 CURRENT SITUATION IN DESIGN ENGINEERING PRACTICE

The majority of engineering designers work in traditional ways, "hop-skip-strategies" or "seat-of-the-pants working," and concentrate on the representation of the constructional structure of the TS(s). They mainly work intuitively, without conscious application of a defined method, based on their experience and "know-how," and with a belief in creativity, but without an explanatory framework. Some published methods are sometimes applied, for example, brainstorming, value analysis, FMEA, QFD, TRIZ, and so forth. Under these circumstances, it is much more difficult for engineering designers to find an optimal solution. Good results can be achieved within an engineering designer's specialty, when the task is routine or consists of small alterations or evolutionary developments. A very good or even optimal solution may be found by chance, but with no guarantee that the solution or its procedure can be applied for the next task. High quality of the organization product can only be achieved in several evolutionary "improvement steps" over a longer time period. Initiated by new conditions of economics, globalization, and advances in general knowledge, rationalizing the DesP has revealed these weaknesses, and since the 1960s has given impulses to the search for new and more effective strategies.

2.1.2 CURRENT SITUATION OF DESIGN ENGINEERING IN
SCIENCE AND EDUCATION

Technical universities in some industrialized countries have integrated rationalizing the design process into their research. Cooperations have developed, for example, internationally at the ICED-conferences since 1981, and interdisciplinary with psychology.

The literature offers procedural models with different strategies (1) in the aspect of *addressees* (for whom): engineering designers in practice, industrial designers, research colleagues, or students, item 3 in the morphology of Section 12.4.1; (2) in their construction, most are based on personal and pragmatic experiences of the authors or of a third party, for example, observed research subjects; (a) some models have been developed based on "pure logic"—for example, the General Design Theory [588–591] in attempts to program "intelligence" into computer aids, especially to capture the "design intent"—the results of design process considerations and decisions that lead to the final designed technical system; (b) some process information has been obtained from other bodies of knowledge, for example, problem solving from psychology, decision theory from mathematics [286,454,540]; and (c) combinations of knowledge sources also occurs frequently.

Some surveys and comparisons of the models exist [315], see Figure 2.2. Most models miss details, and show divergence in terminology. Working up to speed and understanding is needed, especially for the definitions of *termini technici*. For engineering designers in practice, orientation is difficult, and selecting "their" strategy

Basic operations (Problem solving)   (see Figure 1.21, level 3, and 2.7)

VDI-2221 (1987)	Rodenacker (1970)	Pahl/Beitz (1995)	Roth (1995)	Koller (1985)	Hubka/Eder (1996)
Problem assignment	Problem assignment	Problem assignment	Problem assignment	Product planning	Problem assignment
Formulating the task 1. Clarifying and increasing the precision of the task description	Required action relationships	Clarifying of task Clarifying the task description, list of requirements	Formulating Clarifying the task description, main task, instructions, list of requirements	Market analysis Establishing the problem definition	Develop the design specification
Finding principles, conceptualizing 2. Establishing functions and their structure 3.1 Search for solution principles, effects level 3.2 Search for solution principles, form level	Functions Logical action relationships Physical events Physical action relationships Action type Kinematic action relationships	Conceptualizing Developing the solution in principle Functions Action principles Action structures Solutions in principle Variants Technical-economic evaluation	Develop functions Establish functions, general, logical, function structure Develop solutions in principle Functions with effects, develop solutions in principle, spec. functions Develop effect carriers for sketch of principles, technical-economic eval.	Function synthesis Purpose or main functions, division for partial and basic functions, technical-economic evaluation Qualitative synthesis Allocation and variation of effects Variation of effect carriers Represent principle Select solutions for total concept Form-giving Laying out	Establish transformation process Establish technology Establish function structure Morphological matrix Establish organ structure
Form-giving, embodying 4. Division into realizable modules 5. Form-giving of significant modules	Design action relationships	Embodying, laying out Develop the constructional structure Rough form-giving, form, material, calculation Definitive form-giving Technical-economic evaluation Definitive form-giving for constructional structure Weak point analysis, disturbing factors, costs Parts lists Manufacturing instructions	Form-giving of structure and form Form of structure — sketch Contours and sections Materials, strength Assembly, naming Function integration Overall development Technical-economic evaluation	Quantitative synthesis Calculation, sizing Experimental investigations Testing, improvement	Establish constructional structure — 1 Preliminary layout Establish constructional structure — 2 Dimensional layout
Elaborating, detailing 6. Form-giving of total product 7. Establishing the execution and usage values	Manufacturing technological action relationships	Elaborating Execution and usage documentation Manufacturing documentation Assembly, transport, testing instructions	Form-giving for Manufacture Weak point analysis Form-giving for manufacture, assembly, transport, recycling Finalized layout Detailing, tolerancing Manufacturing documentation Assembly, usage and test instructions	Detailing Formulate working plans Manufacturing and assembly documentation	Establish constructional Structure — 3 Detail and assembly drg. Computer models Parts lists and so forth *(see Figure 1.16, levels 1 and 2, and, Figure 4.1)*

Literature is listed in References

See also Figures 1.22 and 4.1

FIGURE 2.2 Comparisons of some procedural models of design engineering.

can be demanding, time-consuming and conditioned by "trial-and-error." Introducing new knowledge into engineering practice can be achieved by a holistic, supportive and sympathetic management approach (see Section 11.5).

Design engineering, solving technical problems, is a personal, individual activity involving availability of information, mood, motivation, cognitive abilities, experience (and mental structuring and restructuring of acquired knowledge), creativity, open-mindedness for suggestions from others, stress, perceived leadership and equity, and so forth. Design engineering is an extremely complex human activity, helped by many tools and methods—individual methods can act as prescriptions for strategic or tactical actions (see also Section I.12.5).

NOTE: Engineering designers have traditionally developed their own procedures in designing novel products, or redesigning existing products, usually without being able to explain their processes. By understanding the theory of TP(s) and TS(s), and of DesPs (closely coordinated within EDS), and using the derived and adapted methods, engineering designers should be able to modify and rationalize their procedures. The DesPs can be brought closer to completion, and errors or omissions detected earlier. This should make the engineering product better for all customers and stakeholders, and help to bring the product to market quicker and more economically.

2.2 TASK OF THE ENGINEERING DESIGN SYSTEM

For a particular problem, several alternative TP(s) and TS(s) can be obtained as intermediate result or as output of design engineering. The TS(s) will exhibit a variety of constructional and other structures. An optimal TP(s) and TS(s) should be achieved under the given circumstances, in the shortest time and at a minimum cost. If it is important to meet any agreed time limits, or if other primary goals arise, the priorities can be revised.

NOTE: "Minimum cost" depends on the boundaries of the *economic and social system* as part of the design environment. The narrow view considers only the economics of the organization, the "bottom line" of annual or quarterly reports. Broader views should include life cycle costs, social and environmental costs, and risks and benefits to the community. They should also include long-term influences and preferably aim for sustainable conditions and changes, for example, survival of the organization.

"Designing" (as *terminus technicus*) covers all products with respect to the "operand of the design process"; see Sections I.7.1 and 6.11.9. "Designing" implies a wide range of products, "design engineering" implies that the product is technical.

Design engineering for any TP(s) and TS(s) must acknowledge the physical principles and laws of nature (and therefore the engineering sciences), and take into account the information and knowledge obtained from practice and experience, the real phenomena of this world, at least at a conceptual level, compare Figure I.3. Any assumptions and the resulting predictions by calculation from scientific or empirical formulae must be documented. The DesP is thus an element in generating and implementing TP(s), and realizing TS(s).

The *quality* of a future TP(s) and TS(s) describes its suitability for the intended task, within the social, economic, and other situations. This quality critically depends on several aspects, see Figures 6.9 and Section 6.7, but in particular on the *origination* phase of TP and TS, that is, planning and designing, the first processes in Figures I.13 and 6.14. Design engineering involves anticipating and interpreting needs, selecting optimal principles of operation, arrangements, elements, and proposals for manufacturing, testing, adjusting, delivering, and so forth.

2.2.1 OPERAND OF DESIGN ENGINEERING

The *operand* of design engineering is *information*. For each individual design project or contract, a customer or sponsor (e.g., a representative of a sales department) provides a design brief, a requirements specification, a statement of needs, which serves as input to the design process.

This starting information is transformed in the design process into the *descriptions of the desired TP(s) and TS(s)*, including information needed by the organization-internal and -external "customers" in each life cycle process. The future TS(s) is described by the elemental design properties, classes Pr10, Pr11, and Pr12 in Figures I.8 and 6.8 to 6.10, consequently design engineering involves a search for suitable design properties. The aim of a DesP, its output, is a complete description of an optimal TP(s) and TS(s), which consists of assembly and detail drawings (and computer-resident representations), parts lists, operating and maintenance instructions, user manuals, and so forth as applicable, see Figures I.16 and 2.1. Therefore the task of the design process is to clarify, interpret, and transform the design brief from the designers' point of view, to establish the TP, and conceptualize, lay out and detail the TS(s) into a complete description, ready for possible manufacture, see level H1 in Figure I.21. Designing is an information-transforming process, and its *technology* is important, see Figures I.16 and 2.1. Mostly, the manufacturing processes can be established at the same time as the TS-constructional structure—*concurrent* or *simultaneous* engineering.

NOTE: Many people include designing in the category of "problem-solving," although problem-solving can also be regarded as a part of designing (see Figure I.21), the hierarchical circularity is only an apparent paradox. Figure 2.1 shows that *designing* may be used as a collective term for a wider range of activities, yet some of their practitioners claim that designing is a part of their process (e.g., city planners [38]). Again, this apparent hierarchical paradox is mainly a matter of perception.

"Project engineering" and "planning" are used for *designing* large-scale and complex systems, for example, road networks, transportation systems, industrial plant, and so forth. The process is interrupted at various stages, for example, for political and legal reasons, to obtain approval from relevant authorities on the basis of documents that record an incomplete stage of development of the proposed system, or for practical reasons to elaborate a smaller subsection of the problem and its solution. They should thus be included in design engineering.

"Integrated product development" is a process led by organization management that intends to develop new (implying mass-produced consumer-oriented) products by using cooperating teams from various specialities in the organization. These usually include marketing and sales, designing, and preparing for manufacture, that is, participants in concurrent or simultaneous engineering. Designing may thus be an integral part of integrated product development, see Figure 7.10.

2.2.1.1 Design Specification, List of Requirements (Input to the Design System) Operand of Design Process in State Od1

The *design brief* as delivered or assigned to the designer by an organization-internal or -external customer, for example, management, administration, or sales (a sponsor), is a preliminary formulation from the viewpoint of these persons. It usually states the direct requirements, and the legal contract conditions. For design engineering, it is usually incomplete, and a better definition from the viewpoint of the engineering designer is needed [189]. The brief must usually be expanded by formulating or "framing" the problem, declaring the deficiency, showing the needs, listing the constraints, and so forth. Assisting inputs are experiences from previous design projects, information from suppliers (COTS, OEM parts), and so forth. Secondary inputs may be changes in the requirements discovered during designing, faulty information, and so forth. This is one of many task definitions for design engineering; each subproblem needs its own task definition.

"A good task definition is already half-way to a solution." The goal for designing should be clear; then the result (the proposed means as suggested solution) can be optimal, and progress toward a solution can be rational and fast. The task definition can use a pro forma, with statements that are specific for the range of design problems, the design situation, and the organization, see Section 9.5. Nevertheless, the task should be understood by the designers, and should be reviewed (especially for changes relative to previous projects) and amended.

2.2.1.2 Representation and Documentation: Operand of Design Process in State Od2

The forms, accuracy and completeness of documentation are decisive for the TS-life phases of "preparation for manufacture" (production planning) and "manufacture." This documentation includes (1) the representations of the designed TS(s), and capture of the "design intent" and (2) documentation for the design process itself, the design report, calculations, information search, and so forth. These may also be critical if a product liability suit is brought against the organization.

2.2.1.3 Secondary Inputs and Outputs

Secondary inputs to DesPs include disturbances to a "normal" procedure, for example, changes in the design specification required by a customer during the DesP, changes in design personnel, in design management or environment (e.g., a takeover of the organization), in priorities and deadlines, in the undesirable need for "crash" programming or "charettes," in available information (e.g., new research results that put previous decisions in question), in software for the task, in the political and economic situations, and so forth.

2.2.2 PROCEDURE IN DESIGNING (TECHNOLOGY OF THE ENGINEERING DESIGN PROCESS) PROCEDURE, STRATEGY, TACTICS, AND METHODS

The total transformation of information in the DesP can be realized as the prescriptions or rules for procedure in each of the partial transformations and operations. The prescription is as a *technological* procedure to answer the question "How should one proceed in a particular situation with a specific problem, in order to most effectively obtain an optimal result?" Prescribing a general *strategic* procedure for designing TP(s)/TS(s), formalizing a procedural model, and presenting it for engineering design practice are primary goals of this book. Among the goals is to indicate application of *tactical* methods. The procedures and the models must be adapted to the design situation, usually by designers.

NOTE: The principles, methods (see Chapter 8), and procedures used in various design situations are different, see Section I.11. Designers describe their working procedures by laconic statement like "an idea came into my mind, and I drew it out." Thought processes used by engineering practitioners usually cannot be applied in a *conscious* and *manageable* way; they are intuitive or in the subconscious, based on experience in practical activity. For discussions about "creativity" and "intuition," see Section 11.1. The goal of the design process is to develop an *optimal* TP(s) and TS(s) with maximum effectiveness. The traditional procedures are not necessarily the best—they can hardly be investigated in any measurable way, and teaching them is difficult, although under favorable conditions some aspects can be learned.

Any TP/TS is artificial, its short-term and long-term behavior is subject to heuristic *causality* (causal determinacy) of an existing ("as is" state) TP and TS (cause → effect). The desired result of a proposed transformation are the operands of the TrfP in their output states, Od2, that is, objects that are subjected to change (Od1): M, L, E, I, at the outputs of the processes, with their aggregate of properties. The effects delivered by the TS that cause the changes in the operand of the TP appear at the end of a causal chain, the effects emerge as *causa finalis* (an original cause, but not in an infinite regression) from the TS. TS-behavior can usually be predicted (e.g., calculated within the limits of precision) to follow the laws of individual TS-"sorts," for example, a machine system follows the laws of the engineering sciences. Many of these laws are statistical, that is, subject to random influences and variations. Some other laws may be as yet unknown, that is, surprising failures may occur that are only later investigated for their scientific contents [466].

The contrast to *causality* is the concept of *finality* (purpose determinacy); see Figure 2.3. The goal for finality is to establish a suitable *causa finalis*, as a future ("as should be" state) TP and TS intended for that purpose. The goal for the design process is therefore to establish a suitable constructional structure (and other structures). Finality plays the role of a compass to show a direction toward an envisaged goal.

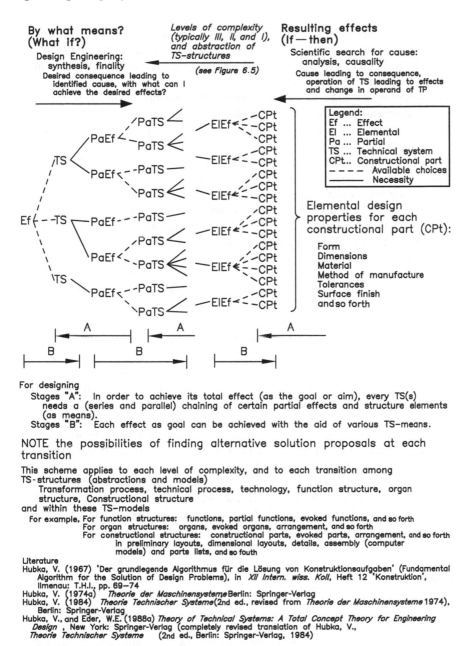

By what means?
(What if?)

Design Engineering:
synthesis, finality
Desired consequence leading to
identified cause, with what can I
achieve the desired effects?

Levels of complexity
(typically III, II, and I),
and abstraction of
TS-structures

(see Figure 6.5)

Resulting effects
(If — then)

Scientific search for cause:
analysis, causality

Cause leading to consequence,
operation of TS leading to effects
and change in operand of TP

Legend:
Ef ... Effect
El ... Elemental
Pa ... Partial
TS ... Technical system
CPt.. Constructional part
– – – – Available choices
———— Necessity

Elemental design
properties for each
constructional part (CPt):

Form
Dimensions
Material
Method of manufacture
Tolerances
Surface finish
and so forth

For designing
Stages "A": In order to achieve its total effect (as the goal or aim), every TS(s)
needs a (series and parallel) chaining of certain partial effects and structure elements
(as means).
Stages "B": Each effect as goal can be achieved with the aid of various TS-means.

NOTE the possibilities of finding alternative solution proposals at each
transition

This scheme applies to each level of complexity, and to each transition among
TS-structures (abstractions and models)
Transformation process, technical process, technology, function structure, organ
structure, Constructional structure
and within these TS–models
for example, For function structures: functions, partial functions, evoked functions, and so forth
For organ structures: organs, evoked organs, arrangement, and so forth
For constructional structures: constructional parts, evoked parts, arrangement, and so forth
in preliminary layouts, dimensional layouts, details, assembly (computer
models) and parts lists, and so fouth

Literature
Hubka, V. (1967) 'Der grundlegende Algorithmus für die Lösung von Konstruktionsaufgaben' (Fundamental
Algorithm for the Solution of Design Problems), in *XII Intern. wiss. Koll*, Heft 12 'Konstruktion',
Ilmenau: T.H.I., pp. 69–74
Hubka, V. (1974a) *Theorie der Maschinensysteme* Berlin: Springer-Verlag
Hubka, V. (1984) *Theorie Technischer Systeme* (2nd ed., revised from *Theorie der Maschinensysteme* 1974),
Berlin: Springer-Verlag
Hubka, V., and Eder, W.E. (1988a) *Theory of Technical Systems: A Total Concept Theory for Engineering
Design* , New York: Springer-Verlag (completely revised translation of Hubka, V.,
Theorie Technischer Systeme   (2nd ed., Berlin: Springer-Verlag, 1984)

FIGURE 2.3   Scheme "goals—means" or "effects—technical systems."

The relationship (intended effect → possible cause → optimal cause) according to
finality is the relationship (goal → means → optimal means).

Designing is always directed toward envisaged *goals*. Engineering designers
look for mainly technical means with which the (intermediate) goal can be reached,

problems (inadequacies, defects) can be eliminated, and needs can be fulfilled. Design engineering consists of a sequence of (goals → means) transitions, a state of finality. Consequently the leading question for design engineering is "With what means can one achieve the necessary effect?"

The relationship of goals to means expresses the "finality nexus" (linkage of finality) and represents the process of synthesis—starting from the goal, a search for suitable means is performed. A finality nexus proceeds in steps in which the main question is "With what?"

Design engineering is concerned with designing technical means to fulfill a purpose, that is, TS(s), and their operational processes, TP(s). TS and TP, and designing them, thus has dimensions in at least three significant axes; see Figure I.15B.

The *causa finalis*, the ultimate cause—the behavior of a TS—is found in the TS-internal properties, classes Pr10, Pr11, and Pr12 in Figures I.8 and 6.8 to 6.10. These are the constructional parts with their elemental design properties, their mutual relationships (e.g., structures of constructional parts, organs, and functions), and the design characteristics and general design properties, which reflect the "design intent." These can be developed in iterative and recursive cycles at various levels of abstraction, see Figure 2.4.

A complete constructional structure that incorporates the TS-internal properties is the precondition for the capability of the future TS(s) to deliver the desired effects to the operand in the TP(s), and for the external properties of the technical system. The "trial-and-error" or "cut-and-try" method used in all sciences is conditioned by previously obtained information—informed trial, not random action—and interpreted as model of knowledge acquisition, see Section I.8. Attempts to rationalize this process must aim at rationalizing the "trial," to avoid and minimize the emerging "errors." Models of TP and TS can be used as assumptions for a "black box," from which (mental, computational, and physical) experiments and simulations can be performed, assumptions verified, and conjectures confirmed.

In the finality procedure, a *hypothesis* (see Section I.5), preliminary to an informal or codified theory, must be proposed as direct answer to the question "With what?" This constitutes a "search for solutions." The quality of each proposed solution must be established by an analytical process, "evaluate and decide." The questions and the hypotheses will be different for other steps of design engineering, given the designers' available information (explicit/codified and tacit/internalized), motivation, mental attitude, and capability.

NOTE: The "trial-and-error" procedure can be immediately rejected as a procedural model for designing—it is a known method for problem-solving (see Figure I.20), but it cannot fulfill the aims. This comment should not *prevent* anyone from using trial and error procedure. Systematic and methodical procedures are more effective, permit better documentation of the procedures, and easier auditing of results.

It is thus necessary to formulate a conscious and transparent procedural model for designing and redesigning TP/TS, design engineering. The model should be based on rational actions (by humans and by TS) and should demonstrate and record progress toward the proposed TP/TS.

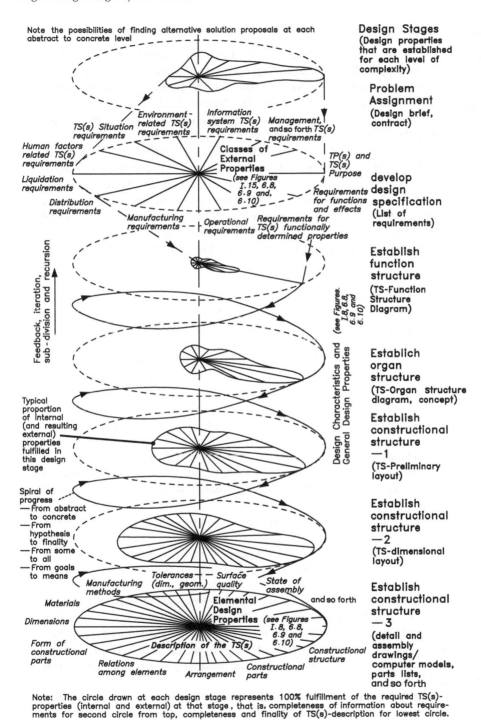

Note the possibilities of finding alternative solution proposals at each abstract to concrete level

Design Stages
(Design properties that are established for each level of complexity)

Problem Assignment
(Design brief, contract)

Environment-related TS(s) requirements

TS(s) Situation requirements

Information system TS(s) requirements

Management, and so forth TS(s) requirements

Human factors related TS(s) requirements

Classes of External Properties (see Figures I.15, 6.8, 6.9 and, 6.10)

Liquidation requirements

Distribution requirements

Manufacturing requirements

TP(s) and TS(s) Purpose

Requirements for functions and effects

develop design specification
(List of requirements)

Operational requirements

Requirements for TS(s) functionally determined properties

Establish function structure
(TS-Function Structure Diagram)

(see Figures I.8, 6.8, 6.9 and 6.10)

Design Characteristics and General Design Properties

Feedback, iteration, sub-division and recursion

Establish organ structure
(TS-Organ structure diagram, concept)

Typical proportion of internal (and resulting external) properties fulfilled in this design stage

Establish constructional structure −1
(TS-Preliminary layout)

Spiral of progress
—From abstract to concrete
—From hypothesis to finality
—From some to all
—From goals to means

Establish constructional structure −2
(TS-dimensional layout)

Manufacturing methods

Tolerances (dim., geom.)

Surface quality

State of assembly

and so forth

Establish constructional structure −3

Materials

Elemental Design Properties (see Figures I.8, 6.8, 6.9 and 6.10)

Dimensions

Form of constructional parts

Description of the TS(s)

Constructional structure

(detail and assembly drawings/ computer models, parts lists, and so forth

Relations among elements

Arrangement

Constructional parts

Note: The circle drawn at each design stage represents 100% fulfillment of the required TS(s)-properties (internal and external) at that stage, that is, completeness of information about requirements for second circle from top, completeness and finality of TS(s)-description for lowest circle.

FIGURE 2.4 Degree of completeness and finality of TS(s)-properties during the design process.

2.2.2.1 Characterization of Procedures in Design Engineering

The partial processes in the structure of the DesP are extremely varied in their complexity, the simplest being the elemental operations that are repeated many times during design engineering (see Figures I.21 and 2.5). The thinking processes that take place range from the creative, to the routine (see Section 2.2.3.1).

Thinking processes are used: (1) *intuitively*, cognitive processes, contemplative and reflective thinking [167,506,507] in which a holistic (partial) solution emerges in a leap of the subconscious or preconscious mind—by creativity [153], based on prior learning, possibly assisted by methods that are intended to enhance creativity—the reasoning that leads to the solution proposal can hardly be explained, except by *post hoc* rationalization or (2) *discursively*, using individual and distinguishable steps, a *methodical and systematic* way, in which an optimal solution is approached in small, conscious, goal-oriented steps—iteratively and recursively.

From this aspect we can distinguish two basic kinds of design process: (1) *intuitive*: a traditional process and (2) *discursive*: a systematic and methodical process, with planned methods.

Methodical and systematic designing can be characterized by subdivision of the total procedure or of a complex task (recursively) into recognizable smaller processes or partial tasks (see Figures I.22 and 4.1), and steps with clearly defined aims—a holistic overview should be maintained in this procedure, to avoid any difficulties arising from relationships among the subdivisions. A recommended, prescribed path leads from the abstract and probable toward the concrete and definitive, for that time and state of progress in problem solving. This path should use a predefined system of abstract models and methods—a procedure based on the *progressive concretization* of the TP(s) and TS(s), that is, moving from incomplete to complete information, and from approximate to definitive values. Individual design characteristics, design properties and features of the TS(s), are progressively established after careful (not necessarily mathematical) optimization, compare Figure I.18. Alternative solution principles and proposals are formulated and evaluated, with decision-making and optimizing at various levels.

The planned and controlled way to achieve the desired quality in a product can only be effectively accomplished by using systematic and methodical design procedures, such as the procedural plan in Chapter 4. Preparing the necessary engineering design report to verify and confirm reasons for various decisions is possible with systematic and methodical procedures.

Design engineering of products can take various forms (see Figure I.20). Form (A) in its pure form is inefficient. Engineering design experiences best use form (C), a combination of forms (B) and (D). Form (D) uses a strategic sequence of checks and iterations (see below), supported by tactical methods, see Chapter 8. The systematic and methodical procedure of form (D) offers the possibility of achieving an optimal TP(s) and TS(s).

If an algorithm is known for the situation, forms (E), (F), and (G) are also possible, mainly as extensions of form (D). The algorithm may be flexible, and can include techniques of artificial intelligence. Form (G) is currently only approached for a narrowly restricted range of technical systems, for example, VLSI chip layout.

In progressing froms (A) through (G), the probability of reaching an optimal solution tends to increase, but applicability to a range of TP/TS decreases.

NOTE: Forms (A) and (B) are regarded as unstructured, although a management structure is needed when applied in industry. The structuring for management purposes (e.g., [24,25,457]) covers the phases of clarifying the task, conceptualizing, embodying, and detailing, see level H1 of Figure I.21, and Figures 2.2, I.22, and 4.1. Between phases, *design audits* can and should be performed, for example, as *phase gates* in which the project is evaluated (formally or informally) by an independent group of practitioners. If the proposals look sufficiently promising further work and expenditure is authorized.

The systematic and methodical approach (D) is also intended to provide functions for rationalizing and control, before and after a design operation, because many people consider that human minds naturally work intuitively. A systematic procedure with record-keeping/documentation is essential for team work, Chapter 11.2, to encourage coordination and cross-fertilization of ideas between team members, even though much of the individual designer's work may rest on subconscious mental processes. The methodical approach provides for a good "preparation" of the problem, and a subsequent "incubation" period that can permit the subconscious thought processes to bring forward latent ideas (see Chapter 8). After-the-event rationalization (procedural form {C}) should then provide a link back to systematic procedures. The need for systematic design methods is based on the inadequacy of intuitive procedures when applied in isolation, and on the need to avoid expensive rethinking to overcome undetected errors during large-scale design tasks performed on high technology systems. Redesigning becomes more expensive as the TP/TS reaches a concrete description. The TS(s) as designed should be "right first time" and optimal. A classification of errors in design [409] has been reported.

2.2.3 STRUCTURE OF THE ENGINEERING DESIGN PROCESS

The DesP contains activities for which designers need specialized abilities and qualifications. These activities are significant in establishing and developing a TP(s) and TS(s), that is, thinking it out prior to implementing the TP(s) and manufacturing the TS(s).

Design activities and their structural elements—design operations—can be categorized and surveyed by (1) a hierarchy of complexity in the activities, in which elements (design operations) in the lower levels are members of each of the elements in higher levels (see Figure I.21) and (2) *activity blocks* that are repeated to achieve strategic aims.

As a technical process, the DesP may be divided into a finite number of partial processes or design operations; see Figures 2.1 and 2.5. The majority of these belong to a few recurring classes of such operations, regardless of the type of design process.

NOTE: Activity blocks follow conceptually from the TrfP and TS models (Figures I.6, 5.1, 6.2, and 6.3). They form logical sequences, which can be used to control the progress of the DesP. In practice, and in observational design research, they can be recognized as orientation points or critical situations [208,209,211,212,248] (see Figure 11.4) in what appear as random and rapid changes of activity within an engineering design procedure. They may also be recognized in reconstructive research on design engineering [435,438].

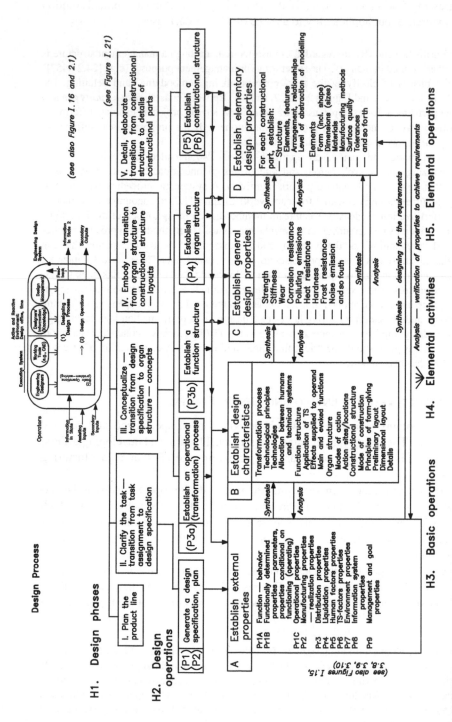

FIGURE 2.5   Structure of activities in the design process.

2.2.3.1 Hierarchical Arrangement of Engineering Design Activities

The degree of difficulty of the engineering design operations depend on the activity itself, and on the specific task—on the "sort" of TP(s)/TS(s), its degree of originality, complexity, difficulty of designing the future TP(s)/TS(s), and so forth. The lower classes of design operations in this hierarchy, Figures 2.5 and I.21, stem in the first place from their complexity and their relationships, and secondly from the kinship among the range of problems to be solved.

Level H1: The activities at level H1 in Figure 2.5 comprise the whole DesP, the administrative phases of (1) clarifying the problem, (2) conceptualizing, (3) laying out, and (4) detailing, as shown in Figure 4.1. According to Figure 2.1B they may also include product planning and customer negotiations about the design problem.

Level H2: Figure 2.5, level H2, presents the *engineering design operations* on two levels. These are discussed in more detail in Chapter 4.1. The upper level contains five relatively complex (classes of) operations, which completely present the procedure of designing, typically for a novel system. The five classes (see Figure 4.1 for numbering of design stages) involve establishing (1) (P1)/(P2): the task description, design brief, design specification, and a working plan; (2) (P3a): an operational (transformation) process—a TP(s), its technologies, and allocation to the operators, if possible with alternatives; (3) (P3b): a function structure of the TS(s), if possible with alternatives; (4) (P4): an organ structure, if possible with alternatives; and (5) (P5)/(P6): a constructional structure, in several stages, if possible with alternatives.

The lower level of the design operations contains four classes of operations that coordinate with the classes of TS-properties. In the sequence according to Figure 2.5, these classes are (1) the external properties of the TS(s), classes Pr1 to Pr9 in Figures I.8 and 6.8–6.10. These have been first formulated as demands, requirements, and constraints for the whole life cycle of the TS(s), as output of (P1)/(P2); (2) the design characteristics, class Pr10, which appear as characteristics in the individual structures of the TS(s) and present the planes (and models) on which variations and alternative solution proposals can be generated and evaluated; (3) establishing the general design properties, class Pr11; and (4) the elemental design properties, class Pr12, which directly represent the constructional structure of the TSs.

Figure 2.5 also implies that the relationships among classes of properties of technical systems must be established by the designers for the specific TS(s), see Figure 6.10. All external and internal TS-properties are causally determined by the elemental design properties.

Level H3: The basic operations (level H3 of Figure I.21) in the hierarchical order shows a cyclic *block of operations* that continually and repeatedly occurs during problem solving. This block scheme is identical to the procedural model of problem solving developed in Section 2.4.1. It may be used (1) as a strategic concept for the entire DesP, see Figure I.21, that is, level H1: clarifying the problem, conceptualizing, laying out, detailing or (2) for its parts, level H2: for example, selecting materials, dimensioning, and so forth, level H3: for example, solving a detail problem. These basic operations are among the most frequently used general activities (compare [398,441,470,522,560,561,579]) and should be mastered by all engineers and designers.

NOTE: The usual term for a procedure such as problem-solving is an *algorithm*. For problem-solving the algorithm must be prescriptive, heuristic, advisory, voluntary, adaptable and flexible, but not (or only partly) capable of being automated. Engineering design technology also covers this kind of prescription.

Level H4: The activities at this level in Figure I.21 are no longer specific for designing; they are general activities that are needed in any organization and situation.

Level H5: These operations form the building blocks for all other operations at the higher levels.

2.2.4 OPERATORS IN THE ENGINEERING DESIGN SYSTEM

Figures I.16 and 2.1 shows the operators. Other factors in the DesP are related to the design situation, see Chapter 3, for example, the task assignment, the input to design engineering. Each of these operators and the technology (design procedures, methodologies, methods, and working principles) is decisive for the resulting quality of the TP(s)/TS(s), and the effectiveness of the DesP. Figure I.23 gives a qualitative assessment of typical influences on some characteristics of the design process.

Routine work (especially on computers) can generally be performed by an individual, and may involve sketching in words and diagrams for feedback communication with one's own mind [438]. Critical situations [208,209,211,212] and novelty usually require cooperation among individuals; see Section 11.2. Consultations, discussing sketches with team members and colleagues, bouncing ideas off other people, and tapping into available expertise are usual, and this also constitutes an aspect of team work.

2.2.4.1 Operator—Engineering Designer(s) (HuS)

Engineering designers (working individually and as team members) are the most important operators in the engineering design system. Yet designing is not their only role [247]. Team activities, competencies [456,458], and physical and cognitive capabilities are important. The job profile of professional engineers engaged in designing is different from that of design technologists, detail designers and draftspersons, other engineers, scientists, and so forth. Problem areas include mutual understanding among the team members, status, motivation, leadership, and so forth. The requirements for engineering designers must be derived from the characteristics of the DesP; see Figure 2.6A ("list of requirements" for designers) and part B (evaluation of designers) [153,287,290,438].

These requirements relate especially to competencies and abilities, available working means (including computers), and (codified, recorded, documented, tacit, and internalized) specialist technical information. Operators "management" and "environment" are reflected in requirements for personal properties of designers, for example, attitudes. Learning objectives for engineering design education can be derived from typical job descriptions, for knowing, competencies and abilities.

(A) **Model of characteristics of ideal designers**

Ideal (engineering) designers possess

Knowledge (knowing), understanding	Abilities, skills, faculties	Personal characteristics, attitudes

General knowledge
  Languages
  Literature
  History
  Geography
  and so forth
 	
Mathematics
Geometry
Physics
Chemistry
and so forth
 	
Technical branch information
  Basic information
  Specialized branch
    information
  Design information
    (design tuition)
  Manufacturing information
    (manuf. technology)
  Materials information
  and so forth
 	
National economics
Legal information
Psychology
Technical esthetics
Ergonomics
and so forth
	Memory
	Logical thinking
	Synthesis ability
	Cost awareness
	Visualization ability
	Combination ability
	Creativity
	Mental flexibility
	Systematic and
	   methodical working mode
	Information procuring
	Decision ability
	Representation ability
	Observation ability
	Concentration ability
	Punctuality
	Leadership
	Organization
	Orderliness
	Personal bearing
	Precise expression
	Persuasive power
	Interactive skills for
	   team work
	and so forth
		Productivity
		Perseverance
		Willpower
		Honesty
		Responsibility
		Duty awareness
		Openness
		Thoroughness
		Conscientiousness
		Care
		Contact readiness
		Broad horizons
		Objectivity
		Critical attitude
		  (including self-criticism)
		Self-confidence
		Enthusiasm, delight in
		  designing
		Readiness for cooperation
		Constant study
		Fairness
		Psychological typology
		Psychic stability
		Collaborative attitude
		Confidence to accept
		  shared responsibility
		  for team decisions
		and so forth

(see also Figure I.3)

(B) **Level of requirements for three typical design team participants**

Knowledge, attitudes, personal properties,	Layout engineer	Detail designer	draftsperson	Skills, abilities	Layout engineer	Detail designer	Draftsperson
Knowledge				**Skills, abilities**			
General knowledge	2	1	1	Logical thinking	3	3	2
Technical fundamental knowledge	3	2	1	Synthesis capability	3	2	0
Specialized branch knowledge	3	3	1	Cost consciousness	3	3	1
				Memory	2	3	2
Attitudes, personal properties				Visualization capability	3	2	2
Productive capability	3	3	2	Creativity	3	2	0
Endurance	3	3	2	Mental flexibility	3	2	1
Sense of responsibility	3	2	1	Systematic working	3	2	1
Thoroughness	2	3	3	Decision capability	3	2	0
Self-confidence	3	2	0	Skills for representation	3	3	1
Enthusiasm	3	2	1	Skills for drawing/sketching	1	3	3
Willingness for cooperation	3	2	2	Leadership ability	3	1	0
Willingness for continuous learning	3	1	0	Organization ability	2	1	1
				Capability for concentration	3	2	2
				Precise expression	3	1	0

Legend: 3 ... High, 2 ... Intermediate, 1 ... Low, 0 ... no requirement

FIGURE 2.6 Characterization of designers.

Whether the DesP should be performed by an individual designer or a design team, plays a large role in evaluating the qualities of engineering designers.

Designing is a human activity, and involves flair, ability, intuition, creativity, spontaneity, judgment, reflection, feel, experience, and so forth. Designing is necessarily heuristic, iterative, recursive, opportunistic, flexible, and idiosyncratic. These aspects are classified in Section I.11.1. The merits of different proposed and executed solutions will also depend on human judgment.

Only a planned systematic and methodical engineering design procedure, as shown in this book, can ensure an optimal solution of the presented problem, approached in an effective process.

Engineers often work on problems of substantial complexity and lack of transparency [425]. The performance of problem solvers is influenced by the *rationality* of their behavior (see also Section 11.1). This is characterized by the designers' abilities to (verbally, pictorially, and symbolically) analyze a situation, learn from experiences, adaptively regulate their progress, adapt proposed methods to the design situation, orient themselves conceptually (holistic understanding), bring their understanding of problems into coincidence with the situation of the problem, structure their information, plan the measures to be taken in suitable breadth, make abstractions in suitable ways, make multidimensional and substantiated decisions, maintain a selected conception, and follow it diligently, and so forth.

Such behavior—also influenced by emotion, motivation, aesthetic capacity, language capability, and other person-specific factors—helps to preserve problem solvers from cognitive emergency reactions that would reduce their performance in problem solving and design engineering.

2.2.4.2 Operator—Working Means (TS)

All objects, devices, tools or systems that aid and support engineering designers during their work are termed working means. The area of working means, TS in the engineering design system, is diverse, and includes means for (1) handling information—capturing, documenting, recording, storing, classifying, relating, cross-referencing, arranging for easy access, searching, and retrieving; (2) modeling and representing—for possible representations of TS and their properties by graphical (sketching and drawing) and other techniques, for example, by TS-structures, Figures 6.2 and 6.3, and by computer modeling of the constructional structure of the TS(s); (3) calculating, simulating, analyzing, and optimizing, for example, by computer; (4) conventional office work—writing (e.g., word-processing), communicating, dictating, storing/filing (e.g., data base systems), and arranging for easy access; (5) reproducing—copying, printing, enlarging/reducing of drawings and documents (e.g., via microfilm), and virtual or rapid prototyping; (6) handling drawings (originals, copies), issuing, transmitting, storing, recording, recalling, and version control; and (7) experimenting, testing, and so forth—measuring instruments, test apparatus, simulation apparatus.

Electronic data processing (computer aided design [CAD], or computer aided engineering [CAE]) shows great promise, including links to computer-aided manufacture (CAD/CAM), numerical control of manufacturing equipment, company-wide gathering and coordination of information (computer integrated manufacturing [CIM]), product data management (PDM, attaching production data to the computer-resident design documentation), and "intelligent" advisory systems (agents) for computer work, for example, [139,180,532,533]. Computer application programs could be considered as part of the operator "information," but when installed on a computer they are part of that TS. These working means are TS, and must therefore fulfill requirements and possess properties as shown in Figures I.8, and 6.8 to 6.10. The class of human factors properties (especially ergonomics), related to the area of psychology (see Section 11.1), is of great importance for these working means.

2.2.4.3 Operator—Active and Reactive Environment (AEno)

The AEnv in which design engineering takes place influences the results of design engineering, in view of the particular character of engineering design work. At a macro level, for the general environment, the cultural, social, political, economic, and other factors can hardly be influenced and changed; one can only attempt to ameliorate or guard against any negative effects. The micro-level can and must be actively formed—it significantly influences the operators in the engineering design system, especially the engineering designers. The major considerations are the physical and organizational working conditions in the design office: position and size of working rooms (areas); equipment and its arrangement, allowing space to hang and view layouts and other drawings; illumination intensity and type; climatic conditions, temperature, humidity, air movement; noise, loudness, type (music); color and texture; relationships within the working group, psychological working conditions, status, time pressures, "crashing" or "charette," and so forth. Each represents important evaluation criteria for the quality of the working environment [289].

2.2.4.4 Operator—Domain and Branch
Information—Information System (IS)

Technical information, the designers' specialist information, is dominant for the quality of design engineering. The engineering designer's quality depends on the information that he/she commands and is competent to use, including engineering science, general information, experience, methods, and so forth; see Figure I.3, Section I.11.4, and Chapter 9.

Engineering designers' working information consists of two parts (1) information about TP and TS, their theory, actual embodiments, and particular TS-"sorts," those for which the engineering designer is responsible, and closely related ones and (2) information about design engineering, especially methodological and procedural information, consciously applied or learned.

The basic information about TS is presented in Chapters 5 and 6; the information about design engineering is the subject of Chapters 2 to 4. For engineering designers, the transformation from the available general information into the specialized information needed for the branch and domain of their tasks is important, as is the classification and form of this knowledge (see Chapter 9), which should guarantee effectiveness in the process of finding and obtaining the relevant information. Information retrieval listings and computers can also help to ease the work with information. The information needs of engineering designers change according to the design situation and phase of development of the TP(s) and the TS(s).

Engineering designers need to be up to date with relevant information; providing engineering designers with facilities for obtaining information from published sources is essential—a management task. Information should cover the available standards, verified empirical and scientific knowledge (e.g., [5,7]), and the state of the art in the specific field of operation, for example, from specialized books, and from scientific and trade journals. A newly recognized area of formalized knowledge is "design for properties," that is, the theory of properties and information for designing to achieve properties; see Figures I.8 and 6.8–6.10—design for manufacture, maintenance, packaging, and so forth, in general "Design for X" (DfX), see Figure 6.15.

2.2.4.5 Operator: Management System (MgtS)

The complexity of the DesP (Figure 2.1) indicates that management is important for the effectiveness of designing; see Section 11.3. Formulating the goals for the organization, (see Chapter 3), and for the design process in each specific project, is critical.

Compared with other management tasks, managing the DesP reveals special problems. The management system (MgtS) should coordinate the measures to rationalize the DesP, directly and indirectly.

Some tasks of the MgtS are also critical to long-term survival of the organization (1) managing information, (2) coordinating design engineering with manufacturing and marketing (concurrent or simultaneous engineering), and (3) ensuring timely delivery of products with appropriate properties (quality). Two main duties emerge, and include managing the interface between the two duties. One is to plan and rationalize the range of products and to define the tasks set for engineering designers, partly included in the DesP shown in Figures I.11 and 2.1 (see Section 11.3.1). The other duty is to manage the process and progress of engineering design work on specific projects, TrfP, TP(s,) and TS(s), including the personnel and the documentation that results (see Section 11.3.3).

2.2.4.6 Overview and Weighting of Factors That Influence Quality

Figure I.23 shows a weighting of the factors that influence the quality of design of TP(s)/TS(s), and for other aims, for a particular design situation. The influencing factors for the quality and description of the TS(s) are (1) quality of input—design specification (see Section 2.2.1.1), (2) technology of design engineering (see Section 2.2.2), (3) engineering designer (see Section 2.2.4.1), (4) working means (see Section 2.2.4.2), (5) AEnv (see Section 2.2.4.3), (6) IS (see Section 2.2.4.4), (7) management (see Section 2.2.4.5), and (8) secondary inputs (see Section 2.2.1.3).

2.3 BASIS OF KNOWLEDGE FOR RATIONALIZING ENGINEERING DESIGN PROCESSES

DesPs can be rationalized by bringing the results of cognitive operations out of the human mind, into records and accessible documentation. Engineering designers interact with these records, and use them for communication. An appropriate system to classify the records should be used. Reflecting on these operations is also necessary [167,506,507]. Rationalization is best achieved if results are brought into the design methodology based on EDS—"systematic and methodical designing," see Figure I.20, level D. Systematic design engineering involves a consistent and heuristic technology of designing—design methodology, preferably based on EDS; see Chapter 12. Design engineering can and should be methodical, that is, using appropriate methods, yet including intuitive and pragmatic approaches.

The DesP contains and uses a wide range of activities (see Figures I.11 and 2.5), which have a basis in various areas of knowledge. The systematic and methodical form of engineering design work is founded on relevant insights from psychology, formal heuristics [425], and other areas of knowledge (including science). The relevant principles of these areas of science may usefully be employed in individual cases.

NOTE: One aim of this book is to propose a valid general model of engineering design procedure. If a flexible and adaptive "algorithm" for design engineering exists, questions to be answered are

1. Are all the phases of the DesP rational, or do irrational phases occur?
2. Is design engineering basically scientific or artistic work?
3. How far can one progress by abstracting from the objects to be designed, so that the general procedure can deliver sufficiently concrete advice?
4. Does a general engineering design procedure (an object-independent method) exist, or only engineering design methods for specific TS-"sorts" (e.g., design method for designing cranes)?

From a rational viewpoint, the word "or" in these questions is misleading, the reality is a continuum between extremes. The DesP is composed of rational operations heuristically applied [365]. Many thought processes can be recognized (after the event) as rational, although they were not under conscious control. Controllable and intuitive (subconscious) thinking are considered rational. Deliberate attempts to use a thought mode or mental process do not necessarily lead to the desired results, especially if the intention is to stimulate creativity [31,119–123,234,425,448,558]. Advocating conscious and rational thinking modes avoids the error of "jumping to conclusions" without investigating the problem and alternatives. It would be an error to rely only on intuitive thought.

Design engineering contains artistic behavior and science "application," yet designing "is" neither an art nor an applied science [152]. In designing of TP(s) and TS(s), constraints must be considered, that is, satisfying customers, fulfilling their needs and appealing to their senses; providing economic survival for the organization; remaining within physical and legal constraints, conforming to standards and ethics, and so forth.

Design problems range from those requiring relatively routine solutions based on well-developed information and existing systems, to those demanding highly innovative solutions. A general engineering design method that is *neutral* with respect to objects can be formulated, that is, independent of particular TP(s)/TS(s), and this is the starting point for the research into design methods since the early 1960s. Intuition is given its place [285]—intuition and method are compatible and can be combined.

Design engineering contains mental activity, that is, thinking, cognitive processes. The *psychology of thought processes* (see Section 11.1), cognitive psychology, investigates human thinking activities. A coordinated group of theories has been proposed, which attempts to explain these thought processes; some terms relevant to systematic and methodical design engineering are as follows:

1. *Association*—takes place by forming mental connections between different concepts (ideas, trigger words); one concept can cause another associated concept to rise into consciousness. New ideas can be stimulated by association, an aspect of creativity.
2. *Thought*—thinking, can be either conscious, subconscious, or emotional/unconscious.
3. *Incubation*—can take place after a problem has been recognized, by allowing subconscious thought processes to progress during other human activities, see Chapter 8. Systematic and methodical design favors conscious thought processes.
4. *Intuition*—experiential thought in which the various stages of thinking are no longer fully conscious. In the sense of point 2, this constitutes subconscious thought.
5. Causes of *thinking errors* are investigated, with warnings against fixations (prejudices), poor problem definition, and solving of problems under time pressure [438].

NOTE: Studies of heuristics and creativity provide a starting point for investigating thought processes. Heuristic procedures ("serving to discover, proceeding by trial and error" [2]) are based on the principles [31,119,234,425,448,558]: (1) ensure motivation, (2) show limiting conditions (expanded, clarified problem formulation), (3) dissolve prejudices (no fixations), (4) search for variants (possibilities of optimization), and (5) reach decisions on the basis of evaluations with maximum objectivity; the design process is impossible without decisions. Some authors state that "design is decision-making." Point (5) shows it is a part of design engineering (see Figure 2.5, level 3), and other applications for decision-making exist.

Management of production-operations, including work study, generally aims to improve the sequences and operations of human work. The basic principles (simplified, see Section 8.2) are (1) create favorable working conditions, (2) ensure clear formulation and understanding of every task or problem, (3) analyze every task and its work content, divide tasks into an appropriate series of partial tasks and sections; give priorities and specify time deadlines, (4) prepare and critically assess all necessary data, (5) choose the optimal solution for the given conditions, which implies searching for alternative candidate solution proposals, (6) carefully prepare each work item in both technical and organizational aspects, (7) prepare a plan for the execution of each larger section of work, (8) supervise, control, and ensure proper organization for executing the work, (9) inspect results and compare with desired values (quality control), (10) consider all insights and experiences (reflect), and prepare a written evaluation.

These principles can be transferred to design engineering. A good range of knowledge of design tasks is needed in order to develop concrete advice about methods and procedures for the engineering designer based on these principles. This is particularly true for item (3), the engineering designer's working methods. Designing needs to act and react flexibly and adaptively to the developing design situation.

2.4 BASIC OPERATIONS IN THE ENGINEERING DESIGN PROCESS

The DesP consists of several activities of different complexity, ideally based on the triad "subject–theory–method." They contain partial processes, operations, and steps, performed by human engineering designers, individually and in teams, with the help of technical means. Several models of problem solving have been proposed, including for single-solution analytical problems [441,470,522,563,579], and considerations of self-motivation, and overcoming panic/fear [398,560,561,583].

Figures I.21 and 2.5 [287,314,315] show a hierarchy of complexity of these activities. Problem solving consists of the *basic operations*, level H3 in Figure I.21, that appear as repeated building blocks of engineering design procedures, with iterations and recursions in the main cycle and frequent calls to the supporting operations. The basic operations of level H3 reflect the whole engineering design procedure (see Figures I.21 and 4.1), and they appear in each design operation, see the inset matrix in Figure I.21. Figure 2.7 shows the basic operations, additional questions, hints, and operations taken from published problem-solving models as extended heuristic prescriptions.

The cycle of basic operations proceeds in four operational steps

Op-H3.1   Determining, defining, and clarifying the task ("framing" the problem).
Op-H3.2   Searching creatively and routinely for likely (and alternative) solutions, principles and means of differing abstraction.
Op-H3.3   Evaluating, optimizing, improving, making decisions, and selecting the preferred or most promising solution(s).
Op-H3.4   Fixing, describing, capturing the "design intent," communicating the solution, transmitting to the records, the next phase, stage, step, or organization function.

These four operations are supported by the results from three additional operations

Op-H3.5   Providing and preparing of information.
Op-H3.6   Verifying and checking (including auditing, validating, and reflecting).
Op-H3.7   Representing (data, solution proposals, and so forth).

NOTE: Various implications are embedded in this scheme of problem-solving. All of the steps and operations are interdependent, and must consider the later or earlier steps or operations. Iterative and recursive working is essential, usually in several rounds of concretization. The

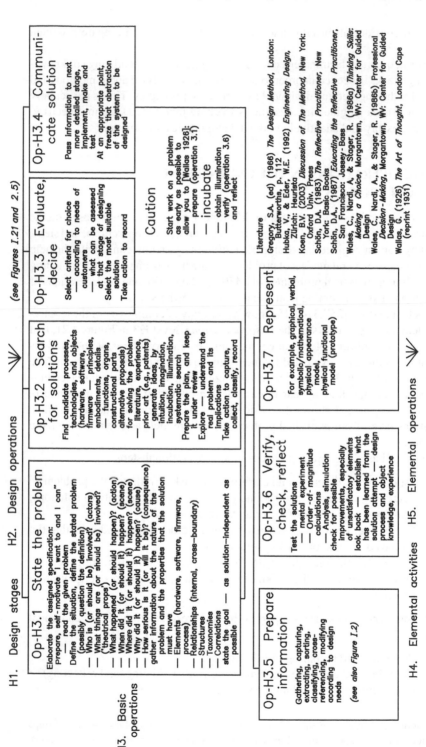

FIGURE 2.7 Basic operations—problem solving in the design process.

sequencing of these operations can be altered. The operations can be performed "on paper" (recorded) and in the human mind—preferably in an interaction between these, keeping in view the mental limitations (see Section 11.1). The general procedure should consist of a "quick-and-dirty" [482] first run through the steps and operations, and then a more detailed treatment in several rounds. Especially before attempting step Op-H3.2, at least a rough and preliminary consideration of Op-H3.1 should have been made. Steps Op-H3.2 and Op-H3.3 should be kept separate, psychology indicates that judgmental attitudes tend to suppress idea-generation (see Section 11.1). Normal human behavior is to mix these steps. Better control and consideration of alternatives uses the steps in the given order, or at least keeps the documented records (the engineering design report) in that order. The problem-solver should be familiar with the use of the component steps and operations, and be aware of this usage.

Each engineering design operation is usually processed in an interchange among the basic operations. Investigating or assuming a part of the problem in another operation (making a *conjecture*) can explain and clarify the task, and lead by iteration and reflection to a more concrete result. The tasks can and should be decomposed (recursively) into subordinate problems, treated in (similar) cycles of basic operations to achieve partial solutions, and incorporated into the main cycle and its results. Conflicts between partial solutions may occur, to be treated as problems to be solved in a cycle of basic operations.

These basic operations of level H3 also appear recognizably in subconscious (preconscious) forms in the intuitive modes of operation. They should be learned in formal ways, and thus become internalized, "second nature" or intuitive. Simpler and more routine engineering design operations are more readily performed in a subconscious way. More complex and difficult engineering design operations may demand a conscious procedure, for which level H3 gives the procedural guidance. These basic operations represent fundamental engineering design activities, although they are not exclusive to designing. The working principles listed in Chapter 8.2 should be respected.

Only the generally applicable elements are listed in the following description of these basic operations. They are expanded and concretized by considering special cases, for example, the step of "elaborating and clarifying the assigned problem" is shown in more detail in the procedural model (see Figures I.22 and 4.1). In these basic operations the engineering designer has the added task of concretizing and adapting the general procedure for the particular situation.

2.4.1 Op-H3.1: Defining and Clarifying the Task ("Framing" the Problem)

This operation can be applied (1) as starting point for the total TP(s) and TS(s) (see Figures I.22 and 4.1, and Section 2.2.1.1), the task definition for the TP(s) and TS(s), a *design specification* and (2) as starting point for each problem or subproblem that arises within the total DesP—task definitions for the partial systems and elements, and for each step in the DesP. This last item is the role of the task definition within the basic operations, as described here. Iterative review and completion of the task definition for the total system during further design work is encouraged, and usually essential.

2.4.2 Op-H3.2: Searching for Likely (and Alternative) Solution Principles and Means

The second basic operation (Op-H3.2) involves a search for candidate solutions, creatively and routinely, that is, alternative principles and means that are likely to achieve the desired state of the operand, or the desired output effect from the TS, or to eliminate defects, and other properties. For the engineering designer the most pressing problems are the *effects* that should be delivered by the TS(s), the considerations concern and influence mainly the action locations and design properties. Engineering designers search for solution proposals in the existing literature, in results of scientific research, in existing TS (e.g., competing products on the market as advertised), by reverse engineering, from suppliers, from the internet, patents, or from self-generated ideas (tacit knowledge, experience, imagination, creativity, and opinions), in order to (eventually) define the concrete constructional structure. They may also need to search for alternative design/solution methods. This operation is the source of creative solutions.

A search for solutions represents synthesis; many methods can help with this search. Methods suitable for the solution search are, see Chapter 8: (1) discursive methods: analogy, aggregation, similarity, morphology; reversal, design catalogs (see Section 9.3), literature search and (2) intuitive methods: brainstorming, synectics, method 6-3-5, and brainwriting.

Selecting a suitable method is guided by complexity, kind and originality of the task (the design situation), and other required properties of the TS(s).

A different situation emerges within the operation "search for solutions," where the goal is to establish individual properties. The solution to this subproblem consists of establishing (synthesizing) certain manifestations and values the relevant property. For example, to establish the strength of a constructional part, select a suitable material (e.g., [7]), and an appropriate size (dimension) to withstand the applied stress. The comparison of the stress (as determined from the engineering science of stress analysis) with a permitted value (the stress resistance capability of the material) for the individual raw materials allows a statement about whether the strength is adequate for the proposed duty, usually expressed as a "factor of safety." Similar considerations are valid for other characteristics that are determined by laws of a certain discipline. Methods for many of the analytical and predictive determinations are prescribed by standards and codes of practice. The operation of comparing an assumed loading with a limiting condition is falsely called "calculation," although mathematical–analytical calculations play a large role. Similar conditions are valid for evaluations (see below).

Measurement scales or mathematical–analytical expressions do not exist for other manifestations of properties, for example, appearance, aesthetics, personal comfort. Judgment of individual persons (or of groups, teams, experts, consultants) is then needed for making decisions.

Experiences define some general principles of conduct, see also Chapter 8. (1) Always try to find several solutions (among the available principles, functions and organs, constructional parts, and arrangements) for the problem, to have the possibility of selection, and to allow rapid reversion or change of direction if needed.

Concretization attempted too early in these considerations can trap thoughts into a certain direction, create a "mind-set" [31] that may not be optimal. Available and proven solutions should be retained in reasonable measure, few steps of design engineering need an invention or innovation, control of risks is a large part of engineering design strategy—and a higher degree of originality (innovation) usually represents greater risks. (2) Cooperation with other branch experts can bring good results in discovering an optimal solution. This usually involves team work, including with people from other design or engineering science disciplines, from outside consultants and customers, and from other departments in the organization—manufacturing, cost estimating, purchasing, sales, customer service, and so forth. Systematic and methodical procedure is recommended for searching, but should be used for reviewing, verifying, and checking. Solutions (partial) should be documented.

Suitable resources and information facilitate the search for solutions, and can lead to improved solutions. Resources include surveys about solutions for functions, that is, design catalogs (see Section 9.3), and masters (see Section 9.2), can help to establish the structures of the TS(s).

2.4.3 Op-H3.3: Evaluating, Choosing the Preferred or Most Promising Solution(s), Making Decisions

The third step (Op-H3.3) in the basic operations is evaluating. Alternative solutions that were compiled in the preceding operations should be evaluated (rated, appraised, at least ranked) to select the most promising (and likely optimal) solution for a TrfP(s), a TP(s), or a TS(s) structure, for the given conditions as established in the task definition, subject to the results of further concretization. Evaluations involve criteria, usually as a comparison of the evaluated subject with a master: (1) a measurement scale, usually transduced; (2) an ideal based on cultural and aesthetic aspects; or (3) other subjects of similar type, for example, a previous or competitor system.

Evaluation criteria should be selected from the design specification, but these should include only those properties as criteria that can be evaluated at the current stage of design engineering. A complete evaluation can only take place when all information is available: (1) at the end of design engineering and (2) after manufacturing (see Figure I.18).

For plant and large capital equipment, where a "request for proposals" has been issued to several potential contractors (Figure 6.18, the lower stream), an evaluation of the offered tenders may be critical. The issuer of the request for proposals should ideally perform some preliminary designing, and select the criteria and evaluation scheme, before the tenders are offered. This is intended to reduce the influence of personal bias of the evaluator, and to avoid adjusting the evaluation to yield an anticipated result. The issuer can thereby anticipate the difficulties of evaluating the tenders, and choose the most appropriate tender for further processing.

Two or more proposed solutions should be carried forward if the choice cannot be made as definitely as desired, for example, if the differences of evaluations do not show sufficient (statistical) significance. In intermediate phases and stages, it is useful to eliminate the weakest ones from further consideration (but keep records of eliminated alternatives, which helps in backtracking), and take forward the strongest alternative (one, or a few, if the decision is not definite enough).

All solution proposals subject to an evaluation should exist at about the same level of concreteness to allow a reasonably valid evaluation and comparison. The remaining criteria from the task definition should not be ignored; they may not be directly amenable to evaluation, but their considerations may influence the choices. The chosen solution should not be regarded as an absolute optimum; this is a relative judgment (by the design team) for the design situation as it exists at that stage of development of the system to be designed. Decisions can then be made to continue, cancel, or to request further work or information on a project.

The initial goal of an *evaluation* is a statement about the expected or realized quality (value) of a TP(s)/TS(s) as a total appraisal, of a partial system, or of one or more properties. Useful considerations may be found in value theory and in decision theory [421,523,528]. An evaluation during designing can use one or more of the following questions

Type I   How good is something? (demands for a total or a partial evaluation)
Type II   Does the solution correspond to the task definition?
Type III   Which of the proposed solutions is optimal?
Type IV   What are the optimal values of some properties?

These types of evaluation are presented in Figure 2.8 by flow charts, with repeated operations

1. Selecting the criteria.
2. Choosing the evaluation scale for each of the criteria.
3. Determining (or assessing) the values for each of the criteria.
4. Processing of the individual values into a total (aggregate) value.

The significance of the results of a planned method and approach for evaluating a system depends on the method. For instance, the evaluation should allow discussions of merits and deficiencies of proposed solutions, and transfer of good features among proposals. Evaluations should then initiate a change of design procedure from *generating* solutions to *correcting* solutions [138,229].

1. *Selecting the criteria for the evaluation*: The quality of a TP(s)/TS(s) is a suitable aggregate of the states of its properties, formulated as a total value, or partial, technical, usage, economic, or esteem values. The evaluation can be based on simple or complex properties, for example, economic, return on investment, profitability, efficiency, energy consumption, and others, which are able to describe an object in a meaningful way. The choice of the criteria for an evaluation is not simple. The object should be described as extensively and universally as possible, but combining different characteristic values can bring relatively problematic results. Design engineering is still a relatively speculative and abstract stage, where determining the criteria values cannot always be objective and clear, and sometimes is not possible. Choosing as criteria those TS-properties for which determination is not reliable at that stage of design engineering would be meaningless.

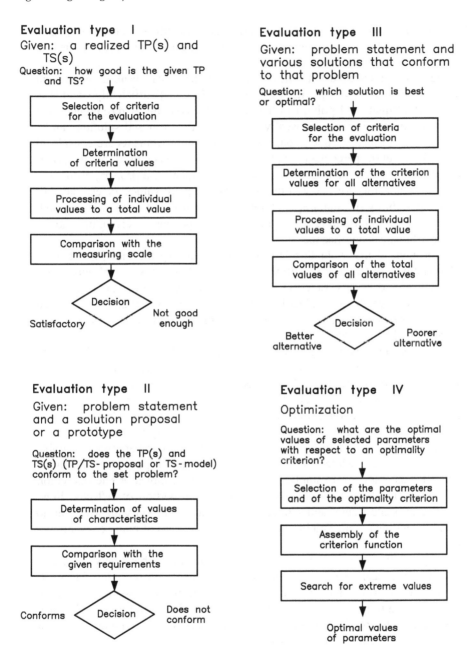

FIGURE 2.8 Types of evaluation—procedural model.

2. *Measures of the criteria for the evaluation*: Where a measurement scale with numerical values and units of measurement is available for the state of a particular property, it should be used to determine values. Otherwise the manifestations of properties can be used to provide an assessed numerical value (see also Sections 6.6.1

and 12.2.1). The evaluative statements should describe the fitness of the TP(s)/TS(s) for its purpose: (1) for the requirements in the design specification or (2) for the task definition or for problem "framing."

For technical practice a usual scale of four quality levels can be allocated as point values: very well suited = 4 points; well suited = 3 points; sufficiently suitable = 2 points; just adequate = 1 point; less than adequate = 0 point. A zero points value given for an important criterion usually means that the proposed solution can be eliminated from further consideration, unless drastic improvements are considered possible. Ten-level (or percentage) evaluation scales have been used, but the possibility and ease of a consistent and meaningful assignment of a property value to a level is less satisfactory for many of the properties. If point values are estimated or found, the possibility of obtaining a "total value" exists—but caution is required; this total value relies on estimated numbers, and the apparent precision may not be justified by statistical significance.

A refinement of the evaluation is offered by weighting the individual criteria (weighted rating, see Chapter 8). Different properties do not have equal importance for the total value of a TS(s) or TP(s). The most important criteria can be described with four (or ten) weight points, less important ones with fewer points—as assessed by the designer or team. The total score is the sum of criteria values from quality points multiplied by weighting points.

A relatively objective rating of a particular (partial) TS(s) or TP(s) for quality demands that the chosen quality point values should be associated to the actual manifestation and values of the properties (see Section 6.6.1). This is relatively simple for quantifiable properties, for example, an amount (e.g., as calculated from a mathematical relationship) of a target value (e.g., 2.1 m) is given 4 points, a largest allowed deviation (e.g., 2.9 m) rates 1 point. A satisfactory objectivity is possible for properties that possess clearly defined manifestations, of which some are more desirable than others, for instance white = 4 points, black = 1 point. It is possible to quantify a property such as color, for example, by defining three variables of hue (the color, about 1200 distinct steps), brightness (roughly the proportion of reflected light, about 300 grey-scale steps), and saturation (strength or dilution with white, up to 100 steps depending on hue and brightness). This is the basis of many color-mixing schemes.

Many properties do not allow an accurate assignment of objectively detectable values to the point scale, for example, safety or friendliness for servicing. The evaluation then has to rest on an investigation that is as objective as possible. In practice it has been shown to be essential that each evaluation of this kind must be justified with reasons, preferably by a written set of comments. Other methods usable for evaluation are listed and described in Chapter 8.

NOTE: The more abstract the structure of the TP(s) or TS(s)—constructional, organ, function, technologies, transformation process structures (see Figure 6.3)—the more difficult it is to judge and verify conformity to the requirements. Evaluations and decisions based only on experience and apparently reasonable assumptions can therefore be wrong. In most cases the total quality of a potential solution can be estimated with sufficient accuracy from a few of the available criteria. If the uncertainty is too large, it is recommended that a definite decision should not be made at a more abstract level, but that a further concretization is performed before coming to a decision. Further theoretical statements and discussions may be found in the literature, which includes some computer implementations [60,69, 79–81,118,129,148,251,253,256,284,349,370,396,401,405,415,457,475,478,490,497,569].

The decision can result in selection of a nonoptimal proposal, a collection of suboptima could result. This danger should be recognized when selecting the evaluation method and the solution. For example, the matrix methods proposed by Starr [523], Morrison [421], and Suh [528] rely on linearity and independence of variables for their validity, see Section 2.4.3. It is possible to depart from the "strict" optimum by a larger or smaller amount, depending on the sensitivity of the properties to such changes.

Statistical significance can be improved by using independent ratings from a few persons [148], and ratings in discussion teams [475,478]. The subjectivity of ratings, if developed in a group discussion, can be adversely influenced by a strong, dominant personality, whose opinions can bias the procedure. It may therefore be useful to modify the form of rating, by using a two-point (+ or −), or three-point (+, =, or −) scale, and comparing one selected solution proposal only with the others, see concept selection, Chapter 8 [475,478]. This comparison can be repeated using the "winner" from the previous round as datum for the comparison. Group discussions about these relative ratings can stimulate suggestions for improvements of the solution proposals (and for procedures). The Art Gallery method is similar. When allocating weights for a numerical weighted point rating, decisions of relative importance among the criteria should be consistent—either by a hierarchical subdivision of weights, such that the weighting in each branch sums to unity [457,569] (see Figure 8.8), or by using a mathematical relationship between each set of pairs from three criteria [81], preferably established (calculated) by computer.

A clearer decision may often be made on the basis of a two-dimensional "relative strength" diagram [349], dividing the evaluation criteria into two separate groups, see Figure 2.9. The abscissa (horizontal axis) shows the weighted-rating sum of technical valuation as a fraction of the maximum theoretically attainable sum (the ideal score): a relative quality of the solution's achievement. The ordinate (vertical axis) plots the economic valuation ratio: anticipated costs, especially for manufacture, compared to anticipated returns, that is, a benefit/cost ratio. Any solution scoring less than 0.7 on either scales is likely to be unacceptable. A variant proposal adds a third dimension, for example, *esteem valuation*, to take the aesthetic and similar values into account [79,80].

3. *Determining (or assessing) the values for each of the evaluation criteria*: Many values of properties of a realized TS or an implemented TP are measurable by experiments, and allow a clear statement. Other properties can only be assessed by judgment. The situation is more problematic during design engineering. Only the elemental design properties can be directly represented in drawings or their computer-resident equivalents. Many of the manifestations and values of properties are not directly represented, for example, in a drawing or analysis, and can only be determined by using mental experiments and models.

Modeling helps the designer to establish facts, informally by mental experiments and by using available materials, or formally by mathematical analysis or by deliberate construction of *physical/tangible* models, for example, by rapid prototyping. Experiments and models include analytical or approximative calculation or simulation models derived from engineering sciences, often implemented as computer

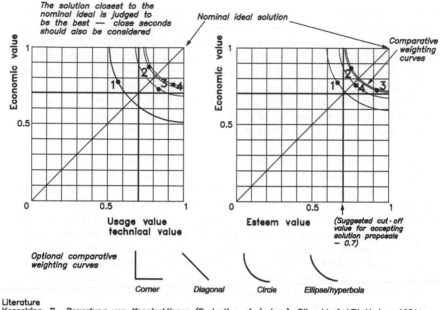

FIGURE 2.9 Evaluation—total quality with respect to two or three groups of properties—relative strength diagram.

applications, and should be the first line of action. For known limiting conditions, a comparison (evaluation type II, Figure 2.9) can lead to a confident decision.

The experience, objectivity, and integrity of the engineering designer (and of the other team members) are the most important elements in situations where different opinions exist about the same problems. After the TS(s) has been realized (manufactured, assembled) and the TP(s) has been implemented, the truth is obvious, but may not be revealed until the system has been in use for some time, for example, fatigue failures have a significant "incubation time."

Even after the TP(s) has been implemented and TS(s) has been realized, the full truth of objective quality and performance is only clear inside the manifestation of the TP(s)/TS(s), not from external observation. Most of this remains hidden from the casual observer, that is, the internal properties. Human opinion about a system can still vary considerably—for a new car in the sales room of a dealer, the reliability can only be estimated by considering the reputations of the makers and the dealer. It would take a long series of well-recorded experiments to obtain an objective value of reliability for the car type, but that value is not applicable to the individual car—it may be one of a rogue population that fails every other day, or it may outlast others.

4. *Processing of the individual evaluations into a total (aggregate) evaluation*: If the values for the selected criteria are available, for example, in point values, the problem of calculating a total value remains. Practice and theory offer many models, from simple addition of the point values, to extremely complicated processing

methods. It has been found that these models do not necessarily deliver reliable results, caution is recommended, especially if broad claims are made [20,60,284,304,396, 405,457].

NOTE: Combining different measures into a single criterion value tends to obscure the interactions of causes, and reflects an assumption that the parameters are linear, orthogonal, and independent. Linearity in behavior implies that point-values are equally spaced, and independence implies that they are "Cartesian" or "orthogonal." The behavior of most TS and their properties is nonlinear. Axiomatic Design, as proposed by Suh [528], gives no advice about the design process; he declares this as simply "creative." Design is defined as a mapping of functional requirements (FRs), to proposed solutions, design parameters (DPs) in the physical space. Each of the FRs and DPs is assumed to be linear and orthogonal. If the numbers of FRs and DPs can be made equal, a square matrix of FRs vs. DPs can be formulated, which can be inverted—implying synthesis as a direct inversion of analysis, but this is necessarily a special case. The axioms (see Section 8.2.3, items 13 and 14) and procedures are intended for evaluation of the "proposed designs" (noun), making decisions about the "best" of the candidates from mathematically solvable criteria by linear algebra, that is, matrix methods. This normally leads to formulating complex FRs, and probably simplistic choices [421,523]. The simplistic mapping of FRs to DPs by Suh, may be compared with the multiple mappings recommended in this book [314,315], in which alternative solutions can be developed: design specification; TrfP; TgStr; FuStr; OrgStr; CStr in preliminary layout: definitive layout; detail, steps in the Procedural Model.

2.4.4 OP-H3.4: FIXING, DESCRIBING, AND COMMUNICATING SOLUTIONS

The result of the cycles of work should be fixed, documented, and recorded. The representation and its content is chosen according to task definition and purpose. The purpose can be: a memorandum notice for internal tasks; communication media for discussions; report generation, for example, communication of the solution proposal(s) to the next phase, stage, or step of design engineering, or technological and organizational preparation for manufacturing (production planning), see Figures I.13 and 6.14, or "stressing"—stress analysis in the aircraft industry.

The display possibilities shown in Section 2.4.7, should not only present solution proposals (results), but also indicate justifications and reasons (see Chapter 3). The representation should be understandable for the designer, and for the addressee, even after a substantial time period. The progress of design engineering should be formulated in the final design report in a clearly traceable form, as required in recent laws concerning product liability, and by standards about responsibilities of designing organizations [9–11].

2.4.5 OP-H3.5: PROVIDING AND PREPARING INFORMATION

Obtaining correct information at the right time is a problem in all branches and specialties; see Chapter 9 and Section 12.1.1. The present situation is similar in designing, research, or management and business administration. Increasingly diffuse presentation of partial information occurs, for example, as "ten-second sound bites"—deliberately short communications that merely summarize a particular point of view, and usually attempt to present it as unalterable fact.

2.4.6 OP-H3.6: VERIFYING, CHECKING, AUDITING, VALIDATING, AND REFLECTING

Verification should determine the accuracy and correctness of previously accomplished work. A statement of evaluation such as: "The TS meets the requirements of the design specification" could also be considered a verification, but this would be false in the context.

A fundamental assumption is that *errors always occur*—conforming to methodical doubt or scientific skepticism according to Descartes (see Chapter 8). A verification process must take place simultaneously with the DesP, for example, as frequent alternation between creating/synthesizing and criticizing/evaluating/deciding. Both processes are interrelated, they influence each other reciprocally, but should be separated for reasons of psychology (see Section 11.1). If a controlling verification is performed only after an extended set of stages of designing, this can cause either large losses, or a long delay such that deadlines cannot be met (see Figure I.18). The view that verification is unproductive is incorrect. Omissions and errors are more likely to be detected and corrected from a verification, savings emerge through avoidance of unnecessary work, and valuable design capacity can be saved. The work following a verification takes place with added knowledge, greater interest, higher motivation, and greater conviction that the work is not futile. It is nevertheless difficult to prove (scientifically, or to a skeptical observer) the usefulness of verification.

The *strategy of verification* suggests that the probability of an error occurring should be estimated: this probability depends on the degree of originality and complexity of the task (e.g., innovation), the quality of the task definition (e.g., design specification), and on the state of the operators in the design process, especially of the designers. The consequences of an error should be considered: errors in conceptualizing, if they are only discovered in the prototype, can mean that design work must proceed again from the beginning at high expense; errors in the dimensioning of a pin are usually easy and cheap to eliminate. The outlook on errors depends on how quickly, easily, and reliably they can be discovered; early discovery of an error is usually less expensive to correct than a late discovery. *Reflection* [167,506,507] is a necessary step that consists of questioning and validating the assumptions, evaluations, methods and results, and finding out what can be learned from the situation for future applications. This should occur toward the end of any cycle of basic operations (problem solving), and after each step and stage of the procedural plan (see Figures I.22, 4.1, and 2.7).

General *working principles* in the checking process are listed in Section 8.2.6, include the following:

1. *Self-examination*: engineering designers should always check their own work, preferably to perform the verification in a different way, and if possible after a longer time interval.
2. Examination by the engineering design (group or project) leader or by a designated checker (especially for detail and assembly drawings): the work

should be checked, reviewed, and audited regularly by the project leader with each engineering designer. Discussions over the drawing board (desk or computer screen) serve for control, and are suitable for the continuing education of the designers. An atmosphere of trust and objectivity in these technical discussions is important.

3. Examination by branch specialists (technical expert, consultant): problems should be presented to the relevant experts, including more senior engineering designers (by scheduling formal or informal periodic meetings). These can be questions of aesthetic form-giving (industrial design), transporting or packaging, and about welding, casting, forging or machining, fixturing, stress and dynamics calculations, simulations, and so forth. Such conversations to examine ideas and proposals should be conducted at the right time (1) when the proposals are sufficiently concrete for the experts to be able to understand, assess, and discuss them and (2) when suggestions for change are still possible without large losses in time, costs, or prestige of the participants.

4. Questioning an expert is also applied for obtaining an external examination (audit, expert opinion), when a completed solution is assessed, and a written report is submitted.

5. Examination by an independent *branch team* (committee, syndicate) is among the most effective forms. Branch specialists of different disciplines and specializations can be invited to take part. Many aspects and viewpoints can be discussed in the conference. This procedure, a form of "design audit," "phase or stage review," has proved useful for examining the results of larger stages (conceptualizing, layout), whereby decisions can be made about the subsequent engineering design steps, for example, release for detailing. Methods such as QFD (see Chapter 8) can be useful at such stages.

6. *Tests and trials* are effective instruments for examinations, if the TP(s) is implemented and TS(s) is realized, for example, as a "proof of concept," a test rig, a physical functioning model, a manufacturing prototype, a preproduction model, or even a final product.

7. Modeling (see Section 2.4.3), provides further methods for the designer to test or transmit various kinds of facts.

8. *Value analysis* is applicable as a complex method that includes economic properties.

2.4.7 OP-H3.7: REPRESENTING

During the DesP, different relationships are needed in order to represent the TP(s) and TS(s) (see Section 2.2.1.2), for supporting visualization, as a note for memory support (e.g., capturing the "design intent"), or for communication with other partners, for example, manufacturing. The represented facts can lie on several levels of abstraction, the displays and representations can be (see Section I.7.1) (1) *iconic* representation: copies of the idea or the original, reasonably true to form (sketches, drawings, photographs, physical models); (2) *symbolic* representations: abstraction using assumed or agreed symbols, for example, flow charts, wiring diagrams, organ structure, function structure, mathematical; (3) *linguistic*: abstraction using words; or (4) *diagrammatic*: graph, diagram to display mathematical and other relationships.

Mixed forms are also possible, for example, a sketched "mind map," concept mapping [92] (see Figures I.3 and I.10), or a hierarchical tree [181,445]. Partial work operations of representing are therefore: sketching, drawing, modeling. The importance of representing for systematic and methodical design engineering is emphasized. Some computer support is available, see Chapter 10. Appropriate working principles are listed in Section 8.2.4.

Design documentation is closely coordinated with the methodical steps regarding their information content—documents should capture the elemental design properties, general design properties, design characteristics, and other information [42,288].

2.5 DESIGN OPERATION INSTRUCTIONS

The triad "subject–theory–method" is of interest for engineering designers, to establish the actions and activities, information, methods and tools, to reach the goal of designing a TP(s)/TS(s) (see Figures I.21 and 2.5). This includes establishing (1) whether an action is possible because operating instructions are available, that is, advice about procedure for acting, especially for activities in Section 2.2.3.1, level H2; (2) what technical knowledge is necessary, which known laws and rules apply in the particular area, both object information and design process information (see Figures I.16 and 2.1); (3) which properties of the TP(s) and TS(s) are to be established or determined; (4) what models, methods and tools can be used; and (5) where corresponding information and data may be found, and so forth.

Figure 2.10 illustrates a typical engineering design procedure, with iterations, recursions and applications of problem solving, for a novel design project—various parts of a TP(s) or TS(s) will exist at different levels of concretization. Broadly linear progress with a wide scatter band of activities is typical [425]. Figure 2.11 shows a similar pattern for redesigning, starting from a design specification, then analyzing from the existing TS into an appropriate abstraction level (e.g., TS[s]-functions) and reconcretizing (see Sections I.1, 1.2, and 4.8).

The object to be designed is normally a TP(s) and a TS(s) (see Chapters 2 to 7). This does not contradict *simultaneous or concurrent engineering* (see Chapter 8), for which "the product and its manufacturing process" are to be designed together (see Figures 2.10 and 2.11). For the designers of the *tangible product*, the TS(s) is their object, considering its TP(s), life cycle (Figures I.13 and 6.14) and operators (Figure 6.15), and properties of the TS(s), see Chapters 6 and 7, including possibilities and economics of manufacture. The designers of the manufacturing process have the manufacturing system as their object to be designed, and their operational process is the manufacture of the TS(s). They must nevertheless consider the appearance and other properties of the final TS(s), and must conform to the needs of the organization (see Chapter 3—Design Situation). Coordination of these two design problems and others is actively encouraged.

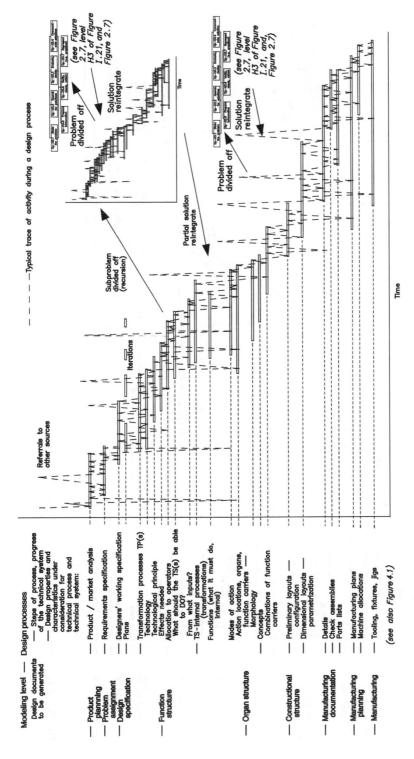

FIGURE 2.10 Design situation—procedural aspects and concurrency—new design.

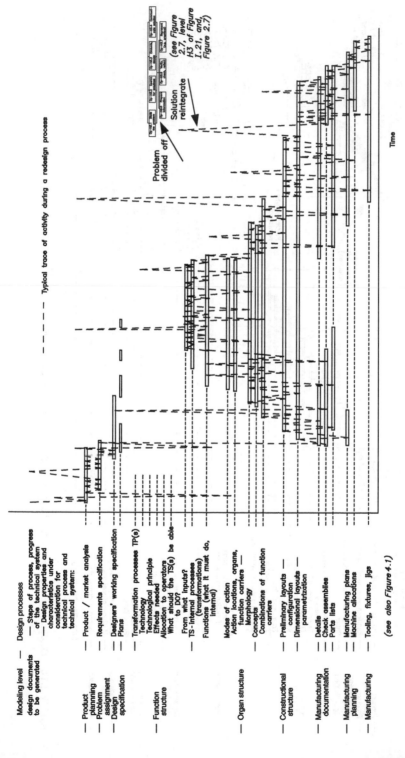

FIGURE 2.11 Design situation—procedural aspects and concurrency—redesign.

The outlines of operating instructions presented below fit into this scheme, and are adapted for establishing internal properties of a TS(s) as directly as possible, see OI/2.6 through OI/2.12. The external properties are established indirectly (by establishing the internal properties), therefore the operating instructions for external properties will be less definite in their advice, and will need further iterations in design engineering, see OI/2.1 through OI/2.5. The first rounds of design engineering of the TS(s) should establish hypothetical assumptions, and from them should develop the rough structures. Only then can an analytical validation be made. Different data is needed for each stage, and therefore different accuracy and precision, and capability for making definite statements about the results is expected.

To facilitate orientation, the *design operating instructions* below are treated as uniformly as possible, and use the following presentation structure

1. *Goal*: the expected output of the design operation is indicated, for example, either how large (what magnitude) is the required property of the constructional part, and the constructional structure (determining a realized value) of an existing TS or a proposed TS(s), or how large should it be (establishing a proposed value) so that the design goal is reached.
2. *Kinds of operation*, or kinds of output.
3. *Requirements*, including demands and constraints, that are connected with the property to be established; see Figures I.8 and 6.8–6.10. Classes of requirements are distinguished according to their source
   a. Requirements that secure the function of the proposed TS(s).
   b. Requirements that ensure fulfillment of the design brief and design specification regarding other demands made on the TS(s), and that consider the special conditions stated in the contract—defined as the external properties.
   c. Conceptually "internal" requirements that are connected with design engineering (class Pr11), or are evoked by needs of manufacturing, assembly, experimenting, adjusting, and other processes of the life cycle phases (see Figures I.13 and 6.14).
4. *Relationships* of the property to be established with other properties—this demands records of the mode of action, and laws and rules of the processes that lead to fulfilling the requirements. These are multiple influences, and need a model of relationships to describe their mutual actions and accurately define the influencing factors.
5. The requirements should be conditional on (and heuristically comply with) natural laws, processes, and realities. According to the status of information, a qualitative or quantitative statement may be available. Where possible, a mathematical equation or formula with measurable values is needed: $Pr_i = f(a,b,c)$, to make reasonable predictions and infer the anticipated behavior of a model of the TP(s) and TS(s).
6. *Insights and measures*: This includes further information and experiences that influence the property under consideration.
7. *Methodical instructions*, references, rules of conduct (see Chapter 8)— the description of a procedure that can be adapted to different situations,

corresponding actions, experiences with individual conduct—developing special patterns for frequently occurring design situations, and experiences with planning statements and use of time.

8. *Information.* Collection of important information sources: literature, patents, and so forth; catalogs of manufacturers, important properties, parameters; standards, norms, legal prescriptions; experts, possibilities of external or internal consulting.

NOTE: The operation instructions do not present a theoretical basis, and do not replace a scientific discipline. A theoretical basis should be as complete and coherent as possible. Operating instructions should be based on such a theory, as indicated by the triad "subject–theory–method." Many statements about the proposed TS(s) are entered as notes on drawings, on calculation sheets and other documents on the basis of a designer's experiences, previous products, and so forth.

Technical information can become of irreplaceable value in the branch or specialty—an organization's treasure—and should be collected and categorized, it should not remain in tacit memory. Concretization of information should be executed with great care, see Section 9.1.1.

2.5.1 DESCRIPTION OF INDIVIDUAL CLASSES OF DESIGN OPERATIONS

A selection of the engineering design operations according to Figure 2.5 is described. Some operations are treated as completely as possible (yet in an abstract and general formulation), because they must appear in all areas. Other typical operations are explained only as examples. The five operations at level H1 present the general progress of designing, see Section 2.2.3.1.

2.5.1.1 Upper Level: Classes I to V

Comments here accord with the connection of operations in the upper level of Design Operations, H2 in Figure 2.5, to the procedural model, see Figures I.22 and 4.1. The last operation, (P5)(P6) "establishing the constructional structure of the TS(s)," is treated in detail in the procedural model, Figure 4.1, because of its importance for engineering practice.

2.5.1.2 Lower Level: Classes A to D

Four classes of operations are shown in Figure 2.5, lower level of H2.

2.5.1.2.1 Class A: establishing the external properties
The engineering design operations are considered with respect to four goals and questions: (1) Which properties should the TP(s) and TS(s) possess? Which requirements are needed in the design specification to make the external properties of the

TS(s) achievable?; (2) With what (primary and alternative) means can the external properties be reached? Which causes and conditions (internal properties) are to be created during synthesis?; (3) When the constructional structure of the TS(s) is established (e.g., preliminary and definitive layouts), what states have the external properties achieved?; and (4) When the constructional structure is realized, what state of the external properties is carried by the TS(s)?—the basis for evaluating the TS(s)-quality.

Question (1) is treated both in the basic operations (Section 2.4), and in connection with the procedural model (Chapter 4).

Question (2) is the central question for the engineering designers during synthesis—design engineering, establishing the properties. The necessary knowledge *originates* from the theories of individual technical disciplines (e.g., the engineering and other sciences) that deal with the range of problems (e.g., with strength, mechanics, thermodynamics, optics, ergonomics, aesthetics, and so forth), and the empirical information on which the theories are based. This is referred to as "property-justified design" and "DfX." The experience of a specialty belongs to the IS that is (or should be) captured, in part in heuristic statements.

Question (3) requires the same information as question (2), and refers to the analytical design operation (a mathematical formulation exists), if the question asks about the state of a property in a known constructional structure. A mathematical formula (or at least an approximative mathematical procedure—an algorithm) is the desired form for the engineering designer. This is the basis for making decisions and selections among alternative proposals.

Question (4) refers to an implemented TP(s) or a realized TS(s). The states are determined by measuring, testing, and observing during operating, experimenting, and developing the TS(s) (see Figure I.18). The results of this operational experience should be conveyed to the designers as positive and negative feedback, especially of maintenance, failure conditions and repair—as full reports, classified summaries, and statistical data depending on the branch and product.

Clarification and definition of the relationships of classes Pr10, Pr11, and Pr12 to the external properties of the TS(s) can contribute to the quality of decisions about internal properties.

Examples for the operation instructions for class A, and especially for question (1), are outlined in the procedural recommendations:

- Establish suitability for safety (instruction OI/2.1)
- Establish suitability for maintenance, service, repair (instruction OI/2.2)
- Establish needs and means for adjustment (instruction OI/2.3)

1. OI/2.1 Design operation instructions—establish suitability for safety

a. *Goal*: increase the safety of TP and TS. If a threat, risk, danger, or hazard can be controlled by suitable measures, or limited to an acceptably small amount, the *safety* of a TP, a TS, and other operators, and of the environment, can be sufficiently ensured.

  b. *Kinds*
     i. Safety of functioning (constructional parts, assembly groups, machines, systems), related to reliability, dependability, availability, consequential damage, hazards.
    ii. Operational safety: safety of industrial personnel; of the TS—damage to the TS; and of parts of the environment, that is, human, animal, and tangible objects.
   iii. Transport and storage safety.
   iv. Environment safety (including personnel not involved in operations, pollutants, deterioration over time, and so forth).
  c. *Requirements connected with safety*
     i. Reliability, dependability, availability, fail-safe conditions, fault hazards, and so forth.
    ii. Ergonomics, access for rescue, and so forth.
   iii. Friendliness to or compatibility with the environment.
  d. *Relationships*
     i. Many general design properties, for example, strength, stiffness, wear and corrosion resistance.
    ii. Elemental design properties, that is, form, arrangement, size, tolerance, surface quality, and so forth.
   iii. Choice of evoked functions, eliminating, protecting, guarding, warning.
   iv. Conduct of service, maintenance, and repair.
    v. Quality of documentation, quality of education.
  e. *Insights—measures*
   Absolute safety can never be achieved—safety measures [58] (see Section 8.2.3).
     Measures according to the principle of safe existence: (a) principle of limited breakdown, include a consciously designed weak point that is easily replaced (e.g., fuse, shear pin); (b) principle of redundant arrangement (series or parallel redundancy); (c) principle of ensuring that a failed system does not present an undue hazard ("fail-safe").
     Potential for danger (hazard): (a) is objectively known (including human breakdown), and can be averted by expedient measures, or can be accepted as conscious risk or (b) is unknown.
  f. *Methodical instructions*
     i. Recognize potential dangers and hazards.
    ii. Analyze the disturbing quantities—human error actions?
   iii. Select measures and accept risk consciously.
   iv. Examine remaining risks: "Is the TS safe enough?"
  g. *Information*
     i. Engineering handbooks.
    ii. Available standards (national and international).
   iii. Fault tree analysis (see Chapter 8).
   iv. Failure modes and effects (and criticality) analysis, FMECA (see Chapter 8).

2. *OI/2.2 design operation instructions—establish suitability for maintenance, service, repair*

   a. *Goal*: simplify maintenance and service—running care, servicing, maintenance of the TS(s) to conserve its value, reduce wear, corrosion, and so forth; implement adjustments and repairs; install upgrades; prevent third-party and consequential damage; implement a service and maintenance plan to systematize maintenance work (observation, supervision, inspections, and controls after prescribed operation time), and to realize completion of repairs and adjustments.

   b. *Kinds*
     i. Supervision by a person.
     ii. Automatic supervision (sensors, controllers/regulators, hazard warnings to supervisor by sound and light signals)—the added TS that implements supervision may also fail.

   c. *Requirements particular to maintenance*
     i. Service life.
     ii. Safety of persons.
     iii. Reliability, dependability, availability, maintainability, repairability (statistical, values).
     iv. Running (direct) and overhead costs, and their allocation.

   d. *Relationships*
     i. Practically depends on all elemental design properties, especially on the arrangement ("configuration") of constructional parts, of organs, and at times even of functions;
     ii. Kind of maintenance implementation, for example, central lubrication, adaptive or proactive maintenance plans;
     iii. Systematic and specialized implementation of supervision, inspections, repairs, and so forth.

   e. *Insights—measures*
   Danger source, recognized from experience, are
     i. For movable parts: wear on sliding surfaces (caution, look also for micromovements, they may lead to fretting damage); lubrication—oil levels, oil changes; bearing heating and noise; corrosion (surface, pitting, crevice); sealing, seal exchange or renewal; brakes, couplings, their adjustment.
     ii. For parts in exposed places: renewal of protective paints; alarms for fire protection.
   Automation of some supervision functions is possible:
     • Central lubrication devices.
     • Smoke and ionization monitors.
     • Temperature (and noise) measurement for bearing locations.
     • Monitoring for metal particles, sludge, and so forth.
     • Analysis apparatus for poisoning protection (e.g., carbon monoxide alarms).

f. *Methodical instructions*
Instructions and schedules for maintenance should include: (1) list of important constructional parts and groups with corresponding documentation; (2) explanation of the functions with all operational data; (3) instructions for starting and operating; (4) instructions for necessary controls and inspections at prescribed time intervals to check for permissible wear, and deviations from normal operating state; (5) instructions for assembly and disassembly, use of jigs, fixtures, and special tools; (6) interlocks to prevent unauthorized operation.

g. *Information*: literature, standards.

3. *OI/2.3 design operation instructions—establish needs and means for adjustment*

a. *Goal*: reduce functional deficiencies caused by dimensional and geometric tolerances in manufacture, tolerance accumulations, changes over time or errors during TS-operation. This goal suggests evoked functions during design engineering, especially in the preliminary and dimensional layout stages.

b. *Kinds*: only the first two are considered in the remaining subsections:
   i. Initial setting of an operating point during assembly of a TS(s), for example, alignment, timing.
   ii. Resetting of an operating point to accommodate wear or other time-dependent changes.
   iii. Monitoring and stepwise input change to maintain steady operation within a tolerance band of an operating point—a two-state "bang-bang" control.
   iv. Monitoring and continuous changing to maintain steady operation close to an operating point (proportional, differential and integral control, analog or digital, and so forth).
   v. Stepwise input change to alter the operating point.
   vi. Continuous input change to alter the operating point—automatic control.

c. *Requirements*
   i. The functional deficiency (or operational parameter) must be measurable, an adjustment point must allow setting to reduce the functional deficiency.
   ii. The measurement indication of the operating point (output location) must be visible from the location of the input change (adjustment point).
   iii. A limitation is given by the precision and sensitivity of measuring instruments at the output location and adjustment point.

d. *Relationships*
   i. Smaller (closer, tighter) tolerances cause nonlinear increase in manufacturing cost.
   ii. Trade-off between cost of (a) manufacture (e.g., to achieve correct function without adjustments) and (b) providing means to allow adjustment, performing adjustments.

    iii. Desired minimum resulting error, and minimum effort to achieve that resulting error.

e. *Insights and measures*

Adjustment can be achieved by

    i. Change of position at adjustment point: (1) fixed by a manufacturing operation; (2) shiftable (e.g., manually) before fixing for operation; (3) shim-settable (settable by fixed incremental inserts) before fixing for operation; (4) settable (e.g., by a setting screw) before fixing for operation; or (5) variable during operation (e.g., by a setting screw and handle) without fixing.

    ii. Change of shape (form) at the adjustment point.

    iii. Exchange of constructional part at adjustment point, for example, selective assembly—sort the realized constructional parts into groups with smaller tolerance ranges, select appropriate matching elements to achieve the desired fit.

f. *Methodical instructions*

Adjustment becomes necessary when

    i. A required dimensional and geometric tolerance is not (economically) manufacturable.

    ii. Manufactured clearances must be reduced.

    iii. Tolerance accumulation (stack-up, build-up) from assembled constructional parts may cause some assemblies to be out of overall tolerance for the assembly, for example, using the program Geomate ToleranceCalc 4.0 http://www.inventbetter.com.

    iv. A common foundation (base) for several constructional parts is only realized at the final location (e.g., during site erection of a TS delivered in assembly groups).

    v. Different functional locations for the same TS are desired, that is, different null positions at different times (e.g., transportable surveying instruments).

Requirements for achieving rational adjustment provisions

    i. The total adjustment (motion range) must be divided into a series of independent adjustment steps, performed in a prescribed sequence, such that (1) each step must be capable of completion; (2) previous steps must not be altered (otherwise an iterative trial-and-error procedure would be required, that can hardly result in a repeatable adjustment of the TS): (a) this may require an adjustment hierarchy: internal adjustment within a subassembly, adjustment of that subassembly relative to other subassemblies, adjustment of the assembly for functional use, operational adjustment by the user; (b) adjustment motions should be orthogonal: an adjustment for one direction should be independent of adjustments for the other two space directions.

    ii. Each adjustment step consists of moving only one element in only one direction such that a definite output result can be observed (measured), and compared to a target value (this rule can be relaxed).

General rules

   i. Good adjustment provisions yield a definite division of subassembled constructional parts into an adjustment unit that should coincide with the physical subassembly (assembly group) as designed for functional purposes.

   ii. Every adjustment step must provide the prerequisites for the following step.

   iii. Adjustment and functional (technological) principle should allow the coarsest manufacturing tolerances, while allowing easiest adjustment.

   iv. Adjustment provisions should aim to provide a minimum of adjustment locations, as close as possible to the observable result position.

   v. A selected functional (technological) principle demands a definite number of adjustment locations.

g. *Information* [252].

2.5.1.2.2 Class B: establishing the design characteristics

Design characteristics, class Pr10, are important internal properties of the TS(s) that are established directly during design engineering—finding alternatives for a proposed TS(s) is also possible in these operations. Establishing the design characteristics is always a subclass of the engineering design phases II to V at level H1 (see Figure 2.5), and of the design operations (P1) to (P6) at level H2. For a detailed specification of the design characteristics, together with the models of the TS, see Chapters 5 and 6.

General design operation instructions for establishing the design characteristics depend on so many factors that they only allow statements of directions or tendencies. Some remarks are contained in the comments to the procedural model in Chapter 4.

The breadth of possibilities (alternatives) for proposals should be established and presented in sufficient depth, and assessed and judged from uniform evaluation criteria. This evaluation (judgment) should limit the initial variety within reason, so that those solutions that are unfit for the given task are immediately excluded. Alternating between searching for alternative solutions, and evaluating and eliminating (see Figures I.22, 4.1, and 4.2), helps to control the possible *combinatorial complexity* of the solution field, which would quickly make the number of available choices and their combinations too large to comprehend. The eliminated variants should still be documented, and reinstated if the initially chosen alternatives prove to be less favorable in their subsequent development.

Nevertheless a balance must be struck—sometimes an innovative solution can be found by accepting a less favorable solution among the retained ones. Compatibility with other solutions must be considered, for example, when several solution proposals (e.g., for subsystems) need to be combined. Engineering design catalogs (see Section 9.3) are useful as assisting tools.

Confining the work to one or a few TS-"sorts," and to certain factors of the organization permits making concrete recommendations for the choice of individual design characteristics. A systematic and methodical gathering of experiences (positive, but also lessons from the negative ones) can increase the quality of this material. The progress of work of establishing the design characteristics is based on the problem-solving cycle of the basic operations, Section 2.4.

2.5.1.2.3 Class C: establishing the general design properties

The general design properties, class Pr11, collect those internal properties that are to be treated as consequences of the established elemental design properties, for example, strength, stiffness. They describe the behavior and quality of the constructional structure, see Chapter 6, and allow finding some alternatives, mainly as variants.

Individual branch areas and specialties (e.g., strength) are explored through the engineering sciences. The relationships are presented in analytical (mathematical) formulae, and an investigation method is recommended, for example, finite elements by computer.

General (theoretical) knowledge is available for class C, yet experience information is indispensable [98,556]—special situations occur in each specialty and branch, and produce difficulties. The experiences should be collected and evaluated.

An example of this class of operation instructions is outlined in the procedural recommendation: establish wear resistance (instruction OI/2.4).

1. *OI/2.4 design operation instructions—establish wear resistance*

   a. *Goal*: decrease wear, that is, decrease undesirable changes on the action surfaces due to mechanical causes (interactions) between constructional parts.
   b. *Kinds*: only the first two kinds are treated in detail as examples: sliding wear; rolling contact wear; shock and impact wear; fluid erosion; jet wear; friction oxidation (fretting wear).
   c. *Requirements connected with wear reduction*: service life; maintenance and servicing; precision; reliability.
   d. *Relationships*
      i. Material surfaces: surface conditions (material, roughness, waviness).
      ii. Form: point, line, or surface contact (mutual conformity, osculation).
      iii. Running conditions and parameters.
      iv. Load: magnitude, time variations.
      v. Kind of lubrication (intermediate material—fluid or solid).
      vi. Kind of movement: continuous, intermittent, reversing, stick-slip; also linear, rotational, helical, spiral, and so forth.
      vii. Temperature.
   e. *Insights—measures*: wear cannot be prevented, only decreased. Possible measures are
      i. More favorable operating conditions.
      ii. More favorable materials, and material pairings (e.g., harder surface paired with softer surface) (1) wear-resistant materials (particular

alloying with Si, Mn, Cr, Mo, Ni); (2) surface treatment: hardening: through hardening, surface hardening, flame hardening, induction hardening, case hardening; surface rolling, shot peening, hammering; nitriding or other diffusion constituents; weld surfacing; spray surfacing: bronze, ceramics materials, carborundum; galvanic covering (electro- or chemical plating); (3) plating with other materials; (4) composite castings: casting in of insert parts;

    iii. More favorable form.

  f. *Methodical instructions.*

  g. *Information*: literature, for example, [7] for selection of materials, standards.

2.5.1.2.4 Class D: establishing the elemental design properties

In designing the constructional structure of a proposed TS(s), the problem is establishing the internal properties, classes Pr10, Pr11, and Pr12 in Figures I.8 and 6.8 to 6.10. Engineering designers must establish all constructional parts and assembly groups, and arrange, configure and parametrize them, whether in a preliminary form (as a "preliminary layout"), in finalized form (as a "dimensional layout") of the TS(s) constructional structure, or in detail. For each constructional part, its form, dimensions, raw material, sorts of manufacturing methods (as general indications and proposals), precision (dimensional and geometric tolerances), surface finish, and so forth, must be established—the elemental design properties. In engineering practice, this design stage is called "embodiment" or "laying out," and "detailing," and is filled with operations of "establishing the design properties." Conceptually an elemental design property can be regarded as a design characteristic at the level of the constructional structure (see Chapter 6), because it is established directly by the engineering designer.

Establishing the many external and internal design properties, is generally a complex and demanding operation. The first rounds take place in the group of the design properties (see Figures 2.5, 2.7, 4.2, and 4.3). Then the group of the external properties of the TS(s) must be recognized, because they exercise the main influence on the properties to be explored. Processing takes place from partial to complete, from simple to complex, and from crude and approximative to definitive statements, to achieve optimal values—with reference to each constructional part and the whole structure. This can be done more efficiently with the help of good "masters" (see Section 9.2). Experiences and available knowledge can reduce the number of iteration loops, if the assumptions and conjectures are good, and if the sequencing of engineering design operations is appropriate (see Design Structure Matrix in Chapter 8).

After establishing the constructional structure of the TS(s), all requirements stated or implied in the design specification and design brief must be examined, checked, and verified to ensure that they have been fulfilled (see class A, goal and question).

Considering the importance of class D, general design operation instructions are given for all operations. These are not complete or exact; they are intended to inspire engineering designers to work out "their own" documents for an area, and are outlined in

OI/2.5	Design operation instructions—elemental design operations—arranging of the constructional parts
OI/2.6	Design operation instructions—establish shape and form
OI/2.7	Design operation instructions—establish material (and raw material, scantlings)
OI/2.8	Design operation instructions—establish kind of production (methods, processes)
OI/2.9	Design operation instructions—establish dimensions, dimensioning
OI/2.10	Design operation instructions—establish tolerances, precision
OI/2.11	Design operation instructions—establish surface quality (finish, roughness, treatment)
OI/2.12	Design operation instructions—form-giving of constructional (machine) parts

1. *OI/2.5 design operation instructions—elemental design operations—arranging of the constructional parts*

a. *Goal*: Arrangement (configuration) of the constructional parts in the given space, so that all requirements are optimally fulfilled.
b. *Kinds*: crude, approximate, preliminary arrangement, preliminary layout; or definitive, finalized, accurate arrangement, dimensional layout.
c. *Requirements connected with arrangement*
    i. All TS-internal functions according to the organ structure.
    ii. Assembly must be possible, preferably using standard tools.
    iii. If needed, disassembly should also be considered for reuse of parts, recovering, remanufacturing and refurbishing, or recovery of materials.
    iv. From the requirements on the TS: appearance, friendliness for transport, service, disposal, liquidation, maintenance.
d. *Relationships*: dependence of the arrangement especially on conception of the constructional structure—mode of construction; and form, and size of the constructional parts.
e. *Insights—measures* about arrangement: clarity, positive and easily detected allocation of functions to organs and constructional parts.
f. *Methodical instructions*, conduct rules: process the arrangement of the construction groups (subassemblies, hierarchy of assembly groupings); integration to total constructional structure.
g. *Information*
    i. Literature, for example [72–74].
    ii. Experts, consultations.
    iii. Standards, guidelines.
    iv. Time planning.

2. OI/2.6 design operation instructions—establish shape and form (of constructional parts)

NOTE: "Shape" and "form" are often used as synonyms, see Glossary. Most statements are valid for shape and form, this range of problems is treated jointly.

Form is one of the most important elemental TS-properties. Different kinds of form exist, depending on which of the (external) properties sets the conditions. Form can be conditional on

i. *Function properties*: aircraft wings, screw thread profile, tooth flank for gear wheel, milling cutter profile.
ii. *Ergonomic properties*: tool handle, lever handle, driver seat, control panel.
iii. *Appearance properties*: form of a smoothing iron (for clothing) or a machine tool.
iv. *Transport and storage properties*: holes or rings to hang and lift (e.g., for a crane), form of a vessel or a forging.
v. *Manufacturing properties*: draft angle for a casting or forging, edge roundings and fillets for castings, runouts for keyway milling and thread cutting (see Figure 2.12).
vi. *Economic properties*: simplest possible form—not always for each constructional part, note trade-offs and synergies between constructional parts and assemblies.
vii. *Strength properties*: beam of equal stress, fillet radius of notches, form transitions.
viii. *Wear properties*: adjustable guideways.
ix. *Safety properties*: no sharp edges.
x. *Material properties*: from available raw forms and scantlings.

Form, established from several requirements and demands, must usually be explored and evaluated from viewpoints of strength, production, assembly, appearance, and economics, especially costs. Questions about form can be directed at all TS of any level of complexity (see Figure 6.5), especially in connection with a constructional part, elemental design properties.

A constructional part (considered as a TS of complexity level I) delivers effects that act internal to a TS of higher complexity, or at its boundary, its action locations. A constructional part must form other spatial features, entities, planes or lines (edges), and combine these into a material unit. Strength, deformation, ability to resist the environment, and other properties of this unit are represented as functionally determined requirements, or evoked functions of a constructional part.

From this general set of requirements, principles for form-giving state that form should be established to (1) provide the required functions and their parameters; (2) accord with arrangement and size requirements; (3) ensure strength, corrosion resistance, temperature resistance, and so forth; and (4) comply with aesthetic requirements.

(A) **Milling a sled-runner keyway**

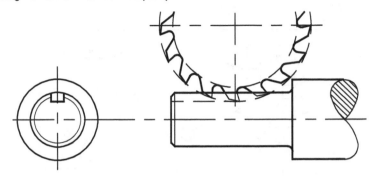

(B) **Tapping a screw thread**

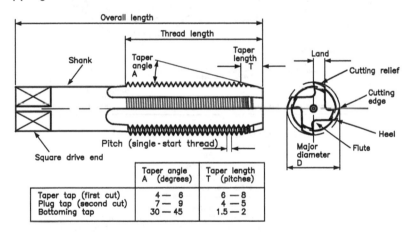

	Taper angle A (degrees)	Taper length T (pitches)
Taper tap (first cut)	4 — 6	6 — 8
Plug tap (second cut)	7 — 9	4 — 5
Bottoming tap	30 — 45	1.5 — 2

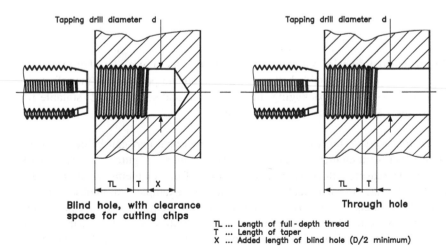

TL ... Length of full-depth thread
T ... Length of taper
X ... Added length of blind hole (D/2 minimum)

FIGURE 2.12　Runouts for keyway milling and thread tapping.

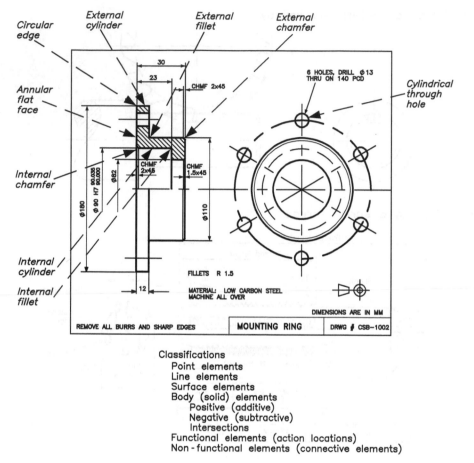

FIGURE 2.13 Form and shape elements—geometric features.

From considerations of realizing (manufacturing) the future TS, other principles are that form should be established to (1) accommodate production (manufacture) and assembly (often also packaging, transportation, and so forth); (2) suitable for materials; and (3) economics (general principle).

(a) Form topology The (total) form of a constructional part can be decomposed into individual form elements, shape elements or *features* (Figure 2.13). Especially for economic reasons, it is expedient to choose form elements as simple geometric bodies that are analytically describable, for example, parametric surfaces in computer models. A constructional part, as a system of form elements that serve either as action locations or as connecting bodies, reveals possibilities of variation of form, see Figure 2.14 (see also [178,370]), (1) form of these elements (e.g., cylinder, prism, cone, parametric surface); (2) dimensions; (3) number of elements; (4) connections among elements; (5) arrangement of the elements mutually and in space.

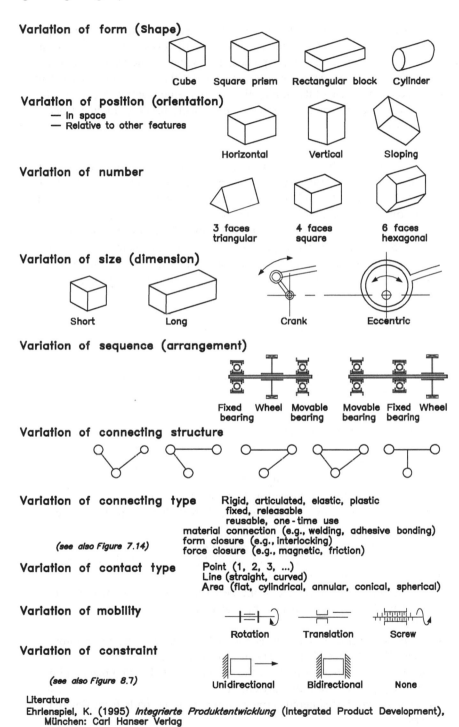

Variation of form (Shape)

Cube   Square prism   Rectangular block   Cylinder

Variation of position (orientation)
— In space
— Relative to other features

Horizontal   Vertical   Sloping

Variation of number

3 faces triangular   4 faces square   6 faces hexagonal

Variation of size (dimension)

Short   Long   Crank   Eccentric

Variation of sequence (arrangement)

Fixed bearing   Wheel   Movable bearing   Movable bearing   Fixed bearing   Wheel

Variation of connecting structure

Variation of connecting type   Rigid, articulated, elastic, plastic
                                   fixed, releasable
                                   reusable, one-time use
(see also Figure 7.14)   material connection (e.g., welding, adhesive bonding)
                                   form closure (e.g., interlocking)
                                   force closure (e.g., magnetic, friction)

Variation of contact type   Point (1, 2, 3, ...)
                                Line (straight, curved)
                                Area (flat, cylindrical, annular, conical, spherical)

Variation of mobility

Rotation   Translation   Screw

Variation of constraint

(see also Figure 8.7)   Unidirectional   Bidirectional   None

Literature
Ehrlenspiel, K. (1995) *Integrierte Produktentwicklung* (Integrated Product Development), München: Carl Hanser Verlag

FIGURE 2.14   Possibilities for variation.

Individual form-elements can be integrated into partial systems, for example, a system of holes in a flange. A form-element can be regarded as a system of planes, and can then be described by: form of these planes; dimensions of the planes; number of the planes; connections among the planes; arrangement of the planes. Similarly a plane can be described as a system of its limiting lines by the same features. A combination of arbitrary manifestations of features (a finite number) in all hierarchical planes creates a form variant.

(b) Standards and unification   Several form-elements or groups of form-elements are often repeated in constructional parts, it is economical to standardize and unify in the form area, for instance by, full or partial reuse of international or national standards, or selection to limit the choice within the organization or among organizations, for example, (1) preferred number series for selection of dimensions (see OI/2.9); (2) preferred dimensional tolerance ranges for fits between components (see OI/2.11); (3) geometric tolerances; (4) national and international standards for (usually purchased) components: keys, splines, fasteners, rolling bearings, electrical insulators, and so forth.

1. *OI/2.7 design operation instructions—establish material (and raw material, scantlings)*

The material for manufacture of the elemental TS (constructional part) is conditioned by external properties, requirements that influence this choice are [7]:

a. *Function properties*: thermal or electrical insulation, heat conduction, chemical resistance ability, oscillation (vibration) damping.
b. *Operational properties*: better quality material for longer duration of operation, or for higher reliability and safety.
c. *Ergonomic properties*: cushioned seats in motor vehicles.
d. *Appearance properties*: woodwork, polished brass, matte-finished aluminum.
e. *Transport properties*: corrosion resistance during transport.
f. *Manufacturing properties*: choice of material (e.g., 1050 steel for ordinary parts, better free-cutting than 1035 steel).
g. *Economic properties*: principle of the best value-for-money material that is sufficient for all requirements, trade-available sizes, delivery time and deadline.
h. *Strength properties*: alloyed steels for highly loaded parts, temperature resistant alloys for the blades of steam and gas turbines.
i. *Wear properties*: guide surfaces of the machine coated with a plastic film or layer (e.g., poly-tetra-fluor-ethylene [PTFE]).
j. *Hardness properties*: sintered carbide turning tool bit, parts from sintered material.
k. *Corrosion resistance properties*: alloyed steels with Ni and Cr or plastics.

2. OI/2.8 operation instructions—establish kind of production (methods, processes)

The kind of manufacturing processes (e.g., envisaged for a constructional part) are closely connected with the external properties, indirectly through the general design properties (see Figures I.8 and 6.8–6.10):

a. *Strength properties*: forged parts tend to be stronger than parts machined by a chip-forming process.
b. *Hardness properties*: hardened parts may be through-hardened or surface-hardened, both produce a hard surface.
c. *Form*: some complicated forms can only be achieved by casting.
d. *Raw material*: each material is limited according to whether or how well it can be machined, cast, welded, and so forth.
e. *Surface properties*: the kind of processing should be chosen according to the prescribed surface quality and state (e.g., painted, see OI/2.9).
f. *Dimensions*: the size (and required number) of the parts influences the kind of manufacturing that can be used.

Manufacturing processes will normally also influence the functioning and functionally determined properties (including performance) of a TS(s). The variability of these properties (from one nominally identical realized TS to another) can be reduced by using appropriate operational values for the manufacturing processes, for example, determined by experimentation, see Taguchi, Chapter 8.

3. OI/2.9 design operation instructions—establish dimensions, dimensioning

Dimensions (sizes) influence practically all properties of the designed TS(s)—functional or strength-determined dimensions. Dimensions are conditioned by

a. *Function properties*: piston diameters and stroke, belt pulley diameters.
b. *Operational properties*: safety factor or allowance added to calculated dimensions (e.g., for anticipated corrosion), larger dimension of parts for better accessibility for maintenance.
c. *Ergonomic properties*: dimensions of operating stand or desk (e.g., control panel, driver compartment, cockpit), spatial arrangement of operating levers, handwheels, buttons, machine displays, and so forth, see Figure 9.3.
d. *Appearance properties*: dimension relationship between individual parts, plane distribution, and proportion (golden section).
e. *Transport properties*: maximum dimensions to the permitted railroad profile, consideration of the standard sizes of crates or packing cases.
f. *Manufacturing properties*: (1) for the constructional parts—minimum wall thickness for castings or forged pieces, minimum bend radius and minimum adjoining "leg" length for sheet metal parts; (2) for representation, placement of dimensioning on drawings to assist manufacturing, inspection, tolerance control, tolerance accumulation (stacking), and so forth.

g. *Economic properties*: smallest possible dimension to reduce material costs.

h. *Strength properties*: dimension of shaft journals and bearings, width and tooth pitch (module) of gears.

i. *Stiffness properties*: mass and dimensions of machine tool bed, vibration behavior.

j. *Wear properties*: allowance for wear of the frictional surfaces.

k. *Material properties*: from the sizes of the available scantlings and raw materials.

l. *Location properties*: prescribed space or connecting dimensions, standardized flange.

The dimensions of constructional parts are often dictated by strength and loading relationships, for example, during manufacturing, transporting, operation, testing. In connection with the loading of material and form, the dimensions (sizes) can be established either in orders of magnitude or accurately. Further criteria for dimensioning are stability, manufacturing, corrosion, and so forth.

4. OI/2.10 design operation instructions—establish tolerances

Dimensional and geometric tolerances, and their accumulation, influence the space relationships between the elements. The choice of tolerances is influenced by higher properties, for example

a. *Function properties*: dimensional and geometric tolerance between bearing journal (shaft) and shell, tolerance between shaft and wheel hub for shrink fit.

b. *Operational properties*: interchangeability of the elements.

c. *Ergonomic properties*: wide tolerances may cause more noise in the mechanism.

d. *Manufacturing properties*: select tolerances of the elements so that assembling is possible without rework or scrap.

e. *Economic properties*: closer tolerances cause higher production costs.

f. *Wear properties*: during shock conditions, wider tolerances cause more wear.

g. *Manufacturing properties*: a kind of machining or manufacturing process allows achieving only a range of tolerances.

h. *Dimensions*: relationship between sizes and the achievable tolerances.

5. OI/2.11 design operation instructions—establish surface quality (finish, roughness, treatment)

"Surface quality" implies the state of the surfaces, including geometric determinacy, precision, waviness, roughness, color, reflectivity, hardness, wear resistance, corrosion susceptibility, and coating. Surface quality is important for achieving the external properties of a TS(s)

a. *Function properties*: guide surfaces, planes for sliding bearings.
b. *Operational properties*: longer usage duration through better surface quality, adequate roughness for oil retention.
c. *Ergonomic properties*: machined surfaces where a person comes in contact; higher friction or roughened surfaces for handles so that they do not slip, yet avoid injury.
d. *Appearance properties*: surface quality is established for the appearance.
e. *Economic properties*: minimum number and sizes of surfaces to be machined.
f. *Strength properties*: smooth surfaces generally increase durability (fatigue resistance).
g. *Corrosion resistance properties*: smooth surfaces are usually more corrosion resistant.
h. *Wear properties*: machined surfaces usually resist wear better.
i. *Tolerance properties*: close tolerances necessitate high surface quality.

6. OI/2.12 design operation instructions—form-giving of constructional (machine) parts

This design operation instruction expands on a part of instruction OI/2.5. The form of (machine) constructional parts fundamentally influences the action locations or action conditions necessary to fulfill the purpose of the TS. These action locations are then joined together into a material unit. Two examples are offered.

The first example shows a lever [251,253], for which Figure 2.15A, geometrically defines the task. The lever should join the action surface B (cylindrical surface) with the action surface C, so that the lever swings around the fixed axle at A (cylindrical action surface). It must also respect the restriction space D. This task follows from an analysis of kinematics, loading (strength), and other constraints. The point of interest in this instance is only the form-giving possibilities in connection with the manufacturing process.

The influence of principles of form-giving is explored in Figure 2.15B. These possibilities are used in form-giving (see Figure 2.15C). The manufacturing processes of the variants were influenced by different considerations, for example, by expected loading, branch of industry (heavy mechanical engineering, instrument engineering, and so forth), number of levers to be made, weight or kind of material (steel or plastics), and show the diversity of possible solutions.

The second example demonstrates 11 typical stages (a development series) during form-giving of a ring lubricated plain sliding bearing [383] (see Figure 2.16). The most important constructional design and form-giving zones and details are established as first priority, then the less important ones in a hierarchical completion of the constructional structure—a change of "windows" as indicated in Section I.7.1 [438]. The task starts from the given diameter of the shaft journal and the height of the shaft axis from the base plate (diagram 1). The needed dimensions (the calculations) have been established, and the form-giving steps are pursued, starting from the choice of the shell length and thickness (diagram 2). The action surfaces (the organ) to achieve the main purpose is thus defined.

(A) **Task definition**

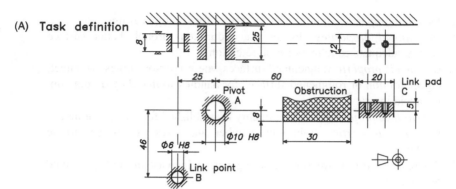

(B) **Organ structure (organs and organ connectors)**

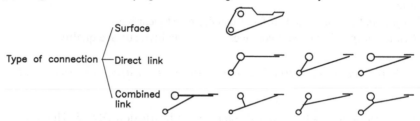

Type of connection ⟨ Surface
— Direct link
Combined link

(C) **Constructional structure (Layouts — typical modes of engineering construction)**

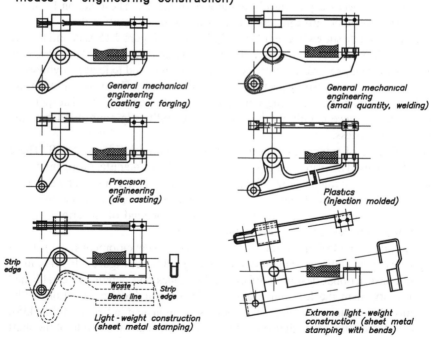

General mechanical engineering (casting or forging)

General mechanical engineering (small quantity, welding)

Precision engineering (die casting)

Plastics (injection molded)

Light-weight construction (sheet metal stamping)

Extreme light-weight construction (sheet metal stamping with bends)

Literature
Hansen, F. (1974) *Konstruktionswissenschaft — Grundlagen und Methoden* (Design Science — Fundamentals and Methods), München: Carl Hanser Verlag

FIGURE 2.15   Structures for lever.

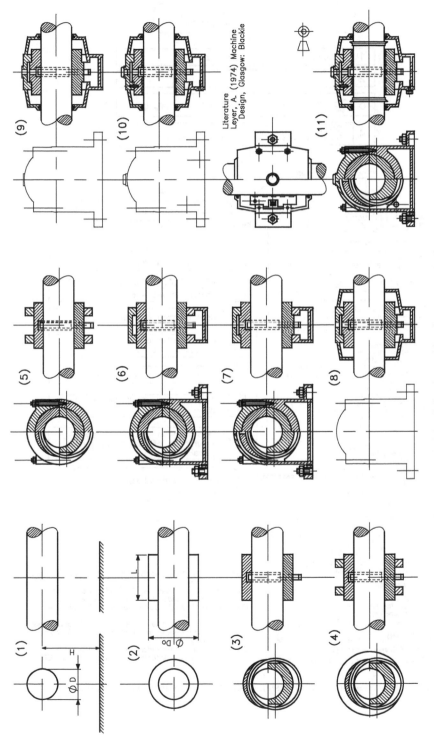

FIGURE 2.16 Form-giving sequence for a ring-lubricated plain sliding journal bearing.

Literature
Leyer, A. (1974) Machine
Design, Glasgow: Blackie

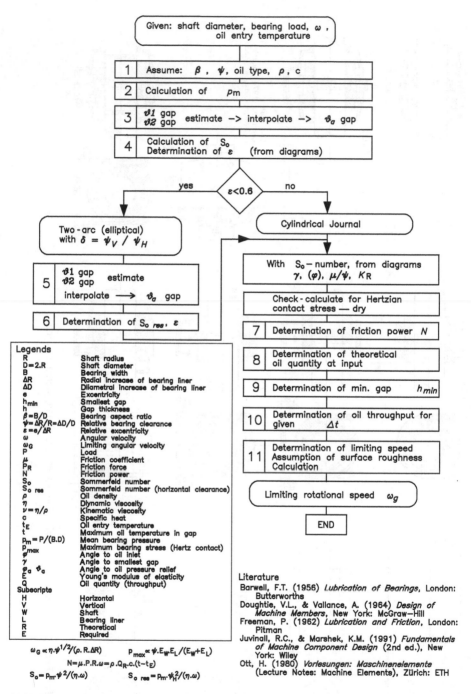

FIGURE 2.17   Procedural master for design calculations of hydrodynamic bearings.

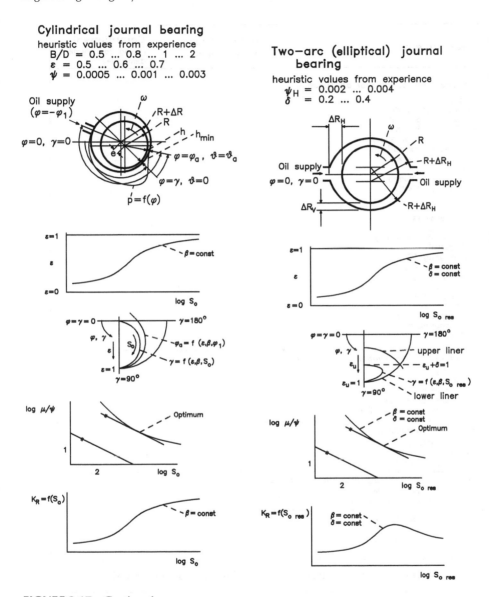

FIGURE 2.17 Continued.

In the next two steps, the zone of the lubricating ring is fashioned (diagram 3). The two shell halves are held together by two robust rings (diagram 4), which for assembly reasons must be of two-piece construction and are fastened by screws (diagram 5). As next action surfaces (organ), the connection flange (to the base plate) is fashioned (diagram 6), and this follows the form-giving guidelines for castings. Then the oil storage chambers (action volumes) are formed (diagram 7). In the next steps the spatial unit under consideration (the housing) is formed to achieve the possibilities given by casting techniques (diagram 8); during this work other action locations (evoked organs) are constructed for the oil filling, level indicating, and draining functions

Requirements	Units, values, tolerances	Commitment (Fixed requirement, desirable)
Requirements conditional on function		
Loading		
— Magnitude radial		
— Magnitude axial		
— Time variation		
— Frequency of shock		
— Possible exceptional loading peaks		
Motion		
— Rotational speed		
— Time variation		
Damping		
Shaft		
— Dimensions		
— Material		
— Positional accuracy		
— Elasticity		
Neighboring TS		
— Interface locations		
— Dimensions		
— Behavior		
— Space restrictions		
Energy losses		
Environment situation		
— Temperature		
— Humidity		
Requirements conditional on operation		
Reliability		
Lifespan		
Maintenance		
— Extent (time, cost)		
— Replacement possibilities		
Ergonomic requirements		
Noise emission		
Safety		
Possibility of monitoring		
Requirements regarding manufacturing properties		
Production technologies		
Batch size / order quantity		
Economic requirements		
Costs		
— Manufacturing costs		
— Running costs		

Master list of requirements for bearing arrangements

FIGURE 2.18   Masters for hydrodynamic journal bearings.

(diagrams 9). Then a form-interlock must be provided to hold the shell against rotation in the bearing casing (diagram 10). Finally the oil circulation must be guaranteed in the bearing through sealing of the side spaces to prevent oil exit (and dirt entry) along the shaft (diagram 10). The completed layout for the bearing (diagram 11) is the final result.

     Engineering designers know that the process as described needs extensive information and knowing, and presupposes some mental and physical models (masters) for the practical. These masters can exist in a rather broad solution space, either close to

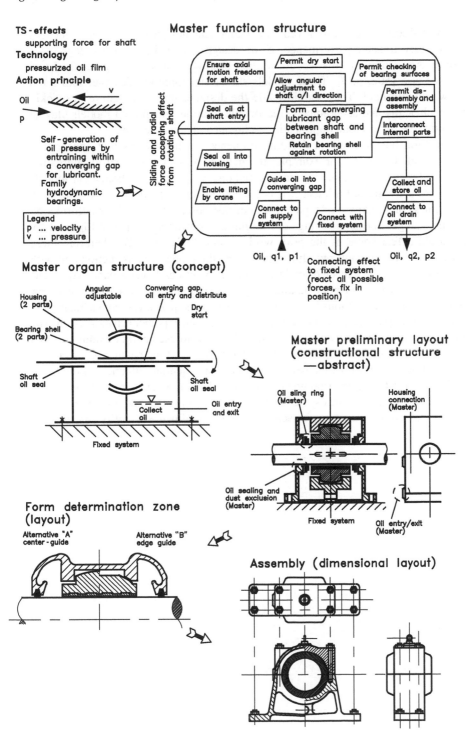

TS-effects
  supporting force for shaft
Technology
  pressurized oil film
Action principle

Self-generation of oil pressure by entraining within a converging gap for lubricant.
Family hydrodynamic bearings.

Legend
p ... velocity
v ... pressure

Sliding and radial force accepting effect from rotating shaft

Master function structure

Ensure axial motion freedom for shaft

Permit dry start

Permit checking of bearing surfaces

Allow angular adjustment to shaft c/l direction

Permit dis-assembly and assembly

Seal oil at shaft entry

Form a converging lubricant gap between shaft and bearing shell
Retain bearing shell against rotation

Interconnect internal parts

Seal oil into housing

Guide oil into converging gap

Collect and store oil

Enable lifting by crane

Connect to oil supply system

Connect with fixed system

Connect to oil drain system

Oil, q1, p1   Connecting effect to fixed system (react all possible forces, fix in position)   Oil, q2, p2

Master organ structure (concept)

Housing (2 parts)
Angular adjustable
Converging gap, oil entry and distribute
Dry start
Bearing shell (2 parts)
Shaft oil seal
Shaft oil seal
Collect oil
Oil entry and exit
Fixed system

Master preliminary layout (constructional structure —abstract)

Oil sling ring (Master)
Housing connection (Master)
Oil sealing and dust exclusion (Master)
Fixed system
Oil entry/exit (Master)

Form determination zone (layout)

Alternative "A" center-guide
Alternative "B" edge guide

Assembly (dimensional layout)

FIGURE 2.18   Continued.

the expected solution or far removed. The engineering designer's situation influences the necessary transfer performance of the designer, and the available degrees of freedom. An example of masters for a plain bearing is shown in Figures 2.17 and 2.18; see also [502, p. 9]. These models should not force the designer into a too narrow solution field. They should guide the designer into a procedure to quickly find a technically and economically reasonable solution. The master models play an important role for computer application in layout and detail design.

The form has generally a deciding influence on the stresses that appear in a machine part. Consider, for example, the notches in a shaft subject to alternating stress—a fatigue condition. For many situations we possess mathematical models that record the expected influence. It is uneconomical if one delays trying to show the unfitness or weakness of a part from considerations of strength or deformation until the strength verification stage of designing, for example, step 8 of Figure 2.16. This could defeat some of the prior work, compare Figure I.18. Therefore these properties should be analyzed after establishing the mode of construction and with the knowledge of the expected loading, using the stress and performance equations for the concrete case, and crude qualitative geometric goals should be formulated. For instance, the stress conditions for a curved beam with unequal stress values (inside and outside of the curve, central plane) allow the choice of an expedient cross-section. The dependence of the stress value of the internal radius (to use another example) leads to the form-giving of the largest possible internal radius of the beam cross-section.

The arrangement of the elements always has an essential influence on the form. This is valid for the parts (arrangement of form-elements), and for the arrangement of elements of more complex systems. The example of the tea machine in Figure 1.9 gives an indication of what can be achieved with the influence quantity "arranging."

3 Design Situation

The designer experiences many different design situations (DesSit) of various kinds [315], which are determined by several factors. The information presented in this chapter is mainly situated in and around the "east" hemisphere of Figures I.4, I.5, and 12.7.

> Organizational position determines the perspective from which the engineer looks at the process s/he is engaged in. The current assignment in turn directs the focus of thinking. The focus can be changed as necessary. The engineer can scan and zoom into or out of the focus areas of the view. It can be thought that the engineer looks at the task environment and the object of design through a restricted 'window' . . . defines the current task within these frames and the task definition embraces an appropriate degree of detailed knowledge [440].

3.1 MODEL OF DESIGN SITUATIONS

Designers in each DesSit should find an optimal solution to the given problem. Therefore, the information about the factors that influence products and procedures (in general, and in designing) are of fundamental importance for designers.

The problem for design engineering may be to establish a future TS(s) and a TP(s), or to find a procedure of design engineering. The (final) design reports, and any changes needed for manufacturing or other down-stream activities, should explain and justify the decisions. This is aided by considering the factors of the engineering design problem and the DesSit.

3.2 FACTORS OF THE DESIGN SITUATION

From the viewpoint of designing, Figure 3.1A shows (1) classes of internal factors that describe the design system, factors FD, and form the design potential of an organization and (2) classes of external factors, the environment of the design system, and the organization potential.

FT   Factors of the design task—planning and thinking ahead for a future organization product, for example, assisted by the classifications of products (Sections I.7 and 6.11).

FO   Organization factors—organization aims, policies, administration structure— manufacturing, selling, servicing, and so forth.

FE Environment factors—of society: local and world economy, culture, politics; and of nature: local and world climate, environmental impacts, pollution sensitivity, and so forth.

See also Sections I.7 and 8.1, and comments to Figure I.17 in Section I.11.6.

This enumeration represents a hierarchical ordering of the individual factors, that is, the design system is a factor of the organization system, which is a factor of a national economic and cultural system, which belongs to the world economy system. These systems are factors of the planet earth and are subject to the laws of nature. It may not be necessary to examine all the factors in each situation, but the connections and a mapping for orientation must be understood.

Figure 3.1B, presents the DesSit; internal factors directly influence the design task in an organization and an environment. The DesSit exerts effects on the engineering design system, Figure I.16, and influences the choice of design technology.

(A)

Class		Designation	Typical examples of manifestation
FE	1	Environment factors Politics	Export−, import−, taxation policies
	2	Society (social system)	Constitution, culture, welfare
	3	Organization policies	State of education, incomes, "fashion"
	4	Science, categorized knowledge	State of research, scientific knowledge
	5	Technology (wide interpretation)	State of the art (engineering)
	6	Market　　for goods	Competition, prices
		for personnel	Employment rate, expertise availability
		for acquisition	Materials, devices, semi-finished goods
		for services	Transport, services, repair
		for financing	Financing availability, interest rates
FO	1	Organization factors Organization aims	Cost recovery, social/ethical goals
	2	Production programme	Breadth of assortment, relationships
	3	Size and location of organization	Small, medium, large
	4	Sorts of manufacturing processes	Single, series, batch, mass, continuous
	5	Potential of organization	Human, working means, financing
	6	Development and market strategy	Long-term market Plans
	7	Employment conditions (general)	Full-time, part-time, out-sourcing
FT	1	Task factors Sorts of product	Breadth, variety, consumer groups
	2	Originality, novelty of product	New products, repeats, variants
	3	Degree of complexity	Plant, machines, groups, parts
	4	Degree of difficulty of designing	Requirements for functions
	5	Particular requirements placed on the product — TS(s)	High Safety, reliability
	6	Deadlines, time to design	Pressure of deadlines, systematic work
	7	Particular requirements regarding procedures	Task, documents, prior permissions, FMECA
	8	Stage of development of the product	Conceptualizing, laying out, detailing
	9	Differentiated submission of data	Foundations, stressing, castings/forgings
	10	Employment conditions of designers	Employee, freelance, consultant
FD	1	Desing internal factors Designers (engineers and others)	Education, experience, attitudes
	2	Working means (tools, equipment)	Tools, equipment, computers
	3	Working conditions, active and reactive environment	Working climate, motivation
	4	Information, branch (Domain) knowledge	Type, form, carrier (medium), reliability
	5	management/leadership	Procedural, hierarchical organization

Survey of factors that influence the design situation

FIGURE 3.1　Design situation—general model.

(B)

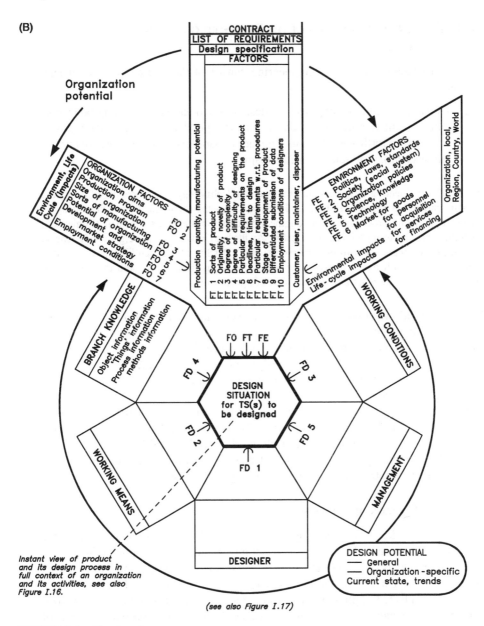

FIGURE 3.1   Continued.

3.2.1 FD—Internal Factors of the Engineering Design Situation

The internal factors for design engineering consist of the operators of the design process:

FD1 Humans—individual engineering designers and teams—including other specialists from manufacturing, cost estimating, purchasing, sales, customer service, and so forth—working in sequence or parallel (simultaneous, concurrent engineering), individually or in a team.

NOTE: The engineering designers choose a suitable **technology** and procedure of designing for their "current" DesSit, that is, tools and methods. They then heuristically and flexibly apply that technology and procedure in the engineering design system to solve their (immediate) problem, which in turn changes the DesSit. At the next review point (e.g., verify, check, reflect), the designers adapt to the "newly existing" DesSit, which then becomes "current." A lagging phase shift exists between the actual DesSit as it develops, and the recognized "current" DesSit under which the engineering designers work.

FD2 Technical systems—working means, including computers, hardware, firmware, software.

FD3 Active and reactive environment—working conditions and other organization factors that directly influence design engineering, for example, security of position of designers in the organization vs. time-limited work of consulting designers under contract.

FD4 Information systems—technical and other branch information of technical process (TP) and technical system (TS), and of engineering design processes, including information, and so forth (see Figures I.10 and I.11).

FD5 Management systems—including leadership, setting goals and objectives, directing, controlling, staffing, resourcing, and so forth.

3.2.2 FT—Factors of the Design Task

The subject of design engineering is the TS(s) (Chapters 6 and 7), and the TS-operational process, the TP(s), for which it is intended to be applied (Chapter 5), that is, the products of the organization. The contracts for TP(s) and TS(s) that an organization accepts are usually limited to a TS-"sort" (see Sections I.7 and 6.11.10). These are established in planning the production and sales program of the organization, usually a management task. Analysis of a task and contract produces a series of features according to which the tasks differ, and can be organized, the class of "factors of the design task" with the symbol "FT." Descriptions in Section 6.11 augment these factors. The most important factors for design engineering are

FT1 "Sorts" of product, range of design tasks. Typical "sorts" of product are described in Section 6.11.10.

FT2 Degree of originality, novelty, of the design task. An important influence on all parameters of designing. The types of design engineering are determined by available experience and clarity of establishing the contents and meaning of the design tasks. Types of design tasks are listed in Section 6.11 and determine whether the engineering designer uses a generative or a correcting procedural manner [138,229]. The demands on the design work are different for each type, and the organization of design engineering is influenced.

FT3 Level of complexity of the system to be designed. The task of the engineering designer may refer to a constructional part, design group, subassembly, machine, or plant (see Figure 6.5). This places demands on the engineering designer and helps to determine the form of organization, including "make or buy" and standardization decisions (Figure 6.7).

FT4 Degree of difficulty, complexity of performing the design task. Tasks place unequal demands on the expert knowledge and the information system of the engineering designer (see Chapter 9). For design engineering, the demands are with respect to the TP(s) and TS(s), and the engineering design processes. Typical degrees of difficulty are a simple container (a storage tank, probably a routine task) and an automatic turning machine (lathe, likely to involve a creative task). Establishing the degree of difficulty must be facilitated and objectified through establishing corresponding classes, and normally not be determined from subjective criteria.

FT5 Particular demands on the product. For a TP(s) and TS(s), the engineering designers' task is derived from a design brief assigned by management. The task is elaborated by the engineering designer in a design specification, which contains prescriptions of many properties to be achieved in the future TP(s) and TS(s), that is, the requirements, including needs, wishes, and constraints. These appear at various levels of priority: as demands (must be fulfilled), wishes (of different intensity according to the possibilities and desirability of fulfilling them), or constraints (restrictions on permissible solutions). More than two levels of priority are possible.

These requirements increase and expand through social and cultural needs, and (local to global) competition on the economic level. Even in a continuous development of requirements, the design specification can be defined as "normal" with reference to particular TS-"sort," in a time frame and conditions, although the tolerance field of variations can be relatively broad.

Unusual demands may be made in a contract, which create a new situation for the designer. For instance, they can involve usual or special safety or reliability values, or very low costs, and can possibly include new properties that up to the present were not called for.

FT6 Design duration, deadlines. The time span within which design work is to be performed is determined by the design process itself, and depends on several factors (e.g., FT1 to FT5). The design duration is an important quantity for product development. The market success of a product can be influenced by the operational properties of a new product and by the location and time of its introduction on the market. According to experience, the design duration tends to stabilize to some "normal" values for a particular product (e.g., a TS-"sort"). If a task demands extreme shortening of time (a "crash program" or a "charette"), then a new situation emerges.

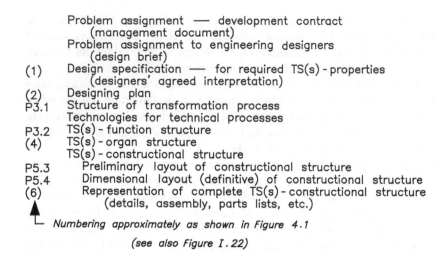

FIGURE 3.2 Main states of existence of the TS during designing.

FT7 Requirements regarding procedures and documentation. Forms and arrangements of documentation may be needed to anticipate liability litigation. Other procedures are prescribed in standards, codes of practice, guidelines, and so forth. Some of these will be given the force of law in ways depending on the constitutional requirements of a particular country.

FT8 Stage of development of the product to be designed—states of existence. For design engineering, the TP(s) and TS(s) reaches different states (abstract to concrete, partial to full definition) in the design process (see Figures I.22, 3.2, and 4.1): task assignment, design brief, design specification; process structure; technologies structure; function structure; organ structure; and constructional structure, at various stages of concretization.

These structures exist simultaneously in a TrfS, and can be set up in several hierarchical planes. They act as substructures, to be recursively designed and incorporated into the more complex structures. They achieve different maturity values and completeness during engineering design, including design transformations into the other types of structure. Different parts of the TP(s) and TS(s) normally exist at different states of maturity, especially as "constructional design zones" and "form-giving zones" (see Chapter 4, Op-P5.3.1, and Figure 2.16). Only the final constructional structure needs to be completed in detail for a TS(s) to be manufacturable; other structures can be incomplete in the design records. These dynamic factors change, influenced by consultations, information, facilities for calculations and experiments, and so forth.

FT9 Differentiated submission of design data. The "normal" case is submission of a solution for the proposed TS(s) and (under concurrent engineering) its manufacturing process at the end of the design duration (see Figures I.16, 2.1, 2.10, and 2.11). The relatively long expected design duration for some TP(s) and TS(s) requires that some important design data are *frozen* and transmitted before the general work is completed. Typical situations are (1) specialized and extensive calculations and experiments, for example, "stressing" for load-carrying structures of aircraft; (2) information for other departments or sections in the organization, for example, definition of castings or forgings, information for the construction and building of special jigs and fixtures (items with long lead-time to support manufacturing); (3) for large stationary plant, definition of the foundations to be built; or (4) mandated by requirements for legal or environmental permissions, or for tendering procedures to compete for contracts. This influences the design program and organization, not always positively. Possible optimization of the "frozen" items, and of their immediate connections, is no longer available in subsequent phases.

For simultaneous or concurrent engineering (see Chapter 8), the influenced departments within the organization (e.g., manufacture, law-conformance, sales) actively advise the designers, especially in the constructional structure, but the engineering designers take responsibility. The personnel from other departments within the organization simultaneously expedite their own work toward realizing the TS(s), for example, by designing jigs and fixtures, planning production processes and methods, preparing sales and advertising literature.

FT10 Employment conditions of the designers. Normally the engineering designers are employees of the organization, are subjected to the conditions of their employment, collect experience, and are loyal to the organization. In some cases (particularly in plant construction and building engineering) designers are operators (owners) or employees of a consulting bureau (industrial designers, design consultancies, architects, etc.), or "loaned" from an employment agency, without loyalty to the organization for which engineering design is being performed.

3.2.3 FO—Organization Factors

Each legal, economic, administrative, and spatial unit that serves to accomplish economic goals is called an organization, for example, private or stock-market-traded businesses, not-for-profit organizations, governmental/administrative departments. An organization can be formed from several organizations (e.g., branch plants, subsidiaries), within one or several countries or continents. Outputs of an organization are products, TS(s) and TP(s), that emerge from information inputs in the first four TS life cycle processes (see Figures 6.14 and I.13), and are marketed. The organization performs by transforming inputs (L, M, E, I) into products, resulting in financial returns for the organization to remain in operation. Important factors are:

FO1 Organization aims. Every organization strives to cover its costs and to obtain a surplus of funds, which usually also requires turnover. Contributions and subsidies help toward this goal. Other goals are noneconomic, including social, ethical, environmental, political, independence, prestige, power, and so forth. Characteristic features of an organization are performance or production program, size and the kind of production process, reputation, market share, and so forth.

FO2 Production program. This determines which products the organization supplies to the market, especially TP(s) and TS(s). The decision on the assortment of product depends on the factors "FE" and the organization potential. In each time period, the organization task is then determined, and its program of deliverables considered, including assortment of product, TP and TS-"sort," and quantities and rates of product(ion), but see FO6. Since design engineering and manufacturing is usually a long-term effort, the individual calendar period should not be regarded as self-enclosed, it must usually feature a long-term planning outlook. The degree of specialization or diversification on a product—TP and TS-"sort"—can be broad or narrow. The assortment has important influences on the stability and success of the organization, economies of scale, organizational complexity, and so forth, and is clearly reflected in the form of organization (see also Section 12.4.5).

FO3 The size and location of the organization. This influences all considerations and solutions in the organization. Size ranges typically between (1) very small organizations—1 to 5 participants; (2) small organizations—up to approximately 50 employees; (3) medium organizations—50 to 200 employees; (4) middle-sized organizations—200 to 500 employees; and (5) large organizations—over 500 employees. Regarding location, typical classifications can be local; regional; national; transnational; multinational; transcontinental (global).

FO4 Sorts of production processes. An important factor and consideration in operational decisions. Large differences in individual areas of the potential may be found, which depend on the production processes (see Section 6.11.5 and Figure 6.18).

FO5 Organization potential. Applying the model of the transformation system (see Figures I.6 and 6.1) allows the preparation of an organization model (see Figure I.7). The transformation of an organization operand from "input Od1 (M, L, E, I)" → "output Od2 (products)" occurs because a technology is applied by effects from the operators: HuS, TS, and AEnv, supported (even conditioned) by IS (including technical knowledge, data, etc.), and MgtS.

NOTE: The deliverable product of an organization may be training and education to humans or other animals, that is, living things as operand (compare Chapter 11.6).

In Figure I.7, the total process is first arranged into four phases (similar to the TS life-cycle phases; see Figures I.13 and 6.14): (1) product planning, including preparing and acquiring of resources; (2) designing, within a DesSit; (3) production planning and production; and (4) sales and distribution, leading to operation and disposal of the product. These partial processes can be considered as main tasks of an organization. The partial processes can help to define the main function structure of the organization, both for its hierarchical composition (organization charts, titles, and responsibility structure) and for its procedural arrangement (duties, job descriptions, prescribed procedures, accountability, formalities, documentation, financial accounting, and so forth).

Operators / Potentials	Design potential	Acquisition potential	Production potential Incl. technological and organizational preparation for manufacture	Sales potential
Human System (HuS)	Designers Designers' tacit knowing of branch and other regions Designers' tacit knowing of design processes Support personnel	Chief executive Upper management personnel Buyers Accountants	Production workers Storekeepers Transport personnel Packers Inspectors	Sales personnel After-sales service personnel Maintenance personnel
Technical object system Working means — tangible means (TS)	Equipment Furniture Computers Computer software (CAD, etc.)	Finance Materials — production auxiliary Energy — primary for manufacture secondary for other uses	Production equipment Storage equipment Transport equipment	Telephones FAX-machines Cars Trucks
Environment — Active and reactive (AEnv) — General (GEnv)	Space Time allocation Motivation Rewards structure	Space Time allocation Motivation Rewards structure	Space Time allocation Motivation Rewards structure	Space Time allocation Motivation Rewards structure
Information system — information, knowledge, data, and so forth (IS)	Recorded branch information Standards Libraries and so forth		Production control and documentation	Product catalogs Operation manuals Service manuals
Management system (NgtS)	Design process (procedural) Documentation recording and control	TQM (see Chapter 8)	ISO 9000 registration ISO 14000 registration	

FIGURE 3.3 Tasks and problems of potentials.

The operators, HuS, TS, AEnv, IS, and MgtS, form the potential of the organization and must have the ability to fulfill the organization tasks. The main functions distinguish four areas of potential (acquisition, design, production, and sales potential), which are composed of individual potentials of process operators (personnel, tangible object assets, management/administration, information, and finance potentials). Figure 3.3 shows the tasks and problems of the individual potentials by examples, attention is directed especially to the design potential.

FO6 Strategic development plan and marketing of the organization. Each organization should act according to long-term business plans. These are founded on estimated and experienced consumption curves, and on development (learning) curves. The risks for individual cases must also be considered, especially the influences of political, local, national, and global economic situations; see factors "FE." The more thoroughly the factors that influence distribution and acquisition markets are known, the more accurately can a forecast be made. Such plans must be continually or periodically reviewed and brought up to date.

FO7 Employment conditions in the organization. This covers all employees and stakeholders of the organization. Factor FD5 is a part of this factor.

3.2.4 FE—SURROUNDINGS OR ENVIRONMENT FACTORS

The environment of an organization can be envisioned as in Figure 3.1B. At the level of the country and world, other organizations, associations (including unions), cartels, persons, markets, among others, are included in the sociotechnical system. From the economic side, the market plays an especially important role for an organization, but also depends on the environment and its influencing factors. At the level of the global to local economy of individual continents, states, countries, regions, and localities, the interest of organizations lies especially in the situation and the trends of further development in the following areas:

FE1 Politics—plans to accommodate the potential situation, policies. Examples are: currency and credit polices; export and import politics (e.g., customs duties/tariffs, subsidies); tax politics; laws and legal policies, environment protection, patents; financing policies, research, subsidies.

FE2 Society, social system: for example, constitution (monarchy, parliamentary democracy, and so forth); cultural, social, welfare, and other systems.

FE3 Company policies: for example, education (including continuing education); empowerment of employees, use of consultants; incomes and work time; interests, "fashion" elements.

FE4 and FE5 Science and technology (including information): for example, general knowledge; research; technologies; (raw) materials; environmental impacts of input materials and products.

FE6 Market, including (1) distribution market: demand (inquiry) dynamics; competition, price, and deadline pressure; internationalization, globalization; (2) labor market: level of employment; availability of trained personnel; (3) acquisition market (including the supply chain): raw materials; production capacity, capitalization; energy; services; and (4) finance market: credit facilities; banking and finance transfer.

These areas of the organization are called environment factors (FEs); the organization is the system under consideration, and lies in the tension field of the FE (see Figure 3.1).

4 Procedural Model of Design Engineering

A center of interest in engineering design research consists of attempts to generate general procedural models, as strategic design methodologies. Many models have been published, for example, [19–21,23,24,27,66,109–111,178,315,369,370,452,457, 498,528,548,588–591], produced in many countries, branches of industry, places of work, with different working conditions, ranges and durations of design and research experience, and objectives and assumptions, in various forms and with their own terminological apparatus. Several publications discussed details of the "schools of thought," for example, [315] and Figure 2.2. Section I.12.3 presents the general model for design engineering of TP(s) and TS(s), based on current knowledge, and integrates many existing models. The information presented in this chapter is mainly situated in and around the "northeast" quadrant of Figures I.4, I.5, and 12.7.

4.1 DESIGN STRATEGY: PROCEDURAL MODELS

Design engineering is needed in the realization processes of TS(s). It includes designing for implementation of TP(s), and is part of integrated product development for technical products.

The strategy of methodical, systematic, and planned procedure in designing is generally defined in a procedural model that has been set up for an assumed state of the design factors consisting of generalized conditions. Mutations of the model are due to the concrete conditions that arise from a specific design problem, for a particular design engineer, and in a particular design situation (see Chapter 3). The plan for the engineering design process should be as *structured* as possible, the details reaching into "design operations" or "steps" where applicable.

The explanations about technical process (TP) (Chapter 5) and technical system (TS) (Chapter 6) show that during the engineering design process the designers are searching for appropriate structures. When designing a transformation process (TrfP) or TP, the process structure should be established. For a TS the search is ultimately for a constructional structure that will be the carrier of the required internal and external properties. This structure should be accurately described by the elemental design properties (see Figures I.8 and 6.8–6.10). The relationship between the given requirements (as input to the design process) and the design properties to be found (as output) is complicated: (1) the number of these relationships is usually large (compare Figure 6.9); (2) quantitative information about some of the relationships is incomplete, many cannot be expressed in quantitative terms; and

(3) all phases, stages, and steps of the engineering design process are mutually interrelated—considerations taken later in the engineering design process usually have some influences on earlier considerations, thus a need for iterative work.

The concreteness (completeness, fidelity, and finality) achieved by the engineering designers in the resulting constructional structure of a TS(s) depends on establishing the internal properties (classes Pr10, Pr11, and Pr12), that is, the design characteristics, general and elemental design properties. The larger the number of design properties that have been established, and the more completely they are determined, the more concrete is the resulting constructional structure.

A direct transition from the list of requirements to a hypothetical concrete constructional structure of a TS is generally inconceivable. Using iteration and the TrfP, TP, and TS structures, engineering designers initially realize the most important properties (in "generating" actions) and then investigate the resulting conceptualization and constructional structure, to determine the adequacy of other properties, and to take "corrective" actions [138,229].

Evaluation (and analysis) of proposed solutions has its problems (see Section 2.4.3). Formulating a general procedure for complicated processes such as design engineering and for problem solving (see Section 2.4) will usually require various additional strategic actions. Strategic principles in design engineering, used between and within the engineering design operations, and between and within the TP- and TS-structures, include:

1. *Iteration* (iterating): used where a direct solution to the problem is not possible, which is almost always the situation in design problems. Complicated relationships exist between the requirements for external properties of the TS and the elemental design properties. Similar to mathematics for solving a system of equations, initial assumptions are made to proceed toward a solution. The results are used as improved assumptions and help to determine a more accurate solution. If convergence is sufficiently rapid, a solution is obtained at the desired accuracy after a few iteration cycles.

2. *Recursion* (subdividing): the problem (or the partially developed solution) is broken down into smaller sections, for example, constructional design groups, subassemblies, modules, organs, form-giving zones. Each section is then developed separately, but coordinated, and the more concrete sections are reintegrated, synthesized, aggregated, recomposed.

3. *Abstraction* (abstracting): attention is concentrated on the important aspects of a problem situation, and less important ones are initially deferred. This principle permits easier entry into a design problem, for example, to first concentrate on realizing the main effects, before considering the assisting and secondary tasks, processes, and means.

4. *Concretization* (concretizing): from rough preliminary and abstract solutions or concepts toward definitive, better defined, and more finely tuned ones, the opposite to abstraction.

5. *Analysis* (analyzing): finding the causes and parameters of the actual or anticipated behavior of an existing or planned structure, and its (detail) values.

6. *Synthesis* (synthesizing): finding suitable means (among the available alternatives) to achieve a goal, for example, a proposed structure that will show a required behavior—this is not a simple inversion of analysis.
7. *Improvement* (improving): starting from an existing solution that shows scope for improvement, achieving a satisfactory solution by constructive criticism, modification, corrective action, and reflection [167,506,507].
8. *Strategy of the problem axis*: strategic progress forward along the problem axis from the problem (or its symptoms) toward physical means may only allow achieving a palliative solution, or one that paralyzes further progress toward a viable solution. Reversal of the search direction toward the causes of the problem can also be attempted, "go back one step." In many cases, this reversal to a more abstract position can permit engineering designers to "get rid of the problem" by avoiding it.

Design engineering, the search for a constructional structure to realize the required main effects to be exerted by the TS(s) on the operand in the TP(s), can progress through many *iterations* and *recursions* on several abstraction levels. The general model in this book shows an optimal procedure consisting of four major levels of abstraction and (at least) four iterative cycles, Figure I.22. In practice, each of the design actions based on these strategic principles merges with one or more of the others. The strategy of the problem axis can be viewed as a combination of these principles. During design engineering, different parts of a TP(s) and TS(s) will exist at different levels of abstraction or concretization. Only when a project is completed can the parts converge to a common state. Managing a design process is discussed in Section 11.3.

4.2 THE RECOMMENDED PROCEDURAL MODEL—SYSTEMATIC AND METHODICAL STRATEGY

The strategy presented in the procedural model is generally valid for design engineering of any TP(s) and TS(s) (see Chapters 5 to 7). This strategy, the *subject* in the triad "subject–theory–method" [351], should be correct, complete, with no significant gaps according to the relevant theory presented in EDS [287,304,315]. The knowledge and laws of these theories are applied to generate a procedure and method, which should then be adapted for the design situation, and heuristically applied to formulate a flexible plan for solving the problem.

The general procedural model offers a search for alternative principles and means at several levels of abstraction (see Chapter 4), and delivers favorable conditions for optimal solutions. Each subproblem or subsystem can be treated recursively in a full (adapted) procedure according to the model. The strategy allows and encourages a transparent design process, structured into clear design operations, which enable comprehensive documentation of the developing system and design processes, and shows effective feedback loops.

The description of the strategy frequently uses graphical models (see Section I.12.6). The nature of the modeled system can be better and more rapidly recognized, especially with the added explanations in the graphical models. The basis for the procedural model (see Section 6.2), is

1. The model of the transformation system, TrfS, Figures I.6 and 5.1, describes the artificial transformation (change) of an operand within its transformation process, TrfP.
2. The model of the TS, Figures 6.2 and 6.3, which in its operational process (TP) is an operator of the transformation process.

NOTE: Both are relative terms, and depend on the viewpoint adopted by the observer. The *operand* of the transformation process is being changed in various *operations*. The operand must generally be regarded as (topologically) "external to" the operators, thus also external to the TS(s). Defining the operand, and the boundary of the TS(s) is therefore important for each design situation.

For instance, food and other things stored in a freezer (TSa) are the operand, OdA, they are external to the freezer, even though they are completely surrounded. The freezer will operate without the stored items. The TSa "freezer" delivers the effect of "removing heat energy from the operand space," when it is connected to an electric power supply and switched on, one of its TS-inputs.

A different viewpoint at a more detailed level exists for the engineering designers (probably in another organization) who are responsible for designing a refrigeration module, TSb, for example, for the freezer, TSa, this designer's "window," see (Section I.7.1) [438]. The TSb "refrigeration module" acts as both an organ and a constructional part for the TSa "freezer." The liquid/gaseous refrigerant is the operand, OdB, even though it is completely contained by parts of the TSb "refrigeration module." The TSb will operate even if it has no refrigerant, that is, rotate the compressor, but not transport heat energy. The TSb refrigeration module consists of the compressor, throttle valve, two heat exchangers, pipes, fittings, electric motor and other parts, and exerts its various effects on the refrigerant (operand, OdB) to compress, cool, expand, and heat it—the operations in the TP "pump heat."

Yet another viewpoint arises for the engineering designers (again probably in another organization) who are responsible for designing the electric motor. This electric motor, TSc, acts as both an organ and a constructional part for the TSb refrigeration module. The operand for the electric motor, OdC, is the compressor, the electrical energy input to the motor is to be transformed from electrical to rotational-mechanical, and the torque and rotational speed is the effect that changes the compressor. The relationship between the TP, its operand, and the TS is further explored in Sections 5.4 and 6.13.

The engineering design system (Figures I.16 and 2.1) is a transformation system. The design process (in this system) transforms the information about requirements (input to designing) into the information about a TP(s) ready for implementing, and a TS(s) ready for manufacture, suitable to fulfill the requirements. This design transformation is accomplished by the effects delivered by the operators of the design process—engineering designer, technical means, the design environment, information system, and management system—following the general model of the "transformation process" (Figures I.6 and 5.1).

The engineering design procedure follows the technology of designing. The applied methods act as prescriptions, instructions, and recommendations in the procedural model, and are mainly derived from scientific methodology, that is, logic, philosophy, and cybernetics. The procedures, methods, and models can be used in various or all levels of abstraction of TP and TS, as selected parts or as a whole, combined with intuitive thinking, and with other published methods.

The procedural model allows effective management of engineering design projects and their complete documentation. The procedure itself develops following the principle of finality—"effects–means–effects" (Figure 2.3), or "function–means–function," a function-means tree (Figure 5.5). This results in a hierarchical chain in which the necessary process operations and effects (what?) are established, and a search for the corresponding means (with what?) takes place. The next more detailed level of working repeats this chain. Numerous possibilities of finding alternatives exist in the area of the means, and in the space of effects, which allow generating, establishing, and selecting an optimal solution. The design characteristics, property class Pr10, representing the classes of solutions, form the backbone of the procedure.

The general procedural model of design engineering should follow the model of the engineering design process (Figures I.16 and 2.1). It should describe and prescribe a systematic and methodical procedure (Figure I.20D) that indicates appropriate operating instructions (see Section 2.5), and operations at levels (H2) and (H3) in Figures I.21, 2.5, and 2.7. A short-form version of the procedural model is shown in Figure I.22. The design phases are coordinated with the structures of TP(s) and TS(s) (see Figures 5.4, 6.2, and 6.3), resulting progressively in more complete and finalized properties for the TP(s) and TS(s) (see Figures 2.4 and 4.3). The engineering design process is divided into stages and steps (Figures I.22 and 4.1), based on several organizational viewpoints. Iteration and recursion, verifications improvement, reflection, correction, and revision are always needed. The procedural model (Figures I.22 and 4.1), includes a repeated block of such operations at the end of every stage, and by implication in every step of designing. This block is especially important in the basic operations of problem solving, Section 2.4, and in concretizing a layout, concentrating on one design zone window at a time [438] (see Figure 2.16).

The TS(s) can be designed (see Section 6.11.2) by establishing a design specification for the revised system, then proceeding: (1) as a novel system, possibly also including a TrfP and TP(s) (see Figure 2.10); (2) for reconceptualizing, by analyzing (reverse engineering) an existing TS to obtain and revise a function structure and thus derive an organ structure (see Figure 2.11); (3) for redesigning, by analyzing (reverse engineering) an existing TS to obtain and revise an organ structure; or (4) as a variant, to change the sizes and functionally determined properties (class Pr1B), with few or no substantial changes in the principles of the constructional structure. From that point it follows the later parts of the procedure as for a novel system.

In designing a novel TS, the theory in Chapters 5 to 7 guides and supports the process with the help of models, for example, for a problem of complexity level III (Figure 6.5), with the consequential problems at complexity levels II and I.

Planning for design work uses the model of design engineering as a transformation system, Figures I.16 and 2.1. This forms a part of the life cycle model (Figures I.13 and 6.14). The process is divided into hierarchically ordered operations of different complexity in the activity structure of the design process (Figures I.13, 2.5, and 2.7), the operations at any one level consist of operations at the lower levels. Design stages and basic operations are composed of repeated activity blocks, usually following the implied sequence fairly closely. Design stages (level H1 in Figure I.21) present the general design procedure, Figure 4.1 (see also Figure I.22), and describe a systematic and methodical design procedure. This represents a rational strategy, especially favorable for "simultaneous or concurrent engineering" and "integrated product development." The systematic and methodical procedure can be performed in six stages, as shown in Figures I.22 and 4.1. The stages are divided into individual steps. Figure 4.1 shows which design properties are established in the operations—partially or completely, approximately or accurately, in preliminary or definitive form.

At each phase, stage, or step (operation), a search for alternative solution proposals is possible (coincident with step Op-H3.2 in the basic operations), which is followed by a step of evaluating, improving and correcting, reflecting, verifying and checking, and selecting, as the end of each procedural stage in Figure 4.1. Figure 4.2 illustrates the repeated procedure of recognizing a goal, diverging to search for alternative solutions, converging by evaluating and deciding, and communicating to the next level—the basic operations in Section 2.4.1. Figure 4.2 indicates a method for controlling combinatorial complexity. The progress toward a fully established set of properties (i.e., especially the design properties) is illustrated in Figure 4.3.

Each of these operations can be described in operating instructions (Section 2.5), and employ tactical methods (Chapter 8), resources (e.g., Chapters 9 and 10), and basic operations (see Figure I.21, level H3, and Section 2.4).

The description of the procedure for a novel TP(s), and TS(s) uses models which deliver general information and have broad validity. These models must be concretized for the branch of industry, the design situation (see Chapter 3) and to derive special "masters" (see Section 9.2). Applying such documents helps to speed up the engineering design process, and increases the probability of reaching an optimal TP(s)/TS(s).

The following subsections sketch the individual stages and steps in a uniform arrangement, that is,

1. *Input* for the engineering design stage.
2. *Output* from the engineering design stage.
3. *Theorem* that formulates the goals and expectations.
4. Applied *models* as resources of the engineering designer.
5. Suggested individual partial *operations (steps)*, specified by key words and phrases.
6. Important *information areas* that are needed by designers (with their "normal" design technical information and knowing) to achieve success. Engineering designers must understand the range of problems and the technical language.

Legend
I., II., III., ... Design phases
(P1), (P2), (P3), ... Design stages
P1.1, P1.2, ... Design steps

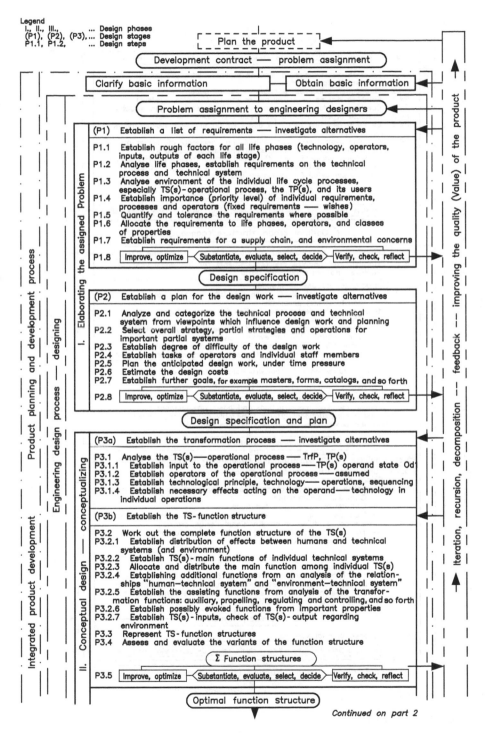

FIGURE 4.1 General procedural model of the engineering design process.

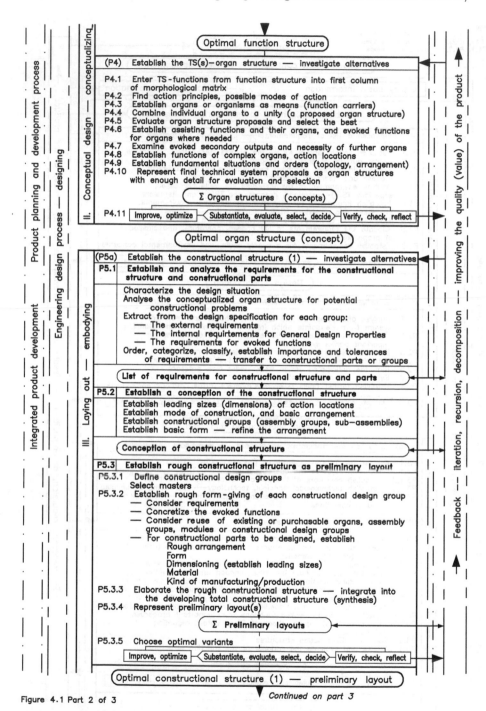

Figure 4.1 Part 2 of 3

Continued on part 3

FIGURE 4.1   Continued.

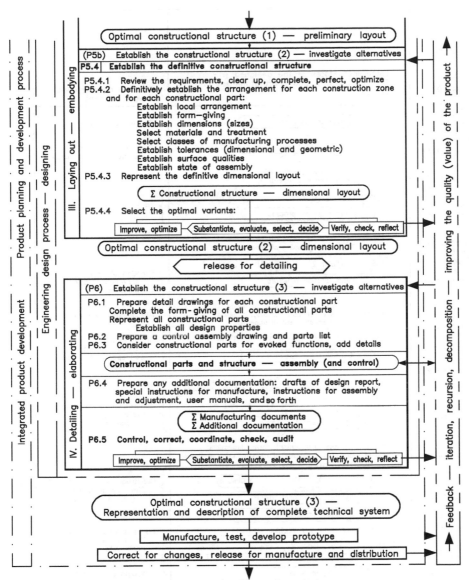

Left side vertical labels: Product planning and development process · Integrated product development · Product planning — designing · Engineering design process — elaborating · III. Laying out — embodying · IV. Detailing

Right side vertical labels: improving the quality (value) of the product · Feedback — iteration, recursion, decomposition

Optimal constructional structure (1) — preliminary layout

(P5b)   Establish the constructional structure (2) — investigate alternatives

P5.4   Establish the definitive constructional structure

P5.4.1   Review the requirements, clear up, complete, perfect, optimize
P5.4.2   Definitively establish the arrangement for each construction zone
         and for each constructional part:
                 Establish local arrangement
                 Establish form—giving
                 Establish dimensions (sizes)
                 Select materials and treatment
                 Select classes of manufacturing processes
                 Establish tolerances (dimensional and geometric)
                 Establish surface qualities
                 Establish state of assembly
P5.4.3   Represent the definitive dimensional layout

Σ Constructional structure — dimensional layout

P5.4.4   Select the optimal variants:

Improve, optimize — Substantiate, evaluate, select, decide — Verify, check, reflect

Optimal constructional structure (2) — dimensional layout

release for detailing

(P6)   Establish the constructional structure (3) — investigate alternatives

P6.1   Prepare detail drawings for each constructional part
       Complete the form-giving of all constructional parts
       Represent all constructional parts
              Establish all design properties
P6.2   Prepare a control assembly drawing and parts list
P6.3   Consider constructional parts for evoked functions, add details

Constructional parts and structure — assembly (and control)

P6.4   Prepare any additional documentation: drafts of design report,
       special instructions for manufacture, instructions for assembly
       and adjustment, user manuals, and so forth

Σ Manufacturing documents
Σ Additional documentation

P6.5   Control, correct, coordinate, check, audit

Improve, optimize — Substantiate, evaluate, select, decide — Verify, check, reflect

Optimal constructional structure (3) —
Representation and description of complete technical system

Manufacture, test, develop prototype

Correct for changes, release for manufacture and distribution

NOTE: The engineering design process summarized in this figure is not intended as a
      linear procedure. Iterative and recursive working is essential, and steps and stages
      overlap, (see Figures 2.10 and 2.11) Various parts of the TrfP, the TP and
      the TS(s) will be in different states of completeness at any one time, this is one
      aspect of the design situation (see Figure 3.1) In all steps of the engineering
      design process, the cycle of basic operations needs to be applied (see Figure 2.7)

FIGURE 4.1   Continued.

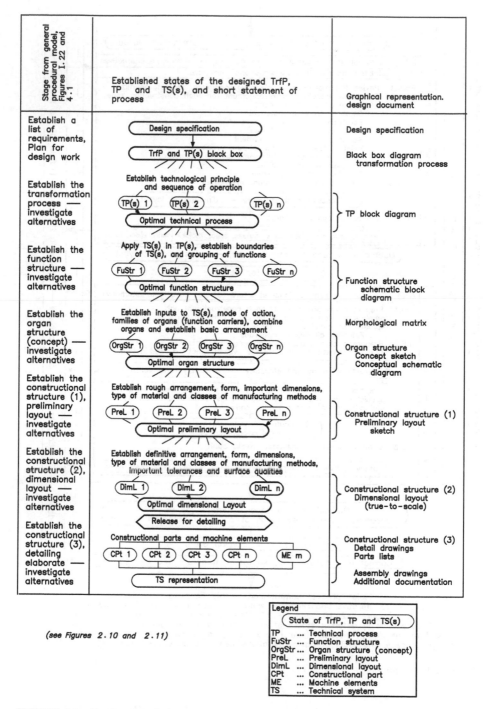

(see Figures 2.10 and 2.11)

FIGURE 4.2 Engineering design process: graphical representation of the states of technical systems during designing.

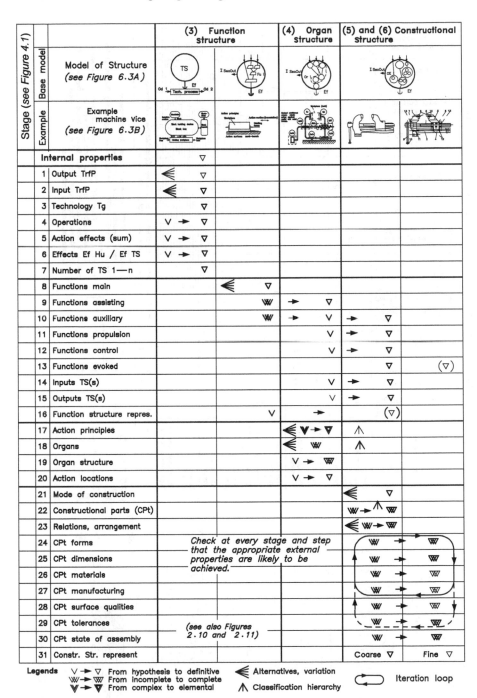

FIGURE 4.3   Iterative and progressive establishing of design properties of Trf P, TP, and TS.

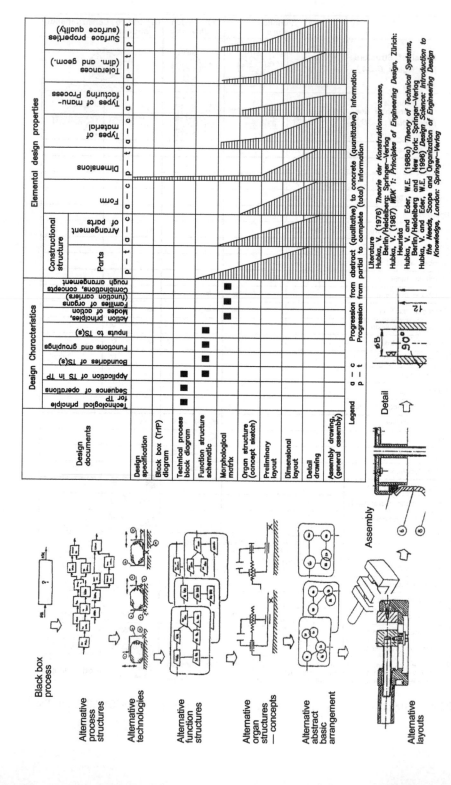

FIGURE 4.3 Continued.

7. Other available *methods and resources* (e.g., computers) for rationalizing the particular stage or step, and their relationship to the presented course of design engineering.
8. Form of representation of the output.
9. *Remarks* about further efforts of organizational or other kind related to this stage, which can be important for successfully handling the stage.

4.3 PHASE I—ELABORATING THE ASSIGNED PROBLEM

Elaborating and clarifying the assigned task proceeds by accepting the assignment of a design brief from management, and exploring it from the viewpoint of the engineering designer, resulting in a "list of requirements," a design specification, and a proposed plan of action.

4.3.1 STAGE (P1): "ESTABLISH A LIST OF REQUIREMENTS"

Clarifying the task offers the possibility to discuss and argue with the problem situation, and to obtain information of the causes. In extreme cases, a solution can be found here, abolishing or avoiding the causes, instead of fighting them.

During establishing the list of requirements, stage (P1) and step Op-H3.1 of the "basic operations," the engineering designers get to know and understand the situation around the TP(s) and TS(s). They develop—consciously or subconsciously—an idea about the possibilities for solutions, and whether there are any completely or partially available TS that will meet parts or all of the requirements. Designers must perform an analysis, to answer all questions—what? how? who? when? where? why? and so forth (see *topika*, Figure 8.7). Thereby a starting point develops for a possible procedural strategy or technology for designing the required TP(s)/TS(s).

The quality of the requirements listed in the assigned task, the design brief given by the organization management, is usually not sufficient for designing. Requirements, wishes, and constraints must be examined and clarified. The goal in this stage is therefore a list of requirements with highest possible quality covering all criteria, a design specification. The design specification should allow multiple solutions at appropriate levels of abstraction. Yet the engineering designers should not be allowed to wander too far for the design situation. This matter needs good judgment. The list of requirements should [503]: (1) *focus* attention on key issues; (2) ensure that all stakeholders (including clients or customers) *agree* on the nature of the problem; (3) *stimulate* innovative thinking; (4) *incur* relatively low administrative overhead; and (5) *facilitate* navigation and searching of information about requirements. A challenge to typical ways of solving the problem can lead to better, more innovative solutions.

During design engineering, undetermined questions are "discovered," even if they are interpreted as "organization-internal" tasks, and must be successively and formally incorporated into the specification. The design specification should be under constant review, and should be updated at intervals, preferably in consultation with the project sponsor.

(1a) **Input**: Assigned design brief—duty booklet, contract for a TP(s)/TS(s) with given constraints (e.g., delivery time), as defined by management, and agreed with a customer.

(1b) **Output**: A full design specification (list of requirements) which has been agreed with the customer or sponsor, with the most appropriate state of anticipated properties in all features.

(1c) **Theorem**: The TP(s) and TS(s) should be capable of effectively transforming the operand and delivering the TS-effects. The TS(s) should also be appropriate for all anticipated life cycle situations, and should have suitable properties (see Section 6.6), for example, for humans, for nonpolluting disposal at "life ended" (see Section 6.11.9), and so forth.

(1d) **Applied Models**: The model of the TS life cycle (Figures I.13 and 6.14) represents individual phases and processes of origination, operation (TP[s], and TS[s] as operator), and disposal is represented as a sequencing of transformation systems (Figures I.6 and 5.1). The model shows only the main processes and phases, which contain several subprocesses. For example, testing can be included in manufacturing and assembling, or distribution can include packaging and transport. During the TS-operational process, the TP(s), additional processes of cleaning, maintenance, repair, and upgrading may be required. The concretization of the general life cycle for a TS-"sort" should contain these particulars, and avoid unnecessary work, iterations and failures for the engineering designer within a specialty (branch of industry).

(1e) Partial **Operations** (Steps):

Op-P1.1   Establish rough factors for all life phases. Determine or assume the properties for the TP(s), that is, purpose, operand, technology, operators, and TS-inputs and outputs of each life phase (see Figure I.18).

Op-P1.2   Analyze the TS-life phases, and establish requirements on the TP(s) and TS(s).

Op-P1.3   Anticipate the environment of the TS-life processes, especially the TS(s)-operational process (the TP) as far as they can be anticipated. A detailed analysis can consider the model of the design situation (Figures 3.1 and I.17), which shows aspects of the environment and the burden on the design potential. This should include a study of technical, financial and other feasibility, possibilities of realization, state of the art, market situation, and other factors.

Op-P1.4   Examine the importance (relative priority level) of individual requirements from the viewpoint of the importance of TS(s)-life cycle processes, and operators.

Op-P1.5   Quantify and state tolerances for the requirements where possible.

Op-P1.6   Allocate the requirements according to life phases, operators, and classes of TS-properties.

Op-P1.7   Establish any anticipated requirements for a supply chain for manufacture, and for environmental concerns.

Op-P1.8   Review the results of this stage, be constructively critical, evaluate and decide (e.g., among alternatives), reflect and improve, and verify and check (including checking for possible conflicts and contradictions).

Enter this information into a list of requirements. This ongoing task covers the information generated in each operation, and reviews and keeps this design specification up to date. Goals and principles can guide the process of generating the list of requirements, as the task definition for the TP(s) and TS(s) (compare Chapter 2), and for each subproblem. As interpreted by engineering designers and team members, after clarification of the task

1. The list of requirements should ideally be *complete*, organized, clear and unambiguous, with reasonable progress to the state of the art, adequate innovation, but controlling risks—requirements, including constraints, should be as explicitly and completely formulated as possible for that design situation.
2. The requirements should be *correct*: the manifestation or measure (quantity, size, value) of all requirements should conform to the desires of the customer and relate to the state of the art, both for the TS-branch and for related regions.
3. The design task and the requirements should preferably be formulated with *process capabilities*, and *effects and capabilities* of the TS(s)—what the TP(s) and TS(s) should be able to *do*—but usually not with means—what the TS should *be*—for example, "a device for regulating flow of a fluid," in preference to "a valve."
4. Requirements should be classified and qualified into at least the two groups of *priorities*, for example, fixed requirement, minimum requirement, desire, wish, constraint, and so forth.
5. Requirements should be *quantified*, and permitted variations (limits of minimum or maximum values, or tolerances) of the values to be reached should be indicated.
6. The list should be *formalized*, for example, according to the classes of properties (see Figures I.8 and 6.8–6.10), and the "Df X" classes and life cycle operators (see Figure 6.15). Appropriate forms of presentation can use pro forma sheets (see Section 9.5).
7. Requirements should *not be duplicated* within the list. Many requirements influence several factors, for example, classes of properties (see Figures I.8 and 6.8–6.10), and operators of the individual life cycle processes (see Figures 6.14 and 6.15). For each requirement that could be classified under several headings, one class should be regarded as primary, the other classes should contain only cross-references to the primary class.

8. The *state of the art* should be clearly established within the design specification.
9. The assigned problem statement and the list of requirements are sets of information, therefore knowledge from the field of information processing is applicable, for example, instructions about establishing the state of the art, information storage and retrieval, and computer science. In this operation, the majority of the technical (branch-related object) information concerning the problem and solutions should be collected. *Study stage* is a frequently applied term, but information is gathered through the whole design process.

Completeness and qualification of the requirements are (and continue to be) points of contention; opinions and conceptions differ. A few hundred statements of requirements must be formulated at the start of the TP(s)/TS(s) life, before and during the conceptual phases of designing. These goals and principles should be appropriate and adapted to the needs of the design situation (see Chapter 3). The working methods used by engineering designers should contribute to reaching a corresponding perfection and quality, both of the task definition and of the TP(s) and TS(s). Suitable methods (see Chapter 8) are given as follows:

1. Analysis of classes of properties of TS: helps to attain clarity, new ideas, quality, quantity.
2. Method of systematic field coverage: aims toward completeness.
3. Method of questioning: strives for perfection, elucidates quantity (see *topika*).
4. Market analysis (market survey): helps to obtain quality of the requirements which correspond to the needs of the market. "Users" (customers) may be [329]: "primary users": operate the TP(s) and TS(s) for the intended purpose; "secondary users": operate the product or operate on the product, but not for its primary purpose; "side users": influenced (positively or negatively) by the product, but do not use it; "co-users": cooperate with the other users, but do not directly use the product. These classes cover the operators of TS-life cycle processes.
5. Method of methodical doubt (Descartes): encourages critical examination of all statements, for validity, accuracy, coherence with known situations, and so forth.
6. Quality function deployment (QFD) (see Chapter 8).

The possibility of realizing the TS(s), and its operational process, TP(s), should be examined (a *feasibility study*), which should indicate whether a solution of the given task is possible, and whether solving it is expedient. A feasibility study will either indicate "a possibility does not exist," or "a possibility may exist"—feasibility can never be proved. Areas to be explored are

1. Technical aspects: Are laws of nature respected? Do the requirements ask for a *perpetuum mobile* (perpetual motion machine)? Are technical means and experiences available?

2. Economic aspects: Are the anticipated TP(s), operation of the TS(s), and its manufacture likely to be economical?
3. Timeline: Is the available time sufficient? Can and should the product be first to be marketed?
4. Environmental and social factors: Are any hazards foreseeable?

An ideal can never be reached, a suitable level of meeting the stated requirements should be satisfactory. The process of evaluation is intended to enable a comparison with the requirements, and to point out deficiencies that must or should be corrected.

(1f) Important **Information** Areas: Engineering designers must know the "state of the art" about their TS-"sort," the development of the branch with its conditions and causes, the market situation and current competitive products (TP and TS) and patents. Information about other areas of engineering, social and cultural aspects, at least at the level of awareness, is also useful, and can help to avoid introducing failures in one's own area for problems that have already been overcome in other areas [572]. Question about possibilities of realization needs well-founded information about (new and existing) materials and manufacturing techniques. An excursion into the sphere of economics, organizations, and management is necessary for designers. Engineering designers are responsible for about 80% of the manufacturing and whole-life costs (see Figure 10.4), they must have information about production, life cycle assessment and engineering [62,160,186,237,244,582], costs and cost management, and so forth. The designed TP(s) and TS(s) must also conform to all applicable laws, standards and codes of practice. This is a further field of information with which designers must become familiar.

(1g) Available **Methods and Resources**: Checklists are a resource for processing the task definition that offer systems of questions and hints, and can deliver new inspiration. An interdisciplinary team is usually an advantage (see Chapter 10). Full records (minutes) of meetings should be kept, especially of any decisions reached. Individual specialists (including engineering designers) will have their own duties to perform between the meetings [247].

(1h) **Form of Representation**: The list of requirements (design specification) can be formalized and unified for a specialty, and used as masters or pro forma (see Section 9.5). Masters and other paradigmatic representations have proven useful in practice, other developments are possible and likely. A graphical representation has been suggested [503], in which the individual requirements (subclassified as product characteristics, functional requirements, constraints, and performance metrics) are coded for importance, and linked into a hierarchical network, preferably in column-and-row format, to improve visualization and understanding of the design problem. The area of quality assurance offers benchmarking, QFD, and other methods for comparisons with competitive products and company procedures. The "feasibility study" method is usable in this stage of designing.

(1i) **Remarks**: This stage demands cooperation with stakeholders, for example, sales department, and manufacturing, so that the established data corresponds to their ideas, or a consensus is reached.

4.3.2 STAGE (P2): "ESTABLISH A PLAN FOR THE ENGINEERING DESIGN WORK"

Planning and organizational preparation of designing, and of all other life phases of the TS(s), is an important section for defining the task. A good vision is needed about the temporal and spatial progress of the solving process and the necessary means, including material, energy and information, resources, staffing, economic, and financial considerations. The plans should be reviewed at suitable intervals, also relevant for managing the design process (see Section 11.3).

(2a) **Input**: First version of the list of requirements, the design specification, output of step (P1).

(2b) **Output**: Complete plan for the contract, that is, (1) goals which engineering designers wish to reach in the development; (2) a solution strategy with significant steps; (3) a timetable with decision deadlines—a timeline; (4) employment and allocation of staff members and branch specialists; (5) economic quantities—design costs, cost goal for the TP(s) and TS(s); and (6) prerequisites that the organization leadership should fulfill to be able to reach the goal.

(2c) **Theorem**: During design engineering, all properties of the TP(s) and TS(s) are progressively established, that is, the elemental design properties. This may be done consciously or subconsciously (intuitively), directly or indirectly. The goal is an optimal quality of all properties, each property should be consciously "designed" and evaluated, which can best be approached with a planned systematic and methodical procedure. Design planning should be recognized as important, the tasks of the engineering designers include *low design costs* and a *short design duration*, which are achievable by an appropriate strategy and procedure.

Planning should consider the degree of complexity of the TP(s) and TS(s), Figure 6.5, that is, plant, machine, assembly group (or module), constructional part—and they are hierarchical. Each level is a design task to be planned and performed, the need for recursion. The degree of difficulty of designing the TP(s) and TS(s), the degree of novelty and originality of the TP(s) and TS(s), and any possible "make or buy" decisions (Figure 6.7) should also be considered.

The majority of design tasks are TS-adaptation, TS-variation, and TS-transformation, that is, redesigning (see Section 6.11.2)—the changes influence the constructional structure, the decisions about design characteristics were already made in the earlier phases, and can be considered as inputs in addition to the list of requirements.

Elements of the design situation that influence design planning are also shown in Figures I.17 and 3.1: (1) environment factors FE, especially the political and economic situation, the market situation in the region, the country, or the world; (2) organizational factors FO, the situation in the organization: branch related, especially in production, economic, and organizational aspects; (3) task (contract) factors FT, the given design problem as clarified; and (4) design factors FD, the design potential.

Decisive for planning is the time situation, for example, pressure of deadlines, work on a design project should start as early as possible to allow time for reflection [167,506,507] and incubation [563], and to avoid "crashing" or a "charette" close to the deadline.

(2d) Applied **Models**: In this stage, the resources are different according to experiences in specific conditions. This range of problems is not treated in general terms. Some remarks about these resources are included elsewhere.

(2e) Partial **Operations** (Steps):

Op-P2.1   Analyze and categorize the TP(s) and TS(s) from viewpoints which may influence the work and planning. Partial systems should be defined, even if only in a crude way for novel design problems.

Op-P2.2   Select an overall strategy, and partial strategies and operations for important partial systems, according to the activity structure (Figures I.21 and 2.5). The complete method, shown in the design stages, is necessary for the novel development of TP(s) and TS(s) (Figure 2.10), that is, valid for all degrees of complexity of TSs (Figure 6.5).

NOTE: Partial systems to be newly conceptualized should be systematically and methodically developed according to the model. For redesign of the TS(s) (Figure 2.11), the constructional structure with its design operations (e.g., dimensioning, form giving) is appropriate, see operating instruction OI/2.12.

Op-P2.3   Establish the degree of difficulty of the design work for the selected design operations, with respect to the available branch information.

Op-P2.4   Establish the tasks of the operators and especially of individual staff members according to the model of the design system (Figures I.16 and 2.1). If the available engineering design potential is not sufficient, either technically and/or in its work capacity, additional assistance and personnel should be planned and secured. Design engineering and other activities (calculation, research, experimentation, supply of OEM and COTS items, special parts, consulting) can be contracted outside the organization (outsourcing), completely or partially. Disadvantages of outsourcing are loss of direct control, difficulties in using accumulated experience and knowledge, losses to the local economy, and so forth.

Op-P2.5   Plan the anticipated engineering design work. Important deadlines already exist, or the design task is accompanied by the demand: "As quickly as possible!" Engineering design work must conform to the time pressures, and compromises must be achieved. Acquisition of information, and deadlines for release (e.g., for detailing, for manufacture) should be planned.

NOTE: One of the strengths of the methodical and systematic procedure is seen here. Using the generated masters, documents and forms, the work can move quickly forward and individual steps can be planned according to experience. This material is rarely available (no time to prepare it), and the methodical procedure is denied or avoided as "too large a time expense." The vicious circle is clear.

Op-P2.6 Estimate the costs of design engineering—this is not usual in most engineering design offices, but is important. Design engineering should be regarded as a productive asset. Proven standard values for time expense need to be available.

Op-P2.7 With only a small expansion of the task, different resources applicable for other tasks can be generated, and the documents for the methodical process can be gradually set up, for example, masters, pro forma, and so forth (see Chapter 9).

Op-P2.8 Review the results of this stage, be constructively critical, evaluate and decide (e.g., among alternatives), reflect and improve, and verify and check (including checking for possible conflicts and contradictions).

(2f) Important **Information** Areas: This range of problems belongs into the design management area (see Section 11.3). Within this frame, it presents a special area, because design planning, like other thinking work, demands a particular interpretation and information.

(2g) Available **Methods and Resources**: The literature about design methodology and product manufacture contains many other models. They have taken different starting positions and paradigms, and are intended for a different circle of addressees. Planning must also proceed differently, especially when a different strategy is utilized as basis. The fundamental course of designing should not deviate too much. Shortening the transit time of a design project can be assisted by a design structure matrix, CPM, PERT, and so forth (see Chapter 8). Data collection involves methods of statistics.

(2h) **Form of Representation**: The plan can be executed in many kinds of display, for example, as a network plan, Gantt-chart diagram, tree diagram, PERT-diagram, and so forth (see Chapter 8). The conventions of the organization must be respected, because the design plan is also an element of the master plan for the organization. This stage and this report should not claim too much time.

(2i) **Remarks**: Within the specialty of design engineering, the stage of design planning is not usual. However, since it is an important activity for engineering designers, its treatment is unavoidable. Such a plan should be periodically reviewed and revised.

4.4 PHASE II—CONCEPTUAL DESIGN—CONCEPTUALIZING

Conceptualizing, or conceptual design, is intended to establish the proposed solution in outline, as a concept, with intermediate results of a TP-structure, a Tg-structure, a TS-function structure, and a TS-organ structure (a "concept") that is likely to be optimal.

4.4.1 STAGE (P3): "ESTABLISH THE TRANSFORMATION PROCESS AND TS-FUNCTION STRUCTURE"

The core of the methodical and systematic process of design engineering begins with establishing the function structure (see Chapter 6). In this stage, except for the main task contained in the list of requirements most other properties are initially deferred, providing that in the next stages all the properties can be successively achieved. A "frontal" procedure considering all (or even most) of the properties is too complicated, because it is impossible to consider the numerous and complex relationships among the properties.

(3a) **Input**: Design specification, output of stage (P1), and plan, output of stage (P2).

(3b) **Output**: List and structure of transformation process, and TS-functions: (1) established operand at input (Od1) and output (Od2), and transformation operations; (2) established technology for this transformation; (3) selected distribution of the effects to be delivered by humans, TSs, and the active and reactive environment; (4) selected inputs to the TS(s); and (5) transformation of TS-input into TS-effect, other output, TS-capabilities, TS-functions. The sorts of functions are described in Section 6.4.2.

(3c) **Theorem**: Mankind has numerous problems, demands, and wishes. The means must be created from the available states of M, L, E, I through transformation processes. The transformation of the operand proceeds by an established technology (see Figure 5.3). In a sequencing of *main operations* (partial processes), the operand is brought from its input state (Od1) through intermediate states into the output state (Od2). Each operation demands effects, delivered by TS, humans or other living things, or the active environment. Information and management systems can also exert significant effects, mainly indirectly through the TS and human systems.

In addition to the *main* operations, it is often beneficial to consider various assisting and secondary inputs to the TrfP(s), especially to the TP(s). This is a choice for a design team, depending on how they see a chance of combining operations in the TrfP(s). Typical operations are shown in the structure of the TrfP (see Figure 5.4).

The TrfP can be sufficiently established, in stage (P3a), subject to any alterations that may be recognized during later stages of design engineering, (P3b) onward. These stages are only concerned with the TS(s), that is, hardware, including specifying and developing any software needed for its operation; see Ref (9 article 3.4.2). The TS(s) with its capabilities should help to realize the TrfP(s) (Figures I.6 and 6.1).

The effects to be delivered by the TS(s) to the operand in the TP(s) represent the required capabilities of the TS(s). The chain of functions that transform the TS-input into the required TS-effects (TS-output) are the main functions, which can be executed by one or several TS(s), some in cooperation with a human, for example, the human manipulates the TS. This shows further variation possibilities (number of TS), which may be useful for complex transformation systems.

The TS-functions must include functions internal to the TS, and across its boundaries. Some of these enable the interaction and cooperation among TS and of humans with a TS—the man–machine interface (see Figure 6.1). A TS must be capable of acting for some parts of its duties, it is "active," at other times it can be reactive, and support, accept, allow, permit, the effects delivered by another TS, a person, or the active environment. The difference between these functions is clear in their formulation, compare Value Analysis in Chapter 8.

The main TS-functions need support from other functions (see Section 6.4.2), for example, *assisting* functions that depend on the requirements (design specification), the input to the TS, its mode of action, its operators, and so forth. The inputs—M, E, I (e.g., signals)—must be taken into the TS by receptors, and processed, regulated and controlled to generate the desired capabilities and effects of the TS. Other functions of the TS can emerge because secondary inputs and outputs are also present. The constructional parts of a TS must form a unit, they must be connected, coupled, joined, supported, guided—supporting and connecting functions. Some additional functions depend on other properties, *evoked* by the active and reactive functions. A typical example is lifting or hoisting—an ability conditional on the means for lifting (e.g., slings)—or the transportability of large TS, made possible by a division into smaller units, and their assembly on site.

The TS-functions can then be realized in TS-organs, using TS-internal *modes of action*. The typical functions and organs are shown in the TS-structures (see Figures 6.4 and 6.6). The *function structure* emerges when the relationships among the individual functions are entered. The model of TS-function structure (Figures 6.3B and 6.4) illustrates the above description, and offers a paradigm for the representation.

(3d) Applied **Models**: The engineering design process develops according to the rule of finality (Figure 2.3) during the search for means (with what?) in the individual design steps. This indicates the possibility of finding several solutions (means) in each step, and to choose the optimal (or at least the most promising) solution (see Figure 4.2). This is important especially in the steps involving the design characteristics, in which the optimal conceptual solution is decisive for the quality of the TS(s). The choices of operational process, technology, mode of action, and arrangement of the TS influence many other TS-properties.

It is not obligatory, and often not possible, to concretize functions in the first round. Much of the information about concreteness, completeness, and finality only emerge in later phases of design engineering. The function structure emerges successively, it need only be established as completely as necessary for further design work. Consider first the main operations in the transformation process, and formulate the TS-functions that are anticipated to deliver the needed effects. Further iterative rounds consider the assisting and evoked operations. Additions from the iterations can be entered into the representations of the TrfP and the FuStr, but mostly this is not expedient—these representations can remain incomplete. The increase in concreteness, completeness, and finality emerges only in the later phases of origination of a TS(s).

NOTE: The *human*, and other living things, can be operand in a TrfP, and in a TP (see Figure 5.3C). A human or other living thing cannot be a part of a TS(s) at complexity levels I, II, and III, that is, they must be external to the TS(s). Nevertheless, a human interacting with a TS of lower complexity can be considered as a "constructional part" of a TS(s) at complexity level IV, plant.

For a *solid material* as operand, the TS-effect acts through a visually obvious technology, for example, in metal cutting, a tool (as constructional part of the TS) with its cutting edge acts to shear material and remove chips from the operand, the workpiece. The TS(s) reacts with TS-internal stress and (temporary) storing of strain energy. A *liquid material* operand often needs to be contained against gravity and pressure. The appropriate TS-effect is a reaction to the presence of the operand—containing, resisting pressure, reacting with TS-internal stress, and (temporary) storing of strain energy. A *gaseous material* as operand usually also needs to be contained, the TS-effect is again a reaction to the operand.

Energy as operand can be treated as an analog to material. The input to the TP(s) is Od1 energy, for example, moment and rotational speed, the output from TP(s) is Od2 energy, for example, at same or different moment and rotational speed. The TS-effect is caused by reacting to the presence of the operand by TS-internal stress and (temporary) storing of strain energy. At zero energy rate of the TP(s), the TS(s) is idling or not operating. An interesting case is electrical conduction "in" a metal wire, usually high-purity copper or aluminum. Electrons, the carriers of electrical energy, repel each other. At high current densities, the electrical current tends to be concentrated in the outer layers of the conducting material. With multiple wires, the inner wires carry less current than the outer ones. This is overcome by suitable interchange of wires among layers along the length of a conductor, Roebel twisting in motor or generator slots, or similar rearrangement in transformer coil disks. At high frequencies, it seems that electrical conduction is no longer needed, hollow wave guides can direct the electrical energy. This confirms the view that the operand can be regarded as "external to" the TS (see NOTE in Section 6.4.4).

Information as operand can also be regarded as an analog to material. Information in the TP(s) is carried by material and energy. The TS provides the operational state in which information can be superimposed as an added signal—see the example of television in Section 4.2.

Chapter 5 discusses the process and its operand. Chapter 6 discusses the tangible TS as it acts on the operand, and the necessary activity and constituents internal to the TS.

(3e) Partial **Operations** (Steps):

Stage (P3a): "Establish the transformation process" (TrfP, TP)

Op-P3.1   Analyze the transformation process (TrfP), Figures 5.3 and 5.4, and the TS-operational process (TP) by establishing or assuming the design characteristics:

Op-P3.1.1   Establish the inputs to the transformation process—operand in state Od1 (if not given as a requirement)—and a sequencing of operations to achieve the operand in state Od2, compare Figure 5.3.

Op-P3.1.2   Establish the operators of the transformation process—assumed.

NOTE: Establishing the operators of the transformation system is important for designing the TS(s), because the performance of the TrfS depends on all its elements—TrfP, HuS, TS, AEnv, IS, and MgtS. Important facts need particular attention: *who* will work with (i.e., use, operate, manipulate) the TS(s), and *where* in the environment is the TS(s) operated. In the first round primarily the main transformation is considered, followed by further iterative rounds to consider the assisting and evoked operations.

Op-P3.1.3   Establish technological principles and the technology in the operations of the TrfP—operations and their sequencing—recognize any evoked operations needed.

NOTE: Different technological principles (TgPc) can be applied to achieve the necessary TS-internal processes. The TS-internal actions and modes of action follow from the TgPc. A possibility exists to obtain new solutions by varying the sequencing of operations, for example, using a morphological matrix.

Op-P3.1.4 Establish the effects which will need to act on the operand—technology in individual operations.

When formulating the transformation process, stage (P3a), a useful viewpoint is to place oneself in the position of the operand, and to follow the change from state Od1 to state Od2. It is therefore essential to clearly identify the operand. Each operation should state a limited change, and the properties that are changed. Alternative operations and processes may be useful.

A good formulation of operations in a TrfP, and TP, can provide a starting point for FMECA or FTA (see Chapter 8). It can also indicate where an engineering science can help to establish the values of some properties, and analysis can (later) verify achievement of these properties.

The TrfP and the TP are now sufficiently well established, that is, the occurrences "external to" the TS(s) have been considered (see the NOTE in Section 4.2). They will need iterative amendment during design engineering. Implementation of the TrfP will probably need to be planned.

Stage (P3b): "Establish the TS-function structure" (TS[s]-FuStr)

Op-P3.2 Generate the function structure of the future TS(s), Figure 6.4 as completely as possible or needed, that is

Op-P3.2.1 Distribute the necessary effects between humans and TSs (one or more), and the environment, that is, allocate operators HuS, TS, and AEnv, individually or in combination, to each operation in the transformation process.

Op-P3.2.2 Establish the TS-main functions of individual TS(s).

Op-P3.2.3 Allocate and distribute the TS-main functions among individual TS(s); check whether a suitable TS exists to deliver the effects for some of the TP-operations, especially the assisting and evoked operations.

Op-P3.2.4 Establish additional "active" and "reactive" functions from an analysis of the relationships "human–TS" and "environment–TS," for example, convert solar radiation for generating electrical energy (TS = solar panels).

Op-P3.2.5 Establish the assisting functions (possibly abstractly, incompletely, temporarily) from an analysis of the TS-internal functions regarding: (1) auxiliary functions: cooling, lubricating, cleaning internal to the TS; (2) propelling functions: motion, power, degree of mechanization; (3) regulating and controlling functions: degree of automation; and (4) supporting and connecting functions. Search for alternatives for assisting TP and TS, which can be treated as new problems at a new problem at a lower level of complexity.

Op-P3.2.6 Establish any possibly evoked functions from important TS-properties.

Op-P3.2.7 Establish the TS-inputs (first assumptions) and check the TS-effects and other outputs with respect to environmental damage. Here too, functions can be evoked to reach compatibility, within the TS and to the environment.

Op-P3.3 Represent the TS-function structure(s)—in the first round this should be performed in a homogeneous way from the degree of abstraction, completeness and finality of the functions (especially assisting functions).

Op-P3.4 Assess and evaluate the alternatives and variants of the function structure—decide on optimal function structures (first progress version).

Op-P3.5 Review the results of this stage, be constructively critical, evaluate and decide (e.g., among alternatives), reflect and improve, and verify and check (including checking for possible conflicts and contradictions).

For the steps of formulating the function structure, stage (P3b), it is useful to imagine oneself as the "black box" TS, that is, "internal to" the TS to be designed (see also the NOTE in Section 4.2). Typical questions are: "What must be done to 'me' or parts of 'me'? What must 'I' do? What inputs must 'I' get and do 'I' get? What outputs are expected from 'me'? What assistance do 'I' need? What physical principles are capable of mathematical investigation?" Alternative functions and their arrangements, and combinations or subdivisions, may be useful.

A good formulation of TS-internal functions in a function structure can provide a starting point for FMECA and FTA (see Chapter 8). It can also indicate where an engineering science can help in establishing the values of some properties, and where an engineering science analysis can (later) verify achievement of these properties in the constructional structure. Initial estimates are usefully "quick and dirty," and may be improved by a static or quasi-static mathematical model, a dynamic model [573–576], and second-order phenomena [411] where needed.

"Generating out the function structure" creates the possibility of forming alternatives (variations), and selecting and developing an optimal solution. Many decisions must be made according to the criteria listed in the design specification. Decisions in this stage are difficult, because the relationships among design characteristics and requirements as evaluation criteria are usually not clear. Yet the decisions in this stage are important for all kinds of requirements, especially costs.

It is usually necessary to work with several variants, and to delay rejecting some of them. The number of retained variants can increase quickly (by *combinatorial complexity*), and complicate the overview. In principle it is consequently advisable to take a "hard" decision position. The report about this stage should indicate as clearly as possible the reasons for rejection, to facilitate control and feedback. Good documentation should retain records of all alternative proposals (especially the rejected ones) to allow a rapid review and retrenchment.

(3f) Important **Information** Areas: The "function structure" is dominated by information about the TrfP and the TrfS, because they determine the role of the TS(s). Effective cooperation with specialists (experts) from the relevant area must be ensured, because this part of design engineering deals with all areas of human activity—transport, manufacture, chemistry, foods, medicine, pharmacology, and so forth. The engineering designer needs good competence in the range of problems and the know-how that leads to success in the project. High value must be allocated to the designers' technical information, even in development or adaptation projects.

(3g) Available **Methods and Resources**: The literature [370,457] mostly describes deriving a "total function" from the requirements, as a "purpose" or "**teleological** function." This process applies a block diagram (black box), which indicates the connection between input and output quantities, for example, as in the IDEF models [18] (see Figure 5.2). The total function is then subdivided into partial functions, which are connected into a function structure, a process called "function decomposition." These recommendations do not differentiate between a TS and its TP, possibilities of finding alternatives (variations) in the TrfP are not recognized—the transformation is considered as given. The disadvantages are obvious. Function decomposition can be operationalized by (1) using a TS-function as TrfP/TP for the next lower level of complexity (see the NOTE in Section 4.2 and NOTE [2] in Section 4.4.2) and (2) using a TrfP/TP as a TS-function in the next higher level.

(3h) **Form of Representation**: Recommendations for representing the function structure can be seen in Figures 6.4 and 2.17. The functions are represented by a special quadrilateral to distinguish them from processes and operations. The advantage of this simple convention lies in the rapid distinction between elements of TS-structures (see Figure 6.3).

A function structure cannot be completely generated (designed) in a single step, because many of the capabilities for partial TS-internal process operations (i.e., functions) can only be recognized during the engineering design process. These capabilities depend on the selection of means (organs and constructional parts) and the conditions under which they operate. Various additional functions are *evoked* from the solution proposals and form a "function–means" chain. These are often not entered into the function structure in a "feedback" mode, the means are selected according to the organ or constructional part which realizes the evoked functions. The function structure thus contains those main functions that are derived from the technology of the TP(s), that is, transformation effects. Additional (evoked) functions are only represented by classes.

Functions should be carefully formulated to state the needed capabilities of a TS(s), as a noun (or noun phrase) and a verb (or verb phrase)—it should be capable of performing this action *with* or *to* that TP-operand, TS-input, TS-throughput, or TS-auxiliary M, E, I, to bring it into its output state for that part of the TS-internal or transboundary process, for example, "hold a cutting tool." Functions should follow in a logical sequencing—the output of one function should be the input to the following function, to progressively process the TS-input into its output effects. Functions *evoked* by the main functions should be established and traced back to their respective TS-inputs or outputs. One function will usually be sufficient for two inverse operations on the TP-operand, for example, insert/remove, open/close, move up/down, and so forth. Functions should be solvable by one or more action principles—research may be needed to make that action principle available. Some functions are needed for completeness of the logical sequencing, but may not be directly solvable as proposals for organs. Action principles should be solvable by one or more organs as means—research may be needed to make that organ available. If a function shows solutions as organs of two (or more) different kinds that could be combined into an organ structure, then the function should be divided to indicate the different kinds. Functions may be further subdivided or combined to explore further possible organs or action principles (e.g., by altering the morphological matrix, see Figure 4.4), or to ease the search for solutions.

Function structures are preferably represented as flowchart networks which show relationships among the functions. A hierarchical tree representation is possible, but cannot show some important relationships among functions in different branches. A hierarchical tree exhibits a minor advantage over a flowchart network, the most significant functions (which have no successor functions and cannot usefully be decomposed) appear at the end of each branch.

(3i) **Remarks**: In engineering practice, the function structure of a TS remains almost unknown or rarely used, because within a certain specialty (e.g., branch of engineering industry) the function and capability of a TS appears almost always unaltered. However the conscious process of establishing the function structure should always be considered as a duty, especially in the area of assisting functions. In addition, masters should indicate possibilities for innovation.

4.4.2 Stage (P4): "Establish the TS-Organ Structure" (OrgStr)

The capabilities of a TS(s), its TS-input, TS-internal and transboundary functions, and TS-effects are now defined. "With what?" can these capabilities be realized—the logical means are organs (see Section 6.4.3). Traditionally, engineering designers offer ideas from memory of screws, bearings, pistons, and other constructional parts (machine elements) as means. If they wish to find optimal combinations of means, they must move in a more abstract sphere.

(4a) **Input**: Optimal function structure (or two or more FuStr that were evaluated as almost equal)—output of stage (P3).

(4b) **Output**: Optimal organ structure—in the most concrete form possible, which connects individual organs into a unit, and to the fixed system (grounding) and other systems in the surroundings. The organ structure should represent the basic (topological) arrangement of organs, usually without dimensions, that is, without wall thicknesses, lengths, and so forth. An organ structure is often called a sketch of principles, a design concept, a conceptual scheme, or an abstract configuration.

(4c) **Theorem**: The path from TS-function to the constructional structure must usually be decomposed into several steps. Concreteness, completeness, and finality must be successively incorporated into the output of design engineering. The number of the steps is not limited, it depends on the available information in the relevant branch specialty. The goal remains, that an optimal solution should be reached with a reasonable number of steps.

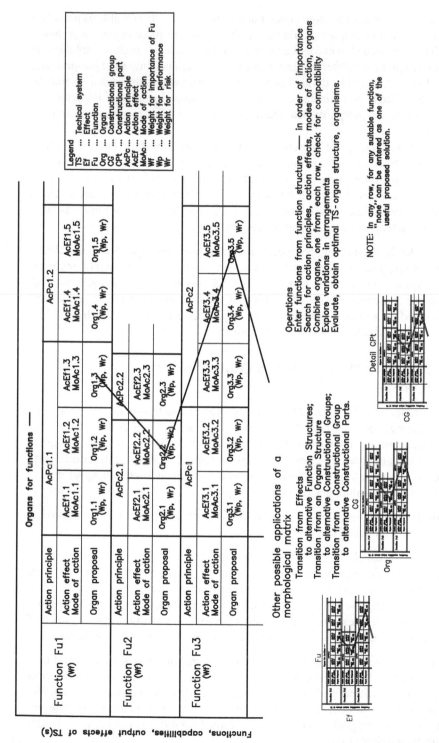

FIGURE 4.4 Morphological matrix—general arrangement.

Organs are chosen as means which realize the desired TS-functions. The cause of the TS-functions is formed through the means—organs. The selection of the means rests on information, and on laws of some sciences (e.g., physics) adapted as engineering sciences.

The individual classes of technical means define the characteristics of the general TS-"sort" to which they belong. They should form the hierarchical planes of a classification taxonomy (see Figure 6.17 and Section 6.4.3). A profitable class hierarchy for designing should be set up according to levels of abstraction. Such a diagram can be valid for all levels of complexity of TS.

In the class hierarchy for TS, the highest plane belongs to the *action principles* that can lead to achieving the desired effects. With this hierarchy we can define the classes of possible organs as means which conform to the laws of a certain information area. TS operate according to various principles: (1) physical (mechanical, hydrostatic and -dynamic, aerostatic and -dynamic, heat transfer, thermodynamic, optical, magnetic, electrical, electronic, etc.); (2) material-internal (stress, strain, plasticity, fatigue, creep, etc., but also metallic, plastic, fiber, etc.); (3) chemical (e.g., combustion, oxidation, corrosion, osmosis, other chemical processes); (4) biological (e.g., fermentation, composting, other biological processes); and others, compare Figures 7.11 and 7.12.

The second plane is occupied by those *actions* with which the required effect can be reached, that is, aggregation of laws in the selected knowledge areas. The third plane contains the TS-"sorts" arranged according to related *modes of action*. The differences in action locations (action surfaces), arrangement or mode of construction are clear, and form essential class indicators.

Other more concrete criteria can lead to emergence of further planes in the hierarchical scheme according to the dependencies of the branch specialty, and degrees of novelty or complexity of the TS (Figure 6.5). Whether it is meaningful to undertake these concretization steps for design engineering must be considered by engineering designers for individual tasks or classes of task.

Figure 6.17 and its example present the hierarchical scheme according to characteristics, and allow a deeper entry into the range of problems.

(4d) Applied **Models**: The procedure follows the finality scheme (Figure 2.3); the steps follow successively in this stage from the classes and the classification scheme (Figure 6.17).

The model of the TS-organ structure (Figure 6.6) is an analog of the model of the function structure (Figure 6.4), with changed symbols. Nevertheless, the function structure and the organ structure do not have a one-to-one relationship: one organ may realize one or several functions, one function can be realized by one or several organs. Individual organs, organ families, or organisms can be entered into the structure with corresponding symbols (Figure 6.6). In some cases the choice of suitable symbolism is prescribed by standards or conventions, for example, for pneumatic, hydraulic and electrical circuits, and chemical "unit processes."

A *morphological matrix* may be used to associate the TS-functions with action principles and organs (see Figure 4.4 and case examples in Chapter 1). Limited computer support is available, for example, [578]. Designers can be helped by design catalogs (see Section 9.3), which collect and organize the known information [178,370,498]. Masters deliver a standard help (see Section 9.2), both in reference to the graphic representation and the tacit experience residing in the designers' mind, for the last steps of working out the organ structure.

(4e) Partial **Operations** (Steps): The methodical and systematic procedure can be decomposed into the following steps, the descriptions include use of a morphological matrix (Figure 4.4):

Op-P4.1   Enter the active and reactive functions from the chosen function structure into the first column of the morphological matrix—these should preferably be in rank order of importance for the TS(s), and initially list only as many as are needed for a set of "quick and dirty" solution proposals for organ structures.

Op-P4.2   Find possible action principles, and (for each of these) the possible modes of action that may be capable of solving each function—record in the matrix.

Op-P4.3   Establish the organs or organisms as possible means (function carriers) within the framework of the action principles and modes of action—record in the matrix; the proposed organs to solve any particular function may include the solution "none" or "accept from outside," for example, from a human or the active environment.

Op-P4.4   Combine individual organs, one from each row of functions, to a unity (a proposed organ structure) whilst checking the compatibility among organs. Several combinations should be attempted, preferably starting with the most desirable organ from each row—different topological arrangements of these organs should also be investigated, see also NOTE (a) below.

Op-P4.5   Evaluate the alternative organ structure proposals and select (or develop by combining) the optimal structure from among the alternatives.

Op-P4.6   Establish the needed *assisting* functions and their organs, and any *evoked* functions with regard to the chosen organs. Investigate possibilities of subdividing or combining (parts of) the organs (especially assisting and evoked organs). This is the first step in which the assisting and evoked functions can be concretized into organs, because now the TS-mode of action has been established and the necessary additions can be concretely visualized. These functions and organs may be solved in a recursive way, that is, treated as a new problem to be subsequently recombined into the main solution(s)—for this purpose, the formulation of a function is adapted as the formulation of a TP for the next more detailed level; see NOTE (b) below.

Op-P4.7   Examine the secondary outputs and the necessity of further evoked organs.

Op-P4.8   Establish the functions (necessary capabilities) of the complex organs, and action locations: action points, lines, surfaces, volumes, action organs.

Op-P4.9 Establish the fundamental situations, topology, configuration, arrangement.

Op-P4.10 Represent the final TS(s) proposal as a TS-organ structure (at first in a more abstract form if needed) or TS-partial system in concrete form (sketch of principles)—this is the output documentation of this stage.

Op-P4.11 Review the results of this stage, be constructively critical, evaluate and decide (e.g., among alternatives), reflect and improve, and verify and check (including checking for possible conflicts and contradictions).

NOTE (a): To reduce the combinatorial complexity from steps Op-P4.3 and Op-P4.4, each organ or organism may be rated in two aspects, for example, potential quality and performance (Wp), and potential risk (Wr) [578], see Figure 4.4. The suggested scale for each rating is:

5	(Wp) Good to excellent solution	(Wr)	No anticipated risk
3	Better than average		One anticipated problem
2	Worse than average		Two anticipated problems
0	Poor		All three anticipated problems

The anticipated problem types are

1. Uncertainties and knowledge gaps that may extend the development time and cost beyond the plan, but not information gaps that can be met without penalties.
2. Potential problems in manufacture that may need extensive investment, cost, time, and so forth.
3. Potential problems in operation or maintenance of the TP(s) and TS(s), too complex, too expensive, too time consuming, especially if customer dissatisfaction may occur.

A suitable computer algorithm can sort the solution proposals in each row according to quality, risk, or a weighted sum of these ratings. In conjunction with an importance rating for each function (Wf), the algorithm could then suggest combining the organs that have the highest overall quality ratings, the lowest risk ratings, or any other criterion. The suggested combinations must still be checked for compatibility of individual organs. Their alternative arrangements must then be investigated and the most promising selected as in step Op-P4.5.

NOTE (b): The terms "function" and "TP" depend on the immediate viewpoint; compare the NOTE in Section 4.2. Consider a *hierarchy of Watching TV*: the most complex level of interest occurs during accepting, setting up, and preparing for operation. The operand is the TV-set with its peripherals. The main operator is the HuS; the TS is the power supply on the wall, the AEnv. The *transformation process* of the TV-set is shown in Figure 4.5A, and results in a watchable TV.

At the second level the "TV is operating, whether watched or not" (Ops 1.7.1.12 and 1.14 in Figure 4.5A). If no signal is applied to the TS receptor (i.e., topologically "external to" the TS), the output of the TV-set (the TS for this next more detailed level) will be only "snow" on the picture tube, and "hiss" from the loudspeaker, that is, the TV-set will still be operational. All operating inputs, outputs and TS-internal processes (functions) can be analyzed and established for each of these operations. For Op 1.9 as the TP, the *functions* required of the TS to deliver the

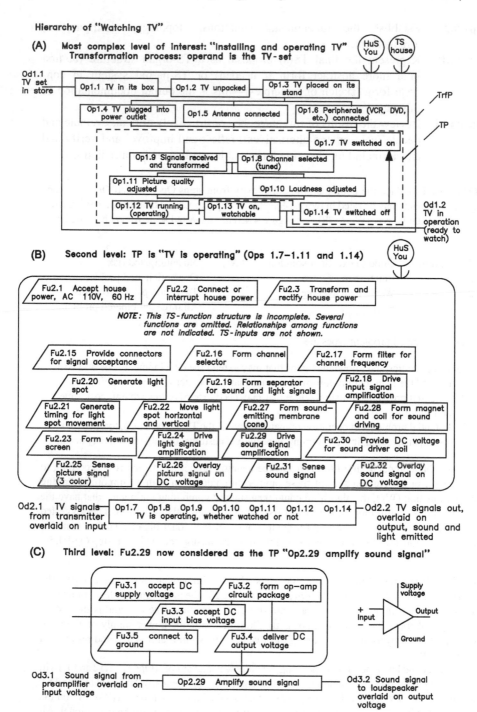

FIGURE 4.5 Transformation process and function structure example "watching TV."

necessary effects are shown in Figure 4.5B. Each of these functions (or groups of functions) can now act as source for the TP for the next level.

At the third level consider Fu2.29 now as the TP "Op2.29 amplify sound signal." The operating TS for this level is an operational amplifier on a circuit board physically inside the casing of the TV-set, as an integrated circuit "component" viewed as a constructional part. The operational amplifier is typically connected to a "supply voltage" and a "bias voltage" when the TV-set is operational (i.e., switched on), whether there is a signal or not. By applying a small modulated sound signal (Od1) overlaid (i.e., topologically "external to" the TS "operational amplifier") on the "bias voltage," a much larger variation (Od2) of the "supply voltage" can be detected, and used to drive the loudspeaker. If the variation of "bias voltage" is too large, the output variation of "supply voltage" will be a distorted replica of the input.

In every case, that is, at every level of abstraction, it is important to recognize the *operand* for the TP, and the transformation that this operand experiences. From this, the tasks of the *operators* can be established, that is, the *effects* that the operators, and especially the TS at that level, should deliver. With the available inputs to the TS, its (TS-internal) *functions* can be established. Each of these functions (or groupings of functions—capabilities for action) can then act as definition for the TP for the next more detailed level of complexity, until all the constructional parts are established.

It is advisable to observe and use the recommended sequence of the operations, but "leaps" and "jumps" (omissions) in the procedures are often found in practice. The recommendation is at least to review the omitted operations within the systematic and methodical design process.

(4f) Important **Information** Areas: The work in this stage demands good information about the engineering sciences—especially those based on physics for conventional TS. For some areas of information, chemistry or biology can provide the conditions for success. In addition, the experiences and a well-grounded overview of the available "means" for a certain branch specialty can be of significant advantage. First considerations about manufacturing costs may now be undertaken. This demands further specialized information and experiences, for example, from the cost area—experience (heuristic) values [365] form an important basis for cost decisions.

Useful information is available for special constructional parts, for example, purchased, delivered from external suppliers, bought out, OEM products, COTS. For such elements, the properties must be understood—they are usually contained in the technical literature (product specification sheets) available from their manufacturer, but cannot easily be compared. Especially the mode of action, connecting dimensions, and conditions for application must be known to the engineering designers. They should understand their own area of expertise, and at least be aware of the phenomena and properties from those "foreign" specialties from which they obtain special constructional parts, for example, electronic control systems for mechanical apparatus, mechatronics.

(4g) Available **Methods and Resources**: This stage is only treated relatively sparsely in the literature, further relevant methods can hardly be found. Methods for some individual steps are available, for example, for cost calculation—HKB (see Section 10.2). Masters and design catalogs (Sections 9.2 and 9.3) are an important information source. Computers have introduced some powerful tools, in representation and analytical investigations.

NOTE: Engineering designers frequently search for information—approximately for every four queries to the Internet, engineering designers query their colleagues 2.6 times, a responsible person 2.3 times, existing documentation once, and other sources once [410].

(4h) **Form of Representation**: The representation of the organ structure (see Figure 6.6) at high abstraction levels follows the paradigm of the function structure (Figure 6.4). The elements are now organs, as means—a "tangible system in principle"—represented by a circle symbol (Figure 6.6). Some possible forms of representation are shown in Figures 2.15 and 2.18, and the case examples in Chapter 1.

Repeated organs can be presented uniformly with simple line diagrams. These symbols may not be standardized, but are intended as a simple and easily understandable language for engineering designers. Such symbolism is also used for explaining the function of a more complex TS, and to describe the relationships and arrangements among individual organs. Other information about formations and relationships are transmitted by means of written remarks.

The organ structure is an engineering designer's shorthand note to describe a TS(s). From the viewpoint of abstraction and finality it does not need to be compiled in a homogeneous form. In the organ structure, a definitive existing constructional unit, a complete purchased assembly group or a module can be delimited as an organ, because each TS carries its organ structure and its function structure as elemental design properties (see Figures I.8 and 6.80–6.10). Other organs can be indicated as a TS-class with an appropriate branch term. Some existing representations can be interpreted as organ structures, for example, electrical wiring schematics, schematic diagrams of hydraulic or pneumatic circuits, "ladder" diagrams of programmable logic, kinematic scheme of (multispeed) gearboxes, and chemical process schematics.

NOTE: A TS of complexity level II or III, acting as a constructional part for a TS of next higher complexity, can be represented as an organ (or organism) by a dimensionless diagram showing only its inputs, outputs (effects and others), and connections to other organs. An electric motor, with its function "transform electrical power to rotational motion and torque," may be represented by the input terminals, the output shaft, and the connection to the fixed system.

(4i) **Remarks**: The organ structure stage is freely characterized and operated. The choice of organs is important for the technical and economic properties of the TS(s). If costs can be saved by a suitable choice of organ structure or of organs, and consequently in layout and detailing, this can result in large overall savings. The quality of the organ structure has large influences on the effectiveness of those organization processes subsequent to designing: ordering of the raw material stocks, scantlings and purchased parts (the supply chain), manufacturing, assembling, and so forth. Some of these tasks can and should be concurrent, to shorten the time to market the product. The disadvantages of concurrent working are that sections (parts, groups) of the designed TS(s) system must be "frozen" (no further changes are permitted), which places constraints on designing other parts of the system. If errors are subsequently found in the "frozen" sections, or major decisions are changed, the financial penalties can be large. Any planned "just-in-time" delivery of constructional parts to the assembly tasks may also create restrictions. This requires a careful consideration of trade-offs.

4.5 PHASE III—LAYING OUT—EMBODYING

Laying out, the phase of embodying, expands the chosen organ structure by establishing materials, forms, and dimensions. The intermediate result is a "preliminary layout," and the final result is a "definitive layout" (a TS-constructional structure) that is anticipated to be optimal.

4.5.1 STAGE (P5): "ESTABLISH THE CONSTRUCTIONAL STRUCTURE (1) AND (2)"

Searching for an optimal solution of the constructional structure is a logical continuation of stage (P4). A certain level of precision to achieve clarity is needed. The main question is "with what?" (tangible/physical) means can a TS(s) be made to operate, the functions and organs be realized, and the properties be fulfilled according to the list of requirements. This expansion of viewpoint complicates this stage. The mutual relationships among properties increase, and some of the requirements may become contradictory. Applying an iterative process is more than ever necessary, in general and in individual steps.

All *properties* listed in the requirements must be achieved, for example, manufacturability and suitability for assembly, safety, reliability, law conformance, appearance, ergonomics, economics, and so forth. The constructional structure must also realize all those constructional, manufacturing, and other functions that are *evoked* by the mode of construction, that is, functions and organs recognized during laying out and detailing. This includes solutions to added connecting, conducting, sealing, isolating/insulating, accessing, supporting, and other functions, and must be achieved by establishing the elemental design properties (see Figures I.8 and 6.8–6.10).

(5a) **Input**: Optimal organ structure—output of stage (P4), one, or several of almost equal "goodness," as concrete as possible, sketches of principles that represent the chosen action chain.

(5b) **Output**: The description of the constructional structure, in which all details are indicated. From this documentation, detail designers can produce the detail drawings, or details and subassemblies can be finished (e.g., by CAD solid modeling, see Chapter 10) and therefore realized. The task consists of establishing all the elemental design properties.

(5c) **Theorem**: The constructional structure, as established in this stage of design engineering, defines the constructional parts that form the tangible system elements. It also defines various combinations of constructional parts for specific purposes, and their relationships (of different kinds), for example, assembly groups, to reveal a hierarchy of assembly operations for a TS(s) (see operating instruction OI/2.3), or modules (of differing complexity) that allow different configurations of TS(s) to be assembled from unchanged subassemblies.

The constructional structure is accurately described in the assembly and detail drawings, and parts lists, or their computer-resident representations. These documents guide manufacturing planning, production, and assembly to realize the TS(s).

NOTE: The constructional structure of an engine lathe is composed of assembly groups and modules of different complexity (e.g., saddle, tailstock, pump, complexity level II, Figure 6.5) that consist of constructional parts (level I—gear wheel, shaft), of different geometric and topological complexity. Constructional parts may be assembled from elemental piece parts, for example, by welding—detail parts (gear rim, hub, gear disk). The hierarchy of complexity is therefore: module; assembly group; part; and element.

A constructional part (an elemental piece part) forms a material unit that is generally not capable of disassembly, but can be (mentally and by computer) decomposed into form elements, defined as bodies, surfaces, edges, points or *features*—combinations of simple geometric forms (see Section 10.3). Some of these are action locations that, in a pairing with the appropriate action location on the interacting part, serve as organs to accomplish a function. Each constructional part (each assembly group or module) is carrier of action locations.

A constructional part is therefore described by its form (system of form elements and their arrangement, relationships), dimensions, material and kind of production, accuracy (tolerances), surface quality, and so forth, the *elemental design properties* (Figures I.8 and 6.8–6.10).

Constructional elements of TS can be classified according to (1) complexity, (see Figure 6.5); (2) function (function groups, modules): assembly groups, constructional parts active with respect to one or several function(s) (monofunctional or multifunctional), or several constructional parts active together for one function; (3) breadth of application: universal or special (various degrees of specialization); (4) degree of unification: standardized, works standardized, unified, typified (see Figure 6.7); (5) origin: manufactured in own facilities, or outsourced (see Figure 6.7); or (6) proven, or new and innovative constructional parts.

An important characteristic of the constructional structure is the relationship of the individual elements to each other: (1) *coupling* realizes in general the functional relationships between constructional parts (e.g., element pairings), and: (a) can be achieved either directly, or indirectly with a coupling member (e.g., clutch); (b) can be realized by different principles: mechanical (solid, fluid), electric (inductive, capacitive), electronic, magnetic, and so forth; and (c) the coupling transfers and processes: M, E (motion, force, power), I—couplings are closely connected with the functions of the TS; (2) *spatial*, topological relationships that depend on the arrangement (configuration) of the individual constructional parts in space—the arrangement is visibly established in the design layouts and the assembly drawings, and the geometrical position of the parts (including tolerances) is defined. Arrangement is a subject of the organ structure (see stage [P4]).

(5c1) **Technical information concerning the constructional structure**: The procedure toward finality to produce elemental, completely describable and manufacturable means—a TS(s) and its constructional parts—is continued in this phase (including optimization) with the goals of concreteness, completeness, and definitiveness.

Answers to the question "with what?" can be obtained from several sources: (1) from experiences, perceptions and experiences of the designers, and which emerge as "ideas," often through association (see Chapter 8 and Section 11.1); (2) from the corresponding class of technical knowledge, for example, machine elements (general) (see Section 7.5), or machine tools (particular); (3) from masters and catalogs (see Sections 9.2 and 9.3); and (4) from the systematic and methodical design information, which recommends methods or working principles, for example, "give form according to force flux or load equalization." Experienced designers organize the knowledge they obtain into "masters" (Section 9.2) that are applied for a certain task or design situation (see Chapter 3).

(5c2) **Primary fundamental relationships**: All properties of the TS(s) depend on, and are established by means of the elemental design properties—elemental design properties are the cause of all properties of the TS (see Figures I.8 and 6.8–6.10). Finality implies looking for the "final" cause (*causa finalis*), that is, the elemental design properties, with which the requirements, and the properties of the TS(s) are fulfilled.

The relationships among individual classes of TS-properties are not always as clearly known as, for example, the relationship: strength $= f$ (elemental design properties). In addition, some TS-properties are not covered by the engineering sciences [98,556]. Several areas of "DfX" have therefore been formulated (see Chapter 6 and Figure 6.15). Information about design engineering for manufacture, transport, operation, costs, and operability (among others) is being collected. Some computer tools are available, for example, DFMA (see Chapter 8).

(5c3) **Mode of construction**: The constructional principle is a design characteristic according to which a constructional structure can be established, and which describes and distinguishes that structure from other structures—the *mode of construction* (see also Section 6.4.4.2).

An integral or differential mode of construction, and a monolithic, composite, modular, or platform-based constructional structure can emerge, according to the chosen distribution of the functions among constructional parts. A further characteristic can describe the typical arrangement of the constructional parts (e.g., an open or closed mode of construction). According to the material and the kind of production the mode of construction can be sheet metal, die cast, plastic molded, and so forth, or in a different field: half-timbered, piled, prefabricated, steel framed, or reinforced (e.g., concrete) modes of construction for technical building systems.

Using the concept of *modular* construction—construction sets or kits—we can create functional or production modules, and "configuration products" (see Section 6.11.10). Repair and maintenance problems may also be solved through suitable choice of modules. A shell mode of construction can be used, the enveloping walls serve as load-carrying components (self-supporting assembly groups—stressed skin).

From a systematic and methodical point of view this provides a further level for variation ("Which other modes of construction are possible?") for generating alternatives, and thereby for possibility of optimization. In addition, variation of number, location, form, and so forth, encourages exploration of alternatives in the constructional structure [178,370] (see Figure 2.14 and Chapter 8).

(5d) Available **Models**: "Laying out" the TS-constructional structure is executed according to the scheme of finality (Figure 2.3)—all TS-properties depend on the elemental design properties; see (5b) above. The procedure for the constructional structure is described in detail in (5e), and presented in Figure 4.1.

An essential help for laying out is offered by masters (see Section 9.2). The experiences abstracted from many details deliver the engineering designer specialized information which is unavailable elsewhere. This influences the assembly groups as functional modules, and groups of high complexity. Classical information and models are contained in the specialty "machine elements," compare Section 7.5. Many models and resources exist for achieving individual TS-properties—among others the area of "DfX" (see Chapter 6 and Figure 6.15).

(5e) Partial **Operations** (Steps): The relationships are complex, for example, the design properties display entangled mutual relationships. Therefore several *iterations* and *recursions* are needed (see Figure 4.3) to arrive at optimal definitive values of the TS-properties (see also Chapter 3). Planning is needed to accomplish the procedure in blocks, to execute the iterations and recursions in the shortest possible time. Different constructional parts, construction zones (action locations, subassemblies, assembly groups, modules) and form-giving zones exist at the same time in different states of completeness of definition—therefore also different methods may be applicable. The experience and knowing of the engineering designers, including their "feel" and judgment, plays an important role.

The individual operations and steps according to Figure 4.1 are given in the following.

Op-P5.1   Establish and analyze the requirements for the constructional structure and constructional parts

A list of requirements is again needed, analogous to the basic operation Op-H3.1. The main task of design engineering at this level of complexity of TS, is to establish *all* the required TS-properties for the immediate design situation, not just a suitable selection. The currently preferred solution proposal (abstract organ structure) has revealed new aspects, which should not be forgotten. This step should be formal: even if the goal is not to generate a formal list of requirements, a complete understanding of the task is required. Agreement with the customers' requirements should again be checked. Individual characteristics of the design situation strongly influence the subsequent decisions of designers, and deliver arguments for their substantiation and justification. With the increasing responsibility of designers before the law, these justifications can in an extreme case be decisive.

The required functions, with their conditions, are taken over and realized by the organ structure—conceptualizing or reconceptualizing. The degree of abstraction at which the organ structure is worked out, its concreteness and quality content, and the division of tasks which were available in the design process determine the continuity of the information flow.

The list of requirements (Section 4.3.1) should be analyzed, and the constructional parts, assembly groups or modules to which the requirements will transfer must be established. Their dependence on the design properties (see Figures I.8 and 6.8–6.10) should be determined, to discover any contradicting tendencies (e.g., increase strength and reduce mass)—finding a satisfactory alternative, trade-off, compromise, or a solution that evades the contradiction by using a different principle, may be difficult. Design properties are often in complicated mutual relationships, which increases the complexity of the task. Methods based on the "theory of invention" (contradiction-oriented innovation strategy, TIPS, TRIZ, and Invention Machine, etc., see Section 8.4) may help to overcome or avoid the contradicting demands.

Then the requirements for general design properties (see Figures I.8 and 6.8–6.10) combining the optimal structure from among the alternatives; should be analyzed for the given design situation, and associated or allocated to the individual constructional parts, assembly groups or modules. Among others, spatial unity of the constructional structure, and its connection to the fixed system (analog to grounding) must be established. Further possible and needed abilities of the constructional structure are usually discovered and postulated (evoked functions), for example, for assembling the constructional parts and assembly groups into the constructional structure.

The task definition for the constructional structure is continuously expanded during the solving procedure, to integrate the requirements for the assisting functions and the evoked functions.

These analyses represent an important phase of preparation for the engineering designers—to establish the requirements (and the evaluation criteria) for important constructional parts, assembly groups or modules, and to recognize those problem areas that influence the quality of the proposed solution, the TS(s), and timelines.

As an organization-internal document, a list of requirements (design specification for this situation) should be compiled, in most cases in writing—even if this is not usual in current engineering practice. This will be limited to the most important constructional parts, assembly groups and modules, and especially the (interacting) features on (mating) constructional parts that realize an organ. The procedure should be shortened and formalized by using preprinted forms (see Section 9.5). A useful form for a branch and field can be presented as a checklist. Any alterations from any previous (routine) projects should be carefully noted and considered.

The minimum that engineering designers must know (and iteratively develop) as a task definition for modules, assembly groups and constructional parts, may be summed up as follows: (1) purpose, function of the constructional part (or of the assembly group or module), needs for reliability; (2) complete environment with details, spatial distribution and arrangement, connectivity to other modules, assembly groups and constructional parts, and to the active environment; (3) loading, such as forces and moments, pressures, electrical current, type and severity of loading condition (static, dynamic, shock, vibration), and permissible deformation (or equivalents) from loading, and so forth; (4) conditions of operation of the TS(s), such as motions, kind of usage, temperature and its variations, humidity, chemical environment, service life, types of regulation and control; (5) requirements for further properties such as weight, size (e.g., limit for transportation); (6) production/manufacturing aspects: which kinds of production processes and equipments are available and so forth—consultation with manufacturing experts is usually needed here; (7) a general estimate of the permitted manufacturing costs; and (8) relevant international and national standards (ASME, ASTM, SAE, DIN, ISO), including standards adopted or developed by an organization for its own use.

Op-P5.2   Establish a conception of the constructional structure

This stage involves decisions of long range significance from technical, economic, and organizational aspects. Concretization of the constructional parts in the assembly groups or modules must be completed, that is, if the assignment of the organs to the concrete constructional parts, assembly groups or modules was not accomplished in the organ structure (operations Op-P4.1 to Op-P4.11), it must happen now, in order to have complete action chains available.

Establishing the sizes of the constructional structure depends on this action chain. This is the basis for establishing the sizes in the constructional structure and of its parts. Designers should begin with dimensioning (establishing the limiting sizes, parametrization) of the action locations, especially those which exercise the central TS-functions and are therefore size determining. For instance, in order to achieve the required performances of a motor or a piston pump, a piston diameter and stroke must be chosen. The performance deciding sizes for a centrifugal pump or for a Wankel motor are different, because a different operating principle was chosen. For many commercial machines, the size of the operands in the TrfP (Figures I.6 and 5.1) decides the main sizes of the TS(s), for example, machine tools, heat treatment furnaces, or forging ovens. The dimensions of other assembly groups or modules are adjusted to these decisive ones.

The mode of construction exercises a strong influence on the constructional structure; see Section 4.5.1, Subsection (5c3). It can determine several engineering design features. Primarily it is the *arrangement*, within the available space, of the constructional parts that make up the assembly group or module which can decide about many properties and characteristics of the TS, for example, at a higher level of TS complexity, compare the choice between front, rear or all-wheel drive for a car. The choice of raw *material* and kind of *production process* can dominate in individual cases, for example, a decision to use fiber-reinforced plastics for the superstructure of a streetcar (frame and body). Close cooperation with manufacturing personnel is essential—the manufacturing processes for the TS can be designed and preplanned concurrently with this stage of designing the TS (see Chapter 8).

A "make-or-buy" decision between *purchase* of assembly groups or modules (including OEM parts, COTS), or their *manufacture* in house or in another organization often demands special measures, mainly the adaptation of neighboring parts. *Societal* decisions about transportation, servicing or disposal, laws about fire protection, and so forth, can give rise to new requirements (see Chapter 3). Pressure on *expenses and costs* exercise a major influence on the conception of the constructional structure, manufacturing costs, and other costs, for example, life cycle costs.

Much cost and time (and possibly vexation) can be saved by realizing these considerations during conceptualizing, instead of in "improvement" and correction of the presented layouts or later life phases, and this must be emphasized by an experienced project leader.

Op-P5.3   Establish a rough constructional structure—preliminary layout

Whether it is possible to reach a complete and definitive (optimal) layout in the three steps (Op-P5.2 to Op-P5.4) is decided by the complexity and the degree of novelty of the TS(s). The general procedure shows laying out in two steps: rough or *preliminary layout*, and refined or *definitive (dimensional) layout*, with corresponding partial design operations.

Only a fraction of the information that should be used to completely define the constructional structure is contained in the abstract organ structure. This "break" or "leap" in synthesis can be solved with the help of "constructional structure masters" (see Section 9.2), which act as guideline pictures (patterns, models) for a part of the constructional structure.

An organ, viewed as a carrier of one or more functions, does not determine clearly or completely its allocation to assembly groups, modules, or constructional parts, or to a form-giving zone (see below). The constructional structure rests on a different principle to that of the organ structure, the correspondence is usually not "one-to-one." Exceptions exist, for instance an electric motor may be viewed as a driving (propelling) organ, a gearbox viewed as an organ for changing the number of revolutions of a shaft or its rotational speed, a coupling viewed as a connecting (or disconnecting) organ, or similar. TS carry their organ structure as constituent of the elemental design properties (see Figures I.8 and 6.8–6.10). These are usually more complex machine elements, which act both as organs for the organ structure, and as constructional parts for the layout. They are mostly purchased or taken over, their organs do not count as exceptions, and consist of simple constructional parts.

This difficulty can be overcome by forming units—modules, assembly groups, partial systems of the constructional structure—which adhere closely to the boundaries of the organs, but in addition retain the needed connecting places (action locations) to the neighboring constructional parts. Experience with the designing of such units probably already exists, these partial systems are treated as *constructional design groups* and *form-giving zones* (see next subsection).

The number of such units, and of the appropriate masters, is usually small in a specialty (e.g., industry sector) for which the particular engineering designers are responsible. This is similar to the number and use of subroutines in computer programming.

The constructional structure must also realize those requirements, assisting functions, and evoked functions that are not present (or not represented) in the organ structure. Some of these requirements can be easily fulfilled—add "something" to the existing (layout) constructional structure, for example, a covering, guard, or shield for safety and appearance. Some requirements can cause troublesome changes. The design operations (see Chapter 2) contain references about achieving the properties, the relevant knowledge is DfX (see Figure 6.15). Most assisting functions and evoked functions can be solved by using the modules, assembly groups, or constructional parts according to Chapters 6 and 7.

In this sense, the rough constructional structure is established (laid out, developed) in four operations: the boundaries of the modules or assembly groups are assumed for this purpose.

Op-P5.3.1   Define constructional design groups

Depending on the complexity of the organ structure, either the total structure, or a recursive subdivision can be processed. The total constructional structure can be divided into individual *constructional design groups*—analogous to the "windows" [435,439], assembly groups and modules chosen to define a suitable assembly sequence of constructional parts.

The boundaries of the constructional design groups are chosen with respect to the organs for establishing the design properties. The set of constructional design groups covers either the whole organ structure with overlaps, or only decisive areas, the *form-giving zones*, to be treated sequentially by one design team, or simultaneously (with problems of inter-team coordination, see also Section 11.2), and combined into a more comprehensive solution (see Figure 2.16).

NOTE: Examples of engineering design procedures that are based on the concept of form-giving zones may be found in the literature, usually using a different designation, for example, [438,545], compare also the case examples in Chapter 1.

The formation of constructional design groups influences the quality of the layout, and the duration of work, for example, iteration loops. Investment of time and effort is useful to propose typical constructional design groups for the situation, and to prepare suitable masters (see Section 9.2). Suitable constructional design groups: (1) if possible, treat complete organisms (see Figure 6.6); (2) are established by a boundary of the constructional design group that will allow determination of as many properties as possible, especially strength (resistance to load), deformations, durability, and so forth, of the constructional parts; (3) establish which masters, unified assembly groups or modules, and unified constructional parts (reused and purchased units, OEM, COTS) are proven in practice; and (4) keep the extent of the constructional design group small enough to maintain clarity, for example, for clear energy and space usage.

The constructional design group usually contains certain "zones" which appear as repeating elements of the constructional design groups. The solutions for these zones can be offered as masters for individual classes of technical means. This deals mainly with the solutions of those assisting functions and evoked functions that are realized with unified components, for example, force transmission by rolling contact bearings and their arrangement. Possibilities for exploring variations are suggested in Figure 2.14.

The form-giving masters (see Section 9.2) can be formulated at different levels of abstraction, as a whole and in individual form-giving zones. They should concretely define the abilities for TS-internal functions, and therefore bring concrete statements about the constructional parts. Depending on the organization-wide policies, the organisms can be established only as a family of means (e.g., rolling bearings), or only as (evoked) functions (e.g., locate in bearings, seal, see Figure 4.6). The resulting degree of freedom allows the engineering designers to strive for an optimal solution for the conditions of the order or contract.

The boundaries of constructional design groups preferably coincide with the boundaries of the main assembly groups or modules. Within the constructional design group, suitable assembly and subassembly groups or modules must be chosen to enable assembly of the TS(s), and to avoid reworking (e.g., adjustment) within an assembled group; see instruction OI/2.3.

Mathematical optimization of constructional design groups may be possible with the help of genetic algorithms, as proposed in the "autogenic design theory" [551,552].

Op-P5.3.2   Establish rough form giving of the constructional design groups

The concept of the solution obtained from the masters is now roughly adjusted to the conditions of the contract, by reviewing the requirements of the constructional design group. Subsequently the individual organs, the organ structure, and the assisting functions and evoked functions must be concretized, that is, the kinds of means established (e.g., deep groove ball bearings). Rough (assumed) estimates of size, strength, and other properties (preliminary parametrization) can be obtained for iterative refinement.

A possible reuse of existing modules, constructional design groups, assembly groups, constructional parts, or organs (i.e., those composite organs that represent composite constructional parts), should be examined—as modules for repeated application, or as purchased parts available on the market. A purchased part (OEM, COTS) is in many cases more favorable than the in-house manufactured one, especially from economic considerations. A reused module, constructional design group, assembly group, or constructional part, implies firm requirements, which must be completely fulfilled by the constructional structure, particularly by the parts in the close environment of the purchased part. In certain cases, disadvantages can accompany such a reuse, in-house manufacture (implying also design engineering, and organizational and technological preparation for manufacture) can become favorable.

The choice of subassembly grouping (and boundaries) can have a major effect on the difficulties and costs of assembly operations. Subassemblies should usually form a hierarchy, individual constructional parts are subassembled (and adjusted), and this subassembly should then not be altered for the next assembly operations (see operating instruction OI/2.3).

In choosing these assembly group boundaries, several considerations may be needed: (1) Are interactions among constructional parts critical? Then they should be included in the same group. (2) Is any change anticipated during the operational life of the TS(s), for example, by wear or other time-dependent influence? Then the change should be localized within an assembly group. (3) Can any anticipated failures or their consequences be minimized? Can a deliberate and easily replaceable "weak link" be incorporated, for example, an electrical fuse, a shear pin? FME(C)A or FTA (see Chapter 8) can be useful in this task. (4) Can the assembly group be standardized

(A)

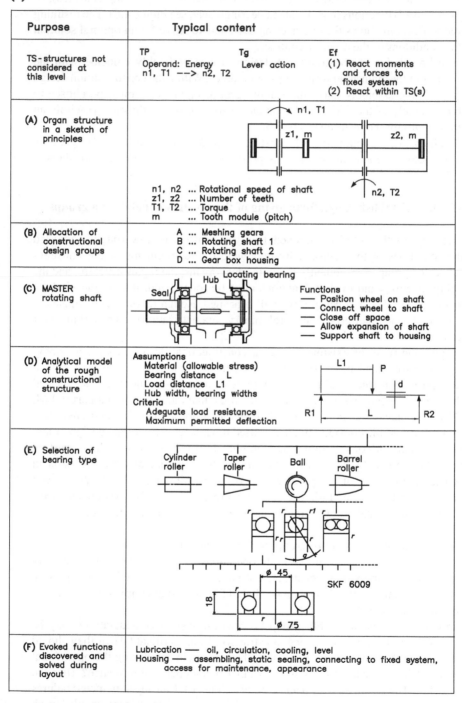

Purpose	Typical content
TS-structures not considered at this level	TP Tg Ef Operand: Energy Lever action (1) React moments and forces to fixed system $n1, T1 \rightarrow n2, T2$ (2) React within TS(s)
(A) Organ structure in a sketch of principles	$n1, T1$ $z1, m$ $z2, m$ $n2, T2$ n1, n2 ... Rotational speed of shaft z1, z2 ... Number of teeth T1, T2 ... Torque m ... Tooth module (pitch)
(B) Allocation of constructional design groups	A ... Meshing gears B ... Rotating shaft 1 C ... Rotating shaft 2 D ... Gear box housing
(C) MASTER rotating shaft	Hub Locating bearing Seal Functions — Position wheel on shaft — Connect wheel to shaft — Close off space — Allow expansion of shaft — Support shaft to housing

(D) Analytical model of the rough constructional structure	Assumptions
	  Material (allowable stress)
	  Bearing distance L
	  Load distance L1
	  Hub width, bearing widths
	Criteria
	  Adequate load resistance
	  Maximum permitted deflection
	
	L1 P d
	R1 L R2

(E) Selection of bearing type	Cylinder roller Taper roller Ball Barrel roller r $r1$ r r r ϕ 45 18 SKF 6009 r ϕ 75
(F) Evoked functions discovered and solved during layout	Lubrication — oil, circulation, cooling, level Housing — assembling, static sealing, connecting to fixed system, access for maintenance, appearance

FIGURE 4.6 Example: gearbox.

(B)

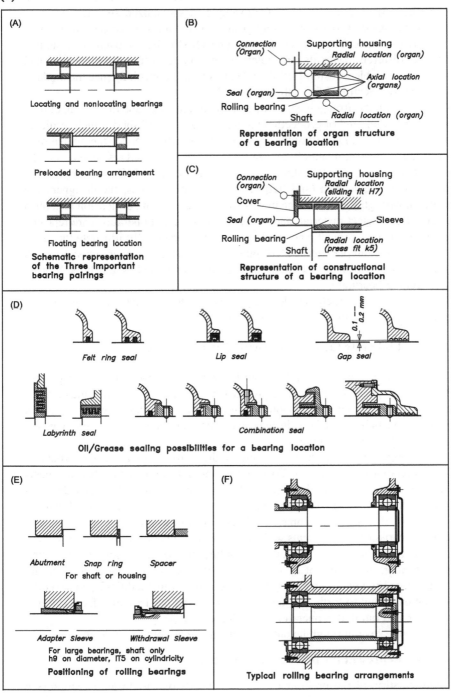

FIGURE 4.6   Continued.

for the organization, ready to reuse in other TS, see Figure 6.7? (5) Are constructional parts available from other sources, for example, a specialized supplier? (6) Can constructional design groups, assembly groups, or modules be chosen, which will allow interchangeable functions within the same TS(s) to be achieved, for example, possibly for different contracts—modularization—modules that can be "mixed and matched" to assemble a customized TS(s)? (7) How could this influence the mode of construction (see Section 4.5.1, subsection [5c3])?

Only then can the arrangement of the constructional parts, and their form, dimensions, material, and kind of production method be fully established, which results in the rough (preliminary) layout of the constructional structure. Since the mutual relationships are extensive, much of the data must first be estimated (assumed), and the first (iterative) round of establishing the means can be started. On the basis of the extent of the available data (and frequent lack of information), the designers can therefore only roughly establish the means, arrange, give form, dimension (establish necessary sizes), select material, and so forth. These design operations are repeated several times in each project, and are described in Chapter 2. To illustrate the range of problems, the example "gearing" (Figure 4.6) indicates two operations: establish the dimensions of the shaft (see model with assumptions) and establish the bearing (only for rolling bearings).

All important constructional design groups are thus brought to "rough layout maturity" by means of the described procedure.

Op-P5.3.3 Elaborate the rough constructional structure

The preliminary layout is assembled from individual constructional design groups. Compatibility among constructional design groups, and assembly groups and modules must be examined. A separate elaboration of constructional design groups may not be needed, and can be carried out during establishing the preliminary layout (e.g., the casing in the "gearing" example).

Coordination of the relative positions of the constructional design groups is important in elaborating the layout. For example, mutual interaction of the gear wheels demands that both shafts must be positioned in a common casing, that is, establishing of location, force reaction and force-flux closure, allowing for differential thermal expansion, possible need for geometric tolerancing, and so forth.

Further (evoked) requirements will emerge during this synthesis, always searching for possible alternatives, for example, cooling (fins on the casing?); lubricating (filling, draining, measuring of oil level, transporting oil to gear faces, and bearing locations?) (see Figures 4.6 and 2.16); transporting (hoisting facilities?); noise reduction (isolation?); and so forth. The function "connecting the TS with a fixed system" must also be laid out in this operation (see Figure 4.6).

In principle, several rough layouts should emerge as solution variants.

Op-P5.3.4 Represent the preliminary layout(s)

The representation of the preliminary layout depends on the design situation (see Chapter 3), especially complexity of the TS(s), degree of concreteness of the documents, and on other elements of the situation.

For simpler TS, a hand sketch, roughly to scale, and in a useful projection, is probably sufficient. An axonometric view (isometric, trimetric, oblique, or perspective) can in some cases present a useful solution. Simple TS can at this point be modeled in CAD systems on computers.

Drawing the preliminary layout for more complex TS can be executed with only the essential space relationships, and few details. Further information, including remarks, key word notes, and explanations belong to the representation. An inconsistent representation is typical. Known zones can remain implied, only interesting new areas can be completed (see Figure 2.16). These "gaps" in the preliminary layout often preclude computer modeling (see Chapter 10).

Op-P5.3.5   Evaluate, select optimal variants, reflect, improve, and correct

Review the results of this stage, be constructively critical, evaluate and decide (e.g., among alternatives), reflect and improve, and verify and check (including checking for possible conflicts and contradictions). This evaluation (see Chapter 5) should check that important requirements (as criteria chosen from the design specification) are fulfilled, and should enable a decision about the optimal solution variant. Methods recommended include the "art gallery" method, "concept selection" [475,478], weighted rating (see Chapter 8). Possibilities for improvement of the chosen solution can be discovered, leading into another iteration loop. The next round in the procedure is producing the dimensional layout (Op-P5.4), thus the suggestions for improvements may be allowed to remain as comments to the preliminary layout.

Op-P5.4   Establish a definitive constructional structure (2)—(dimensional) layout

Op-P5.4.1   Review the requirements, clear up, complete, perfect, optimize

The process "from rough to precise," and "from preliminary to definitive" implies a maturing of the constructional structure, by producing more accurate statements and representations that complete the missing information about the TS(s). In this second round, the preliminary layout should be cleaned up, completed, and perfected. The goal is to establish optimal and definitive data about the constructional structure and to represent it for communication to stage (P6).

The designer's task is to examine the design properties and parts of the constructional structure, to compare them (their anticipated short-term and long-term behavior) with the requirements, and to complete and optimize the structure, in the economic, time, space, and other limitations.

Op-P5.4.2   Definitively establish the arrangement

The goal is to refine, complete, and definitively establish (consulting with manufacturing and other experts) the arrangement of the constructional parts, their form, dimensions, tolerances, surface finish, raw materials, and kind of production process, and so forth—the elemental design properties. Especially the mutual relationships of the constructional parts to each other must be given high attention to finalize the operational TS(s), and to define the intermediate stages of manufacturing and assembly. The preliminary layout usually contains almost no statements about tolerances and surface quality, and few statements about the state and proposed sequencing of assembly or instructions for manufacture (production), standards, and so forth. Important communications about these areas should be included in the definitive layout—indicating the kind of relationships, for example, with texts about tolerances and fit conditions (e.g., geometric tolerancing if needed), degrees of freedom, and others. The design operations (i.e., arranging, configuring, form giving, dimensioning, parametrizing, etc.) are practically repeated at a higher quality level. The design operations are documented in Chapter 2.

Op-P5.4.3 Represent the constructional structure in a definitive (dimensional) layout

The representation of the complete constructional structure, the definitive layout, should normally be accurate to scale. Sectional views are usual to represent the individual constructional parts. Differences can be found in dimensioning these layouts, from almost total lack, to complete dimensioning in details, including many tolerances. Further information is given in different ways with regard to content and formalization in the layouts. This is probably the earliest stage at which computer-aided graphical or geometrical modeling (CAD) can be used to represent the constructional structure (see Chapter 9). The resulting "product models" can record the states of other properties. The trajectories of movable parts should be drawn in the representation, to facilitate control, checking and auditing for collisions or interference. Computer-aided modeling can "animate" the mechanisms to enable and assist checking.

Op-P5.4.4 Evaluate, select the optimal variant, reflect, improve, substantiate, justify

The comments to subsection Op-P5.3.5 may be repeated here. Several variants of layouts, or several variants of some zones, may possibly be expected. They should be evaluated, and decisions made about the optimal solutions according to the chosen evaluation criteria.

Engineering designers are responsible for the solution, in layout and details, and for the manufactured TS(s); they must have good arguments to substantiate and justify the chosen dimensions, form, raw material, proposed production processes, tolerances, and surface finishes. The close connection with "evaluation" (see Chapter 2) is emphasized here. The timing of calculations, considerations and proofs, experiments and tests, and so forth, is distributed over all engineering design phases. A summary should be contained in the representations, and in *design technical reports*. At this point, a management review (e.g., design audit, phase gate or stage gate) and decision to permit further work (e.g., release for detailing) can be taken. The costs involved in the subsequent operations (detailing) tend to be much higher, this is an appropriate point to decide *not* to continue the product development.

(5f) Important **Information** Areas: The comments to subsection (4f) may usefully be repeated here. Questions from the economic area are the main focus—manufacturing costs are decisive criteria in the majority of the cases.

(5g) Available **Methods and Resources**: This stage has close ties to quality assurance—several special methods are appropriate and can be used with advantage.

ISO and national standards have appeared and must be respected. Recommended methods include FME(C)A, FTA, Taguchi experimentation, and Life Cycle Assessment (see Chapter 8), many of which could be improved by incorporating the insights from Design Science [277]. The economic area presents methods of value analysis and value engineering, HKB (see Chapter 10), and others.

(5h) **Form of Representation**: The definitive engineering design layout usually remains an organization-internal document; yet it has vital importance concerning the quality of design, the quality of manufacture, and the perceived quality of the TS(s). Three kinds of layout can be distinguished: (1) According to "maturity": layouts that must be (partially or fully) optimized should not be represented with all details. Layouts selected as "optimal" can be completed. (2) According to "distribution": If an engineering designer continues to complete detailing, information in the layout need not be complete. Completeness is necessary if different staff members should continue the task. (3) Layouts intended for teamwork (e.g., on release for detailing) must contain complete information.

(5i) **Remarks**: The stage of laying out offers very diverse procedural possibilities, the conditions change between organizations, TS-"sorts," and contracts—therefore these discussions should be considered as recommendations. Each organization should "formulate" (and formalize) typical design situations, and offer advice about appropriate conduct for them according to the recommendations in this book. At this point, various instructions for the TS(s) can start to be developed, for example, user manuals, maintenance instructions, and so forth. Regarding the quality of laying out for the TS(s), the comments from subsection (4i) may be repeated here.

4.6 PHASE IV—DETAILING—ELABORATING

Detailing (elaborating) is intended to expand the information established in the definitive layout, to produce a complete representation of each constructional part with information required for planning the manufacturing processes, a complete representation of the total constructional structure, and any other design documentation needed by the organization.

4.6.1 STAGE (P6): "ESTABLISH THE CONSTRUCTIONAL STRUCTURE (3)—DETAILING"

The last procedural phase leads to the complete description of the constructional structure of the TS(s). The definitive layout that serves as starting point completely defines some of the design properties as description features, others are incomplete or not specified. The range of work in detailing changes according to the information content of the layout.

(6a) **Input**: Constructional structure in a dimensional layout of differing quality of statements about design properties, but complete description of the structure, as output of stage (P5).

(6b) **Output**: Description of the constructional parts in all details, that is, with elemental design properties, for manufacture of the parts and their assembly into the total constructional structure.

(6c) **Theorem**: In most branches of engineering, detail drawings are used to describe the constructional parts—graphic representation, dimensioning, texts (notes, remarks), numbers. The completion must be realized according to valid ISO and national standards. Similar requirements are valid for computer-generated information about constructional parts, even when the definitive layouts are "drawn" with computer-aided modelers (see Chapter 10).

The complete constructional structure is represented in assembly drawings, which contain the graphic representation, leading dimensions, remarks for assembly, and parts lists. For each individual constructional part, the parts list usually gives its reference number, title, quantity (per assembly), and initial material (scantling) description, possibly including mass. These drawings have a prescribed format according to the relevant standards. The drawings may need to conform to other instructions and regulations set by the organization, for example, cross-referencing upwards and downwards into a hierarchical system of states of assembly, as used for military documentation.

(6d) Applied **Models and Resources**: Master drawings and documents can supply patterns for choosing material and kinds of manufacture, dimensioning, tolerancing, specifying surface quality, and so forth, and can facilitate the work and increase the quality demanding constructional parts.

(6e) Partial **Operations** (Steps):

Op-P6.1   Prepare detail drawings—detailing

Describing the TS(s) implies representing (defining the form) and complete dimensioning of the individual constructional parts, and statements about initial material, tolerances (dimensional and geometric), surface quality, and state of assembly (e.g., prestressing). Further remarks for manufacture should accompany the description, for example, "drill constructional part x together with constructional part y" (which negates interchangeability of parts), or remarks prescribing fixtures and special tools. Further calculations can be necessary: additional analysis, evaluation and verification of strength, geometric data (e.g., gearing tooth forms, mass), and so forth. Particular attention should be given to the justification for assembly and production methods. The representation should be adapted to the qualifications of the detail designer and the manufacturing facilities, so that information flows rationally. This may involve direct generation of detail drawings from a CAD file, or direct transmission of an NC manufacturing program to appropriate machine tools. Consultations with relevant experts, and team work are probably essential.

Op-P6.2   Prepare a control (check) assembly drawing and parts list

For manual design engineering (i.e., when computer modeling is not used), it is advisable at this point to "forget" the layout, and to "assemble" the individual constructional parts (according to their description in the detail drawings, catalogs of purchased parts, etc.) for checking and control. Detail drawings can be checked for completeness and adequacy of dimensioning and tolerancing (especially to respect the needs of manufacturing and quality control), conformance with standards, cost conformance, and other representation features. At the same time the parts list—list of the constructional parts—can be compiled or checked. The mutual relationships of the constructional parts should be checked in the assembly, and indicated where needed. This defines the complete constructional structure of the TS(s).

Op-P6.3   Consider evoked functions from elemental design properties

During the procedure to establish the constructional structure (see Figure 4.6A), requirements for added capabilities (evoked functions) of the constructional structure emerge. These requirements should be considered in a systematic and methodical way. This implies beginning to formulate the abilities of the TS(s) with a description of capabilities—functions—and then looking systematically for solutions, see Figure 4.6B, for example, in a "function–means" tree (Figure 5.5). The solution is then integrated into the constructional structure.

The solutions for evoked functions usually belong to the class of universal constructional parts, universal assembly groups, or modules, closely related with machine elements (see Section 7.5). As classes of elements they are often repeated in a particular branch of industry.

It is important to prepare engineering design aids such as catalogs, masters (see Chapter 9) and guideline figures, and to regularly record and complete the growing experience in unified ways. Here the computer can provide valuable services (see Chapter 10).

Op-P6.4   Prepare any additional documentation

Engineering designers now "know" the structures that they have designed, and are in the best situation to prepare drafts of the design report—which includes any special instructions for manufacturing, assembly and adjustment, testing the manufactured TS(s) as a prototype and as a finished product, a users manual for operation and maintenance of the TS(s), and so forth.

Op-P6.5   Overall control/checking, verification

The comments to subsection Op-P5.3.5 may be repeated here. The importance of control is high, especially for production and assembly. Control as "Design Audit" (phase gate or stage gate) should deal with the description of the constructional structure as a complete checking of drawings. Checking of drawings should ensure that they are complete, correct, fully and adequately dimensioned, adequately toleranced, with specification of surface quality, complying with the relevant standards, and suitable for economic manufacture with the available facilities, tooling, and fixtures. Checking should also examine other accompanying documentation, such as the design (technical) report, including all design calculations, experiments, and so forth, test and adjustment instructions, storage and transportation requirements, operating manuals, and so forth.

(6f) Important **Information** Areas: Useful information is available for special constructional parts, for example, for purchased parts (see Chapters 6 and 7). For such elements, the properties must again be understood; see subsection (5f).

(6g) Available **Methods and Resources**: The computer has introduced some powerful and novel tools, in representation, in artificial intelligence (AI) heuristic advice, and in analytical investigations. It is to be expected that many changes will be added in time.

(6h) **Form of Representation**: As already mentioned, detail drawings must be executed according to the instructions of the relevant standards and organization-internal rules.

(6i) **Remarks**: The remarks to subsection (4i) may be repeated here.

4.7 SPECIALIZED MODELS OF THE ENGINEERING DESIGN PROCESS

The *ideal general procedural model* (this chapter) is based on generalizing assumptions. In contrast, a *procedural plan* for solving a concrete problem under specific boundary conditions must be in a particular form—preferably derived and adapted from the ideal model, with mutations found in the design situation (see Figures I.17 and 3.1).

Specialized models may be derived from the general procedural model by adapting the recommended procedure step by step as outlined in Figure 4.7: (1) With respect to the environment and *society at large*: standards, regulations, environmental protection and other restrictions, codes of practice, and so forth, problems of fashion and style, general aesthetics, historical preservation, and so forth, and according to the TS-"sort," industry sector/branch, factors FE of the design situation. (2) According to the individual organization and *production*: number of parts (per TS, and for each detail part *vs.* sorts of production method), time deadlines, experimental and manufacturing facilities, traditions, organization, company or institution, factors FO of the design situation. (3) According to the applied design technology; (see Figure I.20), part of factors FT. (4) According to the particular design task, its originality, complexity, difficulty of designing, and so forth, with respect to the task, the TS(s): its degree of complexity, degree of originality, possible variants, number and degree of difficulty of requirements, factors FT of the design situation—and any special procedural requirements that apply for many civil and plant engineering projects. (5) According to engineering designer/design team and their working methods, with respect to the *engineering design process*: state of the influencing operators of the design process—quality (information, experience, creativity, etc.) of engineering designers, state and availability of technical information, working means, management of the design process, working environment and conditions, factors FD of the design situation.

In view of these factors, which clearly influence the strategy of the engineering design process, we must distinguish between

1. The ideal *procedural model*, Section 4.2, contains generally valid information about the ideal flow of work during the engineering design process—in verbal and graphical form, usually as a flow diagram showing *feedback*. The

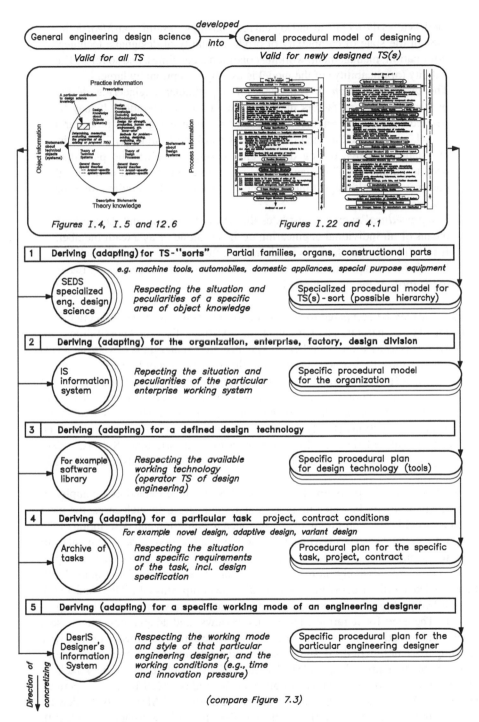

FIGURE 4.7 Derivative path from general engineering design knowledge to specific engineering design knowledge.

     model views the technical and logical relationships, assuming an *ideal* definition of the state of complexity of product, quality of designers, technical information, management, working conditions, deadlines, and so forth. The validity (applicability) of the model is usually exceptionally large.

2. The *procedural plan* is generated for a particular engineering design problem—all given factors may be drawn into consideration. It may be derived (usually by engineering designers) from the general procedural model, and adapted to the circumstances and time limitations. These circumstances change during design engineering, therefore the plan may need to be altered. This plan may be represented in the form of a critical path network, although time estimates for individual operations may be uncertain.

3. The *procedural manner* constitutes the ways of working and the working means of an individual engineering designer, alone or as a team member. It is valid only for the individual and is influenced by personal qualities, knowledge, and practical experience. The procedural manner clearly reacts to the form of the procedural plan.

This scheme may be compared to the scheme for methods presented in Section 8.1.

    From the constructional structures of existing TS, the possibilities for (recursively) dividing these complicated systems into smaller units of varying levels of complexity may be learned. Performing this division according to concepts of functional units (partial systems, assembly groups, modules, constructional design groups) yields a good correlation between these functional units and the individual TS-functions and TS-organs in the operational hierarchy. A division of a given function into partial functions is only possible if the relevant mode of action has been previously determined, that is, a decision is made to use a particular TS-"sort," selected among several possibilities, see the "functions–means" or "goals–means" tree in Figure 5.5.

    In a few cases, the problem of a TP(s) and TS(s) can be treated as a completely novel design problem, in which no precedent systems exist. In such cases, the ideal procedure would give a progressively more concrete definition and description of the TP(s) and TS(s), in broad terms a linear (but iterative and recursive) development from problem assignment to the concrete description of the TS(s), with the design properties becoming progressively more complete—"top–down design" (see Figure 2.10). Other problems will arise, either by subdividing (decomposing) the TP(s) and TS(s), or as evoked problems, each of which can be treated by a similar quasi-linear progression. The need for iterations, recursions, reviews, and adaptations will cause a scatter band of activity around the linear trend, which may be broader or narrower according to the nature of the design problem and the designers' adherence to systematic and methodical procedures [425]. Generally the cycles of iterations, backtracking and feedback corrections are called design–evaluate–redesign (DERD), test–operate–test–exit (TOTE), issue–proposals–argument–decisions (IPAD), activate–detect–advise–improve–record (ADAIR), conceive–design–implement–operate (CDIO), plan–do–check–act (PDCA), analyze–theorize–delineate–modify (ATDM), and many others.

Different sections of the TP(s) and TS(s) will at any one time exist at different levels of abstraction, and in differently established completeness. The properties will be progressively established during design work (see Figure 4.3, part 2). *Concurrent engineering* is normal (see Figures 2.10 and 2.11). If different divisions of an organization work together (e.g., design engineering and manufacturing planning), then concurrent engineering (see Chapter 8) becomes effective.

NOTE: Typical of design work is that an engineering designer works simultaneously at various levels of complexity, and on various sections of the TP(s) and TS(s), to gradually develop the TP-process structure and TS-constructional structure. A single step from the level of machines directly to the constructional parts, is generally impossible. This path toward a solution to the design problem should lead through various levels of modeling (see Figures I.4, I.22, 2.10, 2.11, and 4.1) that are reflected in the quasi-linear progress, supported by numerous cycles of problem-solving (see level 3 of Figure I.21, and Figure 2.7), iteration, and recursion. Actual progress usually appears more random, and changes between activities may be so rapid that they may only be detected with difficulty. The elements of activity are nevertheless present and real, but simultaneous activities make observational research about engineering design extremely complex; see Section I.1.

Iteration paths must be provided within each step of the general procedural model, the procedural plans, and procedural manners, including a possible review of the problem statement assigned to the designers—a managerial review of the product planning phase that usually precedes design engineering. Because at each step the engineering designer has a more complete understanding of the initial problem, all prior stages are subject to (periodic or continuous) review—feasibility checking, refinement, and iteration.

4.8 ADAPTATION OF THE PROCEDURE FOR REDESIGNING

At its most comprehensive, design engineering starts from a problem with no prior solution, as *novel* designing. Nevertheless, the full procedure of designing may be useful in other cases, for example, a recursively defined constructional design group, assembly group or module for which a novel solution is required.

For most other TP(s) and TS(s), variants or adaptations of existing products, *redesign* is the more usual task. Redesigning may be for reasons given in Section 6.11.2. These different scopes will also influence the nature and procedures of the design process [138,229], and the knowledge required to design the TP(s) and TS(s). Depending on the amount of innovation and alteration needed for the engineering design problem, the peak of abstraction can be as high as shown in Figure 2.11, or lower.

A design specification is needed in any case—preferably the assigned problem statement, and a full review and reworking by the designers, so that they understand and are as fully aware as possible of the implications, especially alterations from a previous similar project [208,209,211], which may change a project from routine to novel. Then it is possible either to reuse an existing more concrete structure from the precedent system (this represents a jump in the full procedure), or to abstract from the precedent system into an appropriate abstract TS-structure—"bottom-up design," and concretize again from there (a reversing loop) preferably following the procedural model (see Figure 2.11).

Even for such redesign, it may be useful to recognize the operand in the transformation process for which the TS is intended to be used (see Figure 1.16). For example, for the TS "electrical alarm clock," a human can see the indicated time (TS as operator indicating the time), and the human can set the time and alarm (TS as operand). The electrical alarm clock may also act as a paper weight or an ornament. For the TS "footbridge over a road," the TP has humans as operand, possibly carrying loads (e.g., a bicycle), from one side (Od1) to the other (Od2). Wind, ice, earthquakes, possibly even a tsunami may provide additional loading. A connection to the engineering sciences is clear.

If the alteration concerns the details of constructional parts (their configuration or parametrization), the reversing loop will be short (or the jump very long), and the relevant abstraction will remain in the region of the constructional structure. If larger regions of the constructional structure need to be changed, for example, newly developed constructional parts or organisms are to be incorporated or older ones replaced, then the reversing loop will be extended in time and may reach higher in the abstractions to the organ structure. The reversing loop will be even higher (or the jump shorter), into the function structure levels, if changes are demanded in the organ structure, for instance if some functions of the TS(s) should be changed from mechanical to electronic (computerized, mechatronic). Changes may even be necessary in the technology—then it may be more rational to treat the required TP(s) and TS(s) as novel.

Analyzing and abstracting, working from concrete to abstract (e.g., from a concrete existing solution into an organ structure and a function structure—"bottom-up design"), and subsequent reconcretizing, is likely to be useful for redesigning (see Figures I.17 and 3.1). The further procedure involves revising the function structure for the new conditions adopted in the design specification, and then proceeding through the recommended systematic steps of concretizing to a final solution (see Section 1.2). This form of abstracting and reconcretizing procedure can be useful for all levels of redesigning, to achieve an innovative redesign, an upgrade, a variant, or a small alteration from an existing product. Abstracting need not proceed to the function structure, reconcretizing can be started from any intermediate level.

Part III

Knowledge Related to Designed Technical Systems

Chapters 5 to 7 present the general knowledge about tangible systems, the technical systems and the transformation systems in which they are major actors (operators).

5 Transformation System, Including Transformation Process, Technical Process, and Technical System

The information presented in this supplement is mainly situated in and around the "northwest" quadrant of Figures I.4, I.5, and 12.7. Change is constant, transformations occur naturally, usually with random variations, over long periods of time, assisted by rain, frost, bacteria, and so forth.

Many other transformations are influenced by humans, they are artificial. Human beings have observed and tried to explain nature and the changes in their material environment as a result of natural processes. Then they began to replace some of the natural changes by artificial processes. The *technology* (definite article) for achieving a transformation was based on understanding of known *natural laws*—the abstracted expression of experience (see Sections I.8 and 12.1).

Artificial transformations may be deliberate or accidental, intended or unintended, beneficial or damaging. Avoiding the accidental, unintended, and damaging transformations, or to keep their consequences to an acceptable minimum, and enhancing the deliberate, intended, and beneficial consequences, to make them more efficient and effective, should be a goal.

Humans can use their experience, perceptions, apperceptions, and observations to replace and augment the natural transformation processes by *artificial transformation processes*, brought about with their tools (technical systems [TS]). This results in more rapid, more effectively planned and controlled changes, that produce (partly unwanted) side effects and consequences.

Among the artificial transformations of interest are those that are accomplished with the use of artificial tools, machines, apparatus, devices, and so forth. Artificial tools usually need to be designed (engineered) and made before they can be used—we call them TS.

Interactions among TS, transformations, human culture and society, and so forth are discussed in the Introduction and Chapter 6. The list of symbols shows a *process system* as a rectangular symbol, and a *tangible (object) system* as a symbol with rounded corners.

5.1   TRANSFORMATION—GENERALIZATIONS AND ENTRY TO TERMINOLOGY

Human behaviors and natural, social, societal, cultural (and other) phenomena are directed and guided (teleologically) by certain anticipated recognizable goals—activities act as causes that lead to expected and unexpected results. Humans are driven to *action* by their *needs*, their feelings that something that they regard as useful is missing, important, or even urgently needed to support or improve their lives. The term *needs* is understandably very broad, ranging between objective needs (hunger, thirst) and pseudo-needs (smoking), and also between material needs (hunger, thirst) and non-material needs (social communication, desire for culture and art [264–266,402,403]); see also Section 11.1.

Recognizably, all *goals* (aims, ends) that we set are predicated on certain needs (see Section I.10.1). We have reasons for our (proposed) actions and intentions. Needs are linked with actions, transformations that we wish to achieve, for example, health as outlined in Section I.9.1.

NOTE: Society is constantly developing new needs—either through becoming aware of them, or by being persuaded (e.g., by advertising) that they exist. Such needs have differing *character* and *significance* when compared to the base values of the absolute necessities required for sustaining life; see [204,260,271,380,572] and Section 11.1. Any philosophical discussion about the relative nature of needs and whether they can be "absolutely essential" is undesirable. This book investigates those changes, and their causes, that are relevant to satisfying needs by using technical processes (TP), that is, applying TS to exert the required effects.

Needs (as anticipated goals) are satisfied by appropriate, varied and different, alternative *means*, the most suitable of which should be selected. People must decide what means they wish to use to achieve their intended *goals*, resulting in "*goals–means*" *chains*. They do not have to be clear about the possibilities that may be available, but some choice is preferred. They also do not have to be clear about unintended consequences, but should consider possible outcomes.

In order to progress toward the envisaged goals, everyone uses *methods* for any tasks that are repeated in similar fashion (see Section I.12.5). Tactical methods are included in the description of individual phases, stages, and steps (Chapters 4 and 8). These methods are based on the concepts of causality and finality, as outlined in Section 2.2.2.

In order to achieve goals, the necessary means are not found to be available in the desired (ready-to-use) state. Everything has to be adapted, changed, *transformed* (as operands) in a transformation process (TrfP), from the existing input state to a (usually more desirable) output state. The term *state of an object* contains the operand properties of an object: form (also shape), structure, space (*location*), and current *time*, and others.

NOTE: A related term is the "state of the art," the best currently available embodiment of a product (see Chapter 6.9). This is one factor in the design situation (see Chapter 3).

5.2 MODEL OF THE TRANSFORMATION SYSTEM

Satisfaction of human needs depends on transformations that are supported in the technological sphere, that is, they are artificial. Developments in the technological sphere are accompanied by developments of technologies, processes, and technical means.

The aim (goal) of every TrfP, and consequently of each transformation system (TrfS), is to achieve a particular state of a target object, as means for other changes, this object (and its input, intermediate, and output states) is termed the *operand*. The desired state of the target object (Od2) is the output of the TrfP. If an object capable of being transformed (Od1) is available, it can be selected as the input to the planned transformation.

The model of the TrfS in Figure 5.1 declares

(1.) An *operand* (materials, energy, information, and living things—M, E, I, L) in state Od1 is transformed into state Od2, using the active and reactive *effects* (consisting of materials, energy, and information—M, E, I) exerted continuously, intermittently, or instantaneously by the *operators* (human systems, TS, active and reactive environment, information systems, and management systems, as outputs from their internal processes), by applying a suitable technology Tg (which mediates the exchange of M, E, I between effects and operand), whereby assisting inputs are needed, and secondary inputs and outputs can occur for the operand and for the operators.

A TS-effect is the output (M, E, I) of the chain of TS-internal processes, produced by the TS-action chain, that acts directly through a technology (definite article) to transform the operand (see Section I.9.1.2). Secondary outputs can be emitted by the TrfP, or by any of its operators—the arrow for secondary outputs in Figures I.6 and 5.1 starts from the boundary of the TrfS. "Leaking oil from a gearbox" is a secondary output from a TS, "heat from losses of power transmitted by the gearbox" is a secondary output from the TrfP that is performed by the TS "gearbox" on the operand "rotary power."

NOTE: Various manifestations of the operand, input, output, and effects can be defined:

M   *Material*: gas, liquid, solid; or in special cases a combination of these.

L   *Living things*: only applicable for an operand, includes humans, animals, and plants.

E   *Energy*: according to Figure 7.11, all forms of energy need a state variable (static, "across" variable), and a flow variable (dynamic, "through" variable). Energy can only be transmitted and transformed if both variables are nonzero.

   State variable: force, torque (moment), pressure, voltage, temperature; Newton's law that "action and reaction are equal and opposite" is valid for force, torque, and pressure.
   Flow variable: velocity, angular velocity, volume or mass flow rate, electric current, entropy.

   Energy transfer and its dynamic behavior can be modeled by a sequencing of four-pole elements [575]. One variable, state or flow, can be *active*, it determines the behavior of the system, and can be calculated forward through the sequencing of four-pole elements. The other variable, flow OR state respectively, must be *reactive*, and can only be determined by calculating backwards through the sequencing of four-pole elements; see Section 7.5.

I   *Information*: analog, digital; recorded, tacit/internalized/mental; and so forth.

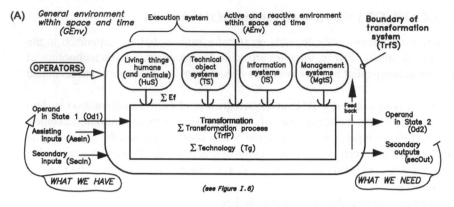

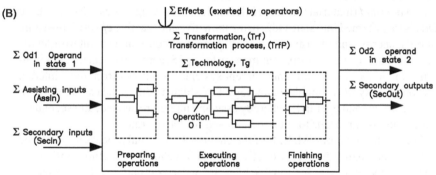

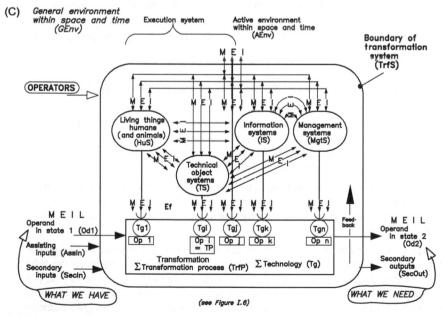

FIGURE 5.1   General model of the transformation system.

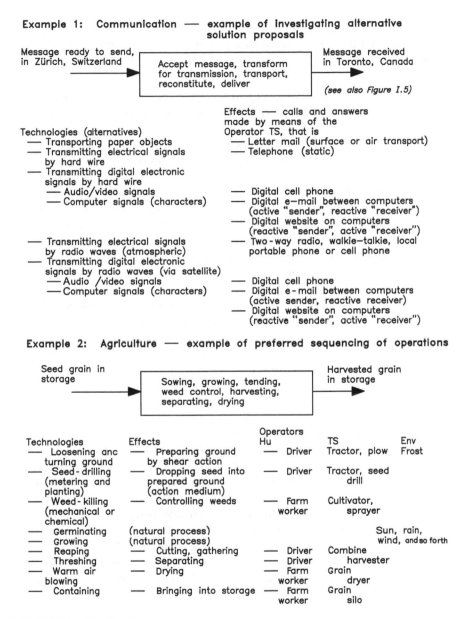

Example 1:   Communication — example of investigating alternative solution proposals

Message ready to send, in Zürich, Switzerland → | Accept message, transform for transmission, transport, reconstitute, deliver | → Message received in Toronto, Canada

(see also Figure I.5)

Effects — calls and answers made by means of the Operator TS, that is

Technologies (alternatives)
— Transporting paper objects
— Transmitting electrical signals by hard wire
— Transmitting digital electronic signals by hard wire
    — Audio/video signals
    — Computer signals (characters)

— Letter mail (surface or air transport)
— Telephone (static)

— Digital cell phone
— Digital e-mail between computers (active "sender", reactive "receiver")
— Digital website on computers (reactive "sender", active "receiver")

— Transmitting electrical signals by radio waves (atmospheric)
— Transmitting digital electronic signals by radio waves (via satellite)
    — Audio /video signals
    — Computer signals (characters)

— Two-way radio, walkie–talkie, local portable phone or cell phone

— Digital cell phone
— Digital e-mail between computers (active sender, reactive receiver)
— Digital website on computers (reactive "sender", active "receiver")

Example 2:   Agriculture — example of preferred sequencing of operations

Seed grain in storage → | Sowing, growing, tending, weed control, harvesting, separating, drying | → Harvested grain in storage

Technologies	Effects	Operators Hu	TS	Env
— Loosening and turning ground	— Preparing ground by shear action	— Driver	Tractor, plow	Frost
— Seed-drilling (metering and planting)	— Dropping seed into prepared ground (action medium)	— Driver	Tractor, seed drill	
— Weed-killing (mechanical or chemical)	— Controlling weeds	— Farm worker	Cultivator, sprayer	
— Germinating	(natural process)			
— Growing	(natural process)			Sun, rain, wind, and so forth
— Reaping	— Cutting, gathering	— Driver	Combine harvester	
— Threshing	— Separating	— Driver		
— Warm air blowing	— Drying	— Farm worker	Grain dryer	
— Containing	Bringing into storage	— Farm worker	Grain silo	

FIGURE 5.1   Continued.

These classes always occur in combination; they cannot be isolated, but one of them will usually be *dominant* relative to the others for any applications or considerations. The laws of conservation of energy and of matter are valid. Energy can be transformed, for example, from chemical to thermal to mechanical. Material can be transformed from liquid fuel to gaseous combustion products.

In general, Σ inputs = Σ outputs = Σ useful outputs + Σ secondary outputs (including losses).

Relativity according to Einstein also permits some transformations from material to energy and vice versa, but only under conditions close to the velocity of light. Information is not subject to conservation.

Explanations about design engineering in this book use this model as starting point, other models are derived by *concretizing*—reformulating in a more concrete way, the opposite to *abstracting*.

The manifestations and values of all properties (e.g., of the operand) are variable to some extent. A permitted maximum variation is a *tolerance*. For example, the input operand in a particular view of "computer printing" is paper in state Od1 (size, weight, strength, color, moisture content, and so forth, in the "in"-tray of a printer, before issuing a "print" command) and a medium (solid or liquid) that can mark the paper; the output is paper in state Od2 (paper imprinted with a selected pattern).

The *effects* received by the operand that change its state (and consequently change its properties) depend on the selected *technology* (definite article), which must conform to the laws of nature. This is usually one of several possible technologies. The technology for the TrfP is an aggregate of the technologies used for each operation; see the wheel example in the NOTE in Section 5.3. According to the type of technology employed, some *assisting inputs*, *secondary inputs* (including disturbances), and *secondary outputs* can occur (pollutants, waste), and also recyclable by-products. For the paper, the effect is a pattern to be imprinted, the technology may be impact through an ink-carrying ribbon, electrostatic transfer of powder and fusing on the paper, or droplets of ink sprayed onto the paper.

Effects can be *active* or *reactive*. Internal damping of a material is *reactive*, energy conversion to heat because of (1) changing strain conditions in the material and (2) resistance to this change of strain—both constitute necessary conditions. The resistance of a material when it contacts a cutting tool arises as a *reaction* to cutting force, and (for metallic materials) uses the technology of shearing the chip material, converting kinetic energy into shear strain energy and heat. The space shuttle, as it reenters the atmosphere, is resisted by air *reactively* moved, accelerating air atoms, exchanging kinetic energy from the shuttle to the air and converting part of it into heat.

Actions by operators can be *direct*, an immediate influence of the TS(s) via a technology to cause a transformation of the operand, or *indirect*, for example, an instruction to a human to use a hammer to forge metal. The environment can deliver various effects to the operand in the TrfP. These can be direct or indirect, active or reactive. Any direct effect causes the environment to be included in the execution system (see Figure 5.1).

NOTE: An effect is an output from an operator directed to the operand in the TrfP. Effects *consist* of material, energy, and information (M, E, I). The symbol for an effect is terminated at the boundary of the TrfP. Yet the tangible outputs of M, E, I from each operator are converted by the appropriate *technology* into transformations experienced by the appropriate operand at the effector location (see Figure 5.1C). All operators interact with each other, exchanging M, E, and I, as also shown in Figure 5.1C.

The manifestations of energy transfer can be: conduction, convection, and radiation. Conduction needs a material contact. Convection needs a fluid contact. Radiation needs no material contact, but need a direct line of travel (including reflection and refraction paths). The manifestations of material can be solid or fluid, and fluid can be liquid or gaseous.

For instance, a very remote operator, the sun, emits energy (electromagnetic waves) and material (particles) in the form of solar radiation. The radiation is only effective if this radiation actually hits the surface of an operand (e.g., your face), that is, the solar radiation is part of the TS "sun" until it contacts the operand. This radiant energy is partly reflected, partly converted into heat to raise the temperature of your skin, as effect Ef1, and partly initiates a chemical reaction to change the color of your skin, as effect Ef2. Physical contact of the radiation is a necessity.

A *technology*, Tg (definite article), describes the nature of the direct interaction between the *main effect* as (active and reactive) output of an *operator* and the *operand* and that causes the operand to be transformed. It is therefore a conversion mechanism that can be described typically by a mathematical expression from an engineering science, or from an empirical relationship, or from any other relevant relationship and its formal or informal theory.

All transformation processes have random variations. These variations are usually small enough to be ignored, but sometimes a transfer and accumulation of small individual errors can result in large variability and can call the strict concept of causality into question. Reducing variability can be useful, for example, helped by Taguchi experimentation (see Chapter 8).

The effects are exerted (delivered) by the operators, especially by the *execution system*—comprising living beings (*humans* with their tacit/internalized knowledge, experience, skills and abilities, attitudes and values, and animals, bacteria, etc.), TS, and the *active and reactive environment*. The transformation can achieve optimal *quality*, providing that the *operators* act optimally, that is, sufficient relevant *information* (recorded/codified knowledge, recorded experience, etc.; see Chapter 12), and adequate process *management* (directing, leading, defining goals, guiding, and controlling toward achieving these goals) are available, and the influences of the *environment* do not exceed certain limits. Effects received by the operand also produce reactions on the operator. The transformation of the operand is therefore necessarily (topologically) external to the operators.

Each TrfS exhibits classes of properties: system behavior, its response to a change of an input or other condition, whether short-term or long-term. Each *system* is characterized by its inputs and outputs, and its structures. Classes of properties for a TrfP consist of the properties (1) of the operand, in state Od1, Od2, and in each intermediate state, (2) of any assisting and secondary inputs, (3) of secondary outputs, (4) the types of transformations or operations that are performed on the operand, and (5) the active and reactive effects exerted by the operators that cause the transformation of the operand by means of the applied technology. The identifiable *elements* of TrfP are the subprocesses and operations. Their *relationships* are the connections (and coincidence) between outputs of one subprocess with inputs of a following subprocess. The relationships in a TrfP comprise the transfer of operand from one operation to the next. Operations or subprocesses can take place sequentially or simultaneously (i.e., in series or in parallel), iteratively, cyclically, interconnectedly, or in any other suitable pattern.

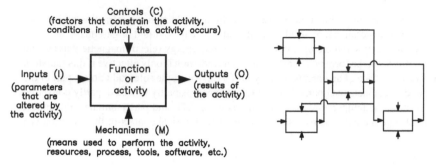

Elemental IDEF0 process model   Typical IDEF0 process structure

FIGURE 5.2   IDEF0 model of a process element and structure.

NOTE: A similar proposal exists in the "set of methodologies," IDEF0 for functions, IDEF1 for information, IDEF1x for data, IDEF3 for dynamic analysis, and IDEF4 for object oriented design [18], that is, especially for computer programming and information processing. IDEF0 is a graphical tool used to produce a functional model of an organization, system, or process as a hierarchical "function decomposition" using only information as operand (i.e., not biological or inanimate materials and energy). As shown in Figure 5.2, the "controls" and the "mechanisms" are equivalent to the effects of the five operators in the model of the TrfS, Figures I.6 and 5.1—but are not differentiated.

5.3 TECHNICAL PROCESSES

Technical processes are those TrfP (partial processes and operations) in which TS find extensive use (see Figure 5.3) either alone, or with additional inputs from humans and the active environment. The explanations about TP are coordinated with the insights about general transformations, each section repeats and expands the above.

A TP is an artificial TrfP in which the *states* of material and biological objects, energy, and information (M, L, E, I, as operands) are intentionally, in a planned fashion, *transformed* by the *main effects received from technical means* (i.e., exerted by TS), and may also be influenced by the *human beings*, an active and reactive environment, and usually indirectly by other operators of the transformation. The *states of operands attained* after the transformation serve to *satisfy human needs*, either directly as goods and services, or indirectly as technical means (see also Section I.9.1.2).

NOTE: The goal of designing a TP(s) is to achieve an optimal output (state Od2) of the operand, within an appropriate time and cost.

The boundaries of the TP(s) and the TS(s) must be made clear. Consider an "automotive wheel." The rim, the tire, the valve, and air are the operands, Od1, at the input to the "black box" TrfP "mounting a tire." Individual operations are Op1: "fix the rim," Op2: "insert the valve," Op3: "mount the tire," Op4: "inflate the tire," Op5: "release the operational wheel."

For Op1: TSa is "tire mounting machine," Od1a is "rim free," Od2a is "rim fixed to Tsa," Tga is "clamping." For Op2: TSb is "tire mounting machine with fixed rim," Od1b is "valve free," Od2b is "valve inserted," Tgb is "valve pulling." For Op3: TSc is "tire mounting machine with rim and valve," Od1c is "tire free," Od2c is "tire mounted," Tgc is "rotational snapping of tire bead over rim," AssIn is "tire lubricant." For Op4: TSd is "tire mounting machine with rim, valve and tire," Od1d is "air at normal pressure," Od2d is "air compressed in tire/rim/valve assembly," Tgd is "pumping and guiding through valve." For Op5: TSa is "tire mounting machine," Od1e is "operational wheel fixed," Od2e is "operational wheel free,"

(A)   General model of technical process (system)

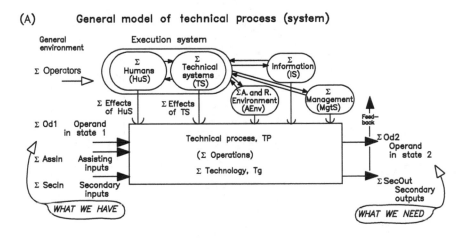

(B)   Example: Drawing round wire

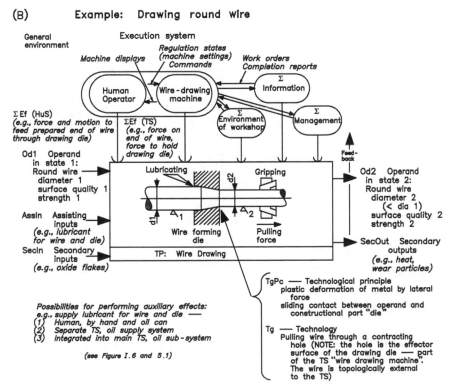

FIGURE 5.3   Model of a technical process within a transformation system.

(C)

Examples						
Operand (Od)			Transformation process (TrfP)	Effects (Ef)		
Type	State		TgPc ... Technological principle PaTP ... Partial process	Exerted in the form of material (M), energy (E) or information (I) by		
	Od-1	Od-2		M	E	I
1. Material steel	Raw material Form 1 Dimension 1	Shaft Form 2 Dimension 2 Tolerance 2 Surface quality 2	TgPc Chip-forming machining PaTP1 Clamp tool PaTP2 Feed, close chuck PaTP3 Rotate workpiece PaTP4 Move tool rel. to wkp. PaTP5 Measure PaTP6 Remove to bin	TS TS TS TS TS 	HuS TS TS HuS HuS HuS HuS	HuS HuS HuS HuS+TS HuS
2. Energy	Chemical energy carried by coal	Heat	TgPc Combustion $C + O_2 =>$ CO_2 + Joule PaTP1 Store coal, ignite PaTP2 Burn, ensure air supply (O_2), remove heat, clean ash and slag PaTP3 Extinguish, clean	TS TS 	HuS+TS TS HuS	HuS HuS HuS
3. Human	Infected	Infection-free	TgPc Antibiotic treatment PaTP1 Prescribe pills PaTP2 Take pills PaTP3 Monitor progress		HuS HuS HuS HuS	HuS HuS HuS HuS
4. Human	Faulty kidney	Functioning kidney	TgPc Kidney transplant PaTP1 Prepare human, achieve sterile conditions PaTP2 Remove faulty kidney PaTP3 Insert functional kidney PaTP4 Observe	TS TS TS 	HuS+TS HuS HuS HuS	HuS+TS HuS+TS HuS+TS HuS+TS
5. Material (water)	In deep well	In storage tank	TgPc Suction — pressurizing PaTP1 Achieve partial vacuum, induct, guide PaTP2 Pressurize, expel PaTP3 Guide to tank	TS TS TS	TS TS TS	TS TS TS
6. Living Object (human, animal)	Place 1	Place 2	TgPc Use transportation vehicle PaTP1 Connect human in place 1 with vehicle PaTP2 (Human/animal + vehicle) —> place 2 PaTP3 Disconnect human from vehicle in place 2	 TS 	HuS TS HuS	HuS HuS+TS HuS
7. Material (machine turning)	Form 1 (starting diameter)	Form 2 (ending diameter — smaller)	TgPc Chip-forming cutting PaTP1 Place workpiece in chuck PaTP2 Move cutting tool PaTP3 Measure PaTP4 Release workpiece from chuck		HuS TS HuS HuS	HuS HuS HuS+TS HuS

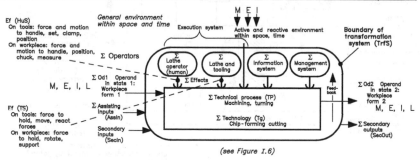

FIGURE 5.3 Continued.

Tge is "unclamping." The technologies Tga–Tge contribute to the overall technology of the "black box" transformation.

It is only when the operational wheel is finally mounted on the axle of a car, TSf, that the car can be operational, and the effects of "transmitting force to the ground" can be realized—the wheel is then internal to the boundary of the TSf.

In Figure 5.3, the separable drawing die of the operational TS "wire drawing machine" must be considered internal to the TS, the inner conical face of the die is the effector that will directly contact the wire and cause its transformation in diameter and other properties.

5.3.1 GENERAL MODEL OF TECHNICAL PROCESSES

The *model* of TP that reflects and integrates all these points is shown in Figure 5.3, which gives detail to the general model in Figure 5.1. Figure 5.3, as with every model of this kind, is intended to give a clear but condensed survey of the subject—in this case the knowledge about TP.

NOTE: This formal representation offers an overview, orientation, and systematization *for* designing, and thereby can help to avoid omissions of some aspects.

5.3.2 THE APPLIED TECHNOLOGY

The technology (definite article) describes how the transformation can be performed. The *change of state* of the operand may occur by various *technological principles* (TgPc), which are available from knowledge of physics, chemistry, biology, and so forth, for instance hydraulic, pneumatic, mechanical linkage, friction, plastic deformation, electron tunneling, radiation, and so on. Depending on which of these TgPc is selected, the technology may be established.

The *technology* (Tg) is the way of implementing a transformation. Tg describes and specifies how the transformation can be performed. A technology may be visualized as the interaction between the TS (e.g., tool) and the operand (e.g., work-piece) that performs the change to the operand. The selected technology (among the available alternatives) for a transformation establishes the operations necessary to achieve the desired change of the operand, and the arrangement and sequencing of these operations.

NOTE: Effects are *exerted* by the operators, especially the TS (although "emitted" might be a better verb for radiation). Effects are *received*, because of the technology, by the operand. Therefore we really only need to talk about the exerted effect and the technology.

This choice of technology permits establishing the structure of the TP, TP(s), the operations and their arrangement, including decision operations that only activate one or other branch of the process structure. The choice of technology also permits establishing the type of effects that must be received by the operand. This leads to establishing the requirements that need to be placed on the humans, the TS, and the active environment, that is, allocating tasks to the executing operators, especially for the effects they must exert. For instance, Figure 5.3 shows how the TgPc of "applying lateral force to achieve plastic deformation" and "sliding contact between surfaces" are applied to the technology of "pulling wire through a tapered narrowing opening to reduce its diameter."

For instance, the technology of hardening a piece of steel prescribes an effect of transferring heat to the item (the operand) to achieve a specified temperature, then rapidly transferring heat from the item (cooling and quenching it) to a lower temperature, and usually reheating it to temper the steel to reduce its hardness from the maximum "glass-hard" state, and restore some of its ductility, followed by slow cooling to room temperature. The technology of radiant heating requires a radiation source, for example, an electric heating element, and a direct line-of-sight to the operand; the radiating energy is considered as part of the acting TS, and is converted at the operand interface to heat.

The importance of the technology for performing the transformation of properties of the operands in TP must be emphasized. The technology determines the achievable quality of the operand in its output state and the effectiveness of the whole process (see Figure 5.1). Major innovations are often connected with new or improved technologies, some of which may be the results of scientific fundamental or applied research. The *information* about technologies available to designers, and likely failure modes, are therefore factors that influence decisions about realizable processes.

5.3.3 CLASSIFICATION OF OPERANDS TRANSFORMED IN TECHNICAL PROCESSES

An operand is subjected to transformations in the TP, and appear as its inputs, throughputs, and outputs (entry, intermediate, and exit states). Four classes of operand can be recognized: M, L, E, I; see the NOTE in Section 5.2, and Sections I.8 and 12.1. These classes always occur in combination.

The state of the operand can be changed by five basic varieties of transformations (see Figure I.20) namely changes of structure, form, location, time, and other *operand properties*: for example, temperature. Time cannot be *arbitrarily* changed, its elapse is a natural process that under normal circumstances only runs forward, but every operand is influenced by the change of time, and in certain operations it is only time that changes.

5.3.4 CLASSES OF OPERATORS FOR TECHNICAL PROCESSES

The general model of the TrfS, Figure 5.1, shows that the operators comprise humans and other animals, technical means (systems), the active environment, information, and management, within space and time. These classes of operators are equally valid for TP. The main distinguishing feature between the general and the TP is that the TP extensively or exclusively involves the application of TS.

5.3.5 STRUCTURE OF TECHNICAL PROCESSES

Following the general definitions in Figure 5.3, the structure of TP can be explained with reference to Figure 5.4.

Every TP can be subdivided into *partial processes* or *operations*, for example, movements, temperature changes, changes in electrical charge or potential, and so forth. Manufacturing (as a particular sort of TP) consists of a large number of operations; they are concerned with preparing the raw materials and tools, and the actual manufacturing, testing, or inspection processes.

The *structure of the TP* consists of partial processes and operations, and their relationships. Two different classifications are used (see Figure 5.4):

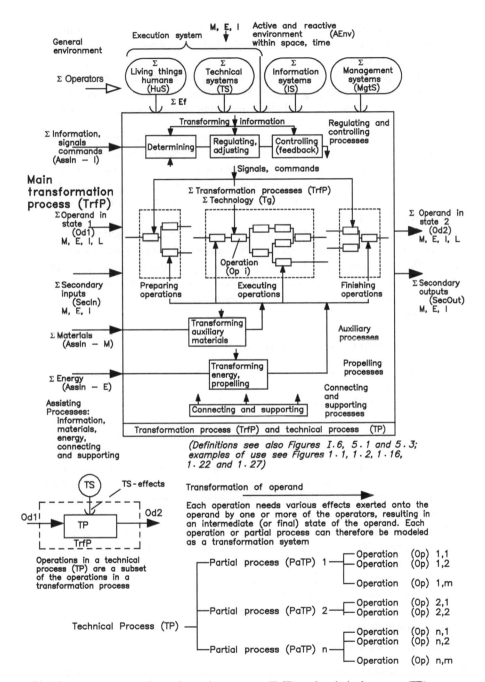

FIGURE 5.4 Structure of transformation process (TrfP) and technical process (TP).

1. A process (partial process, operation, used as a generic term) generally contains a preparing, an executing, and a finishing phase (the horizontal stream of processes).
2. The main processes (conditioned by the purpose of the TP) are always accompanied by assisting processes, that is, auxiliary, propelling, regulating and controlling, and also connecting and supporting processes.

The outputs of prior (partial) processes are the inputs of the following ones. The assisting operations combine with the main transformations to form a total process. These assisting processes are always present (with more or less importance) in a TP, and separating them provides a checklist for designing. They also allow recognition of possible failure modes of the TrfP and therefore an entry into failure analysis, by FMECA (see Chapter 8).

A hierarchical arrangement of TP into different levels of abstraction emerges when various characteristics are progressively concretized, such as the operand of the process (Od), the technology (Tg), the operators (Op), and others. This hierarchy typically moves from abstract to concrete during designing, a concretization hierarchy. Interdependence of processes and technical means is discussed when considering Figures 6.2 and 6.3.

5.3.6 Assisting Processes, Assisting Inputs

Figure 5.4 shows various assisting processes that are usually needed for a TP to be performed. Primary among these are the regulating and controlling processes, and the processes that provide auxiliary materials and energy; see the second point in Section 5.3.5.

Many TP can only run effectively if adequate assisting inputs (AssIn) are supplied through assisting processes. In Figure 5.3, the process of wire drawing can only progress effectively if lubricating oil (AssIn) is supplied to the wire–die interface. The assisting process (TP) "supply lubricating oil" thus fulfills an evoked function, and can generally only be recognized when the next more concrete step for the main TS-functions is decided.

Such an evoked function and assisting process usually needs an appropriate technology and an applicable TS, and can be regarded as a subsidiary design task for a partial TS—but this may be treated as a separate design problem in its own right. This hierarchical development of evoked functions can be shown in a "goals–means" or "functions–means" tree (see Figure 5.5).

5.3.7 Secondary Inputs and Outputs of Technical Processes

Apart from the desired and planned inputs and outputs (the operands), various undesired *secondary inputs* and *outputs* exist.

Secondary inputs to the TP, that is, inputs other than the desired operand and the assisting inputs, consist of material, energy, or information (M, E, I), and can be: *positive*—desired, as additional to the assisting inputs; *negative*—usually undesirable, *disturbances*, mainly as products of the environment, which may be acceptable up to

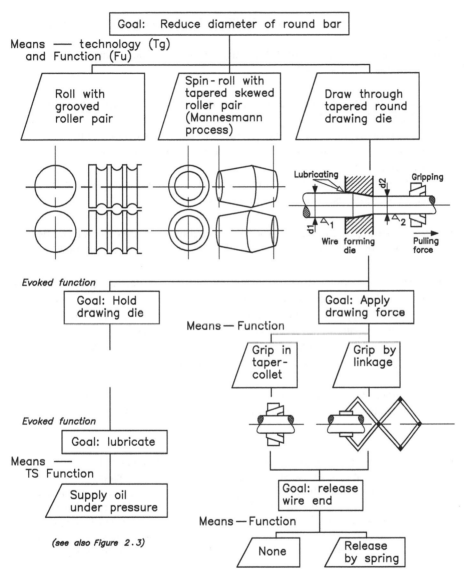

FIGURE 5.5   "Function–means" or "goals–means" tree.

a limit of significant disturbance to the process; above this limit it may be necessary to ensure safe functioning or safe shutdown of the process.

The TP, besides its desired operand in the output state, also produces *secondary outputs*, for example, waste, disturbances of the environment. Ecology and life cycle assessment [237,244] tries to study the results (mainly the adverse ones) from these secondary outputs and their influences on the environment. This demonstrates that most of the actions of operators on operands are really *interactions*; the reactions of the secondary outputs onto the operator and parts of the environment must also be considered. Life cycle engineering [62,160,186,582] tries to avoid or reduce these influences.

Considering the *systematics of TP*, it may be classified according to various viewpoints (see also Section 5.3.3):

1. According to class of operand: processes that transform an operand with predominant focus on its M, L, E, I.
2. According to the category of change: processes that transform an operand with respect to (internal) structure, (external) form, location (in space), time, or other operand property.
3. According to a hierarchy of abstraction: process phylum, class, family, or genus (basic process), specialized processes.

Four basic sorts of process, partial processes or operations are given special names according to four of the varieties of change mentioned in (2) above (see Figure I.20):

1. *Processing*: processes that primarily change the structural (internal) properties of the operand (conversion).
2. *Manufacturing*: processes that primarily change the form (mass, geometry, shape, constitution, tolerances, surface, etc.) of the operand.
3. *Transporting*: processes that change the space location of the operand.
4. *Storing*: processes that change only the time location of the operand.

Combining these types of processes with the types of operands shows the typical relationships of Figure I.20. Changes in other operand properties have not been given a generic name.

NOTE: These terms are familiar when applied to operations and processes that transform materials, but they may also be extended (to make the terminology more precise) to include transformations of energy and information. For processes where the human is an operand (e.g., education and training as experienced by the learner, or the human being as a passenger in a transportation process) such collective terms to describe types of processes are generally not used, but may be applied. For instance, medical operations—surgery, for example, amputation—may be regarded as "processing," fitting of prostheses as "manufacturing."

5.3.8 Trends in Developments in Time of Technical Processes

Developments of the technological sphere are accompanied by developments of a large variety of processes and technical means. In this technical and societal development of human cultures, the means to cause deliberate change, the human was progressively supported (e.g., initially by draft animals and their harnesses) or replaced in various functions as a result of technological developments (see Section I.10).

The indicated directions of development, namely instrumentation, mechanization, automation, and computerization (and its extensions) are historically and currently important. They influence some stages of the engineering design process (designing the desired or proposed TP/TS) in which decisions are made about the scope of the TP within the general TrfP, and distributing the functions of exerting energy (propulsion, drive) and control between man and machine.

According to the distribution of effects (actions) between the human beings and the TS (means), we can distinguish:

1. Manual processes (e.g., craft operations, use of tools) with a preponderance of human (and animal) action for both energy and control effects.
2. Mechanized processes (machine operations) in which the TS takes over the effects of supplying energy.
3. Automated processes in which the TS takes over most of the control effects.
4. Computerized (computer-automated) processes in which routine decisions are made by the TS (includes mechatronics).

The sequence of events of the process, its output and economics are influenced by the effects, which include the current level of information (e.g., the state of the art), the method of control of the process (particularly with respect to its organization, its management and the form of its work procedures), and the environment conditions (in the narrower sense of space and time conditions) under which the process is required to proceed.

An optimal division of tasks between the human and the TS should respect their different capabilities (see Figure 6.1).

Two distinct sorts of technological systems are

1. *Technical process systems*, consisting of *operations* as elements—referred to as technical processes (TP) (see Section 5.3).
2. *Tangible technical object systems*, consisting of *constructional parts*— referred to simply as technical systems (TS).

The requirements for the TS can be found by considering the TrfP/TP and their technologies. The TS is the subject of Chapter 6. The relationship between TP and TS is further explored in the NOTE in Sections 4.2 and 6.13.

According to Figure 6.3, the capability of performing TS-internal processes is described by the functions, and realized in the modes of action. This includes the main functions, and the assisting inputs and their functions, the effects as output toward the operand in the TrfP/TP, and secondary outputs. The connecting and supporting functions in the TS and connections to the fixed system (ground) provide the necessary closures of force circuits—the frames, housings, printed-circuit-board substrates, and so forth.

6 Technical (Object) Systems

Technical systems (TS) conventionally represent the technological sphere. Yet a technical process (TP) to implement transformations is also a system of a technical nature, a process with the character of a system (see Chapter 5).

The information presented in this chapter is mainly situated in and around the "west" quadrants of Figures I.4, I.5, and 12.6. It presents the core of the considerations for the tangible object system, the TS. It sets out the nature and constitution of TS. Together with Chapter 5, it is the theoretical basis for the design process in the "southeast" quadrant, and for the systematic and methodical design process recommended in this book, together with other design methods and heuristic information situated in and around the "northeast" quadrant.

6.1 TECHNICAL SYSTEMS—OBJECT SYSTEMS

The tangible object systems that serve humans include alarm clocks, bathroom sinks, coffee machines, roads, streetcars, bridges, books, motor vehicles, electric motors, telephones, televisions, factory equipment, and so forth. Each of these has to be designed, both for external appearance and functionality (industrial design), and for internal capability for performing (design engineering), with different proportions of these constituents (see Figure I.21). Object systems with a substantial engineering content are called technical systems.

NOTE: TS are "artifacts" that have a substantial engineering content. TS is the collective term for "a tangible technical object that is capable of performing a task for a purpose." *Machine systems*, as special cases of TS, use mainly mechanical modes of action (including fluids and fluidics) to produce their operational effects, mainly products of mechanical engineering. Systems are increasingly hybrids, especially for propelling and controlling, for example, electro- and computer-mechanical systems, mechatronics, robotics, MEMS (micro-electro-mechanical systems).

TS are connected to other systems, and are influenced by the surroundings, for example, moisture, vibration, heat, or light. Every TS is an element of a hierarchically superior system.

6.2 PURPOSE AND TASKS OF TECHNICAL SYSTEMS

Technical system deliver the *exerted effects* that perform the desired transformations of the operand in a TP. Properties and capabilities of TS and humans are shown in Figure 6.1A. An optimal TS exists when (among other considerations)

(A)

Humans	Technical systems
— Are capable of reaching important decisions based on limited information, can tolerate uncertainty	— Work only according to given orders, can only interpret when instructions for interpreting are clear
— Can deliver a correct reaction even under unexpected circumstances, but not consistently	— Unexpected conditions can lead to disasters (i.e., conditions unanticipated by humans)
— Can perform certain operations in various ways, important when the mechanism is damaged	— Number of operations performable by TS is limited, "learning" and "inferring" only within a limited instruction set
— Are flexible in work programming, adapt to changes, capable of learning from own experience and others	— Program changes are usually difficult and costly (exception: digital computers)
— Capacity for receiving information is limited, in quantity and rate	— Capacity of information channels can readily be increased
— Capability of sensing organs (receptors) is limited	— Possibilities of increasing the range of sensing parameters
— Power and attention reduces with time	— Almost constant power and precision
— Working capability requires certain conditions of temperature, humidity, pressure, stress, noise, and so forth	— Can be made for almost any expected environment conditions
— Thinking operations are relatively slow, with large probability of errors	— Rapid execution of (pre-programmed) logical operations almost without error
— Limited capabilities of memory and recall, working (short-term) memory about 7 ± 2 chunks of information	— Almost unlimited information storage, limited by ready access to the stored information
— Information can be complex, can readily combine and interpret, can abstract to formulate theories	— Forms of information basically simple, need full instructions to combine or interpret information
— Are subject to prejudices and bias, form opinions readily	— No bias, except those programmed in by humans
— Are largely governed by emotional factors, feelings, have senses of esthetics, ethics, responsibility, conscience	— No feelings, no emotions

FIGURE 6.1 Characteristic properties of humans and technical systems.

(B)   **Iconic representation**

Environment

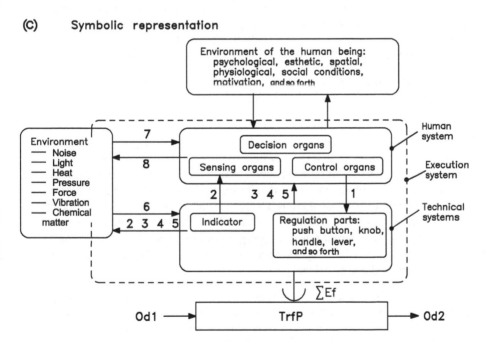

Legend

1 Operation (handling) and
   maintenance
2 Signals about progress
   of TS-operational and
   TS-internal process
3 Physical obstacles
4 Noise
5 Vibration, heat
6 Environmental influences
   on the TS
7 Environmental influences
   on the human being
8 Influence of human beings
   on the environment

(C)   **Symbolic representation**

FIGURE 6.1 Continued.

both the human operator and the TS work in their most appropriate manner, TS and humans interactions should be considered during design engineering (see Figures 6.1B and C).

Exerted effects are output effects delivered by the execution system (operators TS, Hu, and parts of AEnv) at *action locations*, and are at the same time input effects received by the operand. Figure 5.3C shows the main *effectors* and action locations for wire drawing, the gripping device that transmits the pulling action, and the die that causes the lateral forces.

NOTE: In the usual description: "a (manually operated) universal lathe produces rotationally symmetrical parts by a cutting operation of turning" (for a diagram of a lathe, see Figure 6.17, level 1, or a book on manufacturing methods). This expression of the transformation process "manufacturing," "turning," ignores the necessary exerted effects of the human operator— setting the cutting tool, chucking the workpiece, driving the feed motions, and so forth — without which a rotational part cannot result. The lathe, by itself, can only hold and rotate a workpiece as operand, the chuck is active. It can also move a cutting tool in a plane, the tool is an integral part of the TS "lathe." It is reactive when the technology of shear deformation of a small part of the operand produces a different surface, and chips—and only (for a manually operated lathe) when the human operator provides the necessary force/torque (energy) and regulating motions (actions, output effects). The available capabilities of a lathe to exert effects can be used to wind helical springs—a different process from cutting, with a technology of guiding and bending a wire, and a tool to perform a different transformation within the abilities of the lathe (and of the human operator). This justifies distinguishing the TP as (topologically) "external to" its operators.

The *action locations* may be points, surfaces, volumes, and so forth. The action locations of a universal lathe are the conical point of its (live or dead) center in the tailstock, the chuck (or faceplate, live center, and driver), and the cutting edge and faces of the tool—these are the *effectors* of the TS "lathe" that act on or react to the workpiece (the operand), that is, they are capable of performing the holding and cutting actions, the effector functions. The guideways between bed and carriage, and between carriage and top slide are in this view internal to the structure of the TS "lathe" (see the NOTE in Section 4.2), the capabilities are described by TS-internal functions (see Sections 4.4.1 [3c], [3d], and [3h]).

The tasks (and required duties) of the TS are (1) to receive selected main and assisting inputs—M (e.g., assisting material, lubricant), E, and I (signals, commands, data)—and (2) *within the TS at a desired time* to produce the desired output effects exerted onto the operand of the TP at the action locations—movement, force, heating, cooling, protection, warning, and so forth. The exerted effects contain a proportion of the materials and energy supplied as TS-inputs, the remainder must be rejected into the secondary effects and secondary outputs of the TS. The TS-internal process, as described by the TS-internal functions, progresses in *action chains*, according to the *mode of action* selected during designing (see Figure 6.2). The TS-functions are realized by organs and constructional parts.

NOTE: The TS-internal process (for most TS) takes place without the direct intervention of the human. The TS-internal process is distinct from the general TP described in Chapter 5, in which an operand topologically external to the TS is transformed using a selected technology (and a TgPc) by the effects from TS, HuS, AEnv (and indirectly by the other operators), with the purpose of realizing a certain more desirable state of the operand, as is shown in Figures 5.3 and 5.5.

We are, in fact, dealing with *two separate processes* that should be clearly distinguished:

1. As described in Chapter 5, the operand (M, L, E, I), *when present*, are transformed within the *transformation process* (TrfP or the TP) over a period of time from an initial state (Od1) toward a desired output state (Od2) (Figures 5.1, 5.3, and 5.4), which is indicated as the flow through

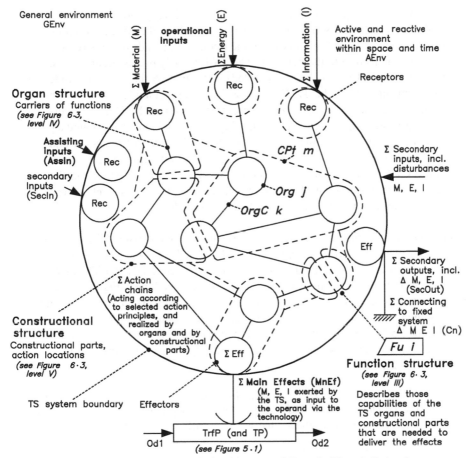

Each technical system exhibits several structures, consisting of different kinds of elements, for example functions (Fu i), organs (Org j) and organ connectors (OrgC k), constructional parts (CP m) and their relationships (see Figure 6.3).

FIGURE 6.2   Model of a technical system—structures.

the rectangular box. This transformation is based on a selected technology (Tg) and its technological principle (TgPc), and is performed with the assistance of and under the combined effects exerted directly or indirectly by HuS, TS, and AEnv, and usually indirectly by IS and MgtS, onto the operand.

NOTE: The wire in Figure 5.3 is the operand topologically external to the TS (wire-drawing machine), and the wire/die lubricant is auxiliary process material. The steam "inside" a steam turbine is the operand topologically external to the TS. The energy transformed by a gearbox is carried by the shafts topologically external to the TS (see the NOTE in Section 4.2).

2. In this chapter, inputs to a TS (M, E, I—signals/commands/data) are taken in by the *receptors*, and are changed by a *TS-internal process* at a certain time into effects (actions, outputs exerted by that TS at the *effectors*) of

the desired type (e.g., movement, force, heat, cooling, protection) directly to the operand. The TS may operate to *actively* deliver the effects, or it may operate *reactively* with effects when the operand demands it—this represents the connection to the TrfP/TP. The *TS-internal process* occurs whilst the TS is operating (or being operated) according to the selected mode of action *internal to* the TS (following the action principle, as shown in Figure 6.2), whether an operand is present or not. When the TS is not operating, it is still *capable* of performing the TS-internal and cross-boundary actions, and this capability is described by its *TS-internal functions* and TS-cross-boundary functions. The effects are exerted onto a process operand (in a TrfP or TP) at the desired place, at the action locations on the selected boundary of the TS, shown by the downward effects symbol from the operators to the TP in Figures 6.1 to 6.4 and others.

The *main TS-internal process actions* are always accompanied by *assisting actions* (*auxiliary, propelling, regulating and controlling*, and *connecting and supporting* actions, e.g., to the fixed system) that must be delivered to the TS or by one section of a TS to another. For instance, the wire-drawing machine in Figure 5.3 needs assisting materials (e.g., lubricating oil for its own mechanism), energy (to change into the pulling force and motion), and information as inputs. Secondary inputs and outputs occur. Each process within the TS demands preparing, executing, and finishing actions. These also constitute some of the classifying aspects for the structures of TS (Figure 6.2).

NOTE: The word *exerted* or *main effect* is used to designate an output task of a TS delivered at its effectors. The term *function* is used to designate the capability for performing a TS-internal action. The range and variety of effects that a TS can deliver are collectively termed its "functionality."

6.3 MODE OF ACTION OF TECHNICAL SYSTEMS—HOW THEY WORK

The capability of a TS to exert its main effects, conditional on its inputs (M, E, I) is determined by a definite *constructional structure* (see Figure 6.2), and embodies the organ and function structures as constituents of the elemental design properties. The constructional structure consists of a collection of structural and manufacturable elements—constructional parts—(including hardware, firmware, and software) that have an active mutual relationship. This relationship, the *mode of action*, determines in which way and by what means the exerted effects occur, and answers the question: "How does this TS operate, function, work?"

Because real TS are largely deterministic, this connection and mode of action results in a causal chain (an *action chain*), from causes to consequences (see Figures 6.2 and 2.4). The actions of the causal chain are based on *natural phenomena*, including: mechanical, electrical, chemical, biological, hydrostatics hydrodynamic, aerodynamic, mechanics of materials, analog and digital electronics, and so forth, and subsections and combinations of these.

All TS-internal processes have random variations. Reducing such variability can be useful, for example, helped by Taguchi experimentation (see Chapter 8).

In many instances an *action medium* assists in transferring actions between constructional parts or subsystems, for example, hydraulic or lubricating fluids, and so forth. For instance, "air" as the action medium "connects" a flying aircraft to the fixed system "earth"; and "gravitational attraction" opposing "centrifugal force" holds a space satellite in orbit.

The *inputs* (their quantities and other properties) of TS consist of M, E, I (Figure 6.2) and are delivered by HuS, other TS, and the AEnv (compare Figure 5.1C).

When designing the desired action chain, design engineers use their tacit, internalized knowing, and recorded information about phenomena, for example, from masters (see Section 9.2) and other records of their experience. A statement such as "a hoist works on a hydraulic principle" indicates from which area of knowledge and science the mode of action has been derived. The mode of action is based on an *action principle* obtained from abstracted experience and science. Generally several classes of modes of action can be found within each action principle.

6.4 STRUCTURES OF TECHNICAL SYSTEMS

Various abstract TS-structures can be formed with respect to differently defined elements, and recognized within a TS-boundary (see Figure 6.2). These structures describe and model some aspects of TS (see Figure 6.3). All structures exist simultaneously, and are elemental design properties (see Section 6.6.2) whether or not they have been deliberately designed.

The purpose of the structures is to provide models and tools for design engineering, especially for obtaining variants/alternatives at these levels of abstraction. Unique one-to-one "mappings" (correspondences) among the effects exerted by a TS, a TS-function structure, a TS-organ structure, or a TS-constructional structure, do not exist, except under some limiting conditions.

A particular effect may be exerted by various different TS-structures. This statement shows that generating alternative solutions or variants should be possible in several steps of the design process, for example, for any TS-structure (and for the TrfP, and its Tg; see Chapter 5). The structure of operations in the TrfP/TP, and allocation of operations between the HuS and TS have major influences on the TS-structures. Level I in Figure 6.3 as a "black box" represents the interactions of a TS with a TP; see also Figures 5.1 and 5.4.

6.4.1 PROCESS MODEL

The process model of a TS shows the TS-internal processes that exert the needed effects of the TS onto the operands whilst the TS is in the *state of operating* (Figure 6.3, Level II). There is necessarily a one-to-one correspondence between TS-internal operations and the TS-internal functions that define the *capability* for performing these operations. Because all these TS-internal processes must be performed by the TS itself, their *functions* (including the capabilities for exerting effects or performing actions) appear directly as tasks for the TS(s). For this reason, this TS-model is

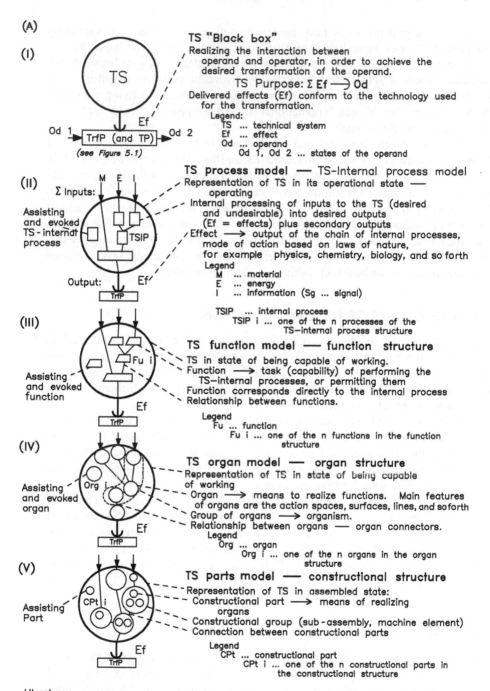

(A)

(I) TS "Black box"
Realizing the interaction between operand and operator, in order to achieve the desired transformation of the operand.
TS Purpose: Σ Ef $\longrightarrow$ Od
Delivered effects (Ef) conform to the technology used for the transformation.
Legend:
TS ... technical system
Ef ... effect
Od ... operand
Od 1, Od 2 ... states of the operand

(II) TS process model — TS-Internal process model
Representation of TS in its operational state — operating
Internal processing of inputs to the TS (desired and undesirable) into desired outputs (Ef = effects) plus secondary outputs
Effect $\longrightarrow$ output of the chain of internal processes, mode of action based on laws of nature, for example physics, chemistry, biology, and so forth
Legend
M ... material
E ... energy
I ... information (Sg ... signal)

TSIP ... internal process
TSIP i ... one of the n processes of the TS—internal process structure

(III) TS function model — function structure
TS in state of being capable of working.
Function $\longrightarrow$ task (capability) of performing the TS—internal processes, or permitting them
Function corresponds directly to the internal process
Relationship between functions.
Legend
Fu ... function
Fu i ... one of the n functions in the function structure

(IV) TS organ model — organ structure
Representation of TS in state of being capable of working
Organ $\longrightarrow$ means to realize functions. Main features of organs are the action spaces, surfaces, lines, and so forth
Group of organs $\longrightarrow$ organism.
Relationship between organs — organ connectors.
Legend
Org ... organ
Org i ... one of the n organs in the organ structure

(V) TS parts model — constructional structure
Representation of TS in assembled state:
Constructional part $\longrightarrow$ means of realizing organs
Constructional group (sub-assembly, machine element)
Connection between constructional parts
Legend
CPt ... constructional part
CPt i ... one of the n constructional parts in the constructional structure

Literature:
Hubka, V. (1984) *Theorie technischer Systeme*, Berlin: Springer-Verlag, Abb. 5.4
Hubka, V. and Eder, W.E. (1988a) *Theory of Technical Systems*, New York: Springer-Verlag, Fig. 5.4

FIGURE 6.3 Structures of technical systems (TS models).

(B) Example: machine vice in operational state — state of capability of operating, or being operated

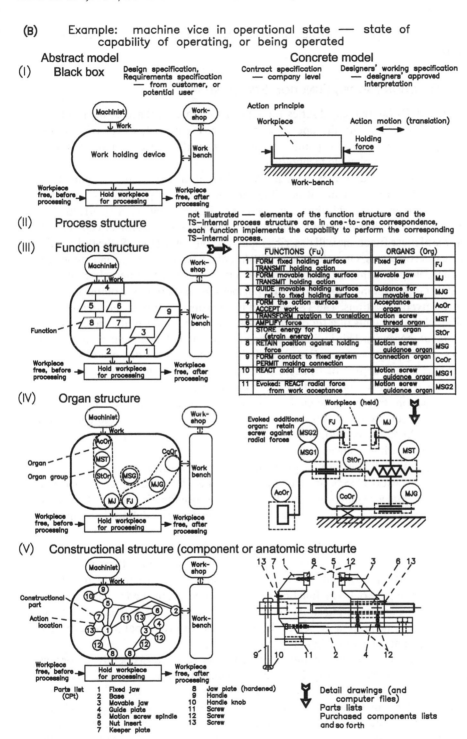

FUNCTIONS (Fu)	ORGANS (Org)	
1	FORM fixed holding surface TRANSMIT holding action	Fixed jaw	FJ
2	FORM movable holding surface TRANSMIT holding action	Movable jaw	MJ
3	GUIDE movable holding surface rel. to fixed holding surface	Guidance for movable jaw	MJG
4	FORM the action surface ACCEPT work	Acceptance organ	AcOr
5	TRANSFORM rotation to translation	Motion screw thread organ	MST
6	AMPLIFY force		
7	STORE energy for holding (strain energy)	Storage organ	StOr
8	RETAIN position against holding force	Motion screw guidance organ	MSG
9	FORM contact to fixed system PERMIT making connection	Connection organ	CoOr
10	REACT axial force	Motion screw guidance organ	MSG1
11	Evoked: REACT radial force from work acceptance	Motion screw guidance organ	MSG2

Parts list (CPt):
1 Fixed jaw
2 Base
3 Movable jaw
4 Guide plate
5 Motion screw spindle
6 Nut insert
7 Keeper plate
8 Jaw plate (hardened)
9 Handle
10 Handle knob
11 Screw
12 Screw
13 Screw

Detail drawings (and computer files)
Parts lists
Purchased components lists
and so forth

FIGURE 6.3 Continued.

redundant for designing, variants cannot be generated at this level compared to level III.

6.4.2 Function Model, Function Structure

The required TS-internal capabilities for actions are closely related to the needs for effects to be exerted onto the operands in a TrfP (Figure 6.3, Level III). During designing, the TS-internal capabilities can be derived from an established TrfP structure using the applicable technologies.

Each TS-function describes the required or desired (internal and cross boundary) *capabilities* of a TS that make it possible for that system to perform its (external) duties, its intended goal tasks, that is, to exert its main active and reactive effects. The main purpose of a TS is realized by the main functions, a sequencing of capabilities.

Some of these functions will describe inputs and outputs of the TS, the tasks of the receptors and effectors. Other functions will describe how the assisting and secondary inputs are changed within the tangible system into the desired TS-internal actions and secondary outputs, namely to convert an input measure into a required output measure under the given conditions. The TS-function is a unique (but usually not one-to-one) coupling of the elements of a set of *independent* input measures to the elements of a set of *output* measures. This is similar to the interpretation of a mathematical *transfer function* for dynamic systems. For example, a spindle with six degrees of freedom is restricted to one degree of freedom (rotation) by the function "hold the spindle rotationally (e.g., in bearings) and provide axial location."

The function structure is defined by its elements, the functions, and the relationships of these functions to each other. The function structure gives the engineer a means to evaluate the operational states of a (future or existing) TS, for example, using the engineering sciences.

The capabilities for performing partial tasks, the *functions*, are thus a consequence of the needs to exert effects onto the operand in the TP (see Chapter 5), and this constitutes the *purpose function* of the TS(s). The purpose function is one of the basic properties of the TS, that is, class Pr1 in the classes of properties defined in Figures I.8 and 6.8–6.10.

The general model of the *TS-function structure* is a collection and arrangement of the receptor functions, main functions, assisting functions, and effector functions of a TS. The basic model and content of this structure (the structure of functions) is illustrated in Figure 6.4. The function structure can be established at various levels of completeness and concretization, depending on the progress of the design work. This should be done preferably before the TS and its constructional parts have been established at that stage of design progress.

Good and complete consideration of functions (and consequently of organs and constructional parts) during designing is frequently decisive for the quality of the resulting system, and can indicate the possible failure modes of the TS, see FMECA and FTA in Chapter 8. These categories of function can act as checklists to verify that the considerations during designing are as complete as possible, leading to "right-first-time" designing. Therefore the types of possible functions should be defined. Many kinds and classifications of functions appear in the literature,

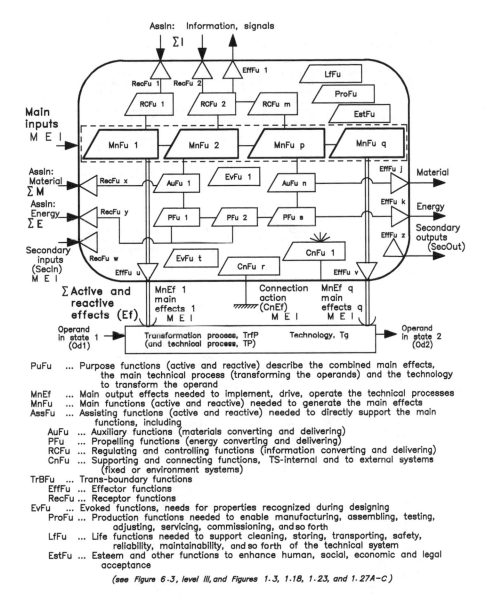

FIGURE 6.4 Complete model of the function structure of a technical system.

for example [228]. Only a few basically important types that will be used in discussions [319] are defined here. Three classifying viewpoints are particularly useful:

1. *Complexity of the function.* Each function may be assigned to a degree of complexity in a hierarchy (compare by analogy with Figure 6.5). The lowest degree is occupied by the elemental functions, those that cannot (usefully) be resolved or decomposed into more limited functions.

Level of Complexity	Technical system	Characteristic	Examples
I (simplest)	Constructional part	Elemental system, usually produced without assembly operations — appears in parts list	Bolt, shaft, bearing sleeve, spring, washer
II	Group, mechanism, sub-assembly	Simple system consisting of constructional parts that can fulfill some higher functions	Gear box, hydraulic drive, spindle head, brake unit, clutch, shaft coupling
III	Machine, apparatus, device, equipment	System that consists of sub-assemblies and constructional elements (components) that together perform a closed function	Lathe, motor vehicle, electric motor, crane, kitchen machine
IV	Plant, equipment, complex machine unit	Complicated system that fulfills a number of functions, that consists of machines, sub-assemblies, groups and constructional parts, and that constitute a spatial and functional unity	Hardening plant, machine transfer line, factory equipment, refinery, underground transportation system

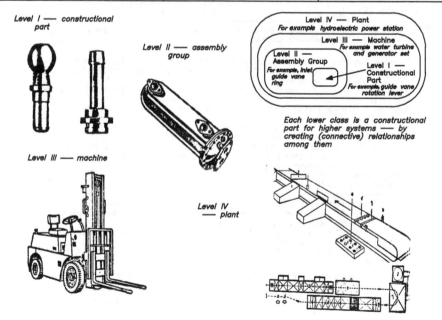

FIGURE 6.5　Technical systems classified by complexity.

2. *Degree of abstraction of the function.* Each function may exist at various levels between "concrete" and "abstract." This influences the number of possible or available organs (function carriers) that can be found (as means) to realize the function (as the goal). For example, if the given function is "change motion of . . . ," then the available means for solving covers a broad

range. As additional data about the function is increased (selected from the ranges of effects, conditions, operations, or operational means), the range of available means of fulfilling a function is narrowed, until a single concrete TS remains. The additional data mentioned here are the design properties, classes Pr10, Pr11, and Pr12, Figures I.15 and 6.8–6.10.

For a "function with i conditions," the order in which the conditions appear is arbitrary. The *behavior* of a real TS can be attained by a different arrangement and sequencing of functions, that is, the states of property classes Pr1A: functional, Pr1B: functionally determined, and Pr1C: operational properties, and their change over time.

The degree of abstraction and the degree of complexity are related. Resolving the functions into partial functions (i.e., functions of lower degree of complexity) is only possible and sensible when a more concrete level of abstraction has been obtained, that is, means (organs) to realize functions. This may be represented in a "function–means" tree, similar to Figure 5.5, one form of "function decomposition." Other characteristics are subconsciously neglected, either because they are implicit in the statement of the task, or they are assumed by tradition to be "fixed variables."

3. *Categories of purpose of the functions.* The various categories of functions that can be useful in designing a system include (see Figure 6.4):

   3a. *Purpose function* (PuFu) of the proposed TS, its essential capabilities, which include those effects that are needed from the execution system, the effects it applies, the chosen technology, and the TP—this is a composite. It is preferable to separate these elements, and to consider each on its own, as Figure 6.4 demonstrates.

   3b. *Main effects* (MnEf) as intended effect of the TS exerted by a technology to the operand.

   3c. *Main functions* (MnFu), internal to and at the boundaries (receptors and effectors) of the TS, as means to fulfill the purpose.

The main functions and main effects, with the selected mode of action that fulfill the purpose of the TS are necessarily accompanied by additional functions. These are essential, and ensure that the main function can be realized, or TS-operation supported. These additional functions are given below:

4a. *Assisting functions* (AssFu) that allow the main functions to fulfill their tasks

   i. *Auxiliary functions* (AuFu) that deliver assisting materials within the TS.

   ii. *Propelling functions* (PFu) that deliver energy.

   iii. *Regulating and controlling* (RCFu) functions that deliver information (data, signals, commands, etc.), perform measurements and comparisons with standard quantities, and give feedback for use within the system.

NOTE: This class of function is now often realized by a combination of transducers, digital electronic devices (computers, programmable logic controllers), and software—which are fully integrated into the TS(s), as required by mechatronics and robotics.

    iv. *Connecting and supporting functions* (CnFu) that keep the TS together in one unit (the local fixed system) and connect it to the fixed system.

4b. Transboundary functions (TrBFu)

    i. *Receptor functions* (RecFu) that allow M, E, I to enter a TS.

    ii. *Effector functions* (EffFu) that exert effects and allow M, E, I to leave a TS.

4c. Evoked functions (EvFu)

    i. *Production functions* (ProFu) that permit manufacture and assembly, concurrent engineering of the TS(s) and its manufacturing processes, and disassembly for maintenance and disposal (e.g., recycling)—the *evoked* functions are discovered during designing in the more concrete structures (i.e., constructional structure, organ structure).

    ii. *Life functions* (LfFu), properties-generating functions other than the production functions, usually *evoked* during designing, for example, those that permit life cycle actions/processes and achieve reliability, stability, and so forth of the TS in its operational process. This includes functions that avoid unintended phenomena [366]—potential malfunctioning may be detected in a function–means tree, a causal chain for that potential failure may be generated (e.g., by FTA or FMECA), and an evoked (supplementary) function may be recognized to overcome the potential hazard. An unintended result may be to reduce an intended input, or to produce an undesired output; which may be countered by (1) distribution functions, (2) inverse functions, (3) consecutive functions, or (4) compensatory functions [84].

    iii. *Esteem and other functions* (EstFu) that may be used to enhance human, social, economic, and legal acceptance of a TS.

For design engineering, a good formulation of the description of a function helps in solving the problem. Functions should be formulated as positives, should relate to only one property, and normally to one level of complexity [503]. Functions are inherited by subfunctions.

6.4.3 ORGAN MODEL, ORGAN STRUCTURE

A structure can be obtained by using the *action locations*, as elements (Figure 6.3, Level IV). These are usually *pairings* of locations that exist on two adjoining constructional parts, and constitute a coupling.

The action locations are arranged in a sequencing to fulfill the functions, an *action chain*, the relationships needed to perform the TS-internal processes. The action chains are realized in "function units," *organs* (or function carriers), an interface (a movable or fixed coupling) between two constructional parts. "Organ" refers to an abstract view of the constructional structure, and can exist at several levels of abstraction: organs can be distributed to form lower function units (suborgans), or functionally collected into *organisms* (see Figure 6.6). In an organ structure, an organ is usually represented without material embodiment and dimensions.

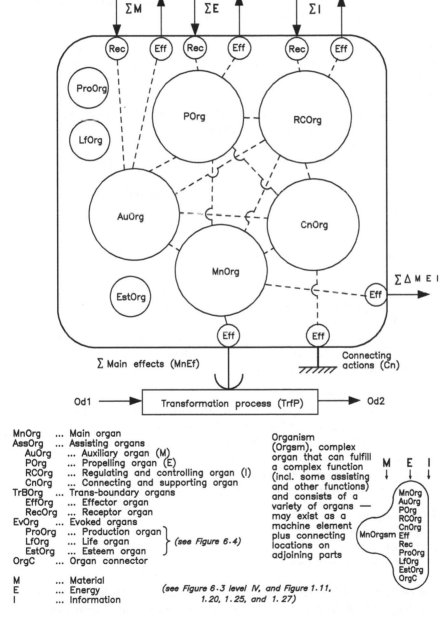

MnOrg   ... Main organ
AssOrg   ... Assisting organs
  AuOrg   ... Auxiliary organ (M)
  POrg   ... Propelling organ (E)
  RCOrg   ... Regulating and controlling organ (I)
  CnOrg   ... Connecting and supporting organ
TrBOrg   ... Trans-boundary organs
  EffOrg   ... Effector organ
  RecOrg   ... Receptor organ
EvOrg   ... Evoked organs
  ProOrg   ... Production organ ⎫
  LfOrg   ... Life organ   ⎬ *(see Figure 6.4)*
  EstOrg   ... Esteem organ   ⎭
OrgC   ... Organ connector

M   ... Material
E   ... Energy
I   ... Information

Organism (Orgsm), complex organ that can fulfill a complex function (incl. some assisting and other functions) and consists of a variety of organs — may exist as a machine element plus connecting locations on adjoining parts

(see Figure 6.3 level IV, and Figure 1.11, 1.20, 1.25, and 1.27)

FIGURE 6.6   Complete model of a TS-organ structure.

The purpose of an *active or reactive* organ is to allow an action to take place, an effect to be transferred. Organs also exist for other purposes, for example, for providing safety, allowing manufacture (e.g., to achieve different properties in sections of a subassembly unit, a vice jaw plate and a vice body), assembly (e.g., a gear box housing divided to allow assembly of the contained mechanisms), or other properties. Some organs carry *evoked* functions.

The various organs of the TS can be classified into typical categories, which are derived from the function structure (Figure 6.4) and its subdivisions (see Figure 6.6):

1. Main organs that lead to exerting the main effects on the operand, that is, that are required to perform the transformation of the operand in the TP
2. Assisting organs especially
   a. *Auxiliary organs* that deliver those additional materials necessary for the main organs (e.g., lubrication systems).
   b. *Propelling or energy organs* that transform and deliver the necessary energy in the desired form for all other parts of the TS, including prime movers.
   c. *Regulating, controlling, and automating organs* that process information, accept command and control signals, or provide a display of information.
   d. *Connecting and supporting organs* that provide connections among the different organs. This duty includes transfer of outputs from one organ or technical subsystem to another within the TS (e.g., energy, motion, electric current, chemical compounds), means to provide the spatial unity of the TS, and means to accept the supporting function as connection to the fixed system (e.g., frame or bed).

   For the user, the TS appears as a "black box," and important are those organs that connect the TS to the environment, the start and end of the action chains formed by the organs (Figure 6.4):

3. Transboundary organs
   a. For the inputs, the *receptors* (Rec) as organs transmit M, E, I (from the supplying devices and operators) inwards to the TS.
   b. For the outputs, the effectors (Eff) with their action locations deliver effects (M, E, I) and secondary outputs outwards to the operand, or connect and support to the fixed system.

   Additional organs that are recognized during designing:

4. Evoked organs
   a. Production organs
   b. *Life organs*
   c. Esteem and other organs

When OEM and COTS items are selected, they can be regarded as complex organisms, they carry their organ structures as constituents of their elemental design properties. The organs of interest are their connection locations to the remaining

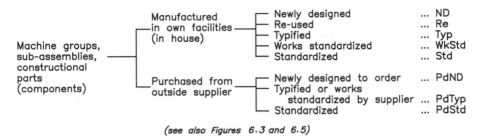

(see also Figures 6.3 and 6.5)

FIGURE 6.7 Technical systems (complexity I and II) classified by manufacturing location and degree of standardization.

system, the receptors and effectors, and their connective dimensions and other product specifications. They are represented on the "outline" drawings published by suppliers in their product catalogs, and are used in selecting and applying assembly groups and constructional parts that are purchased (see Figure 6.7).

The organ structure of a more complex TS almost invariably does not coincide with the constructional structure (see Figure 6.2, and NOTE at the end of Section 6.4.5). For instance, the cross slide of a lathe forms an assembly (a sub-assembly for the whole lathe), but the function "radially guide the tool" and the organ that performs this function are only partly represented in this subassembly, as the surfaces of the bearing pads. The mating surface, the guideway, is situated on the carriage, which is itself an assembly of constructional parts.

NOTE: These are *relative terms*: a central lubrication system is one of the auxiliary organs of a lathe. Lubrication is a system in its own right. Its purpose, an assisting function of the TS "lathe," is to deliver pressurized oil (as operand) to the usage positions (e.g., bearings, gear mesh points, etc.) during lathe operation. Its main organs (e.g., the pump for increasing oil pressure), propelling organs (drive motor), controlling (valves), and connecting organs (pipes) are detectable; this categorization is similar to that used for the complete lathe. A function for a hierarchically superior system acts as the TP of a subsidiary system, compare "watching TV," NOTE (b) in Section 4.4.2 and Figure 4.5.

6.4.4 CONSTRUCTIONAL STRUCTURE

This form is also referred to as anatomic or component structure (Figure 6.3, Level V). When describing the mode of action of TS in Section 6.3, "constructional structure" was used for the totality of its *constructional parts*, which by their *arrangement* (one of the relationships) work to connect and support the action chains of organs to fulfill the functions and cause the effects. Each constructional structure is mapped to the other structures (see Sections 6.4.2 and 6.4.3).

NOTE: The relationship between organs and constructional parts is complicated. Some sections of the boundary of a physical constructional part will constitute an action location, a partial organ (e.g., the bearing diameter and shoulder of a shaft); its form depends on the function of the organ. Others sections will not be restrained by TS-functions, for example, surfaces of constructional parts that are only in contact with a nonfunctional environment (e.g., air), where the forms may be chosen according to other criteria.

Every effect at any level of complexity may be realized by a *variety of constructional structures*. The possibilities of forming variations depend on the TS-internal properties, especially the design characteristics (input, mode of action), and the general design properties of the TS.

During realization (construction, manufacture, and assembly) of the TS, especially assembly, this structure is usually generated in a hierarchical fashion— various (manufactured) constructional parts (e.g., screws or shafts), and composite parts (e.g., electric motors or rolling bearings) are fitted together to produce sub-assemblies (subgroups), assemblies (groups), and modules, which are then fitted together step-by-step (hierarchically) to form higher level totalities. This procedure is clearly observable in assembly line production and leads to the need for some of the evoked functions and organs. Designing provides the information and representations to anticipate and allow these assembly processes, and all other life cycle processes.

Connections between constructional parts can be classified in several other ways according to their nature. For instance, mechanical joints can be classified into those that permit a movement in one or more degrees of freedom, releasable fixed connections (e.g., a bolted joint) and permanent fixed connections (e.g., a welded joint); compare Section 7.5.

Various *modes (or principles) of construction* (see Section 6.4.4) can be used as guidelines for establishing constructional structures, according to experience with similar systems: modular, monocoque, stressed skin fuselage (aircraft), cantilever (bridge), pretensioned (concrete), VLSI (computer chip), integrated, and so forth. A hierarchical arrangement of various constructional structures is related to the degrees of complexity of TS (see Figure 6.5).

The constructional structure of a single TS can thus be defined by various relationships (e.g., hierarchical grouping of subassemblies) according to the type and philosophy of manufacturing and assembling, but also according to the needs of maintenance, transport, storage, disposal, and other aspects (e.g., the mode of construction; see Chapter 4 and Section 6.4.4).

The constructional structure is described in various ways in different documents, mainly in the parts lists, in assembly and subassembly drawings, and in detail drawings, or their computer-generated equivalents (see Chapter 10). These documents serve to provide information to manufacturing (production) and assembling in order to realize the TS(s). Computer-resident documents can, where appropriate, be processed and transferred directly to computer-controlled manufacturing facilities, such as CNC (computer numerical control) machines tools, robotic assembly stations, or rapid prototyping equipment.

It is obvious that establishing and describing the constructional structure is the ultimate task of the designer (see Figure 2.1 and Chapters 2 to 4). Usually this must be done in consultation with manufacturing and other organization sections (e.g., sales, transportation, etc.), as recommended in simultaneous or concurrent engineering (see Chapter 8).

Differences in elements of a constructional structure occur as a result of

1. *Complexity* (see Figure 6.5): (1) constructional parts represent the lowest level of complexity, they are mainly simple parts that are not normally capable of further disassembly, but that can usually only be subdivided by destroying the part; screws, shafts, electrical condensers, resistances; (2) subassemblies of the parts, as intermediate stages of assembly; shafts with gear wheels, keys, bearings, and so forth, mounted; (3) assembly groups, modules of different complexity; lathe saddle, tailstock, headstock; (4) complete assemblies (TS) that can perform a task; a lathe.
2. *Assembly groups* that are formed considering certain aspects: (1) modules or assembly groups with a specified function (functional assembly group, organism) such as pump, gear box, electric motor; (2) assembly groups with other properties, such as assembly groups, module groups, replacement parts groups, construction groups (see below), or groups on which other groups depend regarding their dimensions (e.g., connective dimensions); (3) assembly groups combined into units of higher complexity.
3. Physical *action principle* (mode of action, kind of operand) (see also Section 7.5): (1) mechanical, for example, mechanisms; (2) electrical, for example, switches, transformers, conductors, insulators; (3) electronic, for example, computer controllers; (4) mechatronics, integration of electronic and mechanical functions/components, for example, antiskid braking systems on cars; (5) optical, for example, magnifiers, optical fibers, projection lenses; (6) hydraulic, pneumatic—fluid static, for example, pumps, cylinders, hydraulic motors; (7) thermo-fluid dynamic, for example, heat exchangers, compressors, expanders; (8) measuring elements, for example, transducers, sensors; (9) chemical, for example, osmosis membranes; (10) hybrids of these; and so forth.
4. *Breadth of application* in the TS: (1) universal constructional parts of the assembly groups for broader application (e.g., screws, gears), often standardized; (2) special constructional parts or assembly groups that are only applicable for certain TS or their families or species, for example, special modules.
5. Degree of *unification*, standardization (see Figure 6.7): (1) no unification, but either reused (from a previous application), or newly designed; (2) typified, company standardized; (3) standardized: national or international; (4) legally required.
6. Kind of *realization location* (see Figure 6.7), manufactured within own organization or purchased elements, OEM parts, COTS.
7. Relationships of the individual elements to one another: (1) spatial, *topological* relationships that depend on the arrangement of the individual elements in space. The relationships indicate the geometric situations and motions of the constructional parts relative to each other. The arrangement is visible in the design layouts and the assembly drawings, and is defined geometrically, including dimensional and geometric tolerances. Some computer-resident representations can "animate" mechanisms to

detect interferences and collisions (see Section 10.1). (2) *Couplings* in general realize the *functional* relationships among parts (e.g., element pairings). Several kinds of coupling are distinguished: (a) coupling can be either direct (e.g., electrical contacts), or indirect with the help of a coupling member (e.g., clutch); (b) according to degree of coupling, fixed or loose (e.g., movable) couplings of different degrees of freedom; (c) coupling can be achieved by mechanical, electrical (inductive, capacitive), magnetic, hydraulic (hydrostatic or hydrodynamic), or other effects; (d) according to the purpose of the coupling, we distinguish transfer and transformation of material, energy (power), movement (kinematic), force, and information (signal).

Couplings are closely connected with the functions and related organs of the TS. Construction functions and related organs are often *evoked*, they are needed to enable manufacture and assembly, and to realize other processes and properties. They depend largely on the mode of construction and arrangement, and include interfaces of the assembly groups.

Machine elements (ME) are constructional parts for machine systems, usually with arbitrary boundaries acting as design zones, "windows" [438], form-giving zones [314], and with material realizations of organs (see Section 7.5). Some elements are only found in particular branches of engineering, such as a piston, valve, heat insulator, hook, resistors, capacitors, inductors, transformers, transistors, and so forth.

Constructional parts can be decomposed into form elements (e.g., features) (see Figures 2.13 and 10.16), defined as bodies, surfaces, edges, points, or composite features, and are combinations of simpler geometric form elements, conditional on functioning or manufacturing. They serve as action locations to realize a certain function, and should be defined by their elemental design properties. Features are used in "solid modeling" and "surface modeling" computer representations to define the constructional structure (see Chapter 10). A constructional part is thus described as a system of form elements (features), their arrangement, dimensions, material, kind of production (manufacture), tolerances, surface condition, and other properties—the elemental design properties (see Figures I.8 and 6.8–6.10).

This description, which is typical of mechanical components, can be transferred to other branches of engineering. For instance, a field-effect transistor is regarded (by the designers using or specifying it) as a constructional part, even though the designer of the field-effect transistor may think of the substrate chip, or even the etching and doping layer, as a (constructional) part.

The descriptions till here reflect primarily the required functions (especially the main functions). The other external properties are mainly realized in the constructional structure.

The constructional structure comprises the physical *means* to realize the proposed (abstract) organs. The tasks of the constructional structure can be subdivided into several fields:

(F1)   The constructional structure must realize the (partial) organs, and their functions—the capabilities of the TS. It must physically create the causes for the effects of the TS, delivered by the effectors at the TS-boundary. The action locations should preferably be considered in the organ structure (as general configuration), and even in the function structure; they include energy sources (drives), regulating and controlling, and auxiliary effects, for example, lubricating, cooling.

The constructional structure must physically allow and support the necessary effects to be delivered by the human in the organization, for example, through displays (outputs of the TS, inputs to the human), and inputs to the TS via receptors in the TS (see Figures 6.1B and C). It must also consider and treat the secondary inputs and secondary outputs.

(F2)   The constructional structure should, in addition to those in (F1), fulfill all demands of the list of requirements (the design specification), that is, the external properties of the TS (see Figures I.8 and 6.8–6.10). This includes appearance, fitness for purpose, friendliness for its environment, fitness for producing and assembling, transporting and packaging, servicing and maintaining, repairing and renewal, conformance to laws and regulations (e.g., standards), and the possibilities of disposal at "life ended" (see Section 6.11.9)—many of these are realized through the evoked functions and organs, and their physical realizations.

(F3)   The third class is the internal demand on the constructional structure, especially its integrity, durability, and so forth. Problems emerge in connection with the chosen mode of construction (e.g., modular, fiber-reinforced composite, doped, and etched layer), and with the functions evoked by that mode of construction, for example, transmitting, joining, covering, separating, storing, unifying, sealing, conducting, protecting, securing, and so forth. The spatial unity of the TS must be achieved (connecting parts together within the TS), and the connection with the fixed system (supporting, earthing, grounding) realized.

One of the tasks of the engineering sciences is assisting in parametrizing these capabilities.

NOTE: These tasks of the constructional structure should be the subject of the list of requirements for that design problem (and its subproblems). These tasks, and the classes of properties of TS, indicate the method for generating design specifications (see Chapters 2 and 4).

Fulfilling the requirements depends on the constructional structure, in particular on the design properties of the constructional parts. Designers should find and establish the optimal constructional structure for these requirements and their design situation (see Chapter 3).

The difficulty of establishing individual design properties lies in the large quantity of the requirements that are asked of the constructional structure. Contradictions may be overcome by compromise, trade-offs, or avoidance, for example, using TRIZ.

The constructional principle—the *mode of construction*—is a characteristic that describes and distinguishes that constructional structure from other structures. It is a property of an assembly group (e.g., modular mode of construction), if it is formulated according to functions.

According to the number of constructional parts, an integral or differential mode of construction, or a monolithic or assembled constructional structure may emerge. A typical arrangement of the constructional elements results in an open or closed mode of construction.

From the raw material and kind of production we can distinguish sheet metal, riveted, welded, die cast, plastic transfer molded, plastic lay-up molded, sandwich, and many other modes of construction. Modes of construction in buildings (civil engineering) include half-timbered, pile-founded, prefabricated, fiber reinforced (or prestressed) concrete, and so forth.

From a methodical and systematic point of view, this is another plane for generating, establishing or finding alternatives ("Which modes of construction are possible?"), and thus for possibilities of optimization, especially with respect to appropriateness for the design situation.

In general, the term "mode of construction" is currently not used uniformly and the possible manifestations are not accurately defined. This question should be pursued in the TS-"sorts," to explore the possible modes of construction through a suitable evaluation.

6.4.5 RECAPITULATION

Within any TS, various structures and models can be recognized, depending on which aspects which are considered. Some of these structures can help in finding alternative and variant principles and configurations. The most important for engineers are the function, organ, and constructional structures (see Figure 6.3)—using these elements and showing the relationships among them and to the outside of the system. Examples of structures are shown in Figure 2.18.

Figure 6.3A shows a set of structures, and (Figure 6B) examples to illustrate their interpretation. Starting from the most concrete representations of the TS, the detail and assembly drawings, a movement in this diagram *upwards* through the levels of abstraction (analytically) yields more abstract models of the TS under consideration—a direction of *causality* (Figure 2.3). These more abstract models will progressively include systems with different constructional structure, in configuration and parametrization, and in constructional and action principles. *Configuration* refers to the relative location of surfaces and features of a proposed component, *parametrization* adds specific sizes and dimensions. Movement *from right to left* yields more abstract representations of the same structure. *Designing* a new TS generally implies moving to the right and downwards (synthesizing) through some or most of these structures—a direction of *finality* (see Chapter 2 and Figure 2.3).

NOTE: The mapping between a list of requirements and a transformation process is complex, and almost never one-to-one. The transformation process (TrfP), and its structure, allocation of tasks to humans and TS, technology, required effects, and the TS-function structure may be regarded as supplying the *meaning* (design intent) for a TS. The organ structure and constructional structure provide the *mechanism* to realize the meaning. The mapping between "meaning" and "mechanism" is almost never one-to-one. The mapping between organs and constructional parts can only be one-to-one if both structures are absolutely complete and detailed—then the abstract organ structure (left column in Figure 6.3B) maps inversely onto the abstract constructional structure, that is, the organs are represented by the nodes and the constructional parts the arcs in the *organ structure*, the organs are the arcs and the constructional parts are the nodes in the *constructional structure*. Such completeness is almost never necessary for designing the organ structure, therefore a 1:1 mapping to the constructional structure is hardly expected.

6.5 ASSISTING AND SECONDARY INPUTS AND OUTPUTS TO THE TECHNICAL SYSTEM

The short- and long-term behaviors of a TS is influenced by *assisting and secondary inputs* to the TS which include (compare Section 5.3.7): (1) *positive*—beneficial inputs to the TS, desired and necessary, M, E, I, for example, oil to assist the TS-internal processes and (2) *negative*—usually undesirable inputs to the TS, *disturbances* that are generally not under the deliberate control of the engineering designers; they appear mainly as products of the environment, which may be acceptable until they significantly disturb the functioning of the TS; above this limit it may be necessary to ensure safe functioning or safe shutdown of the TS.

TS cause *secondary* outputs. The danger of damaging emissions and disturbing outputs must be examined for all phases of the life cycle (Figure 6.14).

NOTE: It is usually possible and desirable during the design process to generate a model of the TS-environment in critical circumstances and thereby to establish and model all the likely influences on or from the TS. The properties of the system should then be chosen to take account of the influences of the environment on the system, that is, to resist the negative *secondary inputs*, to avoid operational failure and *disturbances*, assisted by methods such as FMECA and FTA. The behavior of the system must usually be ensured, for example, when disturbances occur, the system should "fail safe."

Noise, vibrations, waste materials, and heat are also usually generated during TS-operation. These *secondary outputs* (pollutants and disturbing effects on the operand and the environment) depend on the mode of action of the TS and other factors. The tasks of engineering designers include assessing these secondary inputs and outputs, keeping records about these considerations (for audits and possible liability rulings) and suggesting solutions.

For this reason, the influence of potential disturbances from inputs or emission outputs in each TS-life phase (Figure 6.14) should be examined and anticipated in all design phases (e.g., by life cycle assessment [62,160,186,237,244,582]), to either avoid them, reduce them to an acceptable minimum (e.g., by additional functions), or use them for a different purpose (e.g., by recycling, inside or outside the TrfS).

6.6 PROPERTIES OF TECHNICAL SYSTEMS

A *property* is anything that is possessed (owned) by an object (a TS). Each *constructional structure* (or TS) is the *carrier* of properties or classes of properties (see Figures I.8 and 6.8–6.10).

(A)

	Symbol	Class of properties	Typical questions about the class	Groups or examples of property class — Emergent properties of TS(s)
External properties — (see Figures I.8, 6.9 6.10 and 6.15) — processes of TS(s)–life cycle — Particular phases — *TS(s)–purpose properties (Pr1)* / *TS-operational process*	FuPr EfPr (Pr1A) – LC6	Functions properties / Effect properties	What does the TS(s) do? What capability does the TS(s) have?	Main function, Assisting functions, Auxiliary function, Propelling function, Regulating/controlling fu. Connecting function
	FuDtPr (Pr1B) – LC6	Functionally Determined properties	What conditions are characteristic of the function?	Power, speed, size; Functional dimensions; Load capacity
	OppPr (Pr1C) – LC6	Operational properties	How suitable is the TS(s) for its operational process (usage, working)?	Operational safety, Reliability, life, Energy consumption, Space occupation, Maintainability, Adjustability, Modularization
	MfgPr (Pr2) – LC4	Manufacturing and other origination properties	How suitable is the TS(s) for manufacture?	Manufacturability, Assemblability, Manufacturing quality
	DiPr (Pr3) – LC5	Distribution properties	How suitable is the TS(s) for transport, storage, packaging, and so forth?	Transportability, Storage suitability, Packaging suitability
	LiqPr (Pr4) – LC7	Liquidation properties	How easy is the TS(s) to liquidate, dispose of, recycle, re-use?	Re-cycling
Particular operators of each Phase — Process of TS(s)–life cycle (see Figure 6.15)	HuFPr (Pr5)	Human system factors related TS(s)-properties	How is the TS(s) to be operated, what influence does the TS(s) have (directly or indirectly) on human beings (esthetic, senses, comfort, danger, endurance, and so forth?	Operator safety, Way of operating, Secondary outputs, Requirements for human attention, Form, color, surface
	TSFPr (Pr6)	Technical Systems factors related TS(s)-properties	What TS can be used for the TS(s)–life cycle process? What other TS can cooperate?	Manufacturing equipment, office machinery, and so forth
	EnvFPr (Pr7)	Environment factors related TS(s)-properties	Are harmful outputs expected? What cultural and societal effects can occur? What laws (and so forth) must be followed?	Cultural norms, societal expectations, pollution, ecological loads, and so forth; Danger of wastes
	ISFPr (Pr8)	Information System factors, "Know—how" related TS(s)-properties	What information and know-how is available? Are instructions sufficient? What laws, codes of practice, standards exist?	Library, publications, standards, patent clearance, legal requirements
	MgtFPr (Pr9)	Management, economics, societal, goals, organization, personnel related TS(s)-properties	What organizational, planning, management influence exist? How economic is the working and manufacturing process? When can the TS(s) be delivered? Manufacturing quantity?	Management procedure, Operating and Manufacturing costs, effectiveness, Manufacturer recommended price, Delivery capability, time, Quantity production
Internal properties — For all external properties of TS(s) — causes of all external properties	DesPr	Design properties	With what means are the external properties (classes Pr1 — Pr9) realized?	Legend: TS ... TS as Operator of (partial) process; TS(s) ... TS(s) as Product of organization
	Pr10	Design characteristics	What technological principles, action sites, and so forth are available?	Structure, Form, Shape, Dimensions, Materials, Type of manufacturing, Tolerances, Surface quality, and so forth
	Pr11	General design properties	Which engineering sciences, and so forth are applicable?	
	Pr12	Elemental design properties	What structures, arrangements, elements/parts are suitable?	

(Sub-classes to classes Pr2—Pr9 are listed in Figure I.8)

FIGURE 6.8 Classes of properties of technical systems.

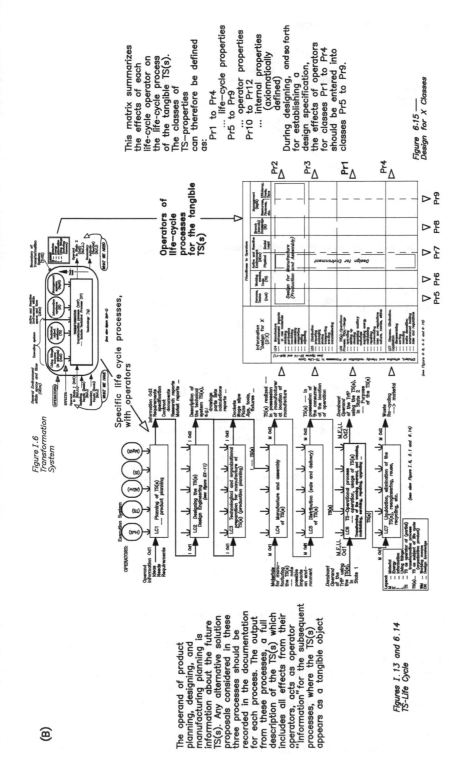

Figure I.6
Transformation System

This matrix summarizes the effects of each life-cycle operator on the life-cycle process of the tangible TS(s).
The classes of TS–properties can therefore be defined as:

Pr1 to Pr4
... life-cycle properties
Pr5 to Pr9
... operator properties
Pr10 to Pr12
... internal properties (axiomatically defined)

During designing, and so forth for establishing a design specification, the effects of operators for classes Pr1 to Pr4 should be entered into classes Pr5 to Pr9.

Figure 6.15 ——
Design for X Classes

Operators of life–cycle processes for the tangible TS(s)

Specific life cycle processes, with operators

The operand of product planning, designing, and manufacturing planning is information about the future TS(s). Any alternative solution proposals considered in these three processes should be recorded in the documentation for each process. The output from these processes, a full description of the TS(s) which includes all effects from their operators, acts as operator "information" for the subsequent processes, where the TS(s) appears as a tangible object

Figures I.13 and 6.14
TS-Life Cycle

FIGURE 6.8 Continued.

(C) **Links for existing systems**

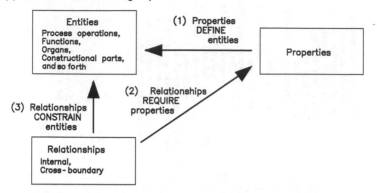

(D) **Links for existing systems and systems to be designed**

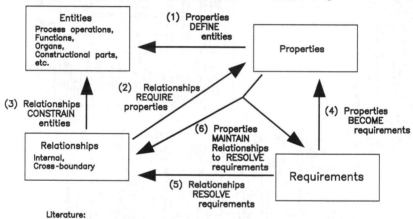

Literature:
Smith, J. and Clarkson, P.J. (2005) "Design Concept Modelling to Improve Reliability",
Jnl. Eng. Design, Vol. 16, No. 5, Oct., pp. 473–492.

FIGURE 6.8   Continued.

NOTE: A clear separation between theory and method is preferable for this classification
(see also Section I.9.2.3). For a theoretically complete classification for an *existing*, or finally
designed, "as is" state TS, "west" hemisphere of Figure I.4, and so forth, the primary classes
of properties are given below :

1. *External properties*, derived directly from the tangible parts of the TS-life cycle
(see Figures I.13 and 6.14—LC4 manufacture, LC5 distribution, LC6 operation, and LC7 liquidation), and the five operators of each of these life cycle
stages (HuS, TS, AEnv, IS, and MgtS), as shown in Figure 6.8B, which deliver
the primary classes Pr1 to Pr9. The Purpose Properties (for LC6) are chosen as the
first class(es)—if these are not sufficiently fulfilled, the TS is not really useful.
2. *Internal properties* can be axiomatically defined, in three classes, Pr10 to Pr12.

In the first three transformation processes of a typical life cycle, LC1 to LC3; see
Figures I.13 and 6.14, the operators of each TrfP establish the properties of the future

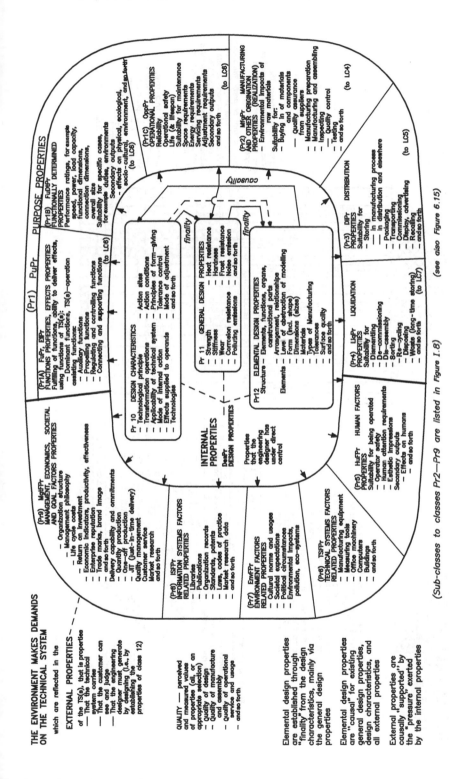

FIGURE 6.9 Relationships among external and internal properties of technical systems.

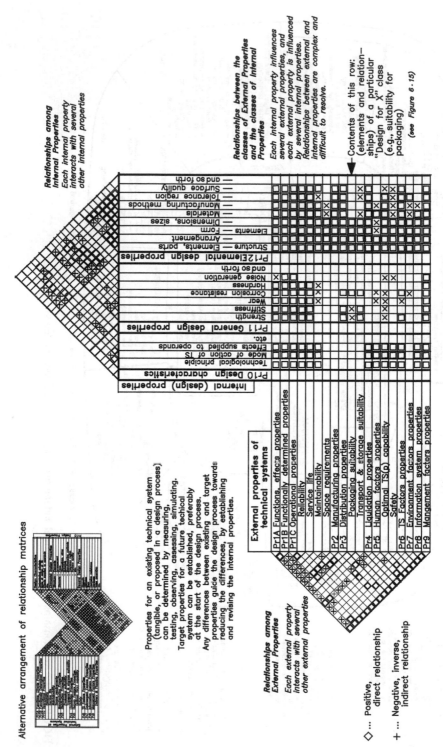

FIGURE 6.10 Technical systems: survey of relationships among classes of properties.

TrfP(s), TP(s), and TS(s), complying with the constraints of the design specification and of the life cycle phases. Designers consider various alternative candidate
solutions, evaluate them, select among them, and transmit the information about the
final choice (and *only* the final choice) to the transformation processes LC4 to LC7,
that is, in which the final choice TS(s) exists in a tangible form. For manufacturing,
the IS is a direct operator (in the execution system) of the manufacturing TrfP, delivering engineering drawings (what is to be made) and production planning documents
(how is it to be made); see Figures I.13 and 6.14, and the NOTE in Section I.9.1.2.
The first three TrfP thus do not need to be considered in the theoretically complete
classes of properties of existing TrfP and TS, nor do the effects of their operators.

From Figure 6.8C, "entities" result from "relationships" and "properties" [517]:

1. Properties *define* entities: Properties are possessed by entities, and
   completely define entities.
2. Relationships *require* properties: For relationships to make use of behavior,
   the entity must have that property. It is through relationships that the use
   of entities is bounded.
3. Relationships *constrain* entities: Relationships in some way constrain the
   use of entities since each entity must maintain relationships with other
   entities.

Entities, properties, and relationships thus belong to the theory of TS.

We also wish to clearly separate an actual transformation system, and design
engineering (compare Section I.13). For a *not-yet-existing* TP(s) and TS(s),
"as should be" state, "east" hemisphere of Figure I.4, and so forth, or an advance
prospect for buying an existing TP and TS, the requirements can be derived from the
same classes. From Figure 6.8D [517]:

4. Properties *become* requirements: Groups can be combined into hierarchies,
   and a property in a parent group is converted into a requirement in a child
   group.
5. Relationships *resolve* requirements: Relationships make use of the entity
   behavior to resolve requirements, such that each requirement is assigned
   to at least one relationship, and each relationship is assigned to at least
   one requirement.
6. Properties *maintain* relationships to *resolve* requirements: Properties indirectly resolve requirements by maintaining relationships that directly
   resolve requirements. Thus the property maintains a relationship such that
   the relationship can resolve a requirement. To determine what properties an entity must have, its relationships must be combined with their
   requirements.

"Requirements" belong firmly to design processes and their theory. Any theoretically based *method* must be adapted by the users to their situation. Any additional
classes of properties belong into methodology and methods. The classes of properties from theory are often not specific enough as guidelines for setting up a list of
requirements. Other considerations are entered pragmatically as the secondary classes

of properties in Figure I.8, and each of these should be consistent and complete. Different arrangements of these subclasses are possible. A distinction according to ISO 9000:2000 [10] into *inherent* and *assigned* properties does not appear particularly useful (see the NOTE in Section I.9.2.4).

In logic, "properties" are treated by unitary predicates. Ordinary language uses predicates to describe properties. In philosophy, properties are termed *attributes*. For technical products, *parameter* is used for those properties (usually measurable) that provide a description of the purpose functions, and information about numerical values and their units of measurement. In engineering, a coordinated collection of measures, or an indication of how one property (dependent variable) changes in response to other properties (independent variables), is termed a *characteristic* (see also ISO 9000:2000 [10], articles 3.5.1 and 3.5.2). "Characteristic" names a class of internal design properties; Pr10.

The *classes* of properties shown in Figures I.8 and 6.8–6.10 are complete, but the examples listed are incomplete, and differ according to the TS-"sort" to be considered (see Figure I.14). The properties need to be expanded for each TS-"sort," and for particular TS. These properties include the societal, cultural, environmental, legal, and many other influences and implications from designing, making, using, and disposing of TS. The terminology is summarized in Figure 6.11. Synonyms are characteristics, attributes, features, and so forth.

The *primary properties* of a TS, those on which all others depend, are termed the *design properties*, classes Pr10 to Pr12. They consist of and are carried by the structures, and their elements and relationships (see Section 6.4). In the elemental design properties, class Pr12, the property "structure" has manifestations (see Section 6.6.1) of function structure, organ structure, and constructional structure—which has sub-manifestations of preliminary layout, dimensional layout, and general assembly with details, parts list, instructions, and so forth.

The constructional structure consists of the constructional parts (characterized by their *elemental design properties*, class Pr12, including form, size, material, manufacturing method, tolerance, and surface finish), the arrangement of constructional parts (the relationships among the constructional parts), and their state after assembly and adjustment (evident in the *general design properties*, class Pr11, and the *design characteristics*, class Pr10).

One implication is that both a bicycle and a digital microcomputer can be described by the same *classes of properties*. Their appearances are different; the property "appearance" does not have the same manifestations in both systems. Their power requirements (values) are different, one requires electrical power and the other human power (different manifestations), one continually needs 200 to 500 W (or zero, if switched off), the other will receive a maximum of about 750 W—the limit of human capability. Individual TS differ in the manifestation and value (magnitude, measured on an agreed scale) of these properties (see Section 6.6.1).

The importance of properties can be seen from their connection with other terms:

1. The *state* of a TS is given by a suitable aggregate of the values of all measurable and assessable properties, and at a given point in time.

Term (symbol)	Definition	Examples
Property (Pr)	That which is possessed (owned) by an object	
Characteristic	Certain properties from which an object can be recognised, that serve to describe it, and that differentiate it from other objects	
Object-specific characteristic	A characteristic that describes the object independent of its nearer surroundings	
Object	Unit (tangible or process) system that can be uniquely described by its object-specific properties and their manifestations and values	Plant, machine, assembly group, constructional part, process
Object group	A group (set) of objects related as a family by its common object-specific characteristics	Machine tools, sort of manufacture
Requirement	Property that is required to be fulfilled during the origination phase of the object to satisfy the requirements. "What is the TS-task?"	Suitability for manufacture
Manifestation	According to the sort of property, this can be an attribute (qualitative statement), and can carry a value (quantitative statement, a scale of measurement exists)	Good, advantageous road fuel consumption, or brake specific fuel consumption
Value	Measure, value, state, condition, goodness of the property	m, kg, s, A, W
Quality	State of an object (unit) related to its suitability to fulfill the given requirements ('how'?). Measured on selected evaluation criteria	Sufficient manifestation compared to requirements
Classes of properties for technical systems (TS) External	Suitability of a TS for individual phases of its life cycle, and its social and economic existence	Adequate functioning and appearance
Internal	Properties that cause the external properties ("with what"?)	
Design characteristics	Characterize the class of the selected sort of realization (embodying experience)	Mode of action (electric) mode of construction (modular) action principle
General design properties	Properties that appear in all designed systems and that decide about other properties, especially engineering sciences	Strength, wear resistance, corrosion resistance
Elemental design properties	Means to achieve all other properties	Structure, arrangement, dimension, material, tolerance, surface
Priority Essential (substantial)	Without which an object cannot exist	
Unessential (incidental)	Which are not essential for the existence of an object	

FIGURE 6.11 Terminology of TS-properties.

2. *Behavior* is the succession of states that the TS assumes in response to a stimulus.
3. A similar aggregate of suitably selected qualities (closeness of a property value to its ideal state) represents the *total quality* of the object (see also Section 3.5.4).
4. Individual properties are usable as *evaluation criteria* for evaluating the object (only if these values are measurable or assessable in that state of the object); see Chapter 9.
5. *Requirements* placed on an "as should be" TS(s) represent its future properties.

To understand the term "property" it is important to note that a TS is the carrier of all properties, independent of whether these properties have been recorded in the list of requirements (the design specification) or not; and independent of whether they were specifically considered during design engineering.

Even if suitability for transportation has not been stated as a requirement, the realized (manufactured and assembled) TS is more or less suited to being transported—its size or its mass may be small enough or too large.

Industrial design (see Sections I.1 and I.10, and Figure 7.10) is primarily concerned with property class Pr5, the human factors properties. These properties include impressions received by the human senses: appearance (visual/aesthetic impression), human–machine interaction (ergonomic—displays and controls), noise (aural), smell, touch (haptic properties [324]), thinking (rational), and feeling (emotional). The haptic properties include tactile (touch), kinesthetic (feeling of motion), cutaneous (skin as sense organ), proprioceptive (position, state of the body), and force feedback.

6.6.1 EXTERNAL PROPERTIES—CLASSES PR1 TO PR9

The ability to exert effects is not the only significant property of TS. Primarily, these effects must be exerted by TS with the necessary operating parameters, for example, power, speed, travel distance, voltage, current, flow rate. TS must exhibit other properties or groups of properties, for example, being able to operate or being operated with ergonomically acceptable facilities, satisfactory maintainability, durability, reliability, appearance, transportability, and so forth, within an envisaged environment, and these are to some extent established by an industrial designer.

Figure 6.8A, contains the classes of properties, a set of typical questions that ask about the specific properties within each class, and examples to illustrate each of these classes of properties. For instance, the question: "How suitable is the TS for the TS-operational process (the TP)?" should bring answers that characterize the quality of the TS with respect to its operational life, behavior, operational safety, reliability, durability (life span, life expectancy), energy consumption, space requirements, maintainability, and so forth. Property class Pr1, and its subclasses, are related to the purpose of the TS; classes Pr2 to Pr4 are derived from the TS-life cycle, classes Pr5 to Pr9 relate to the operators of each of the TS-life cycle processes (see Figure 6.15) and therefore (directly and indirectly) to humans and their society (see Figure I.15).

Relationships among the classes of properties are shown in Figure 6.9 as a visual impression, and the strengths of interaction are illustrated in Figure 6.10 as inter-action matrices. A few sample properties are listed within these individual classes shown in Figures 6.8 to 6.10. The life phases of a TS are illustrated in Figures I.13 and 6.14. The properties of a TS (see Figures I.8 and 6.8–6.10) for these life phases are: suitability for realization, distribution, operation, repair, maintenance, liquidation, and so forth.

6.6.2 INTERNAL PROPERTIES—CLASSES PR10, PR11, AND PR12

The inside area of Figure 6.9, and the right-hand block of Figure 6.10 contain the three classes of internal properties. They are the core properties of TS. These properties are usually not readily visible, interesting, assessable, or measurable by a nontechnical person. Most users of a TS will regard them as a "black box" and will not make contact with the internal properties.

The internal properties are created by the engineering designer using design char-acteristics (Pr10), general design properties (Pr11), and elemental design properties (Pr12). Engineering designers have the task of establishing the individual properties in these classes during their design work.

Class Pr10 of internal properties consists of *design characteristics*, those aspects of the TS-structures that are invariant with respect to any external and internal influences—structural, functional, and technological principles. They deliver know-ledge (including heuristic values and principles) based on experience [98,556]. These are established directly by the engineering designer during design work, their selection is a useful level for generating alternative solution proposals. Figure 6.12 con-tains examples of the design characteristics of TS (TS-characteristics), with their manifestations.

Design characteristics influence most other properties. Proposed solutions to ful-fill the same task (list of requirements, design specification), for which different design characteristics were chosen (e.g., mode of action, or constructional structure), can have large technical and economic differences (e.g., a chain saw driven by an internal combustion engine or an electric motor).

The recommended methodical and systematic design procedure uses the infor-mation about TS-structures (see Figures 6.2 and 6.3). They help designers to establish the design characteristics, and determine the levels at which designers can find alter-native solution proposals and principles of operation. The process of establishing the design characteristics is treated in the design operations (see Chapter 2).

Class Pr11, *general design properties*, characterizes the TS-internal reactions to applied (internal and external) influences and consists of the applications of the engi-neering sciences, of heuristic knowledge about a particular system, of experience information as recorded in the technical literature, of tacit experience knowledge possessed by the individual designer, and so forth.

Class Pr12, *elemental design properties*, directly describe and define the TS-structures, their elements and arrangement (mutual relationships) (see Figure 6.3). The constructional parts are described by their elemental design properties, typi-cally their form (including shape), dimensions, materials, sorts of manufacturing

Viewpoints design characteristics	Manifestations (examples)	Viewpoints design characteristics	Manifestations (examples)
Requirements (external properties, incl. technical, societal, economic, etc.)	Fixed requirements and/or requests, constraints, and so forth from the design specifi cation, can be discretized or negated	Ordering viewpoints:	Parameter types:
		— Number	Single, double, ... multiple
		— Arrangement	Symmetric, unsymmetric, inside, outside, above, below
Transformation Process (TrfP):			
Dominant Input and Output quantities, operands (of black box)		— Form element	Plane (plate, prism), curved (cylinder, cone)
		— Dimension	Numerical value, but also: large, small, compact, long, short
— Material	Transformed material (e.g., dust, workpieces, foods)	— Material	Metal, mineral, fluid
— Energy	Potential energy, heat, electric energy, deformation energy	— Mass, weight	Numerical value, but also: heavy, light
— Information, signal	Mechanical, electrical, optical, digital signal	— Connection type	Form closure, force closure, material closure (see Figure 7.14)
Sorts of processes:		— Position	Coordinates, but also: beside, in front of, behind
— Biological process	Fermenting, decomposing, growing	— Direction	Coordinate direction, but also: horizontal, vertical, radial, tangential, skew
— Chemical process	Oxidizing, reducing, electrolyzing	— Motion	Static (at rest), translation, rotation
Technical system (TS)			
Functions (partial)	Connecting, conducting, converting	— Time relation of motion	Continuous, discontinuous, accelerated, stepwise
Mode of action	Biological, chemical, physical	— Designation	Naming of technical system
Action principle	Mechanical, electrical, optical	— Complexity	Plant, machine, assembly group, component (see Figure 6.5)
Physical phenomenon	Centrifugal force, piezo-electric effect, reflection, gravity	— Mode of construction	Monocoque, fibre reinforced, cut-and-fill, epitaxial
Magnitude:	Speed, electric current, light intensity	— Usage range	Universal, under water, food processing
For example action reality	Space, surface, line, point, location	— Usage magnitude	Continuous, once only
force reality	Tension, compression (pressure), bending	— Technology	Plastic deformation, etching, electro-plating, radiant heating
Characterization of magnitude:	Statement of characteris-tic, continuous, discrete	— Technological principle	Chip removal, fusion, heat conduction, friction to maintain wheel-road contact
For example, value	Statement of value		
Time relation	Constant (static), oscillating (vibrating), intermittent, instantaneous (shock)	— Effects	Connecting, separating, isolating, heating, pressing, rolling, sliding, rubbing
Solution principle	Statement of solution principle, solution (e.g., for partial function)	— Structure	Elements, relationships, couplings
Mode of construction	Modular Lightweight Concrete reinforced Steel-frame building	— Structure elements	Functions, organs, constructional parts
Characterization of solutions	Technical, societal, economic properties		

FIGURE 6.12   TS-viewpoints—design characteristics—examples.

methods (implied or expressly stated on drawings), tolerances (dimensional and geometric), surface properties (finish, roughness, waviness, wear resistance, or other relevant quantities), state of assembly, and so forth. All properties directly or indirectly depend on the "elemental design properties," for example, reliability of a TS depends (among others) on strength and corrosion resistance, which in turn depend directly on form, and so forth, of the constructional parts. These properties are directly established by the designers and are the *causa finalis* for all other properties of the designed TS, especially for all the external properties (classes Pr1 to Pr9). Manufacturing processes and methods can be derived (indirectly) from the materials, form, dimensions, and so forth, and planned to establish the manufacturing process machinery, times, and so forth.

6.6.3 MANIFESTATION AND VALUE OF PROPERTY

In a particular TS, the properties appear in definite *manifestations*, and with appropriate qualitative or quantitative *values* (measures, magnitudes) (see also Section 2.4.3). These values may be absolute (measured on a ratio scale) or relative (measured on an interval scale, a ranking, an assignment to a set, or a comparative assessment).

The quality of individual properties is manifested in their value (measure, goodness, constitution, composition). Some properties allow a unique formulation: length in m, speed in m/s or km/h, and so forth, with value and units, measured according to an internationally agreed scale. Color (a general property) of an object may be manifested as red, blue, white, and so forth. For many properties, quantification is not possible, and even qualitative statements may be questionable—what measuring scale is available for appearance, safety, and suitability for manufacture?

For a consumer or user, the importance of an individual property or class of properties is very variable, Figure 1.14 shows that TS have different mixtures among the classes of properties. Whether a particular manifestation or value of a property is beneficial or acceptable in a specific case can usually be determined. The appearance of a TS (included in class Pr5) can be decisive for buying, when several products are available to fulfill the same purpose and are substantially equal in properties. Cooperation between industrial design and design engineering is needed.

If the design specification quotes a "minimum required value" or "lower bound" for a variable, and possibly an "ideal value," we may be able to achieve or detect (in that order) a "lower constructional value," a "realized value" as the actual measurement on an instantiation, a "nominal constructional value," and an "upper constructional value." Other manifestations of requirements may be for a "maximum value," an "upper bound," for example, maximum permitted fuel consumption (1/100 km), or a "permissible range," a "tolerance."

An overall value for a TS—a quality measure—is an aggregate of individual values, as described in Section 6.7 (see also Section 2.4.3, part 2).

The measure (value) and appropriateness (closeness to requirements) of individual properties generally can change with time. In some properties, variability is low (size, mass), in others it can be significant (strength, surface condition), others can change drastically (by corrosion, fatigue, etc.), and yet others are subjective (fashionable appearance). The changing value depends on several factors and influences the state of an object at a given time.

6.6.4 Relationships among Properties—Model

Properties are related and interdependent. A car must resist corrosion to achieve a certain life expectancy. An object must have a certain (maximum) mass and dimensions for it to be easily transportable, for example, by rail. All properties that we expect of a TS depend entirely on the *elemental design properties*.

Figure 6.9 illustrates the properties of a TS metaphorically as a "balloon" in which the external properties are generated and sustained by the "pressure" of the internal properties. Designers must therefore make some basic decisions about the quality of TS-properties, even in the earliest conceptual phases of designing.

Figure 6.10 shows in principle the interdependence of properties by giving direct and indirect relationships between the external classes of properties (left column), and the classes of elemental and general design properties and characteristics (right column).

6.7 QUALITY OF TECHNICAL SYSTEMS

The quality of the operational TS(s) causally determines the quality of its operational process, TP(s), and the quality of the operand in state Od2—the process output. "Quality is in cycling, not in a bicycle" implies that the use of a product should provide satisfaction, quality as experienced by the user. From a user's point of view, quality "is in the cycling," that is, the perception of quality occurs because of the user's emotional reaction to the TP driven by the TS (and human). Cycling involves the human simultaneously as operand (being transported) and as operator (pedaling, steering, balancing, obeying traffic laws, etc.), with comfort and ease. For the customer to perceive this TP-quality, TS-quality must be built into the TS "bicycle" (and the TS road, traffic signals, etc.) by appropriate (external) properties. This points to a need to consider the TrfP of using a TS during the early stages of designing.

ISO 9000:2000 [9] defines quality as the "degree to which a set of inherent characteristics fulfills the need or expectation that is stated, generally implied or obligatory." This definition connects quality with requirements and properties. The quality of a product can be judged, that is, an *evaluation* can be performed, by comparing the values of properties of a system under development, as designed, or existing, with those of previously defined required and desired values given by the design specification. Reaching "the highest possible technical quality" is problematical; the optimal requirements are different for each case.

The *state* of a TS(s) represents an absolute value. The (total or partial) *quality* is a relative value for a product, and is generally a comparison of an aggregate of all or some of the states of its properties at a particular time with the state of the requirements (including standards), as illustrated in Figure 6.13. The selection of criteria for properties, their manifestations, measures (scales) and the method of treatment of these values determine the validity of the evaluation statements (see Section 2.4.3).

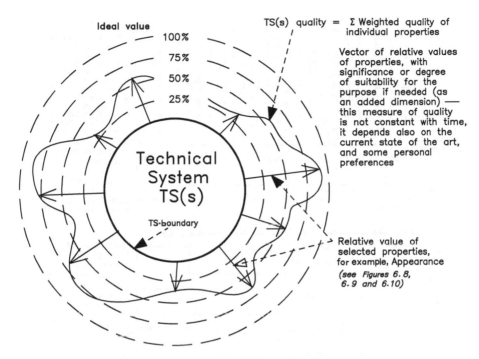

FIGURE 6.13 Quality of TS referred to properties.

In the various phases of establishing the solution (planning, designing, manufacturing, and so forth; see Figures I.13 and 6.14), this selection should ensure optimal behavior and combination of properties.

According to the number and grouping of properties used as *criteria* in performing an evaluation, either an individual value (based on one property), or a single composite value is obtained. This may represent the technical, economic, total, or other kind of value and depends on the place and time of the evaluation, and the evaluators (compare Section 2.4.3). The optimal quality of the object can be characterized either by an ideal or by a desired value.

Combining a number of different values of properties into a total value is not without problems and all known methods have disadvantages. The methods of evaluating (and of producing value statements and characteristic numbers) are the subject of Section 2.4.3.

The *quality of design* (see Figure 6.8A), is established through the internal properties "Pr10—design characteristics" and "Pr11—general design properties," and is incorporated into the represented model of the TS(s) shown in the "Pr12—elemental design properties." It is the quality of the proposed TS at the end of the design process, the relative value of an aggregate of TS-properties, compared to the given requirements for the future TS(s), as determined by simulation and assessment from the output of design engineering—engineering drawings, parts lists, instructions, and so forth. This can be a fairly objective prediction of quality, including time and cost expended up to that point, but only for some life cycle processes and from different viewpoints.

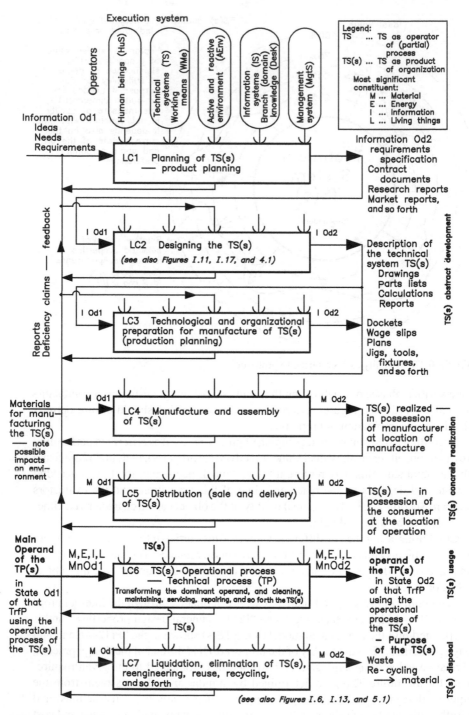

FIGURE 6.14 General model of the life cycle of technical systems as a sequence of transformation systems.

The *quality of manufacture* (or quality of conformance) is the relative value of the aggregate of TS-properties, as determined by measurement and assessment from the output of manufacturing, compared to the requirements for manufacturing stated in the engineering drawings. The most applicable variables are usually stated as "quality, cost, and time" for manufacture. These three belong directly into the TS-properties. Achieving a suitable quality of the TP(s) and TS(s), and accommodating the production system (including production management) is obviously the task of design engineering, and demands a wide range of information (see Figure I.3). The processes of manufacture cannot increase TS-quality above the limit set by the quality of design, it will usually produce a degraded quality, including all expended costs and time up to that point.

NOTE: Production management literature often quotes a triad of "quality–cost–time." Related to *manufacturing*, this triad refers to the achieved quality of production, the cost of production, and the time taken to produce. This book clearly defines:

> *Quality*—an overall assessment of the suitability of all the properties for the purpose of the TrfP and TS, and this includes the properties of *cost* and *time* (see Figures I.8 and 6.8–6.10).
> *Cost*—total or partial cost of implementing a TrfP and realizing a TS, including all the costs of designing, making, distributing, erecting, and possibly liquidating.
> *Time*—total or partial time for implementing a TrfP and realizing a TS, including all the time for designing, making, distributing, erecting, and possibly liquidating.

The triad "quality–cost–time" is therefore *not* as useful for us as it is for the manufacturing managers.

In design engineering, cost and time are classified in the management properties, class Pr9.

The "quality of transport and distribution" will also influence the organization product before it reaches the customer and operator, and can be measured by statistical quality or process control.

The "quality at the point of sale" (as perceived by a stakeholder or other participant) may be a largely subjective and quick assessment based on appearances and reputations, a role for brand names and trade marks [388,389]. The quality perceived by customers or operators of the TS(s) in its operational process, the TP(s), will also depend on their own actions and can be summarized as the "quality of usage and service." Failures (catastrophic and minor) have been caused by users, maintainers, repairers, and so forth. This tends to be more subjective, and depends on many unwritten requirements, for example, the sound, force, and feel of a car door when being closed; feel and feedback to a human of positional response to a control input for "fly-by-wire" controls—haptic properties [97]. The actions (or inactions) of a human operator can influence the quality of usage, for example, the unauthorized experiments that caused the Chernobyl power station disaster, or the lack of knowledge about a valve that caused the Three-Mile Island power station to be destroyed.

The quality of design can be improved if the design problem as assigned to the designers is well formulated to take into account the wishes of the customers, users, environmental concerns, and so forth. The quality of manufacture can be enhanced by suitable and sufficiently early feedback from the manufacturing organization to the designers. Some management techniques help to formalize these aspects (see Section 11.3).

6.8 PHASES OF ORIGINATION AND EXISTENCE OF TECHNICAL SYSTEMS—LIFE CYCLE

The structure of the life cycle of the TS(s) (Figure 6.14), the process that a TS(s) may experience is formed by the sequence of transformation systems, and their processes of origination, operation, maintenance, liquidation, and so forth (see also Figure I.13). By analyzing the factors that influence the individual phases, we obtain information about situations in which the TS(s) exists. In most of these transformation processes, the TS(s) is the *operand*. During actual operation the TS(s) is an *operator* of its operational process, the TP(s). In testing (for development or quality verification), the TS(s) is both operator (of its operational process) and operand (of the testing process). In all these processes it is possible and useful to check for any possible environmental impacts [62,160,186,237,244,582] that may arise from the operands (input and output states), assisting inputs, secondary inputs and outputs, and the operators.

Various additional processes occur as operations within this TS-life cycle. "Product planning" can include tendering for contracts, that is, preliminary designing as a "quick and dirty" estimation. During "manufacture and assembly," processes of testing and adjusting, and packaging for transport and storing may be necessary. "Distribution" will normally involve transporting. In preparation for the "operational process," TP(s), processes of unpacking, erecting, site assembling, aligning, adjusting, acceptance testing, commissioning, and training of human operators may be included. During the "operational process," TP(s), processes of cleaning, servicing, maintaining, repairing, renovating, and upgrading may be needed. "Liquidation" may include decommissioning, dismantling/disassembling, sorting, recycling, and disposing activities. These additional processes are more typical for large and complex systems (level of complexity IV, see Figure 6.5), but may also be applicable to systems of levels I and II that are intended as purchased ME, modules, subassemblies, or constructional parts to be built into larger devices (OEM parts and COTS). Analyzing the operators of the individual processes/operations within this model yields many questions for generating the design specification (list of requirements), and for formulating evaluation criteria (see Figure 6.15).

For instance, by examining the operator "human being" in the various processes, designers must state requirements about the needed quality (and qualification) of workers and operating procedures for the TS(s). This examination should consider all aspects, especially manufacturing, operation, maintenance, and repair of a TS and the TrfP. For example, the physical and operational differences between machine tools and agricultural machinery are caused by the different expectations about the quality

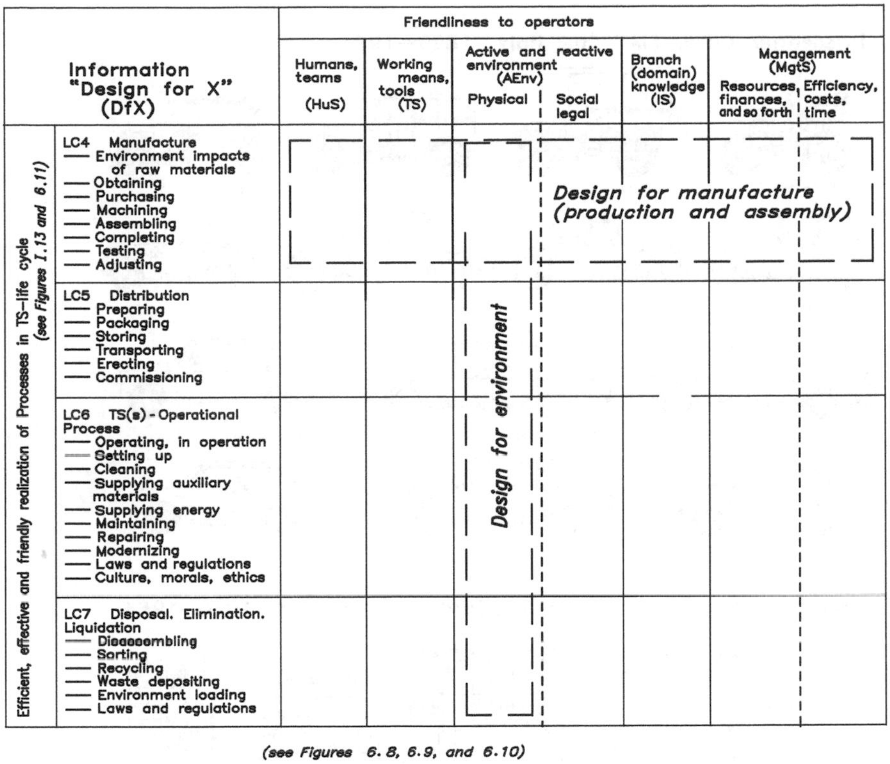

FIGURE 6.15 General systematics of "Design for X" (DfX) classes.

(capability, qualification, knowledge, experience, etc.) of human operators. These considerations must also concern the end user and the manufacturing organization.

6.9 DEVELOPMENT IN TIME OF TECHNICAL SYSTEMS

Development of a TS during a period of time (innovation, evolution, and further development process) leads either to small changes, variations or mutations, with basically unaltered main effects, or to larger changes when a new generation of TS is created (see Figure 6.16). The overall changes of properties are based on individual changes concerning: (1) involvement of the human being as operator in the TP(s): reduction of involvement is accomplished by mechanization, automation, computerization, miniaturization, integration, mechatronics, and so forth (see also Figure 3.1B); (2) main and assisting inputs to the TS(s); (3) mode of action (and other design characteristics) of individual action chains at various levels of complexity; (4) individual constructional parts and their elemental design properties, that is, their form, dimensions, materials, manufacturing methods, tolerances, and surface properties; and (5) arrangement of structural parts, and other general design properties.

Over time, TS have noticeably altered, in the readily visible characteristics of form or size, and with respect to all other properties, for example, maintenance

(A)   Passenger Cars, FIAT Co., Italy, 1899—1971

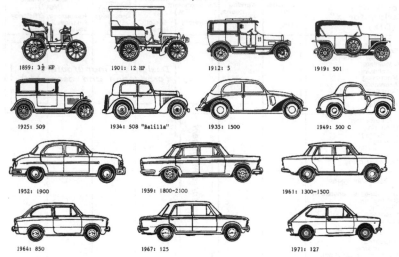

(B)   Ship diesel Motors, william doxford Co., England, 1950—1964

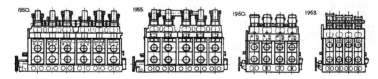

Year of Construction		1950	1955	1960	1963
Power Rating	HP/kW	6800/5060	7000/5220	6640/4940	7200/5360
Length	ft/m	57.5/17.5	52/15.9	33.5/10.2	26/8.5
Mass	tonne	404	395	260	240
Price	$/HP/ $/kW	86/115	64.4/86.4	45.8/61.4	17.4/50.2

(C)   Pressures achieved in laboratory and operation

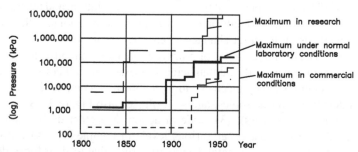

Literature:
Hubka, V. & Eder, W.E. (1988a) *Theory of Technical Systems*, Berlin/Heidelberg & New York: Springer-Verlag, Figures 11.1 and 11.2.

FIGURE 6.16   Developments in time of technical systems.

intervals. For various reasons the innovation process is accelerating. Comparing a car built in 1900 with a more recent model (Figure 6.16A), a diesel motor for ship propulsion (Figure 6.16B), or a calculator made in 1965 with a newer one, substantial changes are obvious [260].

The currently best incorporation of properties in a product is usually called the *state of the art*. The state of the art in a branch of engineering is represented by those TS or available technology (indefinite article) whose properties have reached their highest levels of development at that time. Information about this state (for their own branch, and for related branches) is essential for engineering designers. Keeping pace with the state of the art leads to gradual or intermittent improvements in the TS(s), a static or dynamic design situation [477], a surge-stagnate model [331]. Development decisions usually involve policies at a higher management level or political decisions [203,271,380,571].

This state of the art usually exhibits different levels, according to whether it is observed in research, speculative design and development, production design and development, advanced production, or everyday use (see Figure 6.16C). *Invention* improves the state of the art in research and speculative design and development, mainly by suggesting new embodiments of operational principles into a candidate TS. Invention may or may not lead to a commercially viable TP or TS. If a new TP/TS is implemented, this is usually termed an *innovation*.

NOTE: The time period between a relevant invention and its introduction as a marketable innovative product has generally reduced over time, for example, photographic camera, 1727–1839, 112 years; radio, 1867–1902, 35 years; radar (echo-location in atmosphere), 1925–1940, 15 years; television, 1922–1934, 12 years; atomic bomb, 1939–1945, 6 years; transistor, 1948–1953, 5 years.

Care is advised, the expression "state of the art" is often regarded as a "magic word" that can easily be misused, particularly for advertising a product.

For engineering designers, historic developments of "their" TS-"sort" is a source of experiences and ideas, especially if the designers are able to ascertain the causes for the changes. Further development must be based on these insights, and new requirements must be carefully assessed and justified. Other TS in related fields should also be observed, since developments in their state of the art may suggest improvements for the next generation of the TS under consideration.

6.10 TAXONOMY OF TECHNICAL SYSTEMS—TS-"SORTS"

The large number of TS currently in use demand a systematic classification that can deliver a suitable order and overview. A hierarchical classification can be based on various properties and aspects, and allows generating systems of elements that are in some way related.

NOTE: The process of mentally *abstracting* plays a large role in technology and science. Engineers usually believe that they work at a concrete level, but with many terms and expressions they move on relatively high planes of abstraction. These help engineers to create *hierarchical systems of classification* that can assist them in their work.

The levels of abstraction in TS can be shown on an example from the branch of machine tools (see Figure 6.12). Parts of the terminology for the levels are taken from biology. A larger number of subdivisions is possible by a finer graduation of properties. This arrangement was selected for its clarity of representation, yet the boundaries are flexible, soft, fuzzy. Nevertheless, they represent a source of classification for various regions and branches of industry.

Analogies with other areas of design methodology can be shown, the levels of abstraction are characteristic for the general order exhibited by TP and TS. They represent a useful *series of design documents* that should be established in that sequence during systematic and methodical design work for a novel system. Progress through some or all of these levels takes place during designing (compare Figures I.22 and 4.1). A machine system (a lathe), as a system of the third level of complexity (see Figure 6.5), has been used as an example in Figure 6.17. The same regularity exists without limitation for all degrees of complexity.

Other possible aspects for systematic classification are purpose, mode of action, predominant manufacturing methods, complexity, intricacy, novelty, origin, and so forth (compare Section 6.11).

NOTE: The classifications shown in Figures 6.5 and 6.7 are paradigmatic. Complexity of TS may occur for various reasons:

1. By intricacy, difficulty of manufacture, and so forth, of constructional parts, for example, within level I , different complexity may be illustrated by (1) a simple cylindrical pin cut from a rod, (2) a pin made to close dimensional and geometric tolerances, or (3) a machined casting for an IC cylinder block.
2. By number of components, for example, within complexity level II, (1) a small single-reduction gearbox, (2) a transistor, or (3) a microcomputer central processor with VLSI (very large scale integration) circuit elements with thousands of transistors and other electronic elements.
3. By numbers of relationships—for example, within level III, (1) a manually-controlled lathe, where the consequences of an input are generally close to their cause, (2) a small self-contained computer program, where some output can be hand calculated, and debugging is relatively easy, or (3) a computer application using a knowledge-based (expert) system, a neural network, or a genetic algorithm, where the results (and faults) cannot be anticipated because no human mind can reproduce or follow the intricate pathways that lead to the results.
4. By numbers of alternatives—a frequent design-related problem: if several alternatives exist at each of various levels of interaction (e.g., abstraction), the complexity of choice among proposals increases rapidly, as the product of the numbers at each level—*combinatorial complexity.*
5. By number of participants in a project that is related to (1) the individual person's capability for processing information: 7 ± 2 or fewer chunks of information [101,416–419]; see Section 11.1.1, (2) to the number of communication paths among the participants [453]—$C = N(N - 1)/2$, where C is the number of communication paths, and N is the number of participants: that is, for $N = 2$, $C = 1$; for $N = 3$, $C = 3$; for $N = 4$, $C = 6$; for $N = 5$, $C = 10$; for $N = 6$, $C = 15$;

General hierarchy of technical systems				Example	
Level of Concretization	Designation of level	Definition by ... (graphical model)	Established design characteristics	Designation of level	Graphical representation
0	Technical system (TS)	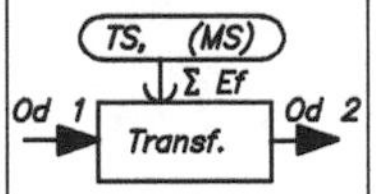 TS, (MS) / ∑ Ef / Od 1 → Transf. → Od 2	TS general MS (as special case) exhibits mainly mechanical mode of action		
0.2	Phylum of TS TS, (MS)	General "black box" 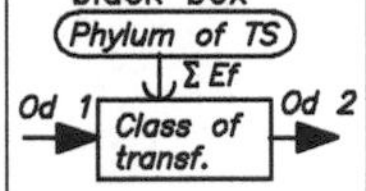 (Phylum of TS) / ∑ Ef / Od 1 → Class of transf. → Od 2	— Phylum of operands — Class of operations	Machine tools	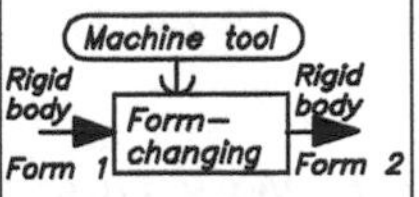
0.4	Class of TS	Detailed "black box" Sketch of technological principle Basic function structure	— Family of operand — Technological principle of transformation — Necessary input effects and from these the basic TS-internal processes and functions	Lathe (metal-working)	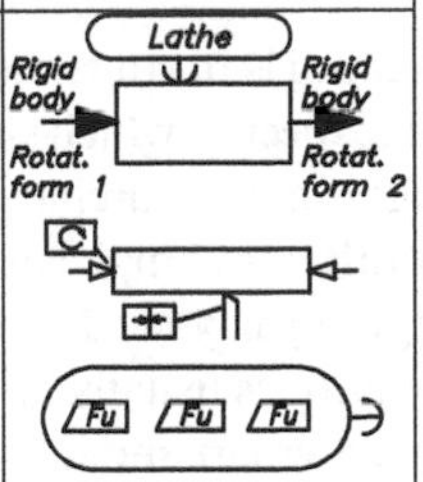
0.6	Family of TS	Detailed function structure Rough constructional structure Concept sketch	— Species of operand for TP — Function structure (functions and their sequence) — Inputs to the TS — Families of function carriers (organs) — Combination and basic arrangement of organs	Universal lathe	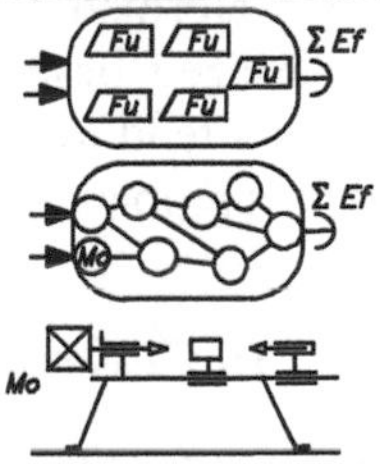
0.8	Genus of TS	Drawing of constructional structure (conditions of similarity will differ according to type series) Drawings of common (re-used) parts and subassemblies	— Complete parts — Arrangement — Partial form — Some dimensions — Types of materials — Some tolerances and surface properties	Lathe type SUR	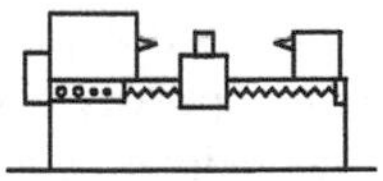
1	Species of TS, or serial size	Complete set of workshop documentation	Complete and definitive specifications for parts and arrangements For all components Forms Dimensions Materials Manufacturing methods Tolerances Surface finish and so forth	Lathe SUR 5	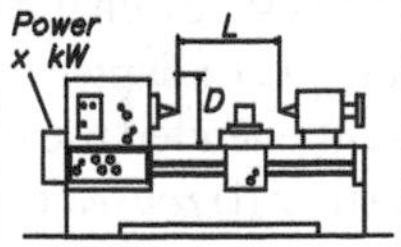

FIGURE 6.17 Taxonomy of technical systems (TS), with examples of machine tools (machine system [MS]).

for $N = 7$, $C = 21$; for $N = 8$, $C = 28$; for $N = 9$, $C = 36$; for $N = 10$, $C = 45$; for $N = 100$, $C = 4950$—the optimal number of participants is probably between $N = 4$ and $N = 10$.

6.11 SYNOPSIS OF CLASSIFICATION OF TECHNICAL SYSTEMS

TS may be classified in various ways. This classification formalizes most of the aspects of TP and TS (as organization products).

6.11.1 WHY THESE CLASSIFICATIONS?

Designing is in some respects different for each combination of these classes (but not in all respects), which should help to characterize the design situation (see Chapter 3). Designing is hierarchical—the design process for each complexity level is similar, but with different emphasis on TS-structures, and depth and speed of problem solving. Designing may need to be innovative or even creative, but in order to control risks such innovation is usually applied for only a smaller part of the whole TS(s), for example, a design group, module, or subassembly. Properties, classifications, modeling levels, and other information about systems should help in designing.

Specialized design situations arise from actual design problems, especially where tasks tend to become more routine because they are in essence repeated for subsequent problems.

6.11.2 TS CLASSIFIED BY NOVELTY

Novelty (see Section I.1) may apply to a whole TS(s), or only to a smaller part of it, compare design situation, factor FA1, in Chapter 3. TS(s) may be: (1) *novel*, with no existing or predecessor systems—most likely in complexity class II for engineering design, complexity class III for industrial design; (2) *redesigned* in several gradations: (a) radical *change* or *innovative*, for example, of functions; (b) *redesigned*, state of the art upgrade; (c) *redesigned*, for example, for cost control; (d) formation of *variants*—change of size and performance magnitudes (e.g., power capability), size ranges; (e) change to *modular* construction; (f) *configuration product* design for customer-adapted assembly on a common platform; (g) *adaptation*—change for different intended use or manufacture; (h) *modification*—change and rebalance some properties; (i) *reused* (needs little or no designing); or (3) direct *adoption* of a previous TS.

The majority of design problems are for redesign tasks. The full procedural model (and its adaptations) may be used for novel designing, it can usefully be adapted for redesign problems (e.g., changes of functions); see Chapter 1 for examples. In many design tasks, risk must be controlled, which implies that a substantial part of previous implementations of TrfP and constructions of TS is retained. Numerical indices for expressing the proportion of commonality among TS and TrfP have been proposed [538].

6.11.3 TS Classified by Usual Stages of Market Development

(1) "First of a kind" on the market, *trendsetter*, market leader, high risk and slow growth, but possibility of high returns (covers about 2% of a potential market); (2) follow up first competitors with improvements, *early adopters* (about 14% of market); (3) follow up competitor aiming for large turn around and substantial market share in a rapidly growing market, diminishing returns due to competition, *early majority* (about 34% of market); (4) competitor aiming into an established and mature market, *late majority* (about 34% of market); and (5) continuing in an established and usually saturated (but possibly declining) market, *laggards* (about 16% of market), expect small returns (per system sold) but may stay in business for a long time satisfying a remaining continuing demand for otherwise obsolescent equipment—extremely low risk for the organization product, but also very low returns, therefore a risk for survival of the organization.

All organizations *need to keep updating*; continuously improving; keep in the market; stay in business, with a progression of products to replace declining products by updated or newer products (see Section 11.3).

6.11.4 TS Classified by Sequence of Demand to Manufacture

(1) TS designed and manufactured in the hope of future sales, and offered on the market, the upper paths in Figure 6.18; marketing involves distributing and advertising an existing product. (2) Order placed by potential customer for a TS to be

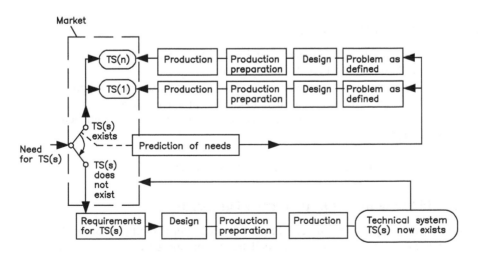

Literature
Hubka, V. & Eder, W.E. (1988a) *Theory of Technical Systems*, Berlin/Heidelberg & New York: Springer-Verlag

FIGURE 6.18 Sequence of market demand, design and production.

delivered later, the lower path in Figure 6.18, which usually needs a "request for proposals," selection of a contractor, and special manufacture; marketing is performed by designers, manufacturers, and organizations—the organization image is important [388,389]. (3) Only advice is offered—consulting.

6.11.5 TS CLASSIFIED BY SCALE OF POTENTIAL PRODUCTION

(1) Single TS, one-off, made-to-order, bespoke, special purpose equipment; (2) small batch, identical discrete systems, may be repeated at longer intervals, which may suggest the use of CAM—computer-aided manufacture; (3) Medium and large batch; (4) Families, small changes among largely similar systems, *flexible manufacturing cells*—may be computer-controlled manufacture including robotics if repeats are needed; (5) Mass produced, long "run" of product, dedicated *production line*; and (6) Continuous product, may be "discretized" later in the production sequence.

6.11.6 TS CLASSIFIED BY SIZE RELATIONSHIPS, COMPLEXITY

Typically in four levels; see Figure 6.5: constructional part; group, module, or subassembly; machine or device; plant.

6.11.7 TS CLASSIFIED BY MANUFACTURING LOCATION AND STANDARDIZATION

Make or buy, the whole or some of its parts; assemble or have assembled; degree of international, national, regional or organization standardization; and so forth (see Figure 6.7).

6.11.8 TS CLASSIFIED BY MARKET

Compare types of artifact (Sections I.7 and 6.11.10): consumer goods; consumer durables; capital goods; OEM goods; luxury goods.

Relationships exist among *suppliers*, *manufacturers,* and *customers*—appropriate quality and reliable delivery of manufactured products depends on these relationships, especially the *personal* aspects, and are partly regulated by ISO 9000:2000 [10].

NOTE: This class should consider TS and TrfP/TP (see Chapter 5).

6.11.9 TS CLASSIFIED BY OCCURRENCE OF LIFE ENDED

The useful life of a product is usually limited, it ends after exploitation of the TS(s) in its operational process (for its intended transformation process), and in some cases even before. Criteria for the point at which the condition of *life ended* occurs ranges between: (1) *wear out*—time dependent, environment dependent, and so forth; life can be extended by repair, partial replacement, refurbishment, remanufacturing, reuse and recycling of parts, and so forth; (2) *catastrophic failure*—time-dependent, environment-dependent, and so forth; (3) declared *obsolescent*—overtaken by development in the state of the art, for example, by a change of operational principle, but still usable or upgradable, or *obsolete*—no longer usable; (4) intended as a *one-shot device*—only a single use is foreseen; or (5) experience *replacement* before completed life expectancy—for example, aircraft gas turbine blading.

See also Introduction, Sections I.9 and I.10. Even after normal "life ended," some artifacts find reuse. A different organization may extend or revive the life (e.g., steam railways run as tourist attractions by volunteers); a partially gutted item may be used as a display item or monument. Otherwise, these items can be (partly) refurbished, remanufactured, recycled into raw material, or sent to scrap or waste disposal.

6.11.10 SOME CONSEQUENCES

Synergy is expected, a system of greater complexity has properties that are more than the sum of the properties of the simpler parts; and there is a *hierarchical relationship* among levels of complexity (see Figure 6.5):

Complexity level IV—plant

1. Usually ordered as single items, capital goods, usually no repeat orders.
2. Involves long lead times to full operation, and slow evolution of state of the art—usually a predesign process takes place before tendering for a contract, regulatory approval may be needed, detail design and construction is allocated to one tendering organization.
3. Designing consists mostly of selecting available items (and procuring them by ordering or having them designed and made by a supplier) from levels of lower complexity (III, II, and possibly I), and establishing their arrangement, connections, behavior, control, and so forth.
4. Large influences of laws, standards, and regulations must be taken into account.

Complexity level III—self-contained functioning system (machine)

1. May be ordered in small to large batches, or manufactured to service a potential market.
2. Designing contains some selection, and some novel structures, arrangements, elements.
3. Importance of properties depends on market to be delivered.

Complexity level II—subsystem, assembly group, composite ME

1. Purchased from a supplier, or a substage of producing a system of complexity III.
2. Designing consists of either selecting, or of newly establishing structures, arrangements, elements, constructional parts.

Complexity level I—constructional part, simple ME

1. Usually for building into a more complex system, that is, a finished component is "raw material" for next level.
2. Designing consists of extracting the *required properties* from the more complex levels, and then establishing *detail*.

The descriptions of products ISO 9000:2000 [9,10] are outlines to those indicated in Section I.7. They are not complete and comprehensive, but provide a rough scale to differentiate TP/TS from other products [164]. The classes are not unique, the boundaries are fluid and overlap. These classifications refer to the TS-operational process, the operand, the TP, the Tg, the effects delivered by the TS, and the complexity of the system. Products may appear in more than one classification. A product from one organization may be an input for another, for example, as OEM items.

At one extremity of the (nonlinear, branched) "scale" of artifacts are *artistic works*. The artist is usually both the designer and manufacturer. Appearance is the primary property, esteem value, and therefore the asking price, tends to be high compared to costs. Usage of these artifacts tends to be relatively trivial. The designed (and made) artistic work frequently consists of only one or a few constructional parts, its structure tends to be simple, but its form can be complicated. Designing can and does take place during manufacture. Evaluation for suitability of the product is continuous, and depends almost purely on the judgment of the artist. There is almost no scientific knowledge base to consider, except for that incorporated into the materials, tools, and manufacture. These artifacts rarely have any engineering content, with the exception of large sculptures and similar works, which rely on the strength of the materials to maintain integrity. Design methods hardly come into consideration. Nevertheless, research activity into artistic design takes place. Typical artifacts are: coins and banknotes, medals, decorations, sculptures, paintings, conversation pieces, jewelry, and works of the performing arts, theater, music, and so forth.

Consumer products are frequently consumable items and materials. Designing and product development involves to some extent the product, but probably more important is designing the packaging and advertising, somewhat artistic works. Again the knowledge and science base is comparatively minimal, especially regarding packaging, unless the product itself is something like a chemical compound that requires design of the composition and of the manufacturing process/equipment. Methods for designing tend to be marginal. Products typical of this group are: packaged butter, motor or culinary oil, bottled water, pharmaceutical products.

Consumer durables must have appropriate appearance and operability. They must project the "right" image, of the product, and of the manufacturing (or selling) organization. This is independent of whether the product is to be used typically in the household, on the road, or elsewhere. These products must be designed and product developed to perform useful tasks, they must function with suitable performance parameters, and be made available at a suitable, usually predefined cost. Maintainability, availability, and reliability are usually important. The product (a TS with its associated TP) tends to have many constructional parts that are frequently not visible or accessible to the casual observer. Two separate (groups of) designers are usually employed. The primary and most advertised design task is that of the "industrial designer" and "stylist," to establish the outward appearance and operability conditions—aesthetics, ergonomics, and customer psychology. Within that "envelope" and usually of at least equal importance for many products, the engineering designers must establish the means for the product to function. Cooperation is called for. There is usually a substantial knowledge base involved, and much of the information can be scientific. Design methods have been discussed, but substantially more for the engineering aspects than for the appearance "styling." Typically such products include: lighting fixtures, domestic appliances, motor cars, personal computers, most of which can be regarded as TS. Other such products have no substantial engineering involvement (e.g., tableware, plates, cups, cutlery), but their manufacturing equipment consists of TS, that is, industrial equipment products.

Bulk or continuous engineering products act generally as raw materials for manufacture in other organizations, for example, bulk lubricating or culinary oils, bulk fuels, metal rolled sections or rod and bar stock, plastics in pellet form for moldings, sheet and foil woven or roving, laminates, and so forth. The processes and tools for manufacturing need to be designed (see industrial equipment and special purpose products, below). Sizes, shapes, tolerances, and material properties of the product are mostly laid down in standards. Imprints or embossings may be added for surface texture and appearance, and may need to be designed—a role for industrial designers and graphic artists.

Industry products are items or assemblies that are purchased by a manufacturing company for assembling into their own products. They include ME, purchased OEM goods (products intended for "original equipment manufacturers"), COTS (commercial off-the-shelf products), other hardware supplies (i.e., TS, usually of lower levels of complexity; see Sections I.9.2 and 6.4.3, and Figure 6.5), and also software. Appearance matters only in some cases. If two items of equivalent functionality, performance and price are offered, appearance (of the product and of its packaging) and reputation of the organization (brand [388,389]) can make a difference. These products include: ball and roller bearings, electric motors and controllers, circuit boards, crane hooks, electronic controllers, high-voltage insulator hangers for power lines, gas turbine engines for aircraft propulsion, and so forth.

Industrial equipment products are self-contained devices (e.g., TS, machines) that can perform a more or less complex function and are intended for use within industry. Functioning and performance (including ergonomics) are primary; appearance is distinctly secondary. Among these products are: personal computers, workstations, machine tools, goods vehicles, turbine-generator sets, chemical reactors, earth moving machines, and aircraft.

Special purpose equipment includes jigs, tooling, fixtures, and specialized manufacturing and assembly machinery, special-purpose robotics, handling and packaging machinery, and ocean-going ships. These *TS* are usually produced to special order, as single items or a small series. Designing and developing takes place specifically for one customer, and in most cases these equipments are newly designed. There is no opportunity for redesign, they must be directly (and usually quickly) integrated into the customer's usage and production facilities. The prototype is the final equipment delivered to the customer—and it must work as ordered. Mechanical, pneumatic, hydraulic, electrical, and electronic hardware, firmware and software components are used in equal importance. Purchased constructional parts are intensively used to control the risks, configuration tasks are more frequent than custom designing of items. Modularity of the structures can be an advantage. Sensor technology, and user or developer software are particularly important. The requirements include extreme demands for cycle times, reliability and availability, ease and flexibility of retooling for a different TS-operational process or its output, and anticipated possibilities for retrofitting to improve performance. The phases of generating preliminary design proposals (conceptualizing) for cost estimating and tendering (to obtain a customer contract) are extensive, and need substantial time and financial expenditure. Tenders have to be offered for complete machinery or retrofit sections of machinery [438], including delivery time and price, before all details of the devices have been worked out. Many of these tenders are unsuccessful. A particularly careful analysis of the task requirements is needed to adjust the design specification to the customer's specific needs. Changes are frequently required during designing and developing, because of progressive increase in the amount of information about the task, and additional wishes of the customer. Software is often developed and adapted during commissioning (tests and trials prior to handing over the TS to its intended owners and users), before the customer's final acceptance of the devices. A major task consists of coordinating deliveries from suppliers, including interfacing with other equipment and components, performance data, drawing and delivery standards, and so forth. The existing know-how of the customer must be used, by providing and obtaining information, and using the customer's personnel as consultants. Training and support for the operating personnel must usually be arranged before and during commissioning. This must especially concern motivation, acceptance, automation psychology and qualification of (operating) personnel. Any future developments and trends that can be foreseen must be thought out in advance and considered. Engineering designers employed in this kind of equipment need high flexibility; the types of tasks change rapidly from one contract to the next. They need good communication capabilities and negotiating styles. Close cooperation between design and production for these machines is essential. Companies involved in this class of artifact are usually small to middle-sized organizations (or subsidiaries).

NOTE: Thanks are due to Professor H. Birkhofer, Darmstadt Technical University, for presenting an explanation of the nature of designing for industrial special purpose equipment to a WDK Workshop at Rigi Kaltbad, Switzerland, from which this summary has been compiled.

Industrial plant usually consists of collections of industrial equipment products, and devices to provide control and connections among them—that is TS. The plant (and some of the connecting devices) is designed to special order, most of the items are bought from other suppliers. This is probably the extreme case of an artifact incorporating other artifacts (organization products from other suppliers), and the (recursive) task of designing them and their components is mainly transferred to the suppliers. Typical are water purification plant for a city, electric power station, oil refinery, telephone network.

Configuration products are items of special purpose equipment and industrial plant for which the components are quantity-produced and standardized industrial equipment products (OEM, COTS)—designed deliberately as modular interchangeable TS (as organisms and constructional parts for the product), often to be mounted on a common platform. The configuration products are then assembled to the customers' requirements, without further designing or modifications—designing means configuring the product by planning the assemblies and their interconnections (e.g., wiring harnesses).

Infrastructure products provide means for supplying services according to ISO 9000:2000 [9], such as transportation, power delivery, water, and fuels.

Intangible products are typically documents such as contracts, insurance policies, and so forth, as tangible items that typically record a specification of the services provided by an organization.

Software products are intangible products presented as computer programs of various kinds and for various purposes, including mechatronics and firmware. They may be delivered on a transportable medium (floppy disc, CD-ROM, DVD-ROM, etc.), by downloading from a computer source, or loaded into a programmable controller.

Designers generally work within a particular branch, and they design a well-defined genus or a particular type of TS. In this case all statements should be concretized according to the relatively lower level of abstraction, and they probably follow well-defined function structures and organ structures. Instead of stating a general formulation of a requirement, a particular manifestation of a property is specified, for example, instead of "a sonic echo location system," the specification is for "a type XYZ submarine sonar equipment to a requisition with a specific reference number."

Figure 6.17 shows clearly the progressive concretization of the example. For instance, at the first level (in the TP) the operand and its transformation of state are established, that is, machine tools work on pieces of material (it would even be possible to specify only a limited range of materials, e.g., "metals" or "aluminum"), and are used to change their form (especially their shape and geometry). At the next lower (more concrete) level, only a rotational form is allowed as subject of the transformation, and this effect on the operand is based on a particular technological principle. The general statements can be therefore concretized for the particular branch, and the resulting possibilities of solution can be represented on the basis of the chosen classifying aspects to provide a good overview.

This path represents a strategy that can lead to success in designing. General insights for the level of TS cannot be directly applied in engineering practice because designers in their special circumstances must rapidly reach decisions about their problems. Every branch of engineering (e.g., mechanical, electrical, civil, etc.) and every subsection (e.g., in mechanical engineering: machine tools, transportation equipment, mining, domestic appliances) demands specialized information. This information must be concretized for a particular organization and even for a particular design engineer (see Figure 5.4 and Chapter 9). The basic hierarchy of systems is shown in Figure 6.19, the levels of complexity combined with the levels of abstraction.

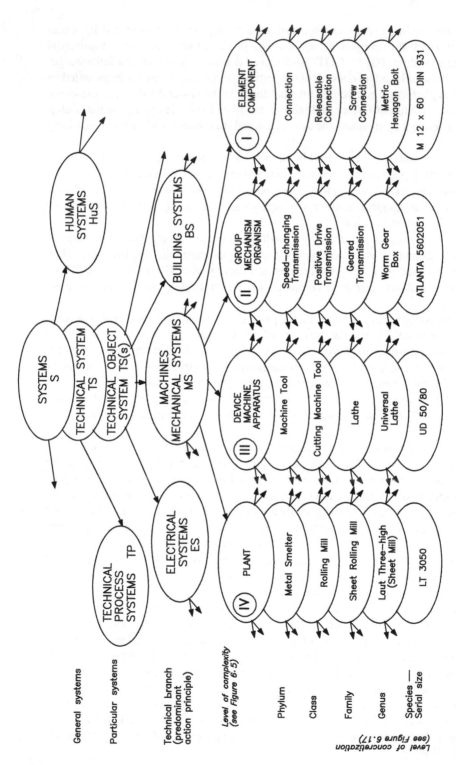

FIGURE 6.19 Hierarchy of technical systems (TS-hierarchy).

6.12 TP-OPERAND, TECHNOLOGY, TS-EFFECTS

According to Section 4.2, the TP-operand is situated external to the TS(s) as operator. The TS(s) exerts its TS-effect through a technology onto the TP-operand. Some reaction to the TS(s) is expected. A TS(s) can still operate or be operated even if no operand is present, or it is idling. The TS-main effect then does not change the (absent) operand; the TS-effect is a potential, just as the TS-functions describe a capability of the TS(s) to perform TS-internal actions.

This relationship of an operand considered external to TS(s) needs to be interpreted for different constituents of operand (see Figure I.9 and the NOTE in Section 4.2).

7 Specialized Engineering Design Sciences, Specialized Theories of Technical Systems

7.1 ENGINEERING DESIGN SCIENCE FOR THE TS-"SORTS"

In general, engineering designers work only within a certain limited range of technical process (TP) and technical system (TS), a branch of engineering. Their involvement is with one particular TS-"sort"; see Section 6.11.10. Their knowledge must include TS of lower complexity (see Figure 6.5) that provide constructional parts, and often to TS of higher complexity in which "their" TS find application. The information presented in this chapter is mainly situated in the "north" hemisphere of Figures I.4, I.5, and 12.7.

General Engineering Design Science (EDS) [315] (see Chapter 12), is associated with the generic plane of TP and TS, the structure of contents is known—Figures I.4, I.5, and 12.7 reflect the four fundamental quadrants. Each Specialized Engineering Design Science (SEDS) can be derived from this model and will contain the same parts. Information about *existing* TP and TS ("as is" state) resides in and around the "northwest" quadrant. The Theory of Technical Systems (TTS) [304], which resides in the "southwest" quadrant, was derived by abstracting from the information in the "northwest" quadrant. Using the triad "subject–theory–method." The coordinated Theory of Design Process [287] that resides in the "southeast" quadrant was developed from both TTS and the information about design processes that resides in the "northeast" quadrant. Heuristic and experience information that can be used during design engineering, both with respect to the TP(s) and TS(s) ("as should be" state) and the design processes, is collected outside the "northeast" quadrant. Both quadrants in the "south" hemisphere have been continuously developed since their inception.

NOTE: Figure 7.2 illustrates the relationships among the quadrants with respect to various models presented in this book, compare also Figure 12.8.

In this way a general design process was found; see Figures I.22, 4.1, and 4.2. However, each actual design process depends particularly on the TS(s), and its TS-operational process, TP(s), and demands concrete and specific technical

343

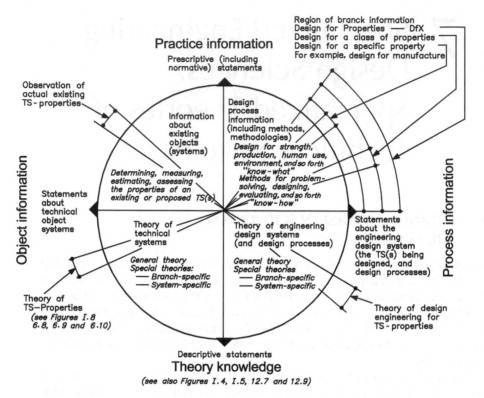

FIGURE 7.1   Categories of engineering design information.

(branch, domain) information; compare Figure 4.7 and Chapter 6. SEDS is used as a *terminus technicus* for an information system at a specific complexity level and in a branch.

Engineering designers must acquire an information system, consisting of tacit knowing and recorded information, formed by classifying the existing information according to the SEDS. The SEDS should be built on one of the planes of the TS-taxonomy (Figure 6.19), probably the "phylum" or the "species" plane (see Figures 6.17 and 6.19), where the designers work.

Figure 7.1 shows a possible construction of an information system with an SEDS. The conventional "map" according to Figures I.4, I.5, and 12.7 represents the general EDS on the plane of the TP/TS and is projected (Figure 7.3) onto the structure of the SEDS, inheriting and transmitting the general information. The segment in the "northeast" quadrant of Figure 7.1 shows as an example "design for manufacture," based on a theory of design for manufacture in the "southeast" quadrant. The next level in Figure 7.4, the "phylum" level (e.g., machine tools; see also Figures 6.7 and 6.17), is chosen as SEDS, which should be as complete as possible, for example, in a fundamental book or a data base. All information systems below this level, on the levels "families" to "species," expand and concretize the information through corresponding "completions," compare Figure 11.6. The figure shows the completions from the TS-plane, and those originating from the concrete organization and even

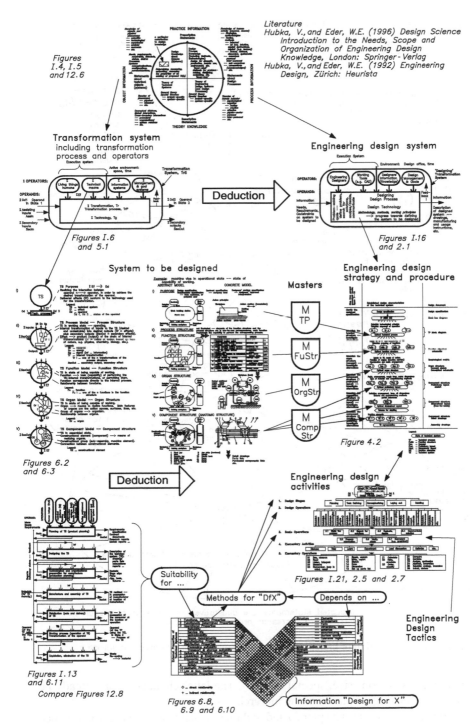

FIGURE 7.2   Derivation path for design strategy and tactics.

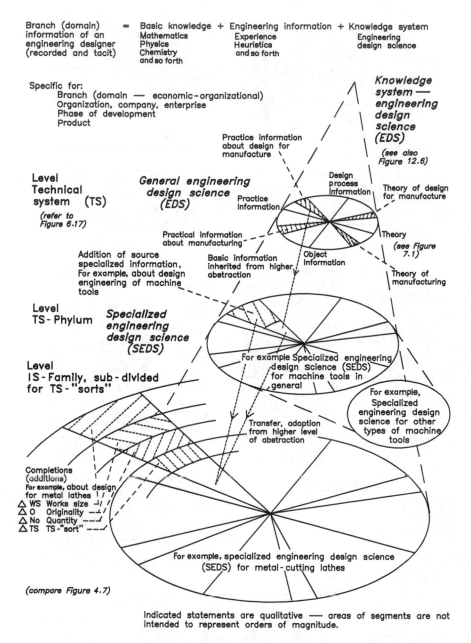

Branch (domain)   =   Basic knowledge + Engineering information + Knowledge system
information of an   Mathematics   Experience   Engineering
engineering designer   Physics   Heuristics   design science
(recorded and tacit)   Chemistry   and so forth
                                and so forth

Specific for:
    Branch (domain — economic-organizational)
    Organization, company, enterprise
    Phase of development
    Product

Knowledge system — engineering design science (EDS)
(see also Figure 12.6)

Practice information about design for manufacture

Level Technical system (TS)
(refer to Figure 6.17)

General engineering design science (EDS)

Practice Information

Design process information

Theory of design for manufacture

Practical information about manufacturing

Basic information inherited from higher abstraction

Object Information

Theory (see Figure 7.1)

Theory of manufacturing

Addition of source specialized information, For example, about design engineering of machine tools

Level TS-Phylum
Specialized engineering design science (SEDS)

Level IS-Family, sub-divided for TS-"sorts"

For example Specialized engineering design Science (SEDS) for machine tools in general

For example, Specialized engineering design science for other types of machine tools

Transfer, adoption from higher level of abstraction

Completions (additions)
For example, about design for metal lathes
△ WS   Works size
△ O   Originality
△ No   Quantity
△ TS   TS-"sort"

For example, specialized engineering design science (SEDS) for metal-cutting lathes

(compare Figure 4.7)

Indicated statements are qualitative — areas of segments are not intended to represent orders of magnitude.

FIGURE 7.3   Constitution of branch (domain) information for design engineering.

the concrete case (series size, special wishes). The latter elements could already be considered on the higher planes of the hierarchy.

Technical process (see Chapter 5) and technical system (see Chapter 6) influence the essence of the design processes. In the area of TS, the particulars and peculiarities of an individual design process are caused especially by (see also Section 6.11)

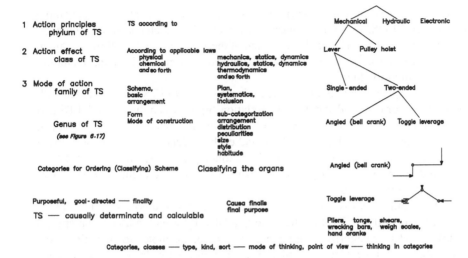

FIGURE 7.4 Hierarchical analysis of categories for engineering design information.

function and mode of action (mode of operation, action principle) of the TS(s); features of individual specialties or industry branches; degrees of complexity; plant to individual constructional part; level of originality of the TS(s), novel design, adaptation, variant, reuse; manufactured number of TS(s), mass production to single item "one-of-a-kind" production; and size and other features of organization where TS(s) are manufactured—TS(s) manufactured in small, medium, or large organizations.

Some examples should illustrate these characteristics:

1. When designing internal combustion engines or Kaplan turbines, conceptualizing will rarely be needed, because the TS within each "sort" almost always use the same mode of action. The organ structure is the "prior art" of existing TS. The situation may be different for the partial organs, which can be newly conceptualized.
2. For material-processing and special TS, the phase of conceptualizing forms the decisive phase of the design procedure in which the quality of the TS(s) is established.
3. The structure of plant exists predominantly of available and purchasable parts. The main task for design engineering of plant ("project engineering") is selecting the elements and establishing their relationships.
4. The engineering design process for a car in mass or series production, is different to design engineering for "one-of-a-kind" or "made-to-order" manufacture.
5. Only the form-giving phase (definitive layout and detail; see Figures I.22 and 4.1) is applied if a variant is to be designed; only some of the information of a TS is changed.
6. Large organizations mostly possess better technical and financial possibilities. The division of labor reaches deeper, and the available information is more widespread.

If each of these different categories of design engineering is taken as an example of a potential SEDS, different suggestions for classifying characteristics emerge, depending on which criteria are chosen. Each of these categories can allow setting up a specialized science, and therefore a specialized information system, including "know-how" in individual areas.

SEDS and Specialized Theories of Technical Systems (STTS) are concerned with individual categories of TP and TS. This relationship between the general TTS and an STTS at the TS-phylum and TS-family levels is illustrated in Figure 7.3. Each branch of engineering, each organization, and each product (TP and TS-"sort") will have its own set of STTS. According to the taxonomy presented in Figure 6.17, an STTS could be formulated at any level of abstraction in the hierarchical arrangement of TS (see also Figure 6.19).

STTS have the task to concretize the general statements (theories) of EDS [315] for a particular "sort" of TS, to collect the available information about that special TS-"sorts," and to categorize it in suitable ways, for example, for machine tools (or for milling machines), for agricultural machinery (or for dairy machinery), for hydro-power equipment (or for Pelton-type water turbines). The advantages are as follows:

1. Adapting the existing scientific theories and other information into the framework of the general TTS should expand and consolidate the information.
2. Cross-fertilization between different branches of engineering should be much easier.
3. Experiences should be more easily transferable between individual areas, and between practice and research—in both directions.
4. Some major failures may be avoided that are attributed to "lack of appreciation for one technical area of information that is available in other areas" [572].

From this point of view, the following hierarchy of ordering characteristics appears appropriate:

1. Complexity of TS: see Figure 6.5.
2. Branch, field of expertise: see Section 6.11.10: complexity level III "machines" is most applicable for this consideration.
3. Level of originality of TS: see Section 6.11.1, and compare Figure 6.3.
4. Kind of production: see Section 6.11.5.

A "branch EDS" can be considered as an element of the partial quantity of sciences within each complexity level; and an "originality EDS" is an element of the partial quantity of each branch, and a "kind of production EDS" is an element of the partial quantity of each originality level.

This is illustrated (see Figure 6.19) by subdividing TS from the degree of complexity, and associating the TS-"sorts" with the individual levels, with examples (see Figure 7.4).

Regarding the completeness of EDS in individual areas, the individual manifestations of the ordering characteristics make possible a general information system that is broadly applicable and differs only in certain special aspects from the other information systems in the region. Information systems that belong as elements of several levels and kinds of sciences can emerge in this way, for example, classes of "design for properties" (compare Figure 6.15), a part of "DfX." Thus a complete SEDS can also be assembled from several "independent" elements, including existing publications on specialized engineering topics (see also Section 7.2.6), which are joined to the kernel of the branch (domain) information.

7.2 SPECIALIZED THEORIES OF TECHNICAL SYSTEMS AND THEIR IMPORTANCE

If all the information about a particular "sort" of TS is collected, and structured according to the general EDS [315] (see Chapter 12) and its constituent TTS [304] (see Chapters 5 and 6, and Figure 7.5), then a more uniform and usable structure of all technical information for that TS-"sort" can be achieved. The resulting information system (see also Chapter 9) is then based on a STTS, in the "southwest" quadrant of Figures I.4, I.5, and 12.7. This STTS describes the particularities of the TS-"sort," and inherits many of its characteristics from the general theory. The STTS then provides the framework (scaffolding) and classification criteria into which the branch information can be sorted, recorded, and cross-referenced.

The quality of this knowledge and collected information about a TS-"sort" is decisive for the work of engineering designers. The structure of the general TTS can be incorporated into the existing (but restructured) knowledge to generate an STTS, whereby completeness, uniformity, and ready understanding can be achieved. The STTS should then be able to provide a source for instructions, information, recommendations, examples, masters, both for the execution of the TS design task, and for the procedures that can be followed. Further adaptation to the design situation (see Chapter 3) is still needed, including *completions* to the information system (see Figure 7.3, compare also Figure 11.6).

A comprehensive theory that addresses these questions will be different from the theories available in technical books. Parts of the existing theories will be directly applicable, parts will need adapting or transferring between areas, and still others need to be newly developed.

The importance of this concretization step is very high, from several viewpoints:

1. For acceptance of EDS in general—designers are overtasked when they are asked to use the general statements for their particular work: they prefer "their tried and tested" (and usually intuitive) working styles and procedures.
2. It is mainly by applying the theories, and other information, to a real design task that the goals and possibilities of EDS begin to be understood.

The specialized theories about the subject of the TS-"sort" will lead to recommendations about appropriate methods for design engineering (see Figure 12.10).

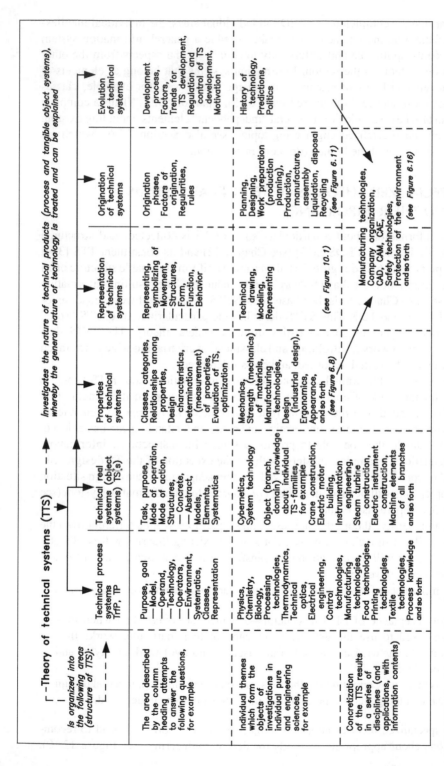

FIGURE 7.5 Regions of the theory of technical systems and information content.

The STTS are therefore faced with a complex of questions (see Figure 7.5):

1. "Sorts" of TrfS where a particular TS-"sort" may be employed—analysis of the TrfS, including operands, technologies, operators, and so forth.
2. Output effects that may be selected as tasks for TS-"sorts"—their typical function structures, variations, and modifications.
3. Modes of action and operating principles of TS usable or available for use for individual functions, as presented in some design catalogs (see Section 9.3), and some books, for example, about machine elements.
4. Typical organs (function carriers) and their relationships—organ structures, for example, some of the information available in the appropriate design catalogs (see Section 9.3).
5. Typical TS-properties, including general design properties and design characteristics, and their manifestations in previously designed and conjectured TS—comparisons, evaluations, quality problems, achieving the desired properties.
6. Problems of generation (ontogenesis) of the TS-"sort," including generating the required properties by design engineering and manufacturing the TS.
7. Experiences of development, and possible trends.

The general TTS provides systematic questions, yet even experienced engineering designers find difficulty answering them spontaneously from their experience. Additional analysis and interpretation of experience and theory is necessary, during which new insights and terminology arise. This forces new thought and brings deeper understanding and new insights.

Information that is not readily available must be searched out and verified in order to obtain a practical and reliable information system that serves the engineering designers. Individual categories of specialized theories built from different viewpoints exhibit overlaps, as shown in Figure 7.6 for mechanical engineering systems. It would be possible to find a "theory of steam turbo-machinery" either in a "theory of thermal machines" or in a "theory of turbines," and consequently the related prescriptive information may be available from various sources.

When concretizing the design characteristics (e.g., technology of transformation, mode of action, or mode of construction; see Figures I.8 and 6.8–6.10), classifications will emerge that offer a source for further possibilities. Using the triad "subject–theory–method," Figure I.2, indications about suitable masters (Section 9.2) or forms (Section 9.5) can be obtained. These items belong to the designers' information system; see Section 9.1.

In analogy with Figure 6.17, a starting point for formulating a hierarchy of special theories can be found in the levels of abstraction, for example, a theory of machine tools ("phylum" level), a theory of metal-cutting lathes ("class"), a theory of universal lathes ("family"), and so forth (compare Figure 7.3).

Formulations of specialized theories may be developed at lower levels of complexity of TS, for example, theories of mechanisms, linkages, machine elements, special theories of constructional parts, subassemblies, and modules of machine systems. Other areas of TS lead to similar examples.

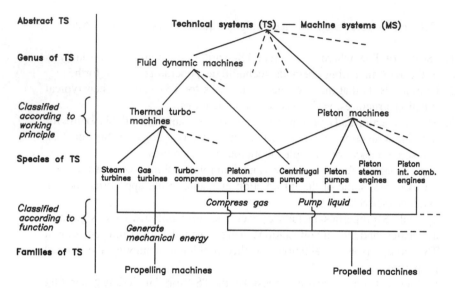

FIGURE 7.6 Regions of technical information—partial hierarchy.

To develop an accurate idea of the contents, these fundamental information elements are described, and machine tools serve as example of the TS-"sort."

7.2.1 Theory of Phylum of Technical Systems

Starting from the "map" in Figures I.4, I.5, and 12.7, the contents of the "southwest" quadrant can be specified through those part segments that are shown in Figure 7.5. Machine tools (e.g., phylum) are analyzed and described, typical functions and features are found, and solutions cataloged. For the classes within the phylum, for example, lathes, milling machines, or planing/shaping machines, the new information is captured in the completions and would generate the SEDS. Particular attention should be given to specific properties of the phylum (or the class), for instance, to precision, stiffness, or stability. The SEDS could also be set up at two levels, for example, phylum and classes (see Figures 6.17 and 6.19).

7.2.2 Theory of Design Processes for TS-Phylum

The theory in this "southeast" quadrant should describe the special design process models for the phylum and the classes (see Figures 6.17 and 6.19), based on the treatment in the general EDS, and emphasize the specific conditions of the field—designers, working methods, representation, working means, management, or working conditions.

7.2.3 Technical Information about Designed Objects

The statements of information in and around the two "northern" prescriptive quadrants are more concrete and application oriented in comparison with the two "southern" quadrants. The contents of the "northern" quadrants are directed toward providing practical and heuristic information and are concerned with three identifiable aspects— the first aspect is the "northwest" quadrant, descriptions of existing ("as is") TS, analytical knowledge about them, and experience with them.

7.2.4 TECHNICAL AND HEURISTIC INFORMATION ABOUT DESIGNING

The "northeast" quadrant asks: "With what abstract or concrete tangible elements (e.g., organs, constructional parts) can the TS be realized?" (second aspect of Section 7.2.3), and "with what procedures and methods can the structure, property, and so forth, be achieved or realized?" (third aspect).

With respect to the second aspect, suitable instructions and heuristics should support efforts to establish fitness of the proposed TS for different life phases (see Section 2.5). The instructions as "know-how" of the engineering designers are generated as individual information segments. Some are fairly well known, such as design for manufacture and assembly (e.g., [45,46,73,74]), and may be computer supported [73,74]. Others wait for adequate formalization, for example, design for maintenance, or the environment [582], where only partial information exists, either in other areas or in a form that is not directly usable or understandable for designers. For example, "design for manufacture" should be derived from manufacturing technology (see Figure 7.7).

The distribution of this kind of information on the individual hierarchical planes of Figure 7.3 is formed in a different arrangement from the relevant theory. Concrete statements can be made on the general levels, because the majority of the information is universally valid. On the lower planes only specialties (completions) are added, for example, instructions for machine tool beds of reinforced concrete as part of the "design for material and manufacturing" area.

This "northeast" quadrant also contains practical information about *design engineering* (third aspect from Section 7.2.3), that is, about the design processes. Predominant information is given by methodical references: procedural models, methods that support the individual design steps, practical references for application of the working means (e.g., computers; see Chapter 10), management, or working conditions. The general references and rules of conduct need to be adjusted for the branches, domains and specialties, and for well-defined conditions of a certain organization, or even for one particular design group or certain designers, compare Figure 3.1. This adjustment is not easy. The relationship of the working methods with the object and with the level of the branch (domain) is evident in Figure 12.7, and a derivation path (compare Figure 4.6) is shown in Figure 7.2.

7.2.5 SUMMARY

State of the Knowledge: The model of the EDS at different levels of abstraction, and as elements of the holistic information basis for designers, shows clearly the advantages of such an information system. Many of the information elements are

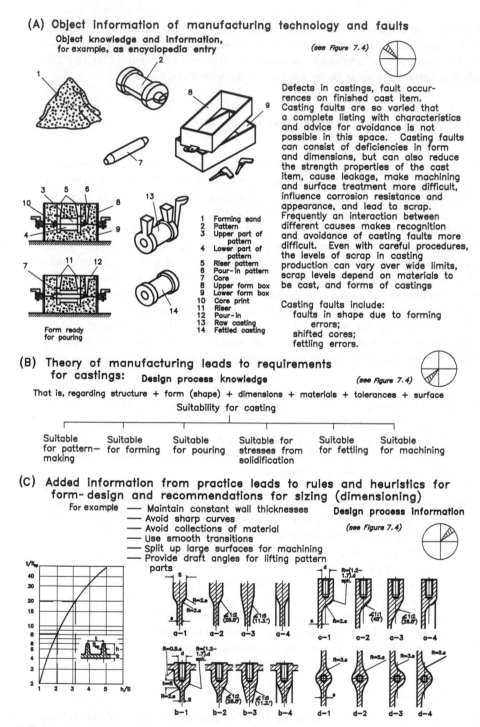

(A) Object information of manufacturing technology and faults

Object knowledge and information,
for example, as encyclopedia entry   *(see Figure 7.4)*

Defects in castings, fault occurrences on finished cast item. Casting faults are so varied that a complete listing with characteristics and advice for avoidance is not possible in this space. Casting faults can consist of deficiencies in form and dimensions, but can also reduce the strength properties of the cast item, cause leakage, make machining and surface treatment more difficult, influence corrosion resistance and appearance, and lead to scrap. Frequently an interaction between different causes makes recognition and avoidance of casting faults more difficult. Even with careful procedures, the levels of scrap in casting production can vary over wide limits, scrap levels depend on materials to be cast, and forms of castings

1 Forming sand
2 Pattern
3 Upper part of pattern
4 Lower part of pattern
5 Riser pattern
6 Pour-in pattern
7 Core
8 Upper form box
9 Lower form box
10 Core print
11 Riser
12 Pour-in
13 Raw casting
14 Fettled casting

Form ready for pouring

Casting faults include:
   faults in shape due to forming errors;
   shifted cores;
   fettling errors.

(B) Theory of manufacturing leads to requirements for castings: Design process knowledge   *(see Figure 7.4)*

That is, regarding structure + form (shape) + dimensions + materials + tolerances + surface

Suitability for casting

| Suitable for pattern-making | Suitable for forming | Suitable for pouring | Suitable for stresses from solidification | Suitable for fettling | Suitable for machining |

(C) Added information from practice leads to rules and heuristics for form-design and recommendations for sizing (dimensioning)

For example — Maintain constant wall thicknesses   Design process information
           — Avoid sharp curves
           — Avoid collections of material   *(see Figure 7.4)*
           — Use smooth transitions
           — Split up large surfaces for machining
           — Provide draft angles for lifting pattern parts

a-1 a-2 a-3 a-4   c-1 c-2 c-3 c-4

b-1 b-2 b-3 b-4   d-1 d-2 d-3 d-4

FIGURE 7.7 Relationship among practical object and prescriptive design information—design for casting.

(A) — **Requirements and typical application**

For example, crankshaft and connecting rod bearings

(B) — **Dimensions — calculation of bearings see Guideline Sht. SB ...**

Select for example

A =   *(see Figure 2.17)*
B =

(C) — **Form - giving**

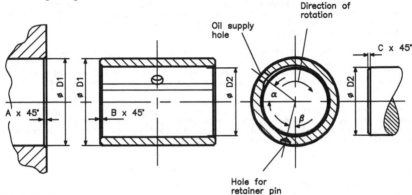

(D) — **Material**

For example

Phosphor bronze   2B 8   for ...
Lead bronze   LB9   for ...

(E) — **Machining**

For example, surface

(a) ..... grind, Rs = 0.4
(b) .....

(F) — **Tolerances, fits**

Diameter   D1

Bore   H6
Bushing   r5

(G) — **Type of manufacture**

FIGURE 7.8   Engineering design guideline—sliding bearing (scheme).

available, their displacement and transformation brings order, which is indispensable for theory, education, and practice.

Form of the Knowledge: In constructing the SEDS, its form is important, an aspect that is validated with examples (see Figures 7.7 to 7.9, and Section 9.3).

7.2.6 Evaluation of the Current Branch Information

Specialized branch information is currently contained in books, for example, about steam turbines. Analyzing representative branch books [45,59,213], features of contents can be detected (e.g., for steam turbines). The theoretical basis from physics or other knowledge areas receive particular and deepened treatment for TP and TS-internal processes of the TS-"sort," for example, thermodynamics. Operational processes of the TS-"sort" (and their main technology) are explored in detail, for example, transformation of thermal into kinetic energy, gas flow, and heat transfer.

Engineering design catalogs: A collection of useful information for engineering designers, external to human memory. They are usually presented in the form of tables, which are arranged according to methodical viewpoints, with information as complete as possible (within a given framework) and systematically categorized. They permit targeted search for particular content. Most catalogs contain parts that categorize, present, and extend the information, and appendices if needed. Operation catalogs do not usually contain extension parts.

Categorization part: That part of an engineering design catalog that determines the systematic arrangement and completeness of its contents. It should be finalized during development of the catalog, such that it also delivers a framework for theoretically possible contents, even if such contents are unknown at the time (they yield empty fields in the presentation parts). The categorization part contains exclusively the characteristics for classifying potential contents.

Presentation part: That part of an engineering design catalog that presents the content, that is the objects, solutions, or operations. The contents are represented by terminology, laws, symbols, formulae, sketches, drawings, and so forth Those contents that are known do not necessarily fill all available fields in the presentation part, they can be completed at a later date.

Extension part: That part of an engineering design catalog which contains searchable information that is not classifed by the categorization part. The extension part (a one-dimensional catalog) is adapted to the needs of the user, can be extended as desired without changes to the categorization or presentation parts.

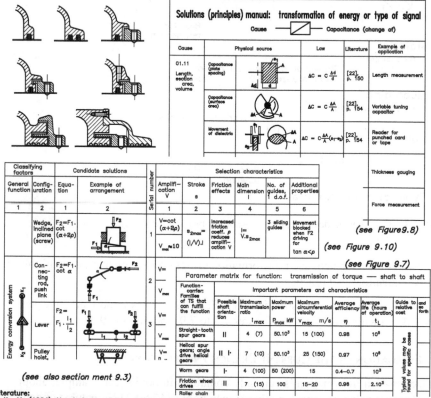

Literature:

Roth, K. (1994) *Konstruieren mit Konstruktionskatalogen (Designing with Design Catalogs)* 2nd ed. 2 vols., Berlin/Heidelberg: Springer-Verlag
Koller, R. (1985) *Konstruktionslehre für den Maschinenbau (Study of Designing for Mechanical Engineering)* 2nd ed., Berlin/Heidelberg: Springer-Verlag
Ehrlenspiel, K. (1953) *Integrierte Produktentwicklung — Medthoden für Prozeßorganisation, Produkterstellung und Konstruktion* (Integrated Product Development — Methods for Process Organization, Product Realization and Design Engineering), München: Carl Hanser Verlag

FIGURE 7.9   Forms of design catalogs.

Branch information is presented for the specific area, for example, problems of strength, raw materials for high temperatures, control and regulation. Theories of these problems are the subject of different books, for example, university text books for courses, which treat special constructional parts of a TS-"sort," for example, steam turbine rotors, high-speed sliding bearings. They discuss total systems, application, structure, description, for example, of energy systems. Design methodology related to the TS-"sort" is only rarely mentioned, examples may be found in some specialties (e.g., water turbines).

The present status shows that currently available technical information is not coordinated, and not complete. Books are different to each other with regard to content and terminology, and are redundant: in several specialties the same sections of information emerge. "Factory knowledge" often forms a special part of the authors' experience [98,556]. Only some specialties are processed in branch books. The isolated origin and application of branch information makes mutual cross-fertilization among branches difficult.

7.3 CONTENTS AND STRUCTURE OF THE STTS

The goal is to build up a paradigm of an information system for designers in a branch of engineering [315]. This information system, including existing elements, should offer a complete basis to classify its contents and relationships, giving reasons for recommendations.

The procedural model with its individual phases, stages, and steps is recommended as structure to apply this technical information in design engineering. Because individual steps are mostly bound to certain models, these are concretized in the STTS for a certain TS-"sort." The possibilities for concretization are different on the individual hierarchical planes.

As in each other activity, a task should be defined at the start—as a list of requirements, a design specification. The next activity concentrates on establishing the *purpose function*, class Pr1. The remaining requirements for properties of the TS must be respected and established during the progressive (iterative, recursive, and problem solving) design procedures. If the design procedure is systematic and methodical, the considerations about functions are paralleled by those about TS-properties, compare Figures I.22 and 4.1.

Designing for TS-functions is described fairly comprehensively in the procedural model.

In the area of TS-properties, the influence of the chosen design characteristic, Pr10, general design property, Pr11, and elemental design property, Pr12, on other individual properties (e.g., external properties) should be investigated (properties as evaluation criteria), and any further measures (evoked functions) necessary to achieve that property should be listed.

Coordinated with the systematic procedure of design engineering (Chapter 4) and the operating instructions (Chapter 2), the STTS knowledge is arranged into six coordinated levels, the STTS structure: (1) information to generate a list of requirements for a certain TS-"sort," equivalent to stage (P1) of the procedural model, Figure 4.1; (2) information to establish (design) a transformation process (TrfP), and the necessary Tg and TS-effects, including consequences for the operand and for the environment [62,160,186,237,244,582], if this is needed for the problem, equivalent to stage (P3a) of the procedural model; (3) information to establish the needed functions—capabilities and tasks of the TS-"sort," including any available masters, equivalent to stage (P3b); (4) information to the means—organs—to achieve the main functions of a TS-"sort," including action principles and mode of action, equivalent to stage (P4); (5) information to constructional parts to optimally realize the chosen organs, equivalent to stage (P5); and (6) information to achieving individual properties, and their dependencies, "DfX."

These six levels are related to the classification of novelty of TS (see Section 6.11.2). All manifestations of novelty should use level (1), list of requirements. Novel design engineering can use all the remaining levels (see Figure 2.10). Redesigning can start at or abstract to an appropriate level, for example, level (4 or 5) and work toward the higher-numbered levels (see Figure 2.11). Innovative redesigning may need to explore one or more lower-numbered levels.

The same structure is observed in the individual levels as described below, with the exception of level (6), compare Figures I.22, 4.1, and 4.2:

(a) **Introduction**: Short characteristic of the procedure with statement of the properties to be established, that is, tasks of the phase.

(b) **Model**: The model forms the basis of information in the individual phase and should be concretized for the relevant specialty—the TS-family or the TS-"sort."

(c) **Design for TS-functions area**: From an analysis of the concretized model, recommendations for the stage can be established and justified. These are mostly internal properties. All ideas and experiences from the manufacturing and organization area (among others) should be entered as information sources in this section. Thereby a "treasure trove of experience" can be generated.

(d) **Design for TS-properties area**: Statements of measures that should be introduced in the stage, to reach the required range and mixture of TS-properties. These include requirements for evoked functions, for example, safety, reliability.

(e) **Form**: The possible or expected form of the results for this stage are indicated, for example, masters (see Section 9.2), pro forma (see Section 9.5), or guidelines/standards (see Section 9.4).

Producing such information is also the goal of the area "DfX" (see Chapter 8).

Level (1): List of requirements (design specification) for a TS-family or TS-"sort"

(a) **Introduction**: The task of this level is to provide information for analyzing the assigned engineering problem, the life cycle of the TS-"sort," and the requirements for TS-properties. A list of requirements should be obligatory for all design engineering.

(b) **Model**: Concretization of the TS–life cycle model (Figures I.13 and 6.11) to typical masters for the TS-"sort" is the most likely starting point. This means the concretization of the inputs and outputs, secondary inputs and outputs, technology, and operators.

(c) **Design for TS-functions area**: This area is not yet relevant, the functions of the TS-"sort" are derived from the required TS-properties and the TrfP.

(d) **Design for TS-properties area**: The model of the TS-properties (Figures I.8 and 6.8–6.10) shows the classes and relationships among external and internal properties. This model can be used as a "checklist," together with the "DfX" classes of Figure 6.15.

The analysis of the life phases and their operators (see Figure 6.15) determines the important properties that the TS-"sort" should display, for example, hygienic measures for food machines, safety for transport of humans, or a changed operating method for the TP. Transport of TS can demand a series of particular properties, for example, compliance with rail transport profiles.

(e) **Form**: Most appropriate for effective analysis of individual life phases is a sequence of TrfS (Figures I.6 and 5.1), in several variants of masters adapted to each life phase, based on Figures I.13 and 6.14. The list of requirements could be executed as a master or a pro forma (see Figure 2.18, and Sections 9.2 and 9.5), the requirements and their number are relatively stable for a TS-"sort" (see Figure 6.17). For a full understanding of the problem, the engineering designer should study the compiled list of requirements.

Level (2): Transformation process (TrfP)

(a) **Introduction**: The task is to provide information about the TP(s) for the TS-"sort," and to support establishing the TrfP, its technology, and effects from the operators.

(b) **Model**: The transformation process that the TS(s) is intended to operate (Figures 5.1 and 5.4), the "TS-operational process," TP(s), should be concretized and analyzed to typical masters for the TS-"sort." This is one of the most important phases for a novel TS(s) and for design engineering, because decisions have far-reaching consequences.

(c) **Design for TS-functions area**: In most cases the operand (Od2) is demanded as output of the transformation. Decisions should therefore be made about the state of the operand as input (Od1), a suitable technology (Tg), and operators. The TrfP is quite stable for the abstract level "phylum" (Figure 6.17), and is thus a model for its lower levels. The concretized TrfP is an adequate source for the definition of the capabilities of the TS-"sort." The most important decision influences the TgPc and the Tg, which demands that well-defined effects act on the operand. A full description of the effects should be recommended at this level, to obtain an accurate description of the tasks of the operators "human" and "TS" for each technology.

(d) **Design for TS-properties area**: This is not relevant, no task is defined for the TS(s).

(e) **Form**: This is not generally standardized, but should define as clearly and accurately as possible the necessary effects to be delivered by the operators to the operand.

Level (3): Functions, function structure (FuStr) of the TS-"sort"

(a) **Introduction**: The established necessary effects to be delivered to the operand (see level (2) above), are now distributed among the executing operators, the HuS, the TS, the AEnv, and (for manufacturing processes) the IS, or combinations of these. This decision is closely connected with any perceived need for instrumentation, human inputs, automation, and so forth. The effects that are assigned to the TS(s) now formulate its tasks, the TP(s), as the goals of the designer. The TS(s) must be capable of fulfilling these tasks and to respect the human as partner. The receptors and effectors also include the active and reactive functions of the "human interface," for example, enable human inputs to the TS, protect them from moving parts, allow human observation of the TS, and so forth. These main functions must be accompanied by assisting functions.

(b) **Model**: Concretizing the TS-function structure (Figure 6.4) to typical masters for the TS-"sort" is the most likely starting point. The model of the TS-internal functions presents the main functions that deliver the effects to the operand, and the other classes of TS-internal functions, that is, assisting functions, input/output functions, evoked functions, and their relationships.

(c) **Design for TS-functions area**: At this level it is only possible to define concrete main functions of the TS-"sort." Other classes of TS-functions may be indicated if they are considered useful, mostly only in abstract terms.

(d) **Design for TS-properties area**: Since the requirements on TS-properties are already known, some assumptions of the evoked functions can be listed.

(e) **Form**: The TS-function structure should contain a good formulation of the main functions and the "TrfP interface," and assumptions for the active and reactive "human interface" functions, assisting functions, evoked functions, and abstract formulation of other classes of functions.

Level (4): Organ structure (OrgStr) of the TS-"sort"

(a) **Introduction**: The abstract means to realize the main functions can be developed in three successive steps. The first should decide about the mode of action of the TS(s), with an action principle (e.g., mechanical) and a mode of action (e.g., gearing). The mode of action sets main and assisting functions that are realized by appropriate organs (function carriers). In this way the main functions structure can be formulated. One or several functions is/are realized with one or several organ(s) because of an established mode of action. The complete organ structure as organism emerges from individual organs, but their mutual compatibility must also be examined.

NOTE: Computer program subroutines are equivalent to constructional parts, their "call" and "return" statements are equivalent to effector and receptor organs in this sense.

For example, the purpose function "heating to temperature" can use an action principle "electrical," with a mode of action "resistance heating." The main function "form an electrical resistance, supply current" can be realized by an organ "electrical heater bar" with its action surface to emit radiant heat. The task at this level is to provide information about suitable organs, to explore candidate organs for the TS-"sort," and to combine them into candidate organ structures.

(b) **Model**: Concretization to typical masters for the TS-"sort" is the most likely starting point, using the function structure (Figure 6.4) as developed in level (3) above. This consists of (1) the morphological matrix (Figure 4.4) to assign the organs to the main (and assisting) functions and combine the organs into an organism, whilst examining compatibility and (2) the concept sketch to display the concrete organ structure.

(c) **Design for TS-functions area**: Items to be established include (1) masters for action principle, mode of action, catalogs; (2) masters for main functions, assisting functions; and (3) masters for organs, combinations of organs to organism, with alternatives and variants. Information about machine elements (Section 7.5) can be useful, they can be considered as complex organisms that interact with the adjoining constructional parts at their action locations, that is, the connecting and transforming organs.

(d) **Design for TS-properties area**: Items to be established include (1) evaluation criteria for all kinds of functions, including evoked functions where possible and (2) evaluation criteria for other properties for which manifestations and values can be estimated at this point.

(e) **Form**: TS-organ structure with accurate representation of the main, assisting, input, output, and other functions. Abstract representation of organs for other classes of functions.

Level (5): Constructional parts, constructional structure (CStr) of the TS-"sort"

(a) **Introduction**: The task of this level is to concretize the selected organ structure (e.g., concept sketch) into the constructional structure. The constructional parts and their arrangement determine the constructional structure using the selected mode of construction (e.g., modular). The elemental design properties are established gradually in at least two phases. Favorable construction zones (form-giving zones) are defined in the preliminary layout. The definitive (dimensional) layout establishes the complete and accurate form, dimensions, and material, from the bare assumptions to a coordinated and optimized whole.

(b) **Model**: Concretization of the constructional structure in preliminary layouts, definitive layouts, details, parts lists, and so forth, to masters for the TS-"sort" is the most likely starting point.

(c) **Design for TS-functions area**: Items to be established include (1) masters for mode of construction, (2) masters for constructional parts: purchased (bought out, OEM, COTS) and manufactured, (3) masters for arrangement, (4) masters for construction zones, form-giving zones, and (5) masters for design properties.

(d) **Design for TS-properties area**: Ensure manifestation and values of all achieved external properties as criteria for establishing the design properties.

(e) **Form**: TS-constructional structure with representation of all elemental design properties, or at least form, dimensions, and materials; and some design properties as standards.

Level (6): Knowledge about TS-Properties

The range of problems about achieving individual properties is covered, that is, their relationships and measures (means) to guarantee their existence (see Chapter 8), "DfX."

7.4   SORTS OF SPECIALIZED THEORIES OF TECHNICAL SYSTEMS

Technical system are tangible products, classes of TS can be formed from different viewpoints (see Section 6.11), for example, required purpose and TP (cranes, machine tools), action principles (mechanical, pneumatic, electrical), complexity level (see Section 7.4.1), and degree of difficulty, among others.

The TS-"sorts" and their levels of abstraction are useful for formulating information systems, resulting in a hierarchy of layered concretization. The possibility of forming specialized information systems according to other viewpoints exists (see Figures 6.17 and 6.19).

7.4.1   SEDS AND STTS FOR INDIVIDUAL LEVELS OF COMPLEXITY OF TECHNICAL SYSTEMS

The hierarchy of TS can be erected according to the degree of complexity (see Figure 6.5): level IV, plant; level III, machines (instruments, devices); level II, assembly groups; and level I, constructional parts, and each of these subdivisions (units) can consist of additional levels, for example, according to Section 7.3.

The best known units of this hierarchy are machine, instrument, device, apparatus (complexity level III) that deliver a complex of effects (outputs) and that form a material unit, for example, locomotive, vacuum cleaner, television set, or measuring instrument. These can be decomposed, depending on the complexity of their structure, according to different criteria, into partial groups that again consist of assembly groups, modules, down to constructional parts (see Figure 7.4).

The specific features of the individual levels reveal how advantageous a SEDS and information system can be, and at which level of the concretization it could optimally be situated.

7.4.2   SEDS AND STTS FOR PLANT—LEVEL IV

Plant, major projects, and infrastructure are TS of the highest level of complexity (level IV). They consist of machines (complexity level III) down to constructional parts (complexity level I). This kind of complex TS includes manufacturing plant, energy-generating plant, chemical plant, traffic, and utilities infrastructure, which serve different purposes and functions. The modern kitchen, in which food is prepared and stored, and which performs other functions for a family or a social unit, could be called a small-scale plant, because the individual items of equipment are brought together into a unit that is unlikely to be repeated elsewhere. The situation is in some ways similar to special purpose equipment (see Section 6.11.10).

Compared with other TS, a plant displays several characteristic features (but these must be regarded as relative terms). It is technically complex, costly to very costly. It contains several kinds of TS (building, machine, electrical systems, and so forth) and combines many purchasable elements, only part of any plant is manufactured to order. In planning and designing, the emphasis lies in conceptualizing and establishing the TrfP/TP, and the arrangement of the chosen subunits, whereby the technology of the TrfP is the dominating element in assessing the anticipated or realized quality. The operators "active, reactive, and general environment" play an unusually important role, for environmental impacts, socially and politically.

Design planning or project engineering of plant is therefore a particular process. The planning process is broadly formalized (e.g., [429,486]) and reacts to particular demands. Questions about potential costs are asked very early in the process and should be contained in the tenders offered to a "request for proposals." The location of the proposed plant must be established immediately, and the solution must react to the location and its conditions. Because a plant generally influences a wider active and reactive environment, often in ways that are not clearly obvious, building a plant is subject to several national and local laws, regulations, and permissions, for example, building inspection, fire department approval, trade supervision, safety regulations, work protection, consideration of transport accessibility (for equipment and operational personnel), environmental impact, life cycle assessment [62,160,186,237,244,582], and so forth. The planning engineer (or architect) is often at the same time the realizer of the project, supervising its construction, erection, testing, commissioning, and so forth.

Planning proceeds as a rule in several stages: (1) preplanning, investigative project, feasibility study, in which an approximate idea of the process, the plant structure, and also the initial costs and expected revenues should be established, usually as a business plan; (2) elaboration of the project—this delivers accurate statements about TrfP (and Tg) of all corresponding streams (M, E, I; forward working and feedback control); layouts of machines, apparatus; plans for electric installation, piping, heating, ventilating, roadways, maintenance areas, and so forth; a complete estimate of all costs; and (3) granting of building permits, and of construction contracts and building project subcontracting (including some more detailed designing).

Plant engineering is demanding engineer work that requires specialized technical information and is concerned with many risks. Special knowledge was developed to collect and classify the information, for example, cost calculating and the characteristics connected with costs and calculation.

Judging the economics of a project, especially of the cost emerging in the first project engineering phases, is extremely difficult. A method for rough cost estimating consists of creating a simple model to obtain an idea of the dimensions of the buildings, apparatus, and so forth, to establish the enclosed space, and multiply this by experience numbers (e.g., cost per unit volume). Costs can be added for further equipment (e.g., for electrical, heating, sanitary installations). The method and numbers form the technical knowledge of the project engineer.

Plant is realized in single-item production, although many of its constructional parts can be made in larger quantities. With regard to the effectiveness of such information systems as goal, the selected level to construct the SEDS and STTS (see Figure 6.19) should be as high as possible. Here the context to concreteness and similarity aspects (e.g., the representation of a plant) tends to be lost. The information system for design engineering can then be completed with the broadest special information for individual levels (compare Figure 7.1).

7.4.3 SEDS AND STTS FOR INSTRUMENTS, MACHINES, APPARATUS—LEVEL III

The complexity level III of TS—machine, apparatus, instrument, device, house, bridge—is best known to the users, manufacturers, and engineering designers. For an organization, these products form an output that necessitates their existence. General explanations about design processes in Chapters 2 to 4 are valid as analogs for all complexity levels. The description of this complexity level can be established as a TS with capabilities (especially TP) that are required by the consumer (technical unit). Although the requirement is not necessarily presented as a direct contract (Figure 6.18, lower section), the supplier will anticipate the customers' requirements and speculatively market a TS (Figure 6.18, upper section). This TS forms a spatial unit, with an extensive diversity of types, each with different requirements, and manufactured mostly in large to very large numbers of pieces. They show continuously increasing requirements (functionality) in depth and breadth, are found mostly in a development series, have many predecessors, and at the same time are accompanied by members of a size range and by competing products.

When designing this complexity class, the phase of conceptualizing (compare Figures I.17 and 4.1) is rarely needed, the new TS(s) retain the existing mode of action, that is, the organ structure remains unchanged. The goal is mostly a development, by redesigning, adapting, transforming, or generating variants. The emphasis lies in the preliminary layout, in which the individual organs and their forms are established. Innovation usually takes place in individual organ(ism)s, modules, or assembly groups, when higher abstractions may be useful.

In the layout stage, designers strive to realize the new requirements for the TS(s) to become marketable, to increase fitness for manufacture, and to lower the manufacturing costs. Design engineering adjusts to the quantity of the product to be manufactured and must consider the high demands of production and assembly. In most cases a prototype (or a set of prototypes) is first prepared, realized, and tested, after which the documentation for series and mass production is executed. The degree of automation of the manufacturing plant must also be considered, because raw material costs, rational production, and assembly play an important role. Design and production use "simultaneous" or "concurrent engineering," TQM and QFD (see Chapter 8), "integrated product development," and "industrial design" where applicable, so that

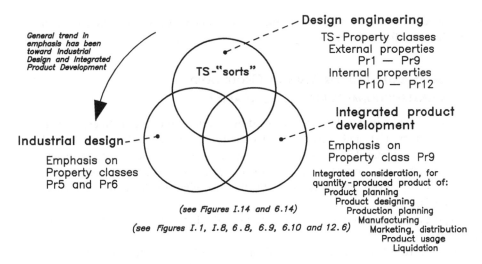

(see Figures I.14 and 6.14)

(see Figures I.1, I.8, 6.8, 6.9, 6.10 and 12.6)

FIGURE 7.10   Scope of sorts of designing.

problems of manufacturing and marketing are anticipated and avoided in the prototype. The last two have much in common and also substantial differences (see
Figure 7.10 and Section I.7.1).

Time to design is shortened as much as possible; all work proceeds under time
pressure. The design process in its structure and procedure depends on the size of the
organization, and on the available design technology (see Chapter 10). The design
specification is an important element that influences design engineering and establishes the complexity of the TP(s) and TS(s). Some further elements are seen in the
design situation (Figures I.12 and 3.1).

Few methodical problems exist in this class. The biggest concern is the branch
(domain) information, in and around the "northwestern" quadrant of Figures I.4, I.5,
1.6, and 7.1, which describes existing TP and TS ("as is" state), and the "northeast"
quadrant with object information about designing the products ("as should be" state)
and design process information. The situation is different for new products, within
"integrated product development."

In individual branches, particular demands that require special information as
peculiarities of these areas may emerge. Due to the severe hygienic regulations, the
food, pharmaceutical and medical equipment industries, and their machines require
the possibility and need for ease of cleaning and sterilizing. The TS(s) must have a
state and condition of the effector surfaces, which make cleaning possible, surface
quality with no porosity that could support the growth of bacteria, accessibility of the
areas to be cleaned and disinfected. Disassembly and reassembly of the machines is
to be done by personnel with little technical training. Suitable coverage (enclosure)
from outside must be provided, so that disturbances are avoided. No touching of
sensitive metals by aggressive materials (especially foods) can be permitted. Special
sealing and containment organs are needed.

This information is mainly available, but its range and accessibility in the records
varies from maker to maker. The SEDS and STTS need revision, and a taxonomic

level selected at which the range of problems should be processed, to efficiently and economically fulfill its function.

7.4.4 SEDS AND STTS FOR ASSEMBLY GROUPS AND CONSTRUCTIONAL PARTS—LEVELS II AND I

The lowest complexity level of TS is formed by assembly groups, modules, mechanisms, organisms (II), and constructional parts (I). A 1:1 coincidence "elemental function = elemental TS" does not exist, organizational aspects (e.g., sequence of assembly) can predominate over functional aspects. In mechanical engineering, both complexity levels are represented by the discipline of "machine elements" (Section 7.5).

Varying and abundant manifestations of individual features are typical (see also Chapter 6).

These TS fulfill simple tasks (i.e., TP) in many different conditions but are not necessarily designed for a specified block of functions (therefore not as organs, function carriers, or functional units). They occur as "species" in many different machines, that is, gears, bearings, shafts. They occur either in mass production (screws, rolling bearings, seals) or in single item or small series production (special casings or shafts). Their form, raw materials, dimensions, tolerances, surface qualities, and arrangement are elemental design properties that determine all other properties (see Figure 6.9).

7.5 MACHINE ELEMENTS—DESIGN ELEMENTS

Machine elements are conventionally defined as TS that are frequently used as known and proven solutions for functions. They appear as constructional parts (e.g., shafts—level I), or as assembly groups, modules, subassemblies, or mechanisms (e.g., rolling contact bearings, couplings, brakes, power transmission units—level II). Their action locations interact with adjoining constructional parts (as organs), and many of these organs also constitute "machine elements" in the conventional literature [133,188,342,521], for example, press fits, splines and serrations, bolted joints, and so forth. Mechanical principles such as hydrostatic, hydrodynamic, thermal, or control elements usually do not appear [33]. Similarly recurring parts exist in other engineering disciplines, for example, steel-reinforced concrete columns (civil engineering), distillation columns (chemical engineering), diodes (electronic engineering), and so forth. A wider definition to include these parts is needed, for example, as *engineering design elements* [573].

The main process of engineering design elements concerns energy. Materials and information need energy for their processing. Each type of energy is characterized by (1) a static, "across," tension property—a "state variable" and (2) a dynamic, "through," rate property—a "flow variable" (see Figure 7.11). Power is the product of a state variable with a flow variable.

Engineering design elements, as carriers of a function with active connections to other constructional parts, can be regarded as *organisms*. The TS-functions are formulated as generic verbs (i.e., as flows) applied to nonspecific energy (as noun).

Forms of energy	State Variable (static, "across" variable)	Flow Variable (dynamic, "through" variable)	Typical physical quantities	
			Power (= state × flow variables) $P = dW/dt$	Work $W = \int P.dt$
Mechanical translational	Force F	Velocity $V = ds/dt$	$P = F.ds/dt = F.v$	$W = \int F.v.dt = \int F.ds$
Mechanical rotational	Torque $M = F.r$	Angular velocity $\omega = d\alpha/dt$	$P = M\omega = F.d\alpha/dt$	$W = \int M.\omega.dt = \int M.d\alpha$
Fluid	Pressure $p = F/A$	Volume flow rate $\dot{V} = dV/dt$ Mass flow rate $\dot{m} = dm/dt$	$P = \triangle p.dV/dt$ $P = \triangle p.\dot{m}/\rho$	$W = \int \triangle p.dV$ $W = \int \triangle p.m.dt/\rho = \int \triangle p.dm/\rho$
Electrical	Voltage U	Electric current $I = dQ/dt$	$P = U.I = U.dQ/dt$	$W = \int U.I.dt = \int U.dQ$
Thermal	Temperature T	Entropy $\dot{S} = dS/dt$	$P = \triangle T.\dot{S} = \triangle T.dS/dt$	$W = \int \triangle T.S.dt = \int \triangle T.ds$

Literature:
Weber, C. and Vajna, S. (1997) "A New Approach to Design Elements (Machine Elements)", in (see also Figure 9.12) WDK25 — Proc. ICED 97 Tampere, Tampere University of Technology, pp. 685–690

FIGURE 7.11 Forms of energy and quantities.

These verbs include (1) transmit, (2) reduce/increase, (3) connect/disconnect, (4) store, (5) divide/unite, (6) transform, (7) distribute/combine, (8) distribute/collect, (9) separate/mix, and many others.

Compared with Figure I.9: (5) "divide/unite" is a transformation of *structure*; (6) "transform" is a transformation of *form*; (2) "reduce/increase" is a special case of "transform," from one form to the same form; (1) "transmit" is a transformation in *location* and is often a special case of "reduce/increase"; (3) "connect/disconnect" is a special case of "transmit"; (4) "store" is a transformation in *time*; (2) "reduce/increase" can also be accomplished by serial application of two successive "transform" (6) operations, for example, mechanical rotation to hydrodynamic flow to mechanical rotation, as in fluid couplings and torque converters. A systematic classification of design elements according to function verbs [573] is shown in Figures 7.12 and 7.13 and includes several elements in addition to the conventional mechanical machine elements.

These functions, by which the TS, the operand, and the operational situation interact, indicate the analytical methods from the engineering sciences that can help to evaluate a proposed TS(s), and to establish the needed sizes (and sometimes forms) of constructional parts. At first a "quick and dirty" estimate is made [482], followed by a "static" view, for example, maximum stress from static loading. If needed, a "dynamic" investigation and simulation can follow, for example, vibration behavior, for which a four-pole simulation model can be employed [575].

More extensive classifications of design elements are too complex for human searching (compare design catalogs [178,370,498]) and are better implemented using computer hypermedia [64]. Force transfer (in couplings and elsewhere) takes place by different manifestations of closure as shown in Figure 7.14, and these form an *information unit* in the hypermedia classification system. Existing couplings can be classified as shown in Figure 7.15, similar to a morphological matrix (Figure 4.4).

New principles can be found by systematic variation (see Figures 2.14 and 8.7) of active organs, and checklists [178,179] (e.g., Figure 2.14) Systematic classification and variation can assist (individual and team) creativity, and can produce many possible alternatives [176]. These methods increase need for suitable selection and evaluation of alternatives to find combinations that show technical and economic merit.

7.6   SEDS AND STTS FOR OTHER TS-"SORTS"

Degree of complexity and product "sorts" were the most important ordering features for SEDS and STTS. Three other classification characteristics are: level of originality, number of pieces, and organization size.

The range of problems and task of *processes* as systems of operations, and the range of problems of designing of processes, is distinct from designing TS-object systems. Especially in connection with plant, we mentioned the importance of the TrfP, its operand, and the technology.

In chemical engineering, "designing the process for transforming the operand" has been acknowledged and integrated into education [239,241].

Forms of energy	"Function" Verbs		Verbs of TP operand transformation, or of TS-internal function (see Figure I.9)	
	Transmit	Reduce/increase	Connect/ disconnect	Store
Mechanical translational	Rods, links, cables, belts, chains, connection elements (e.g., connecting rod), guidances, ...	levers, wedges, pistons, ...	Ratchet mechanisms, traction couplings, mechanical flip-flops, ... (connect) welding, soldering, adhesive bonding, riveting, ...	Springs ("static" strain energy), counter-weights ("static" potential energy), inertia masses ("dynamic" kinetic energy), ...
Mechanical rotational	Shafts, keys, splines, serrations, cotters, clamp connections, force and shrink fits, couplings, bearings (sliding, rolling), guides, ...	Gears, belts, timing belts, chains, friction drives, flat and vee belts, ...	Clutches, brakes, ...	Torsion springs ("static"), flywheels ("dynamic"), ...
Hydrostatic Pneumatic	Pipes, tubes, fittings, ...		Valves, ...	Pressure vessels, hydraulic accumulators, ...
Hydrodynamic Aerodynamic	Vanes, guides, wings, ...	Diffusers, ...	Valves, ...	
Electrical	Wires, fuses, ...	Transformers	Insulation, switches, ...	Capacitors ("static"), accumulators, inductors ("dynamic"), ...
Electronic (analog, digital)	Conductors, ...	Amplifiers, attenuators, chokes, inductors, ...	Transistors, diodes, ...	Magnetic memory, ..
Thermostatic	Heat pipes, cooling fins, ...	Heat exchangers, boilers, condensers, ...	Thermal insulation, fractionators ...	Heat storage units (fire-brick), reactor vessels
Thermodynamic	Combustors, spray nozzles, ...	Heat pumps, ...	Diverter channels, ...	
NOTE:	Special case of "reduce/increase" if value and form of output = input	Special case of "transform" (see Figure 7.12) if form of output = input	Addition to 'transmit'	

NOTE: Functions should normally be formulated as a combination of a verb (or verb phrase) and a noun (or noun phrase). Both the verb and the noun should be chosen appropriate to the specific TP and TS-"sort". See also Figure 9.9.

Literature:
Weber, C. and Vajna, S. (1997) "A New Approach to Design Elements (Machine Elements)", in WDK25 — Proc. ICED 97 Tampere, Tampere University of Technology, pp. 685–690

FIGURE 7.12 Classification of engineering design elements for some "function" verbs.

7.7 WHO SHOULD PROCESS THE INDIVIDUAL SORTS OF SEDS AND STTS?

Ideally, a few representative branch areas would be presented in engineering schools, with an outline and explanation of the STTS for some classes of TS (e.g., machine tools, compare Figure 6.17). Within an organization, the information system should be generated according to the speciality of the TS-"sort," and within a design office the concentration can be on the TS-genus or species. Engineering designers can start from the "species" level and complete their contract.

From these examples it is possible to derive some significant criteria about TS from which special theories may be classified, particularly: (1) their function, "pumping a fluid," "cutting metal," and so forth; (2) their mode of action: for

Forms of energy — Transform to: / Transform from:	Mechanical translational	Mechanical rotational	Hydrostatic Pneumatic	Hydrodynamic Aerodynamic	Electrical	Electronic	Thermostatic	Thermodynamic
Mechanical translational	(see "reduce/increase" in Figure 7.11)	Wheels, rack and pinion, slider-crank, motion screws	Cylinder/piston		Linear generators	Linear transducers piezo, lvdt, strain gage	Linear brakes	Reciprocating gas compressors
Mechanical rotational	Wheels, rack and pinion, slider-crank, motion screws, cams	(see "reduce/increase" in Figure 7.11)	Hydrostatic pumps	Turbo-pumps	Rotary generators	Rotary transducers	Rotary brakes	Rotary gas compressors
Hydrostatic Pneumatic	Cylinder/piston	Hydrostatic motors	(see "reduce/increase" in Figure 7.11)			Pressure sensors, specific gravity sensors		
Hydrodynamic Aerodynamic	Water jet	Turbines	Pitot-static tubes, fluidic control units	(see "reduce/increase" in Figure 7.11)	Magneto-hydro-dynamic generator		Throttles	
Electrical	Linear motor, actuator, solenoid	Rotary motor, rotary actuator	Magneto-hydro-dynamic pumps		(see "reduce/increase" in Figure 7.11)		Electrical resistances	
Electronic	Mechatronics	Mechatronics				(see "reduce/increase" in Figure 7.11)		
Thermostatic	Linear heat engines, combustion engines	Rotary heat engines, combustion engines			Heat sensors, temperature sensors	(see "reduce/increase" in Figure 7.11)		
Thermodynamic	Turbojet engines, turboprop engines, ramjets	Gas turbine engines						(see "reduce/increase" in Figure 7.11)

(see also Figure 9.12, matrix of transformations among physical quantities)

NOTE: Functions should normally be formulated as a combination of a verb (or verb phrase) and a noun (or noun phrase). Both the verb and the noun should be chosen appropriate to the specific TP and TS—"sort". See also Figure 9.9.

Literature:
Weber, C. and Vajna, S. (1997) "A New Approach to Design Elements (Machine Elements)", in WDK25 — Proc. ICED 97 Tampere, Tampere University of Technology, pp. 685–690

FIGURE 7.13 Classification of engineering design elements for "function" verb "transform."

the function "pumping a fluid"—piston, centrifugal, axial bladed, gear action, screw compression, and so forth; for the function "cutting metal"—rotating or reciprocating tool, rotating or reciprocating workpiece, geometrically defined or undefined cutting edge, and so forth; (3) their degree of abstraction: for the function "pumping a fluid"—pumping machinery, gas compressor, roots type diesel supercharger, and so forth; for the function "cutting metal"—machine tool, lathe, turret lathe, and so forth; and (4) their degree of complexity.

	Type of closure principle		
	Form closure		Friction or force closure R
	Rigid contacts F	Flexible contact E	
Iconic Representation	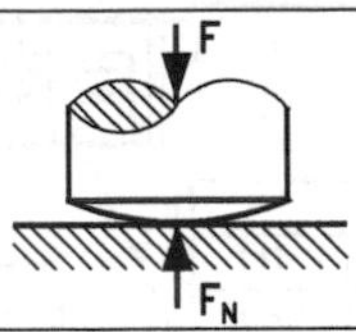	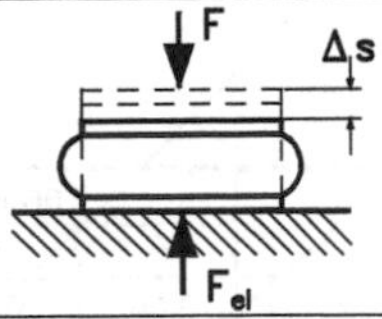	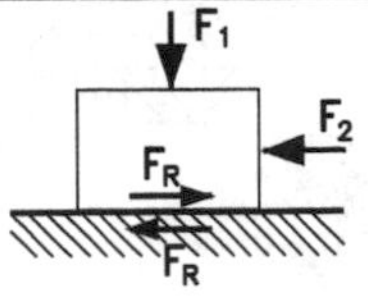
Equation	$F_N = F$	$F_{el} = F$	Static $$F_2 \leq F_R = \mu_S \cdot F_1$$ Dynamic $$F_2 \geq F_R = \mu_D \cdot F_1$$
Types of stress on materials and surfaces	Surface pressure, Hertz contact stresses, Sub-surface shear stresses.	Surface pressure, Elastic/plastic deformation	Surface pressure, Surface shear stresses, Static friction (limited), Wear
Properties	With clearance: shock, plastic defor-mation. Without clearance: pre-load stresses, high precision of force trans-mission.	Relative motion, Rebound (elastic strain energy), Vibrations, Damping, Compensates peaks of force appli-cation	Limited force trans-mission parallel to contact surface, Micro-movement ul edge of contact (danger of fretting), Sliding when horizontal force exceeds static limit.

(see also Figure 2.14)

Literature:
Birkhofer, H. (1997) "The Power of Machine Elements in Engineering Design — A Concept of a Systematic, Hypermedia Assisted Revision of Machine Elements", in *WDK 25* — Proc. *ICED 97 Tampere*, Tampere University of Technology, Vol. 3, pp. 679—684

FIGURE 7.14   Closure principles as classification criteria for machine elements.

The broadest application for the SEDS and STTS is on the hierarchy level from TS-"sort" downward, these are normally in the manufacturing program of an organization. The interest and technical information in an organization provide motivation for developing an STTS.

Engineering education should contain such considerations and expert knowledge, but in a more general form than for the concrete SEDS and STTS needed in an organization. It would be necessary to adopt the plane "phylum" or "class" as a suitable educational level. In technical universities in continental Europe, subjects such as machine tools, turbines, pumps, and cranes are integrated in the educational program—English-speaking regions tend to believe that these subjects should be left to industry. A uniform view should allow a transfer possibility into other speciality regions, it would therefore be necessary and profitable that the schools work out such STTS. Engineering practice would also see an advantage; a concretization of this plane is made considerably simpler. For disseminating EDS [315] into engineering practice, it would be decisive to work out this view of the STTS on such higher levels at the universities. The whole system of these STTS

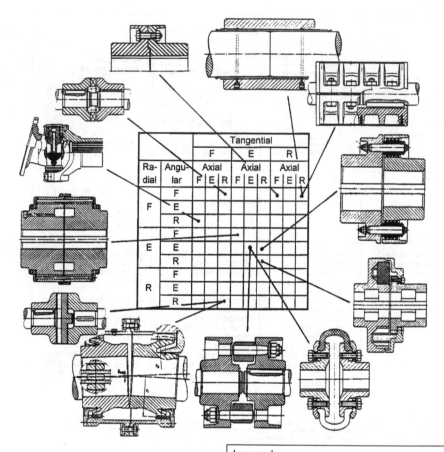

The following table appears within the central figure:

		Tangential								
		F			E			R		
Ra-dial	Angu-lar	Axial			Axial			Axial		
		F	E	R	F	E	R	F	E	R
F	F									
	E									
	R									
E	F									
	E									
	R									
R	F									
	E									
	R									

Legend:
F ... Form closure, rigid contact
E ... Form closure, flexible contact
R ... Friction or force closure
(see Figures 2.14 and 7.14)

Types of mis-alignment:

Angular, intersecting axes
Radial, parallel axes
Skew, non-intersecting angular axes
　combined angular and radial
Axial displacement

Static, no substantial change during operation
Dynamic, substantial change of offset and/or
　angle during operation

Literature:
Birkhofer, H. (1997) "The Power of Machine Elements in Engineering Design — A Concept of a Systematic, Hypermedia Assisted Revision of Machine Elements", in WDK 25 — Proc. ICED 97 Tampere, Tampere University of Technology, Vol. 3, pp. 679–684.

FIGURE 7.15 Classification of couplings according to closure principle. (Reproduced from Birkhofer, H., in WDK 25—Proc. ICED 97 Tampere, Tampere University of Technology, 1997, Vol. 3, pp. 679–684. With permission.)

(coordinated by the general TTS), and the knowledge and experience collected in them, will also satisfy computer experts, because data can be made available in suitable form, categorized and cross-referenced.

Engineering designers are responsible for developing their own information systems, in small steps, starting with the most important areas of TTS. The first most important of these should be the design characteristics and properties, and the TS-structures (see Chapter 6).

Part IV

Support for Design Engineering

This group of supplements relates to *operators* of the design process. Human designers (operator HuS) can be supported by developing, offering, and prescribing methods that are suitable for various design operations presented in Chapter 8. Information systems (operator IS, Chapter 9) provide support in several ways, including masters, design catalogs, standards, and prepared forms (proforma). Technical support (operator TS, Chapter 10) predominantly concerns the use of computers in design engineering. Human resources (operator MgtS, Chapter 11) relates to psychology, team work, management, education, and so forth.

8 Design Methods

Individual design operations (see Chapters 2 and 4), elements of the design process, are often relatively complicated, and need a planned procedure, rules for behavior, *design tactics*, to be used by engineering designers in carrying out those smaller tasks that have relatively concrete problem formulation. The engineering designer's tactical instructions, activities, and behavior in design operations can be controlled by suitable *methods* or *working principles*. The information presented in this chapter is mainly situated in and around the "northeast" quadrant of Figures I.7, I.8, and 12.7. *Representation* has high priority if the solution process is to be successful. This is related to the technical means used to record the results of thinking, even if they are preliminary and transient [438], to allow interaction between the mental and external representations (see Chapter 10).

8.1 DESIGN TACTICS: METHODS AND WORKING PRINCIPLES

"Method" is used to describe large complexes of activities (see Figure I.24), for example, modeling techniques, market, or value analysis (and they contain their own methods). "Method" is also used for more limited procedural instructions, for example, stepping backwards, or the Descartes method, which could also be regarded as working principles, or guidelines.

A method is likely to be most suitable if it is based on and derived in a systematic way from a well-founded theory (see Figures I.2, 4.7, 7.2, and 7.3). A method needs to be adapted, usually by the user, to the design situation (see Chapter 3). Some initial definitions are

1. *Working principles* are general guidelines and *formalized heuristics* for engineering designers, and acceptability of solutions, and so forth, usually without prescribing a procedure.
2. *Methods* are prescriptions of recommended (or sometimes mandatory, normative) procedures, and their steps and operations, *procedural instructions* that can be adapted and flexibly applied to perform a task, for example, operating instructions in Chapter 2.
3. *Tools* are assisting devices (TS, including software) that can help toward implementing the methods and their procedures—they realize parts of the methods.
4. *Techniques* are local and individual adaptations (ways of applying) of the methods to conform to regulations, preferences, and the design situation.
5. "Best practices" are mainly pragmatic methods that are accepted in some organizations.

From the viewpoint of the engineering designer

1. A *method* is a consistent system of formal rules that define classes of possible procedures that should lead from a given starting position (usually with iterations) to a desired final situation. The validity of methods is usually broad. A *methodology* is a set of methods with some logical context (e.g., recommendation for consecutive use) and structural or procedural relationships, that is, a system of methods.
2. The existence of a method or methodology permits the creation of a *plan* to define the sequencing of activities for a particular case. Each method allows creation of several plans, according to the concrete conditions of the design situation.
3. A *personal working mode* is based on the method, suitably and flexibly adapted for the use of (and usually by) the engineering designer, and it can often be used by only one person; it is suitable for that designer's procedural manner, compare Section 4.7.

According to the particular goals, methods may be *classified* into those for finding solutions, evaluating, determining properties, eliminating thinking errors, and so forth, and combinations of these. An important criterion for methods is the human ability or property that is addressed or stimulated. For example brainstorming [448] and synectics [234] use association (see Section 11.1); systems thinking gives a guide toward planned procedures and order; and questioning brings about a consciously discursive thought mode. "Trial and error" and random searches can be regarded as methods. Some bodies of information (Chapter 9) are related to appropriate methods, for example, "Design for X" (DfX) and life cycle engineering [62,160,186,237,244,582].

The designer has a tactical tool in the working principles, short prescriptions that give important *general instructions for appropriate attitudes*. They include rules for behavior, general solution guidelines, and codes of ethics. Some working principles may be regarded as requirements for engineering designers; others present technical information or guidance for design management.

Engineering designers should produce such lists from their experience, and thereby create for themselves an important personal working aid: a *procedural checklist*.

8.2 WORKING PRINCIPLES FOR DESIGNING

Working principles can be classified by the steps and stages of the engineering design process.

8.2.1 GENERAL WORKING PRINCIPLES

1. *Critical acceptance*: do not accept any information without examination and verification.

2. *Control principle*: every result must be examined, using an advantageous control strategy and tactics, and examining important requirements, for example, function, feasibility, economics.

3. *Principle of effectiveness and efficiency*: striving in every process to maximize them.

4. *Principle of economy*: the operation of the TP(s) and TS(s) should be attained in the cheapest fashion. Depending on organization policy and on ethical considerations, this "cheapest fashion" could be the lowest first, running, whole-life, or environmental cost.

5. *Optimization principle*: aim for an optimal solution or best compromise for the given conditions, with optimal care and accuracy, design time, and design cost.

6. *System and totality principle*: every simple or complex object and every process is simultaneously a system in its own right and a system element in a larger system, that is, systems are hierarchical. Connections, interactions, and relationships should be considered.

7. *Principle of recording information*: human memory is unreliable. Every important item of information should be recorded and classified (cataloged) in an economic fashion.

8. *Ordering principle*: every area of knowledge should be classified and cross-referenced.

9. *Overview principle*: create a usable and comprehensive survey, to show elements, relationships, structure, and so forth, of the subject.

10. *Principle of methodical and planned procedure*: guide the systematic and methodical progress of activities in a planned way.

11. *Professionalism in design*: follow ethical rules of behavior, aim for best possible personal performance, and admit faults and limitations of competence (knowledge).

8.2.2 PRINCIPLES DURING "SEARCH FOR SOLUTIONS"

1. *Orientation principle*: carefully determine the state of the art. Think of possible solutions to the problem, using creativity, intuition, and imagination, before applying this principle. Searching for the state of the art can then proceed with better knowledge of the questions to the specific problem, with less danger of mental set or fixation; but consider also the opposite danger of "jumping to a conclusion," and defaulting on this orientation principle.

2. *Accept good existing solutions* (to control the risks), but only after critical examination.

3. *Principle of accurate problem formulation*: for each subdivision (phase, stage, or step) of the problem, produce a sufficiently precise formulation of the problem to be solved.

4. *Principle of abstraction*: abstract from the existing circumstances to find new paths.

5. *Division principle*: divide every complex problem into reasonable subdivisions, define suitable boundaries for the system and subdivisions—see "windows" [438], constructional design zones, and form-giving zones.

6. *Variations–combinations principle*: combine suitable elements into a totality, vary the elements that can perform equivalent functions—the synthesis part of recursion.

7. *Incubation principle*: permit the subconscious to work. Alternate productive phases with rest periods, and allow time for incubation [563] for creative actions.

8. Always find at least *a few meaningful solutions*, in order to have the opportunity of selecting between them—and find these solutions before evaluating and selecting.

9. *Defer judgment*: concretization too early can direct the considerations into a particular direction, resulting in prejudice, mental set or fixation (mental block) [31].

10. Adopt *existing solutions* in sensible proportions: not every step in the design process can result in creativity, and creative solutions and inventions usually carry higher risk.

11. *Cooperation* with other experts can bring good results in finding the optimal solution.

12. *Systematic procedure* is recommended in the step of searching for solutions, but is particularly important in all steps of review and revision.

13. All solutions should be briefly *recorded*, with *comments* where appropriate.

8.2.3 Principles Concerning Quality (Properties) of the TP and TS

1. Design with respect for *function*: then "form follows function" (Walter Gropius), that is, if it is right, it also looks right, but not necessarily the converse.

    a Design for the *market*: fulfill all customer requirements.

2. Design for acceptable *performance*: consider all operating conditions.

    a Design for the *environment*: consider impacts on material, physical, chemical, ecological, social, cultural, societal, political, and other environmental components.

3. Design for *operation*: especially for the TS-operational process, its purpose and misuse.

    a Consider operational safety and human operator conditions, a safety hierarchy [58] states, in approximate order of effectiveness (mostly all five priorities are needed)

        i First priority: *direct*—eliminate the hazard and risk in the TP(s)/TS(s) itself (design the danger or hazard out).

        ii Second priority: *indirect*—apply safeguarding means, precautions (guard or protect against the danger or hazard).

iii Third priority: *indicate*—warn against dangers or hazards, especially for persons, whether users or observers: (1) *directly*, use warning signs (in easily visible places) that are easily understandable in the cultural environment of TS-operation, applied as labeling on the TS(s) or (2) *indirectly*, written into the operating instructions for the TP(s)/TS(s), but note the human attitude of overconfidence or complacency: "if all else fails, read the instructions," and then forget them.

    iv Fourth priority: *train* and instruct, supervise.

    v Fifth priority: prescribe *personal protection* (and ensure that it is useable and used).

  b Aim for minimum space consumption, minimum dimensions.

4. Design the TS for *manufacturability*: achieve the most economic realizability of the product with available manufacturing systems, aim for optimal manufacturing methods. (1) Design the TS for ease of *assembly and maintenance*: examine assembly methods for the TS, avoid special tooling. (2) Design the TP for ease of implementation.

5. Design for *packaging, storage, and transport*: create favorable conditions.

6. Design and plan for *timely delivery*: consider lead-times for tasks, and the supply chain.

7. Design for *liquidation*: disassembly, disposal, recycling, refurbishing, and remanufacturing.

8. Design for the *human being*: offer maximum protection, avoid causing difficult or monotonous human work, aim for minimal human fatigue.

9. Design for *appearance*: keep in mind the aesthetic influence of the product.

10. Design in accordance with *regulations*: take into account applicable standards, codes, laws. Do not copy products protected by patents, trade marks, design registration, and so forth.

11. Design for *economics*: aim for minimum manufacturing and running costs.

12. Design for adequate *strength*: ensure appropriate strength and stiffness of the TS, considering also fatigue, creep, resonances, wear, and other failures or deterioration.

13. Aim for independence of functions, uncoupling of influences, simplicity [528].

14. Aim to reduce information content of the TP(s) and TS(s) [528].

8.2.4 PRINCIPLES CONCERNING REPRESENTATION

1. Aim for clear, complete, and unambiguous representation of the TS(s) in its structures.

2. Aim for economic and optimal representation.

3. Aim for purposeful representation (considering the receiver of the communication).

4. Consider recording, manipulating, storing, handling, distributing, revising, upgrading, archiving, retrieving, and so forth, of documentation.

8.2.5 Principles for Lists of Requirements (Design Specifications) and Criteria for Evaluation

Compare Section 4.3.1

1. The design specification should be complete, orderly, clear, and unambiguous.
2. The requirements should be *formulated* by stating transformations and *effects*, not means.
3. Requirements should be *classified* and prioritized.
4. Requirements should be *quantified* and, where possible, tolerances should be given.
5. The assigned problem statement and the list of requirements are sets of information.

8.2.6 Principles for Verification, Checking, Auditing (see also Chapter 4)

1. The path of verifying and checking should be different from the way of finding solutions.
2. Verify after a suitable time interval, especially if self-examination is to be performed.
3. Examination, checking, or audit should be done where possible, by persons who did not take part in generating and establishing the solution.
4. Estimate and assess the importance of individual parts for the whole, and distribute attention according to this importance.
5. Estimate consequences, probability of errors occurring, and possibility of their discovery.
6. Use systematic procedures for verification and checking, keep full records.
7. Use suitable resources and aids (including outside opinions, computers, etc.).

8.2.7 Heuristic Principles

Heuristic principles describe problem solving as a battle, fought under strict (tactical) rules, which cannot be universally valid; they depend on the subject. To learn them, each person (and team) must work through examples, and abstract successful heuristic principles from them, to be able to reuse and adapt them. Typical principles according to Polya [470] are

1. Introduction and solution of an auxiliary problem—auxiliary problems are substituted, because they may assist toward solving the original problem.

Solving the original problem is the goal; the auxiliary problem is a means by which the goal may be reached.

2. Search for analogies—analogies are present in many thoughts, in daily conversations and in conclusions. They also appear in artistic expressions and high scientific achievements, whereby all kinds of analogies can play a role in discovering a solution.

3. Generalization of the problem—generalizing leads from the consideration of an object or process to the consideration of a larger collective of the objects or processes, or other levels of abstraction, which contain the goal object.

4. Specialization, detailing of the problem—specialization leads from the consideration of a given set of objects (objects) or processes to the consideration of a smaller subset.

5. Induction—induction is a method by which general laws can be discovered by observation and combination of particular cases; see also Section 12.1.9 [174].

8.2.8 PRINCIPLES OF THINKING BY DESCARTES (1596–1650) [127]

The following four rules should be sufficient, *under the condition of a firm and persistent resolution to always follow them*: (1) never assume a thing as (axiomatically—addition by the authors) *true* that would not be recognized as true with certainty and conviction, that is, to carefully avoid undue haste and prejudice, and to only understand as much as it presents itself, clearly and distinctly in the mind, so that absolutely no doubt exists; (2) *decompose* every difficulty into as many parts as would be possible and desirable to achieve a better solution; (3) *organize* thoughts; begin with the simplest and easiest to comprehend objects, and advance gradually and in easy steps up to the recognition of the most complex; and even assume that things are somehow organized, which by nature show little precedence; and (4) produce everywhere complete enumerations and comprehensive *summaries* that nothing is omitted.

8.2.9 WORKING PRINCIPLES FOR ORGANIZATION

Working principles related to Total Quality Management (TQM) [10,108,125,549] advise that members of an organization should be aware—know what else goes on in the organization, what duties are being performed by others, who has relevant expertise and information, who can be called on as a partner in discussions. Contacts need to be formalized (scheduled) to avoid disturbances for the work of others; see Section 11.1. The following references may be useful: (1) Deming's 14 Points for Management [125,126] (two slightly different listings); (2) Juran's 10 Principles [339–341]; (3) Crosby's 14 Principles [108]; (4) Albrecht's Customer Orientation [36,37] of *Total Quality Service*—TQS, with a 17-point TQS Action Menu; (5) Nadler and Hibino's Principles of Breakthrough Thinking [431,432]; and (6) Axiomatic Design [528], but see Section 2.4.3.

8.3 BASIC CONSIDERATIONS ABOUT METHODS—GENERALIZATION OF "METHOD" AND "METHODOLOGY"

Procedural instructions for methods are available, and tend to increase the probability of success. Methods must be adjusted and adapted according to the ever-changing task and design situation (see Chapter 3). The methods may then no longer be clearly recognizable. Employing a design methodology or method implies that the intuitive modes of operation are also needed (see Section 11.1). Indirect employment of a design method can prove useful. Intuitive results should subsequently (retrospectively, reflectively) be incorporated into the systematic procedural plans and models; the procedure can act as control, checking and verification, and support to completeness of the considerations. The goal-guided search for optimal solutions for a design problem, which is formulated at a suitable level of abstraction of the TP and TS-models, can be supported by using an appropriate (pragmatic, heuristic, or theory-based) method. Most kinds of method can be supported by appropriate computer programs. According to kind of the procedure, *discursive* and *intuitive* methods are available.

Discursive methods may be classified as [425]: (1) *algorithmic* methods that follow strict instructions to reach a predictable result with certainty; (2) *heuristic* methods that, if applied adequately, usually allow reaching a reasonable result, and cause the procedure to become more effective, certain, and goal-oriented [365]; and (3) *post hoc* (after the event) analysis, which may show patterns from an intuitive search with no prior instructions. "Discursive" implies that the method recommends a stepwise procedure that leads systematically to one or several solutions. Through application of discursive methods one can approach a previously defined goal by following a plan. Even though the steps of procedure are defined, the content of information being processed is not prescribed. This means that the content (the object-related information) can be collected in any suitable way, including by creative leaps and intuitive approaches. Methods and creativity are not contradictory (see also Section 11.1.7). Through different combinations of known and conjectured elemental solutions, new relationships can be discovered and new solutions found. Examples of discursive methods are morphological matrix (see Figure 4.4), laws of similarity; analogies; structuring methods; reversal [597,598].

"Intuitive" problem-solving methods may lead to new solutions through intuition and spontaneous, unconstrained, opportunistic "second nature" actions. A solution appears to come suddenly into consciousness, and often the fact that previous work has led to the thoughts about solutions is denied by the problem solver. Examples of intuitive methods are: brainstorming; method 6-3-5 (brainwriting); synectics; induced creativity; contradiction-oriented procedures (e.g., ARIZ, TRIZ, Invention Machine, COIS); heuristic principles (Polya) (see Section 8.2.7); formal heuristic methods (see Section 8.3.3); heuristics in general (see Section 8.3.4).

Both discursive and intuitive methods can and should be used together. A conscious, stepwise processing and application of the method is possible, but (1) if the method has been learned well, a subconscious (or unconscious) use can be observed, an intuitive procedure results (see Section 11.1), with possible feedback into the systematic and methodical record keeping; (2) derivation, change, and adaptation of a method to the task and situation are necessary; (3) combinations and connections of the methods, and transitions and intersections between the methods are possible; and (4) in time, the name and essence of a method often pass over into the vernacular, for example, the word "brainstorming" [111,335,448] that was normally operated according to the given procedural instructions is now used for any kind of free discussion or conference to find ideas, and even as a substitute for "problem solving."

Combinations and connections between methods can be expanded up to logical, theory-based sequences, which are then referred to as *methodologies*. The delimitation between "method" and "methodology" is accordingly fuzzy, yet relationships can be found among the methods and methodologies, and to the design steps. Computer-based method catalogs are in development, which also indicate how methods can be usefully combined, for example, [78,527].

Some methods are commercially marketed and can be purchased [149,151]. Marketing is often aimed at management, and methods are advertised with unfounded assertions. References to QFD [75,151,259,269] show statements that "the team designs the product," and it is implied that QFD is the only method necessary for design engineering. The importance of design engineering, the necessary knowledge and experience of engineering designers, and the special nature of engineering design work are thereby lowered.

Engineering Design Science [287,304,305,314,315] has the advantage that the methods and types of modeling of TP and TS are based on a theory. Influences on designing are brought into context (cultural, social, legal, economic, financial, business, in reference to manufacture process and sales office, etc.), for example, see Figures I.8 and 6.8 to 6.10.

8.3.1 Process (Action) Information

Process-oriented engineering information (action knowledge) deals with information about general actions that engineers (including designers) should accomplish. This includes organization and management, diagnosis of errors, analysis and synthesis, and therefore also contains the design process information (see Figures I.4, I.5, and 12.7). The existing knowledge about design processes originates from normal, traditional engineering practices. In part through design research (see Chapters 1 to 3 in [315]), action instructions (methods) were proposed and derived (e.g., [25,457]), as prescriptive information for design engineering.

Process-oriented engineering information consists of two parts, accessible recorded information, and that internalized by the human designer (see Chapter 2). The known and published methods [111,149,151,315,335] represent recorded information. The actually executed stereotypical procedures that can be observed are indications of the internalized, "tacit" knowing. This internalized knowledge can only in part be synthesized from this observation, as a procedure diagram; see also the research approaches in Section I.1—which limits the possibility of setting up a design theory through research from observation protocols.

A hierarchy of design operations [314,315] is shown in Figures I.21 and 2.5; see Section 2.4.1. These steps cannot be completed in a linear, sequential procedure, but must be processed iteratively, recursively, interactively, and with a problem-solving attitude and procedure; see Chapter 4. Parts of the task, and the TP(s)/TS(s), will *at any one time* exist at different levels of abstraction, in different states of concretization, and will therefore need application of different parts of the methodology.

8.3.2 APPLICATION IN DESIGN STEPS

The recommended sequence of administrative phases for designing a TP(s) and TS(s), for example, following VDI 2221 [25], is (1) clarifying the given task, (2) conceptualizing, (3) laying out (embodying), and (4) elaborating (detailing). A subdivision of these phases into stages and steps is used in this book; see Chapter 4, and Figures I.17, 4.1, and 8.1.

All methods have limited application, and can be characterized by their applicability in the different phases, stages, and steps of designing. Published methods can be allocated to design engineering according to several classification criteria:

Allocation to basic operations of problem solving; see Section 2.3.1 and Figure 2.7.

Allocation to main operations, see Section I.12 and Chapter 4, and Figures I.22 and 4.1.

Allocation to modeling of TS (see Chapters 5 to 7): list of requirements (properties, life cycle and its operators, "DfX" classes), transformation (operands, technologies, effects, process structure), function structure (TS-internal process, effects, additional and evoked functions), organ structure (contents—additional and evoked functions and organs), constructional structure (evoked functions and organs, layouts, details, evoked functions and organs for manufacture—jigs, fixtures, tools).

Allocation to analytical application of the engineer sciences—for example, for evaluations, simulations, predictions, optimization of parameters, and so forth.

Allocation to management of the design processes—for example, product planning, product development: QFD; general management: TQM, PERT, CPM; cooperation with production: concurrent or simultaneous engineering; methodical computer support: HKB, CAD, finite elements, boundary elements, CFD (see Chapter 10).

An allocation of some methods to the main stages and the basic operations (first and last columns in Figure 8.1) are shown in Figure 8.2.

8.3.3 FORMAL HEURISTIC METHOD

The formal heuristic method, defined by Klaus [352] as the "science of the methods and rules of discovery and invention," is a special case of trial and error. It differs from the *deductive* methods by operating with conjectures, working hypotheses, provisional models, and so forth; see also Sections 8.2.7, 8.2.8, and 12.1.9, and [174,470].

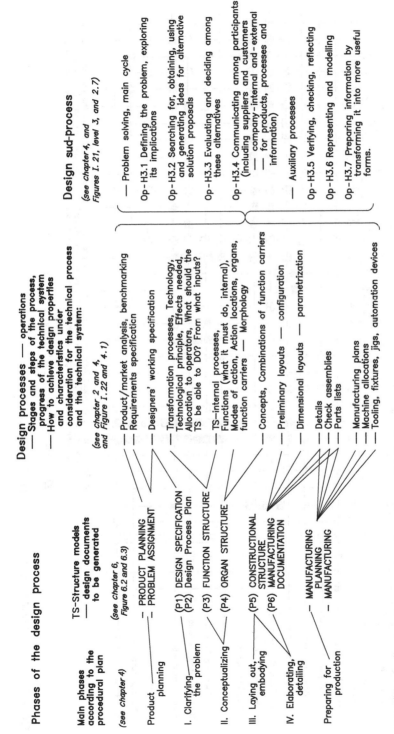

FIGURE 8.1 Design strategy and tactics—phases of designing.

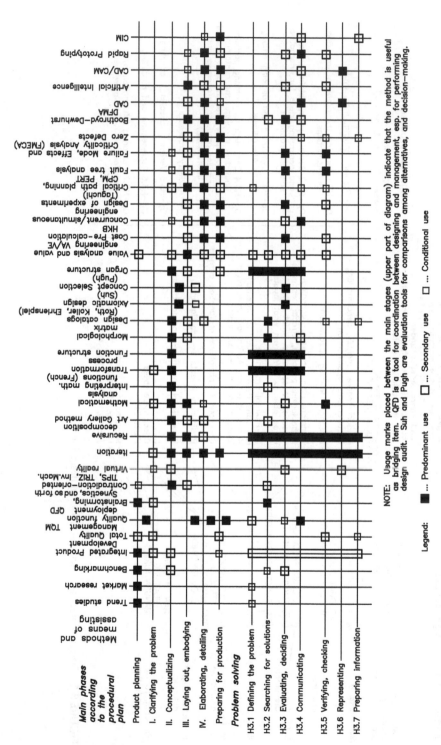

FIGURE 8.2   Usability of methods in phases of the design process.

The heuristic method is a process that assists in the search for an aid toward reaching a proof, not a method of rigorous proof. The heuristic method studies actually occurring discoveries and inventions, and tries to derive general laws of processes for discovering and inventing which do not depend on the concrete task, for example, TRIZ [39,40] is claimed to be an "empirical science."

The formal heuristic method uses results and processes of experimental psychology, information theory, information psychology, and neurophysiology. Heuristic methods can be simulated on computers. "Heuristic" machines work like chess game machines, that is, they are programmed with general strategic rules, and are used according to the situation combined with "trial and error." One example is "systematic heuristics" proposed by Müller [296,298].

The formal heuristic method is an important element of dialectic logic. If a proof is found with the help of heuristic methods, the proof can in general be represented as strictly logical and the heuristic considerations become superfluous; see Section 8.2.7, first principle by Polya [470].

8.3.4 Heuristics in General, Heuristic Data

Heuristics are any statements, guidelines, experience values, items of data, and so forth, that contain scientifically unverified (or even unverifiable) information that is useful for obtaining a specific solution to a design problem [365]. Inverting or rearranging a scientific law or formula does not lead directly to a singular solution—multiple or undefined solutions are the norm. Some of the heuristics and heuristic values may be found in international, national, regional, and organization standards, as voluntary consensus agreements. Examples include maximum and minimum water pressures and maximum recommended water velocity for domestic water supplies.

8.4 CATALOG OF METHODS (ALPHABETICAL)

This catalog cannot be complete, due to development of new, revised, and reinvented methods. Each entry follows the pattern: (1) Title (abbreviation, cross-reference to other entries); (2) *D*—short description of the procedure; (3) *P*—purpose; (4) *L*—literature references (name or reference number); (5) *A*—allocation to steps of the procedural model and basic operations.

Cross-references may be shown as (1) s.—see . . .—descriptions are given only under the alternative title, or included in more extensive descriptions of the method; (2) s.a.—see also . . .—other entries in this list to this description; and (3) rl.—related methods.

These entries may be scanned into a computer to allow search for method according to envisaged usage. Surveys of methods are [31,66,111,141,142,149,151,335], usually with no sequencing.

ABC Analysis (s. Pareto distribution).
Abstraction *D*—Derive a more abstract form of representation of a problem or system from a concrete form, for example, formulate TS-functions from an

existing layout; *P*—Open a solution field to search for alternatives in more concrete representations; *L*—; *A*—All main stages.

Adaptation *D*—Modify or partially transform an existing TP and TS for different conditions; *P*—Obtain a reliable solution for new conditions; *L*—; *A*—General project, any step.

Aggregation *D*—Combine TP and TS subsystems into a single system, or combine functions of several organs into one function or organ; *P*—Achieve new properties, simplified structure; *L*—; *A*—Organ structure, constructional structure.

Alexander's component method *D*—Identify contributing components to a complex design problem; *P*—Allow suitable sequencing of design considerations and decisions; *L*—[31,335], rl. AIDA, Axiomatic design, Design Structure Matrix; *A*—Transformation process to preliminary constructional structure.

Analyze Theorize Delineate Modify *D*—Recommend a problem-solving sequence with forced iteration; *P*—Historic formulation of a theory of design; *L*—[562]; *A*—Basic design operations, s. Section 4.7.1 for other formulations.

Analysis of Interconnected Decision Areas (AIDA) *D*—Identify components of a problem and their relationships; *P*—Grouping of components of the problem to reduce influences of relationships; *L*—[397]; *A*—TrfP to constructional structure (preliminary layout).

Analysis of Properties (s.a. Attribute Listing) *D*—Analyze every TS-property, list all attributes (properties, characteristics) of the thing to be designed, consider each separately to decide whether any can be usefully changed; *P*—Improvement of an existing TS; *L*—[107] rl. Attribute Listing; *A*—General redesign project, any subsystem.

Analysis of Variance (ANOVA) (s.a. Design of experiments) *D*—Mathematical-statistical treatment of experimental data to discover influences of environment properties on the behavior of a system; *P*—Identify those properties that have the greatest effect on behavior; *L*—[201,202] rl. Taguchi experimentation; *A*—Constructional structure (of prototype, or of an experimental test rig).

ARIZ (s. Computer-aided invention search).

Art Gallery Method *D*—Display several concepts or layouts with alternative solution proposals, discuss advantages and disadvantages of competing layouts by viewing the displays in a team situation; *P*—Select and synthesize the most favorable constructional structure from several proposals; *L*—[178,262] rl. concept selection, Pugh method; *A*—Organ structure, constructional structure preliminary and definitive layouts.

Artificial Intelligence (AI) *D*—Simulate some aspects of human intelligence by computer algorithms, for example, to speed up or make more reliable; *P*—Provide computer-resident expertise for diagnostics, give advice for solving problems for which solutions are known to exist, and give warnings during computer-aided design for known fault conditions; *L*—[532,533]; *A*—Mode of action, constructional structure.

Attribute Listing (s. Analysis of Properties).

Axiomatic Design *D*—Use simple axiomatic statements to guide selection among proposed alternative solutions, see also Sections 2.4.3 and 8.2.3; *P*—Apply linear matrix algebra to decision making (but caution, such a selection may be simplistic and unreliable); *L*—[528], rl. Matrix algebra/decision theory [421,523]; *A*—Evaluating and deciding, from limited considerations.

Benchmarking *D*—Provide comparisons among competitors about products and organizational procedures and structures; *P*—Improve own products and organizational procedures and structures; *L*—[154,569,580]; *A*—All stages.

Biological analogy *D*—Study structures of natural phenomena, living organisms and biological matter to derive analogies; *P*—Use patterns of structure as probably optimized by natural selection for the purpose; *L*—[215,297]; *A*—Constructional structure.

Black box (s.a. Function decomposition, Abstraction) *D*—Define inputs and outputs of a process, and its transformation; *P*—Observe and modify the transformation in isolation, independent of the mechanisms that cause the transformation; *L*—; *A*—Transformation process, ergonomics considerations.

Blockbusting *D*—Overcome mental blocks and prejudices by freeing the mind; *P*—Improve search for solutions by avoiding fixation on existing solutions; *L*—[31,111,335] rl. Lateral thinking; *A*—Basic operation "search for solutions."

Boundary Element Method (rl. Computational Fluid Dynamics, Finite Element Analysis, s.a. Chapter 10) *D*—Approximate iterative computer solution of systems of equations with assumed boundary conditions, by dividing a continuous geometric boundary into discrete elements, to model physical phenomena; *P*—Find a close analogy to the behavior of a physical phenomenon; *L*—; *A*—Basic operation "evaluation" (analytical).

Boundary search/shifting (s.a. Systems Approach) *D*—Investigate effects of redefining the boundaries of the system; *P*—Find better distribution of tasks among systems and operators; *L*—[335]; *A*—Any design stage.

Brainstorming *D*—Collect ideas in free discussion without criticism; *P*—Find many solutions to a given problem; *L*—[111,335,448] rl. Brainwriting; *A*—Basic operation "search for solutions" in problem defining and conceptualizing.

Brainwriting (s.a. Method "6-3-5") *D*—6 participants, each write down 3 ideas within 5 min, then pass on to next person for similar 3 ideas; *P*—Find many solutions; *L*—[272,273,297]; *A*—Basic operation "search for solutions" in problem defining and conceptualizing.

Breakthrough Thinking (s.a. Section 8.2.9) *D*—Use guidelines to generate ideas for future products; *P*—Gain advantage over competition; *L*—[432]; *A*—Product planning.

Case-based Reasoning *D*—Derive heuristic advice (not necessarily explained) from previous design decisions by computer-automated capture (using AI techniques); *P*—Provide advice to guide new decisions; *L*—; *A*—Constructional structure.

Cause and effects analysis (rl. Black box) *D*—Analyze consequences ("effects") attributed to causes (causally or statistically); *P*—Clarify relationships; *L*—; *A*—All stages.

Characteristic number method *D*—Use characteristic numbers derived from previous experience to predict expected behavior of a system; *P*—Estimate values of variables that will result in expected behavior; *L*—[72,73,178], rl. regression analysis; *A*—Product planning, conceptualizing.

Check lists *D*—Use suitable lists of items as guideline for considerations; *P*—Completeness of task; *L*—[111,140,335]; *A*—All stages.

Computational Fluid Dynamics (s. Boundary Element Method, s.a. Chapter 10) *D*—Divide flow volume and TS form into discrete elements; *P*—; *L*—; *A*—Flowing fluids in or around the physical form (geometry) of a TS.

Computer-aided design/drafting (CAD), computer-aided engineering (CAE), and computer-aided manufacture (CAD/CAM) (s.a. Chapter 10) *D*—Using a computer to represent a TP and TS being designed (CAD), perform suitable analyses to assist designing (CAE), and convert data about constructional parts for CAD/CAM; *P*—Capture the "design intent" for a designed product (geometry, tolerancing, all considerations toward design properties) for retrieval and modification; *L*—; *A*—Constructional structure.

Computer-aided invention search (s.a. ARIZ, Contradiction-Oriented Innovation Strategy [COIS], Invention Machine, TIPS, TRIZ) *D*—Present object information as analyzed from numerous patents, to suggest a procedure of invention that can be applied to design work; *P*—Overcome apparent contradictions (improvement in one variable leads to deterioration of another, 39 variables and their pairwise interaction; see Figure 9.12 for physical quantities) in requirements, by selecting alternative principles (40 developed by Altschuller in the 1940s) from existing patents, indicate possibly viable new solutions; *L*—[39,40,182,389,536,544]; *A*—Conceptualizing.

NOTE: Contradiction has been questioned [327], by developing a method to prioritize the inventive principles for a problem.

Computer Integrated Manufacturing (CIM) *D*—Computer control of manufacturing plant and inventory; *P*—Management supervision; *L*—; *A*—Constructional structure.

Concept map *D*—Show a set of concepts, and by labeled links describe their relationships, as a hierarchical or network representation; *P*—present a graphical image of related concepts to improve understanding of a situation; *L*—[181,444,445], rl. mind map, storyboarding, s.a. Figures I.2 and I.4; *A*—conceptualizing.

Concept Selection (Pugh Method, s.a. Section 2.4.3) *D*—Pair-wise comparisons of proposed solutions according to selected criteria, performed in teams; *P*—Select an optimal solution among given proposals, and improve it; *L*—[475,478]; *A*—All structures.

Concurrent engineering (s.a. Simultaneous engineering) *D*—Perform design work on the product and the manufacturing process at the same time; *P*—Best trade-off between design features and manufacturing cost/difficulty; *L*—; *A*—Constructional structure.

Continuous improvement (s.a. TQM and Section 8.2.9) *D*—Search for ways to improve the operation of the organization or its sections; *P*—Many small improvements over a period of time can result in a large change; *L*—; *A*—Product planning, and all stages.

Contradiction-Oriented Innovation Strategy (COIS) (s. Computer-aided invention search).

Cost-Benefit Analysis *D*—By analyzing costs and benefits in monetary terms, select a solution among the proposals; *P*—Find an optimally economic solution; *L*—[217,379,595]; *A*—Product planning.

Cost calculation (s.a. HKB, Section 10.2) *D*—Calculate costs of products from regressions on data, or from empirical formulae; *P*—Obtain estimate or accurate value of costs for manufacture; *L*—[91,177,178]; *A*—Constructional structure, layout and detail.

CQuARK *D*—Consider issues, **Qu**estion data/specifications and "is it worth pursuing?," **A**ware of reasons/limitations/trade-offs, **R**efer to past projects, **K**eep options open; *P*—general heuristic advice on procedures (s.a. Section 8.2); *L*—[32]; *A*—All stages, see also the review cycle in each step of the procedural model, Figures I.17 and 4.1, and in the problem-solving cycle, Figure 2.7.

Critical Path Network/Planning/Method (s.a. Program Evaluation and Review Technique [PERT]) *D*—Graphically represent the envisaged activities and their duration; *P*—Create an overview of sequence and timing and find the critical path, shorten times; *L*—rl. Gantt chart; *A*—Time planning of projects.

Decision Tree *D*—Keep records of decisions about a TP/TS, as a hierarchical tree, Figure 8.3 [140]; *P*—Allow retracing of decision steps; *L*—[111,335,401]; *A*—All stages.

Delphi Method *D*—Poll expert opinions (repeatedly) about a problem for the future; *P*—Predict a trend or future development; *L*—[296]; *A*—Product planning.

Descartes *D*—Four principles: criticism, division, ordering, create overview; *P*—Correctness and effectiveness of the thought process; *L*—s. Section 8.2.8; *A*—All stages.

Design catalog (s. Section 9.3) *D*—Collected and categorized information about possible solution principles to fulfill a function, including mathematical relationships; *P*—Present information to assist creative search for modes of action; *L*—[178,370,498]; *A*—Basic operation "search for solutions," Organ structure.

Design for Manufacture and Assembly (DFMA) *D*—Using computer representations of the constructional parts and their arrangement, generate

(A)   General scheme of decision tree

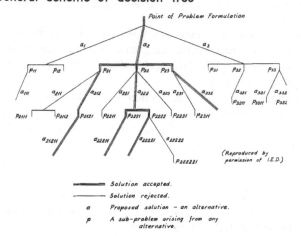

(B)   Example of decision tree — private car

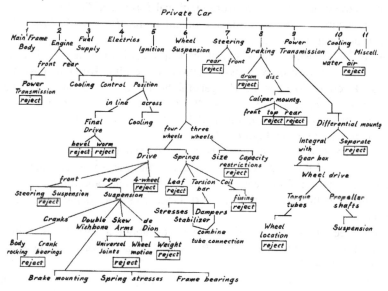

(C)   Concept sketch (preliminary layout) of private car

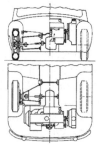

Literature:
Marples, D.L. (1960) "The Decisions of Engineering Design," London: Inst. Engng. Des.; and IRE Trans. Engng. Management, EM—8 1961, pp. 55—71.
Eder, W.E., and Gosling, W. (1965) *Mechanical Systems Design*, Commonwealth and International Library # 308, Pergamon Press, Oxford

FIGURE 8.3   Decision tree.

some characteristic numbers that can guide redesign to optimize manufacture and assembly with respect to cost; *P*—Reduce cost and complexity of a (mechanical) system, reduce number of constructional parts, simplify forms, ease assembly, and so forth; *L*—[72,73,74]; *A*—Constructional structure.

Design for properties and life cycle *D*—Collect information about favorable principles, forms, and arrangements that will optimize the system to be designed for each property (cost, functional integration or separation, assembly and disassembly, testing, maintenance, reliability, safety, serviceability, ergonomics, environment, and so forth—see Figures I.8 and 6.8–6.10), each phase (process) of its life cycle (see Figures I.13 and 6.14), and each operator of these life-cycle processes; *P*—Make knowledge available to and directly usable for designers; *L*—rl. DFMA; *A*—All stages.

Design of Experiments (DoE) (s. Analysis of Variance [ANOVA], s.a. Taguchi experimentation).

Design retrieval (s.a. Precedents) *D*—Reuse previous design schemes as proven parts of new systems; *P*—Reduce risks of failure, and time to complete designing; *L*—; *A*—All stages.

Design Structure Matrix *D*—Generate and process a matrix of precedence relationships (e.g., mathematical) among anticipated steps and variables in designing a particular system; *P*—Reduce the needed number of iterations in a decision and optimization sequence; *L*—[83,184,185,330,525]; *A*—Conceptualizing.

Engineering science calculation *D*—Use engineering sciences to calculate values of variables (e.g., dimensions, sizes) and characteristics of behavior, for properties of strength and performance; *P*—Establish minimum dimensions of constructional parts for safe and reliable operation of the TS; *L*—[12]; *A*—Organ structure, constructional structure—analysis, evaluation, simulation, optimization.

Enlarging the search space *D*—Revise the boundaries of the TP(s)/TS(s), search for principles and configurations that could help to solve the problem; *P*—Allow consideration of alternative solutions; *L*—[111]; *A*—Conceptualizing, constructional structure.

Experimentation (s.a. ANOVA, Taguchi experimentation) *D*—By measuring and testing, obtain desired statistical values and trends; *P*—Determination of the properties of a realized TS, prototype, or test rig; *L*—; *A*—Constructional structure.

Failure analysis *D*—Observe and diagnose failures to establish causes and progresses; *P*—Obtain information toward "design for properties"; *L*—; *A*—Realized TS (prototype, test rig).

Failure Mode, Effects, and Criticality Analysis (FMECA) *D*—Analyze possible failure modes in a proposed system, establish possible consequences ("effects") from each failure (as a network of interactions), estimate statistical probability of occurrence; *P*—Estimate reliability, durability, dependability of a proposed system, find ways of improving them; *L*—[554], rl. Fault Tree Analysis, see Figure 9.16; *A*—Conceptualizing, constructional structure.

Fault Tree Analysis (FTA) *D*—Analyze possible failure modes in a proposed system, establish possible consequences from each failure (as a hierarchical tree structure of interactions), estimate statistical probability of occurrence; *P*—Estimate reliability, durability, dependability of a proposed system, find ways of improving them; *L*—[554], rl. FMECA; *A*—Conceptualizing, constructional structure.

Finite Element (or Difference) Analysis (FEA) (s. Boundary Element Method, s.a. Computational Fluid Dynamics) *D*—Dividing the volume of a constructional part into discrete sections; *P*—; *L*—; *A*—.

Fishbone diagram (s.a. Ishikawa) *D*—Diagram main influences and causes on the behavior of a system, see Figure 8.4; *P*—Clarify subproblems arising in solving a design problem; *L*—[325,326,549] rl. Design Structure Matrix; *A*—Conceptualizing.

Focus Group Interview *D*—Conduct interviews with potential customers as targets for a product; *P*—Establish customer requirements; *L*—[36,37, 111,335] rl. QFD, design specification; *A*—Product planning.

Force (and moment) flux *D*—Trace the transfer path of any externally applied and internally generated force (and moment) to ensure that action and reaction are equal, and preferably determinate; *P*—Detect any force unbalance, uncontrolled forces (e.g., constrained thermal expansion), redundant force paths, to establish minimum sizes of constructional parts; *L*—[178] rl. Force equalization; *A*—Constructional structure.

Function costing *D*—Perform regression analysis of the acquisition costs on a size range of a purchasable constructional parts (OEM, COTS, as complex organs for reuse in a system); *P*—Obtain estimates of the costs of a proposed system; *L*—[384] rl. HKB—Herstellkostenberechnung, Section 10.2; *A*—Organ structure, constructional structure.

Function decomposition *D*—Divide a more complex function into smaller and simpler functions (see Figure 2.3), detect evoked functions (function–means tree, Section 6.4.7 and Figure 5.5); *P*—Redefine and concretize functions to allow easier solution, s. Section 4.4, subsection (3g); *L*—[457]; *A*—Function structure.

Fundamental Design Method (s.a. Figure 8.7) *D*—Use guidelines for goals to start a design process; *P*—Enhance creativity; *L*—[335,404]; *A*—Clarifying the problem, conceptualizing.

Gallery method (s. Art Gallery Method).

Gantt chart *D*—Show the time taken (planned or estimated) for a set of activities (s.a. Figures 2.10 and 2.11); *P*—Plan activities, with implied consideration of relationships among activities; *L*—rl. Critical Path Network; *A*—Time planning of projects.

Granta—Cambridge Engineering Selector (CES) (Materials, Process, Shape) *D*—Select materials of construction and form (shape) from considerations of several material properties, and manufacturing processes; *P*—Optimize

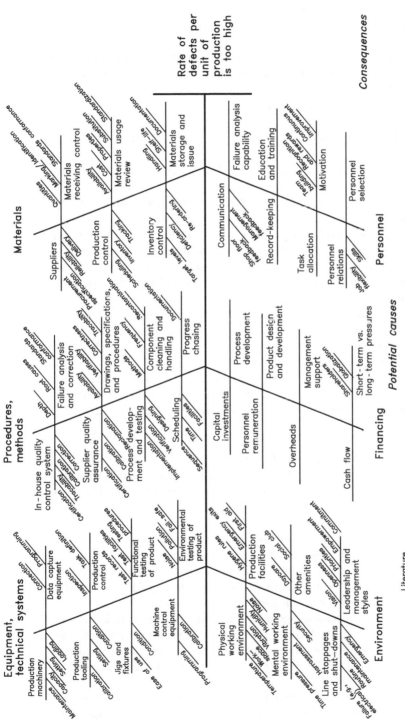

FIGURE 8.4 Fishbone diagram—example: product defects, analysis of situation.

Literature
Ishikawa, K. (1982) *Guide to Quality Control*, Tokyo: Asian Productivity Press
Ishikawa, K. (1985) *What is Total Quality Control? The Japanese Way*, Englewood Cliffs, NJ: Prentice-Hall
Underwood, A. (1993) *TQM Handbook on Total Quality Management* (Canadian Aerospace Industry), Ottawa: Industry Canada

materials specification and usage, optimize manufacturing sequence, establish expected properties; *L*—[7] was Fulmer Materials Optimizer; *A*—Constructional structure.

Herstellkostenberechnung—HKB (s.a. Cost calculation, Section 10.2) coupled with manufacturing planning and data of manufacturing methods available in the organization.

Hitachi method (s. Fishbone diagram).

Ideals concept *D*—Generate an ideal solution, degrade it to be able to realize it; *P*—Improve over existing systems by a leap; *L*—[431]; *A*—Conceptualizing.

Incubation (s.a. Section 11.1 and Figure 2.7) *D*—After thorough preparation of the problem, take a break to allow the subconscious to work; *P*—Find solutions by intuition; *L*—[563]; *A*—Basic operation "search for solutions."

Integrated Product Development—IPD (s.a. Figure 7.10) *D*—Develop a product (including designing), market, manufacture, and so forth, in an integrated procedure that simultaneously takes all aspects into account; *P*—Organizational coordination of activities; *L*—[44,178] rl. Concurrent engineering; *A*—All stages.

Interaction net/matrix (s.a. Design Structure Matrix, AIDA) *D*—Find interactions and relationships among features and variables for a system; *P*—Obtain an efficient sequence of establishing properties; *L*—[111,335]; *A*—Planning the design process.

Interpreting mathematical functions *D*—Set up a mathematical model of the anticipated behavior of a system, interpret it to find the most important variables and their best values; *P*—Optimize a system; *L*—[214,384]; *A*—Conceptualizing.

Interviewing users (s. Focus Group Interview).

Invention Machine (s. Computer-aided invention search).

Investigating user behavior *D*—Observe users of a system to find out what they actually do; *P*—Modify the system more appropriately to suit the users; *L*—[111,335]; *A*—Redesign, clarify the problem.

Ishikawa (s. Fishbone diagram).

ISO 9000 series *D*—Establish organizational systems of quality control, assurance and audit, and design audit, for the product and procedures of the organization; *P*—Provide visible and independently assessed evidence that an organization can be trusted to deliver consistent quality of products; *L*—[9,10,113]; *A*—Product planning.

ISO 14000 series *D*—Establish organizational systems of environment assessment and protection; *P*—Provide visible and independently assessed evidence that an organization can be trusted to consider the environment and its protection; *L*—[11,93,113,328] rl. Life cycle assessment/engineering; *A*—Product planning.

Iteration (s.a. Section 4.1) *D*—Starting from assumed values, obtain progressively closer approximation of values; *P*—Find solution for a system with complicated interactions; *L*—; *A*—All stages.

Just in Time (JIT) *D*—Plan for delivery of materials and constructional parts with minimum lead time before they are needed for manufacture or assembly; *P*—Reduce work-in-progress inventory; *L*—; *A*—Product and production planning.

Kaizen (s. Continuous improvement, TQM, see also Section 8.2.9).

Kanban *D*—Use an inventory control system of documents; *P*—Acquire materials and constructional parts with minimum lead time; *L*—rl. Just in time; *A*—Product planning.

Kansei *D*—Capture the emotional factors of users, feelings about a product with respect to aesthetic, ergonomic, and psychological value; *P*—Identify design specification statements for a product; *L*—[592]; *A*—Clarifying a problem.

Lateral thinking (s.a. Blockbusting) *D*—Use unconventional ways of thinking; *P*—Find unusual possibilities for solving a problem; *L*—[119–123]; *A*—Conceptualizing.

Life cycle costing/assessment/engineering *D*—Take all phases of the life cycle (Figures I.13 and 6.14) into consideration, by formalized procedures; *P*—Minimize life cycle costs and influences on the environment; *L*—[62,160,186,237,244,582] rl. ISO 14000 [11]; *A*—Product planning, conceptualizing, final evaluation.

Market Research *D*—Systematically collect and classify market information; *P*—Establishing marketing conditions; *L*—; *A*—Product planning.

Materials Resource Planning (MRP) *D*—Plan the usage and storage of materials (i.e., give restrictions to some design decisions); *P*—Optimal use and cost of materials of construction for product; *L*—; *A*—Constructional structure.

Mathematical analysis/modeling (s. Engineering science calculation).

Mental Experiment *D*—Observe an idealized mental model at work; *P*—Testing of an idea, establishing an expected behavior; *L*—; *A*—Conceptualizing, constructional structure.

Method "6-3-5" (s. Brainwriting).

Methodical Doubt (s.a. Scientific Skepticism, Descartes) *D*—By systematic negation of existing solutions, search for new solution paths; *P*—Find new solutions; *L*—; *A*—All stages.

Method of Invention (s. Computer-aided invention search, s.a. ARIZ, Contradiction-Oriented Invention Strategy, Invention Machine, TIPS, TRIZ).

Mind map *D*—Map various concepts and their relationship to a central concept; *P*—Improve understanding of a related set of concepts; *L*—[92] (originated in the 1960s) http://mind-map.com, http://visual-mind.com, http://smartdraw.com, rl. concept map; *A*—conceptualizing.

Modular construction *D*—Set TS-boundaries to limit the functions of each section, provide standardized interfaces between TS sections (complex organs), establish a common platform for variations; *P*—Allow easy interchange of TS assemblies or subassemblies to vary their performance capabilities; *L*—; *A*—Constructional structure (probably needs to be planned in an organ structure).

Morphological matrix/scheme *D*—Enumerate possible organs (function carriers) to solve partial functions, in matrix form; *P*—Obtain many solutions by combinations of function carriers and variation in their arrangement; *L*—[111,134,335,443,597,598], see also Figure 4.4, and Chapters 1 and 5 to 7; *A*—Transition from a function structure to an organ structure (also for transition from organ structure to constructional structure).

Objectives tree (s.a. Decision tree) *D*—Generate a list of objectives and sub-objectives (requirements and their priorities), diagramed in the form of a hierarchical tree; *P*—Obtain a prioritized design specification, show relationships among requirements; *L*—[111]; *A*—Clarify the problem.

Optimization *D*—Find an optimal solution to a problem, use mathematical techniques to find an optimal set of values according to a single criterion, or multiple criteria; *P*—Selection and improvement of solution proposals; *L*—; *A*—All stages.

Order of magnitude calculation *D*—Use values expressed only as powers of ten to calculate according to an equation; *P*—Obtain a rough numerical value to get a feel for sizes, and to verify the adequacy of computer output; *L*—; *A*—All stages.

Pairwise comparison (s. Concept selection).

Pareto distribution/diagram (s.a. ABC analysis) *D*—Arrange observations according to magnitude from largest to smallest, give careful consideration to the largest; *P*—Focus on the most important issues; *L*—; *A*—All stages.

Part count reduction *D*—Reduce the number of constructional parts, by combining, eliminating separate fasteners, and so forth; *P*—Reduce cost of manufacture and assembly (caution: this may lead to an increase in the costs of disassembly, recycling, disposal); *L*—[72,73,178] rl. DFMA; *A*—Constructional structure.

Participative design *D*—Consult and discuss with (potential) customers to allow them direct input to design decisions; *P*—Make the product suitable for the user; *L*—; *A*—All stages (most important in clarifying the task).

Plus–minus comparison (s.a. Concept selection, Pugh method, s.a. Section 2.4.3) *D*—Compare solutions according to selected criteria among each other and with an ideal, mark only with "plus" (if better) or "minus" (if worse); *P*—Find optimal solution proposal; *L*—; *A*—Basic operation "evaluating and deciding."

Precedents *D*—Search among prior art, existing realized systems, previous proposals; *P*—Find solution possibilities; *L*—[71]; *A*—All stages.

Problem Analysis by Logical Approach (PABLA) *D*—Use a set of prepared charts to explore the design problem; *P*—Generate a design specification; *L*—[378,537], see Figure 9.17; *A*—Clarify the problem.

Program Evaluation Review Technique (PERT) (s. Critical Path Network) *L*—[246].

Pugh method (s. Concept selection).

Quality Circle (s. Quality Service Action Team).

Quality function deployment (QFD), from Japanese characters—*hin shitsu* (attributes, features, qualities) *kin no* (function) *ten kai* (development, diffusion, deployment)—preferably not *atarimae hinshitsu* (quality taken for granted) D—Develop a set of charts according to recommendations to relate two viewpoints about a proposed product, a "house of quality" (see Figure 8.5A) (survey) and 8.5B (example); P—Capturing the "voice of the customer" and making it heard throughout the product realization process; L—[75,111,151,259,269]; A—Clarifying the problem, release for detailing and manufacture.

Quality Service Action Team (QSAT) (s.a. Quality circle) D—Collect information by discussion about a product, and its manufacturing and assembling processes; P—Improve the quality of a product and production organization; L—[37]; A—Constructional structure.

Quirk's Reliability Index D—By entering criteria into a set of prepared charts, find numerical estimates about potential reliability of a proposed TS; P—Estimate and improve reliability; L—[335,479]; A—Constructional structure.

Rapid prototyping D—Produce a tangible model of a constructional part or subassembly in a plastic (or metal) material under computer control, from a computer representation (solid modeling); P—Obtain a solid model that can be handled, assess its suitability; L—rl. under various other names, mostly proprietary; A—Constructional structure.

Recursive decomposition (s.a. Function decomposition, and Sections 4.4.1, 5.2 and 6.4.2) D—Divide a larger or more complex problem and system into smaller units, perform the necessary operations, and recombine into a larger unit; P—Subdivide the problem to make it easier to solve, or allow solution by different teams (needs good coordination); L—; A—All stages.

Reverse engineering D—Disassemble a realized TS (usually from a competitor) to generate a set of manufacturing documentation (detail and assembly drawings); P—Copy a TS, or modify it in minor details; L—; A—Constructional structure.

Rhetoric D—Analyze the rhetorical situation (audience, affected organization), generate ideas and arguments, organize ideas, write, revise (reflect, iterate); P—Guidelines for preparing a convincing argument; L—[50]; A—All stages.

Science of Generic Design (s.a. Warfield) D—Use political and operational supervision of design engineers to keep error events (faults, failures, recalls) within socially (economically) acceptable bounds, managing the design process and the product is important; P—Reduce design errors; L—[565,566]; A—All stages.

Scientific Skepticism (s. Descartes, Section 8.2.8).

Selection chart D—Produce a list of alternatives and their characteristics relative to selected criteria, with decisions on further processing (see Figure 8.6); P—Maintain a clear record; L—[178,457]; A—Basic operation "evaluating and deciding."

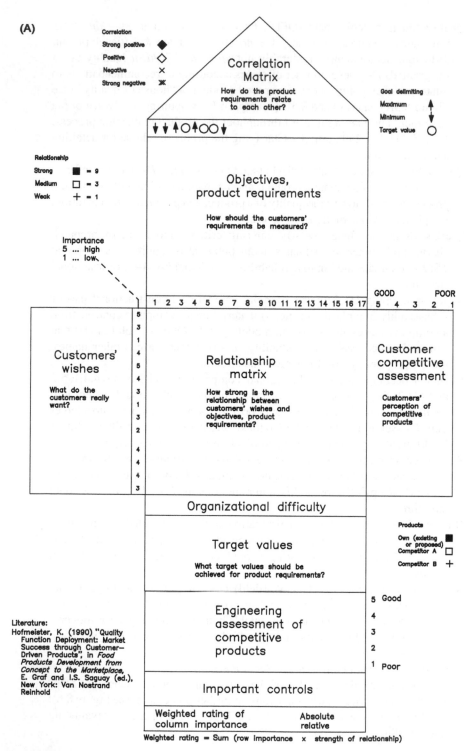

FIGURE 8.5   House of quality—concept survey and example "tape winder."

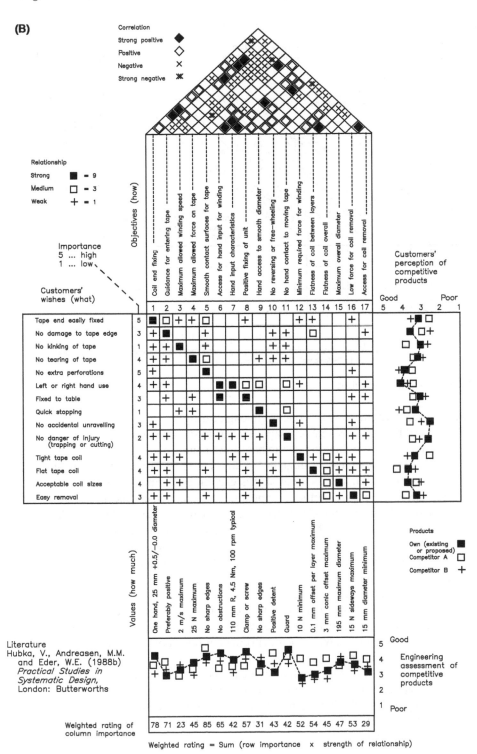

FIGURE 8.5 Continued.

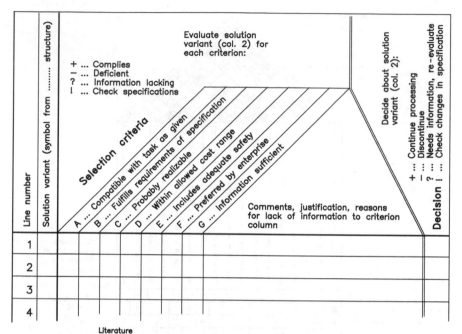

Literature
Pahl, G., and Beitz, W. (1993) *Engineering Design* (3rd ed.), London: The Design Council
& Berlin/Heidelberg: Springer–Verlag

FIGURE 8.6   Selection chart.

Simultaneous engineering (s. concurrent engineering).

Six colored hats *D*—Use real or imaginary hats to signify (show or encourage) different modes of thinking: white hat shows neutral and objective thought; green hat shows creative thought; yellow hat shows optimistic, logical, and positive thought; black hat shows caution and critical judgment; red hat shows feelings, intuition, hunches, and emotions; blue hat shows thinking about the thought processes; *P*—Separate the thinking modes (see Section 11.1); *L*—[123] rl. Lateral thinking; *A*—All stages.

Step Forwards/Backwards (s.a. strategy of the problem axis, Section 4.1) *D*—Attempt both solution directions, from "is" to "should be" and reverse, from more abstract to more concrete and reverse; *P*—Find the most favorable path to a solution; *L*—; *A*—All stages.

Storyboarding *D*—List each item to be considered on a separate card (e.g., sticky notes), rearrange cards into plausible sequences; *P*—Find alternative sequences (e.g., of transformation operations, functions); *L*—; *A*—Conceptualizing.

Strength diagram *D*—Plot relative evaluations for technical and economic (or other) criteria onto a Cartesian coordinate graph—s.a. Section 2.4.3 and Figure 2.9; *P*—Find adequate and best solutions among proposals; *L*—[296,305,349] rl. Evaluation, Weighted rating; *A*—Basic operations "evaluating and deciding."

Strengths–Weaknesses–Opportunities–Threats (SWOT) Analysis *D*—Analysis of a situation, by verbal analysis and quantitative evaluation; *P*—Formulate a vision and mission statement for an organization; *L*—[514]; *A*—Product planning, general management.

Structured Analysis and Design Technique (SADT) *D*—Obtain flow chart of process to be investigated or to be designed, with chronological chart of alterations to keep track of developments, using interviews to gather information (facts, problem identification, opinions on solutions, etc.); *P*—Define transformation process, record progress of the design process; *L*—[400] rl. IDEF0 model (Chapter 5), Transformation process applied to information (Chapter 5) considering all operators as "control," technology as "mechanism," operations (or functions) as "subject"; *A*—Transformation process structure, possibly also for function structure.

Synectics *D*—Formally chaired team analyzes problem and searches for new solutions through analogies; *P*—Discover new solutions; *L*—[111,234,335], rl. brainstorming; *A*—Conceptualizing.

Systems Approach *D*—Work systematically in every situation requiring a solution or decision; *P*—Complete investigation of an area (as far as possible); *L*—[250,335]; *A*—All stages.

Taguchi Experimentation (s.a. Design of experiments, ANOVA) *D*—By performing a set of controlled statistical experiments, find the main influences that can disturb the TS-operational process, or manufacture of the TS; *P*—Find robust solutions that withstand the disturbances; *L*—[391,468, 496,530,531]; *A*—Constructional structure.

Taguchi philosophy *D*—Reduce variation (variance, standard deviation) in properties, aim to achieve a consistent variation (statistical variance, standard deviation) of values around an optimal mean; *P*—Make the product insensitive to variations in external and manufacturing conditions; *L*—[391,468,496,530,531]; *A*—Constructional structure (at times also conceptualizing).

Target costing *D*—Use cost data for proposed solutions; *P*—Obtain a proposed solution that satisfies a maximum permitted cost criterion; *L*—[87,509]; *A*—Constructional structure (at times also conceptualizing).

TIPS (s. Computer-aided invention search).

Topika (Topics) (see Figure 8.7) *D*—Ask a formal sequence of questions about the problem or a proposed solution; *P*—Prompt appropriate answers; *L*—[49,56,57] rl. Six questions, Work study questions (contained in Topika); *A*—All stages.

Tolerance analysis *D*—Computer-aided analysis of tolerance accumulation and six-sigma limit determination; *P*—Find influences of tolerance contributions; *L*—Program Geomate ToleranceCalc 4.0 http://www.inventbetter.com; *A*—Constructional structure.

Total Quality Management (TQM) (see Section 8.2.9) *L*—[10,108, 125,549].

Total Quality Service (TQS) (see Section 8.2.9).

Topika (topics) — inquiry scheme according to aristotle (summary)

Latin	English	Comment
Quis	Who?	Question about the person or object that is to be treated or described
Quid	What?	
	Why?	Proof derived from the considerations
Contra	Opposed to	Antithesis, contrary of proposition, disproof
simile	Similar to	Explanation based on parallel cases
paradigma	Example	Explanation based on examples, patterns, paradigms
	Testimony	Witness of experts
conclusione	Conclusion	
Ubi	Where?	
Quomodo	How?	
Quando	When?	
Quibus auxilium	With what means?	

Ensuring completeness of the considerations

Roulettes (as in gambling, playing) according to Matchett:
  Roulette (1) Concerning any one, two or more features (surfaces, components, groups, organs, etc.) or functions, how can we:

    ELIMINATE   or   INCLUDE / SUBSTITUTE
    COMBINE   or   DIVIDE / PARTITION / SEGMENT
    STANDARDIZE   or   DIVERSIFY / CONSOLIDATE
    TRANSFER / REARRANGE
    MODIFY / ADAPT
    SIMPLIFY
        the whole, or a part of ...?
  Roulette (2) What effects, demands, restrictions
    Will "A" have on "B" ?
    Will "B" have on "A" ?

Systematic variation of forms according to Koller and Erhlenspiel *(see also Figure 2.14)*
— systematically try to vary, change or alter one or more of the following
    characteristics of features (e.g., surfaces, components, groups, organs, etc.)
    or functions:
MAGNITUDE, DIMENSION, SIZE
FORM
NUMBER
PLACE, LOCATION, POSITION, ORIENTATION
TYPE OF MOVEMENT
SEQUENCE, ARRANGEMENT (e.g. INVERSION)
STRUCTURAL CONNECTIVITY, TYPE OF CONTACT or CONNECTION
DEGREES OF FREEDOM, STATIC DETERMINACY
ELASTIC / PLASTIC DEFORMATION
TYPE OF MATERIAL, MANUFACTURING METHOD, ASSEMBLY METHOD
TOPOLOGY, REFERENCE SYSTEM
TIME, TIME SEQUENCE OF MOTION   *(see also Figure 2.14)*

Literature:
Aristotle (384–322 BCE), Organon, section Topika (Topics)
Matchett, E.D. (1981) "Fundamental Design Method", in *Proc. ICED 81 Rome*, Milano: technice nuove, 1981, pp. 221–228
Koller, R. (1985) *Konstruktionslehre für den Maschinenbau* (Foundations of Engineering Design for Mechanical Engineering), Berlin/Heidelberg: Springer-Verlag
Ehrlenspiel, K. (1995) *Integrierte Produktentwicklung* (Integrated Product Development), München: Carl Hanser Verlag

FIGURE 8.7 Topika (topics)—exploring a theme from general viewpoints.

Trend studies *D*—Study time-trend of a variable for a product; *P*—Obtain predictive estimation of near-future developments; *L*—; *A*—Product planning, conceptualizing.

TRIZ (s. Computer-aided invention search).

Value Analysis/Engineering *D*—Analyze and criticize the existing or proposed solution (usually layout and details) from the viewpoint of economics; *P*—Improvement of the economic and other properties of the TS; *L*—[222,261,414,424]; *A*—Redesign, conceptualizing, and constructional structure.

Value Management *D*—Use a further development of value engineering to include the viewpoint of product success as main criterion; *P*—Improvement of economic and other properties of the TS; *L*—[372]; *A*—Redesign, conceptualizing, and constructional structure.

Virtual Reality *D*—Use a dynamic computer representation to provide a visual impression of a design proposal; *P*—Check potential operability and comfort for a user; *L*—; *A*—Constructional structure.

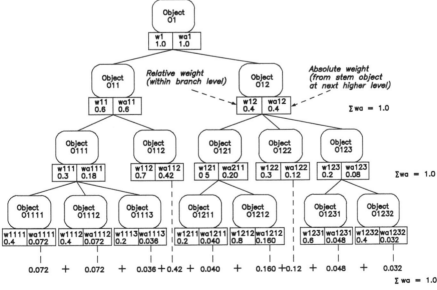

Scheme for allocation of consistent weights to decision objects.

Decision objects can be:
  design criteria (e.g., selected from a design specification)
  alternatives for operations, functions, organs, components
  and so forth

For each tree-branch of a hierarchy, the relative weights allocated to the members should add to unity (1)
For each member of a hierarchy level, the absolute weight is the product of its relative weight (within a tree-branch) and the absolute weight of the stem object of that branch

Literature:
Pahl, G., and Beitz, W. (1993) *Engineering Design* (3rd ed.), London: The Design Council and Berlin/Heidelberg: Springer-Verlag

FIGURE 8.8   Hierarchical allocation of weights.

Weighted rating D—Select criteria, assess weights for each criterion, evaluate goodness of each solution proposal with respect to each criterion, multiply weight times goodness, add to obtain an overall rating or a percentage rating; P—Find relative goodness of each solution compared to a nonspecified ideal; L—rl. weighting (see Section 2.4.1.3); A—Basic operations "evaluating and deciding."

Weighting—assigning weights (see Figure 8.8) D—Assign weights for weighted rating, consistently distributed among a hierarchy of criteria; P—Entry to weighted rating; L—[111,178,457]; A—Basic operations "evaluating and deciding."

Why? Why? Why? (s.a. Topika, Figure 8.7) D—Persistent and repeated use of the question "why?" on each successively given answer; P—Reach in-depth of understanding; L—[111,335]; A—All stages.

Work Study Questioning (s.a. Topika, Figure 8.7, six questions) D—Ask the six questions what?, where?, when?, who?, how?, why?; P—Achieve understanding of a problem; L—; A—All stages.

Zero Defects D—Aim for motivation toward achieving complete correctness; P—Improve the product and the organization; L—rl. TQM; A—All stages.

9 Information and Formal Support for Design Engineering

As outlined in Sections I.8 and 12.1, information is available from various sources. The information presented in this chapter is mainly situated in and around the "northeast" quadrant of Figures I.4, I.5, and 12.7, and presents additional details about information systems (Section 9.1), masters (Section 9.2), engineering design catalogs (Section 9.3), standards, codes of practice, guidelines (Section 9.4), and forms (pro forma) (Section 9.5).

9.1 INFORMATION SYSTEMS

Chapter 2 contains definitions of the design process as a transformation process in which information is processed. The *operand* in the design process is information, yet little attention is paid in engineering practice and education to organizing and documenting information.

The most important question for each person searching for information is: "Where do I quickly find the needed information of most appropriate quality?" The expert asks: "For what kind of information am I searching? Which information carrier (medium) can contain it? Which information banks are available? How is this information arranged?" A brief discussion of terms is presented in Sections I.8 and 12.1.

9.1.1 INFORMATION QUALITY

Important questions that characterize the quality of information are as follows:

1. Are the contents correct (adequate vs. wrong), accurate (within an acceptable tolerance range) and intelligible?
2. Are the contents complete (cross-referenced, no isolated sections, all sections in context)?
3. Are the contents reliable and verifiable (source references, are sources traceable)? What is the range of validity of the information?
4. Is the format clear, unequivocal, precise, easily understood, formal, logical, easily retrievable (sharpness of information)? Is the form lucid and usable (density), short and comprehensible (without repetition, redundancy)?

409

Is the form and arrangement suitable for the needs of design work (or for archiving)?

5. Is rapid orientation possible (volume and complexity of information, availability at the right time)?
6. How old is the information (is it still valid)?
7. What type of information carrier or medium is used (e.g., memory, literature, internet, knowledge-based/expert/neural-network system)?
8. Who are authors, maker, compiler, producer, and publisher of the information?
9. Is the information readily accessible? Can information be taken over? Is it protected by intellectual property rights (patents, copyrights, trade marks, product registrations); see Section 11.3.2? What are possibilities (and permissions) for use?
10. For which potential user group is the information prepared, addressee circle?

Too often information is wrongly applied. Different properties of the information are important for its quality, according to the situation. For designers the accuracy and availability of the information is decisive. For documentation, rapid orientation is important.

The relationships among individual properties give rise to conditional statements about qualities of a class of information. With documents from a known author or publishing house, it is possible to assume accuracy and reliability without knowing the content of the publication, because the name is a guarantee for the content. The age of the information can be decisive for the accuracy. Over time, some information becomes obsolete. Younger information areas, such as computer sciences and engineering, nanotechnology, and so forth, change quickly.

9.1.2 INFORMATION CARRIERS

Information should be fixed (recorded, stored), usually in written or graphic form, but this depends on the kind of information. The possibilities increase with progress in technical information processing. A book (e.g., [54,198,199,225,238, 407,463,513,594]) is only one form of information carrier (medium); see Figure 9.1. Decisions about the most suitable information carrier must be made with respect to purposes, advantages, and disadvantages (see Figure 9.2).

Designers collect information for themselves, or it is collected by special organization personnel, usually for concrete application. Carriers for this information are, for example.

1. Excerpts (statements) from primary information carriers for individual need, recorded on secondary information carriers.
2. Groupings of frequently used data (Figure 9.3).
3. Guidelines, for example (morphological) matrix with assignment of organs to the functions (Figures 4.4 and 9.4) — design guidelines (Figures 7.8 and 9.3).

Information carriers	Information banks and information systems
Engineering books Text books Reports — Theses — Conferences — Visits — Consultants — Design projects — Experiments — Repair, damage — Environmental impact Patent applications Drawings Parts lists Data (magnetic) tapes, diskettes, CD–ROM Film, video tape Microfilm Photographs Samples, models (physical), products User manuals, operating instructions Descriptions (narrative) Advertising leaflets Product (enterprise) catalogs Advertisements Price lists Memoranda, newsletters Tenders, requests for proposals Business correspondence Contracts, orders Standards specifications Regulations, codes of practice Appraisals, expert opinions Guidelines Note books (design records) Surveys Masters, proforma, design catalogs Schemes Diagrams, figures, graphs, tables Data carriers (punched cards, tapes) and so forth	Conferences, seminars, workshops Courses Libraries — Own — College — Enterprise — Central (public) Data banks: — Enterprise — Central (public, government) — International Archives: — Drawings — Film — Correspondence (e-mail, surface/air) Museums: — Technical — Scientific — Others Card files Disk files (audio, video, computer) Internet World-Wide-Web

FIGURE 9.1   Information systems and carriers.

4. Masters (see Section 9.2); Pro forma (see Section 9.5); Engineering design catalogs (see Section 9.3); Computer applications (see Chapter 10).
5. Constructional parts (see Chapter 7).

Some magazines and periodicals deliver an overview of contributions in other publications, for example, Engineering Index; year-end surveys in trade journals; Current Contents; Reviews. Information is available from the Internet and "task-specific bibliographies," for example, patents.

9.1.3 CLASSES OF INFORMATION

Information can be organized into several classes according to various criteria (see also Section 1.1.4). Particular viewpoints for this organization can be: finality (see Figure 2.3); questions of designers; questions concerning finding means to solve problems. Classification is usually performed as hierarchies, for example, knowledge disciplines, but the relationships among items of information in different branches

Goals of engineering design operation	Suitable type of information carrier
Basic knowledge within the branch region; overview, orientation	Text books, engineering books of broad interest, compendia, monographs
Depth of branch (domain) knowledge	Engineering (technical) books, monographs, reports, drawings, tenders
Encyclopedic orientation	Reference books, engineering books, survey articles in journals
Revision of knowledge	Text books, reference books, excerpts
Keeping up-to-date in branch	Journals, lectures, exhibitions
Orientation about suppliers	Enterprise publications, exhibition catalogs, advertisements in journals, tenders
Search for solutions (conventional or innovative/novel)	Guidelines, journals, engineering books, monographs, Internet/WWW, computer aided invention search
Decisions	Guidelines, decision tables
Workshop drawings	Design guidelines, results tables, diagrams, handbooks, engineering books

FIGURE 9.2   Relationship of information carriers to goals of design operations.

are lost. Yet classes interact, for example, "statistics" is applicable to science and other activities — should books on statistics be in a separate class, or distributed among other disciplines, or both? How can books on statistics be found? Better classification systems can result from a "flowchart" or multisubject matrix approach. The degree of fixation is described in Section 12.1.3. For working with information, see Section 12.1.5.

Some known classification systems are:

(a) **Library of Congress Cataloging-in-Publication Data/Dewey Decimal Classification**: The Library of Congress system of "call numbers" consists of one or two (capital) letters and a group of (usually three) numbers that define the subject area, the initial letter of the first author's family name, an accession number, and the year of publication. Typical of the Dewey classification, the range of information is divided into (e.g., decimal) classes and subgroups, into a hierarchy of arbitrary depth. The symbols are internationally understandable, therefore the classifications are used in many libraries, and publishers print the chosen classification into their books. Books and periodicals are given ISBN or ISSN identifications (International Standard Book or Serial Numbers), the base number is allocated to the publisher, and each published book is given its own extension, with a check digit added. These classification systems are unique for single-subject publications, cross-linking among branches of the chosen hierarchy are difficult.

(b) **Key Word Systems**: The contents of each information carrier can be described by one or more unified key words, organized in a list of terms, which simultaneously references the relationships, super- and subordination, and synonyms. Most libraries use key word systems in addition to the Library of Congress, Dewey Decimal, or

ROTARY CONTROL

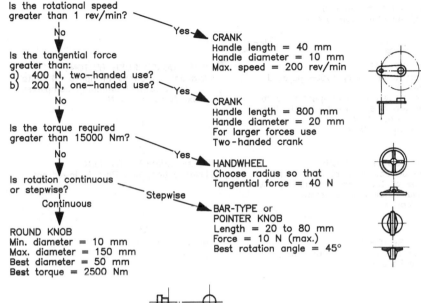

Is the rotational speed
greater than 1 rev/min?

No

Yes ► CRANK
Handle length = 40 mm
Handle diameter = 10 mm
Max. speed = 200 rev/min

Is the tangential force
greater than:
a)　400 N, two-handed use?
b)　200 N, one-handed use?

No

Yes ► CRANK
Handle length = 800 mm
Handle diameter = 20 mm
For larger forces use
Two-handed crank

Is the torque required
greater than 15000 Nm?

No

Yes ► HANDWHEEL
Choose radius so that
Tangential force = 40 N

Is rotation continuous
or stepwise?

Continuous

Stepwise ► BAR-TYPE or
POINTER KNOB
Length = 20 to 80 mm
Force = 10 N (max.)
Best rotation angle = 45°

ROUND KNOB
Min. diameter = 10 mm
Max. diameter = 150 mm
Best diameter = 50 mm
Best torque = 2500 Nm

LEVER CONTROL

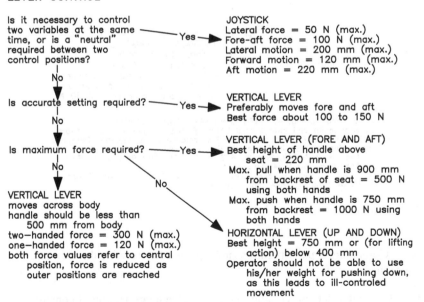

Is it necessary to control
two variables at the same
time, or is a "neutral"
required between two
control positions?

No

Yes ► JOYSTICK
Lateral force = 50 N (max.)
Fore-aft force = 100 N (max.)
Lateral motion = 200 mm (max.)
Forward motion = 120 mm (max.)
Aft motion = 220 mm (max.)

Is accurate setting required?

No

Yes ► VERTICAL LEVER
Preferably moves fore and aft
Best force about 100 to 150 N

Is maximum force required?

No

Yes ► VERTICAL LEVER (FORE AND AFT)
Best height of handle above
　　seat = 220 mm
Max. pull when handle is 900 mm
　　from backrest of seat = 500 N
　　using both hands
Max. push when handle is 750 mm
　　from backrest = 1000 N using
　　both hands

No

VERTICAL LEVER
moves across body
handle should be less than
　　500 mm from body
two-handed force = 300 N (max.)
one-handed force = 120 N (max.)
both force values refer to central
　　position, force is reduced as
　　outer positions are reached

HORIZONTAL LEVER (UP AND DOWN)
Best height = 750 mm or (for lifting
　　action) below 400 mm
Operator should not be able to use
　　his/her weight for pushing down,
　　as this leads to ill-controled
　　movement

FIGURE 9.3　Selection of control input means.

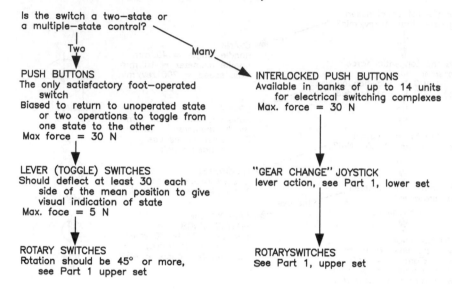

SWITCHES (DISCRETE STATE CONTROLS)

Is the switch a two—state or a multiple—state control?

Two

Many

PUSH BUTTONS
The only satisfactory foot—operated switch
Biased to return to unoperated state or two operations to toggle from one state to the other
Max force = 30 N

INTERLOCKED PUSH BUTTONS
Available in banks of up to 14 units for electrical switching complexes
Max. force = 30 N

LEVER (TOGGLE) SWITCHES
Should deflect at least 30 each side of the mean position to give visual indication of state
Max. foce = 5 N

"GEAR CHANGE" JOYSTICK
lever action, see Part 1, lower set

ROTARY SWITCHES
Rotation should be 45° or more, see Part 1 upper set

ROTARYSWITCHES
See Part 1, upper set

PEDALS

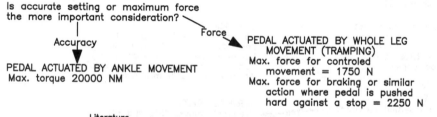

Is accurate setting or maximum force the more important consideration?

Accuracy

Force

PEDAL ACTUATED BY ANKLE MOVEMENT
Max. torque 20000 NM

PEDAL ACTUATED BY WHOLE LEG MOVEMENT (TRAMPING)
Max. force for controled movement = 1750 N
Max. force for braking or similar action where pedal is pushed hard against a stop = 2250 N

Literature
Eder, W.E., and Gosling, W. (1965) *Mechanical Systems Design,*
    Commonwealth and International Library # 308, Pergamon Press, Oxford
Murrell, K.F.H. (1957) Data on Human Performance for Engineering Designers',
    Engineering vol 184 in five parts: Part I, 16 Aug., pp. 194—198; Part II,
    23 Aug., p. 247—249; Part III, 6 Sept., p. 308—310; Part IV, 13 Sept.,
    pp. 344—347; Part V, 4 Oct., pp. 438—440
Woodson, W.E., Tillman, B, and Tillman, P. (1992) *Human Factors Design
    Handbook* (2nd ed), New York: McGraw-Hill, 1992

FIGURE 9.3 Continued.

similar systems. Computers usually permit searching for a key word, either in a title or file name, or within a text.

(c) **Thesaurus**: Computer data processing makes possible a description of contents with several "descriptors"; the list is called a "thesaurus." The contents of a publication can be described and stored, for example, with 3 to 20 descriptors. Usage of words, and finding synonyms and antonyms, can be aided by a language thesaurus, for example [137,520].

Property	Type of regulator / controller			
	Without assisting energy	With assisting energy		
		Hydraulic	Pneumatic	Electrical or electronic
Calculation operations	Only simple operations possible	Analog and digital operation possible, but expensive	Analog and digital operation possible, suitable	Analog and digital operation possible, very well suited
Possibility for networking (cascade control, disturbance detection)	Not possible	Possible, but expensive	Good possibility	Very good possibility
Setting motor	Not available, setting (control) work must be small	Setting force practically unlimited	Setting force limited by normal available air pressures, F_{max} 30 kN	Setting force limited, setting speed also limited
Consequences of loss of energy source (power)	Not applicable	Short-term emergency operation possible with accumulator	Short-term emergency operation possible with air reservoir	Emergency power supply (uninterrupted) possible, but may be expensive
Price	Cheap	Relatively expensive	Relatively cheap	Relatively expensive, to cheap
Dangers, hazards	Very small	Fire hazard from leakage	Low	Sparks may cause fire or explosion, danger to life in humid spaces
Repairs	Simple means (mechanical technician)	Simple means (trained technician)	Simple circuits (trained technician)	Specialist repair, easy if modular units can be interchanged (plug-in replacement)

FIGURE 9.4   Guideline for selection of a means: controller from viewpoints of apparatus technology.

(d) **Patent Classification**: Patents are organized and labeled according to contents. A unified system of patent classes is used. The specialty and the product species are described.

(e) **Internet**: URL and hyperlinks allow rapid searching for information such as forms and availability of constructional parts (Original Equipment Manufacturer [OEM] parts, commercial off-the-shelf [COTS]). Internet assistance is relatively comfortable for acquiring information. Down- and uploading of data files makes information available among users. The quality of this information is often questionable.

As a first priority, the primary classification scheme should be according to the properties of TP(s)/TS(s); see Figures I.8 and 6.8–6.10, and the relevant life cycle processes and their operators (see Figures I.13 and 6.14). A secondary ordering can be according to the contents of relevant books and other information sources.

The hierarchy of design operations (Figures I.16, 2.5, and 2.7), can be used. Relating the branch and domain information to individual design operations (e.g., the basic operations, H3) can fulfill some purposes of information storage and retrieval at the designer's workstation, for example, to formulate categories of information for a retrieval system.

The existing information contained in the different kinds of information carriers are collected, organized, and made accessible (retrievable) in information systems; *information banks*.

Each library has an alignment, categorizing system, information bank, and information carriers. For engineering designers, a survey of libraries and classification systems is important.

An information system (IS) is set up according to the quality of the superior systems. In an organization, we can presuppose that the IS follows a hierarchy (variants are possible): (1) Organizational information system (OIS), for example, a particular selection (subset) of standard fasteners (sizes) to be used as organization standard; (2) Departmental IS; (3) IS of a design team; (4) Personal information system (PIS), of individual designers.

An OIS should be comprehensive, and contain information about a product assortment from all aspects, especially practical experiences of use of the products. The individual sections of the organization should collect their special information — for example, the design section about designing in general, the applicable object information (see Chapters 5 to 7) and the relevant design process information (see Chapters 2 and 4). This should refer to designing the products that are the task of the section.

The IS of a design team should collect information that supports the duties of that particular team and helps to achieve a high effectiveness of the design program. The IS of the design team should include the specialized engineering design science (SEDS) for the product "sort" to be designed (see Chapter 7). It should also contain resources such as design catalogs, masters, pro forma, methods, and so forth. Figure 9.5 shows examples of IS carriers of a design team.

Public libraries cannot by themselves satisfy the information need of individual designers. Designers have different education, experience, talent, personal working mode, and interests, and their need for information should be adequately supported [32,53]. The PIS must be adapted to their characteristics, much of the information is not fixed in the PIS, and survives only internalized in the designers' (tacit) memory. The IS of a design team exerts a large influence on PIS, and decides whether a PIS needs to be built at all.

A PIS differs from a public or organizational IS regarding contents. Excerpts, and documentation for finding primary information carriers are secondary to a PIS. The lists of literature in one's own "primary" information carriers (e.g., study books, lexicons) can be completed and augmented with new sources. Secondary information can be recorded in notebooks or card files.

A PIS can probably best be built up by classifying using a key word system. Key words can be chosen individually according to the designers' (users') feel for language. Classification systems should be adjusted to the individual information areas of designers (see Figures 9.6 and I.3).

For instance, some optimization in selecting materials can take place at an early stage of designing, for example, by using suitable optimization/selection charts [51] or programs [7] relating material properties to loading conditions.

General Engineering Design Science (EDS) is a consumer of the information of other sciences, and serves as source for deriving, ordering, classifying, and providing the information for other disciplines. A deductive process can obtain SEDS for individual TS-"sorts" (see Figures 7.1 and 7.3). Derivations from EDS can be useful from at least two characteristics according to the morphological scheme in Section 12.4.1.

(01) Treatments of basic knowledge and engineering sciences. Preferably engineering designers should use their own handbooks, which contain much of the needed information. Each discipline that is not represented in these handbooks can be covered by suitable specialized books (e.g., enterprise management). Particularly important areas can be covered by monographs (e.g., for hydraulic machines).

(02) Information about working processes, for example, machine tool construction and production engineering.

(03) Excerpts and commentary to groups (01) and (02).

(04) General theory of technical systems — reference books.

(05) Standards issued by the country of origin (manufacture), and of the destination countries.

(06) Specialized theories of technical systems: important sorts of technical books and teaching texts that treat the systems in the relevant areas, including their auxiliaries and partial systems.

(07) Particular works about light-weight construction, reliability, product life and other characteristic properties of the technical systems to be designed.

(08) Enterprise-specific standards: selections from the country standards, standards developed for the particular enterprise, typifying series.

(09) Regulations, laws, codes of practice relevant for the area, for example, safety, explosion and fire prevention.

(10) Patents that are important for the area.

(11) Packaging regulations, recommendations and systems (influence on the form of the product).

(12) Transport regulations and possibilities. Load capacity of cranes, dimensions of doorways, rail clearance profiles, ship transport conditions, cost penalties for excess weight and dimensions.

(13) Storage regulations and possibilities, shelf-life considerations.

(14) Machine elements and other elemental systems that contribute to the final system, branch books, monographs, papers.

(15) Manuals about own technical systems: descriptions with important properties (preferably on unified pro forma), representations (outline drawings, schematics), short operating instructions, claims procedures, reference numbers of assembly drawings, evaluations about the technical systems, possibly their development over time.

(16) Manuals about competing technical system, similar format to group (15).

(17) Experiment and test protocols of the designed products, including claims against them, evaluations from external institutes.

(18) Enterprise publications about purchased partial systems. Recommended is a unified file-card system (self-generated) with all necessary parameters, representations (footprint, overall dimensions), and comparisons of technical and economic properties.

(19) Catalog of repeat parts, preferably classified according to function, with suitable details.

(20) General design process (Engineering Design Science) information, technical books about designing and design theory.

(21) Specialized design process (Engineering Design Science) information, concretized insights about designing in the specialized branch area under the given working and environment conditions, in papers, reports, notes.

(22) Procedural plans and models for typical tasks in the working area with important and compulsory control stations, accurate specification of deliverables, cooperation of specialists (team work).

(23) Particular calculation methods and procedures, flow charts, support documents.

(24) Procedures and documentation for rough calculation of manufacturing costs (e.g., VDI Guideline 2225).

(25) Results tables for frequently performed calculations.

(26) Typical function structures of technical systems in the branch area.

(27) Selection charts for means (organs) for usual partial functions in the branch area, preferably as matrix of functions vs. means and conditions (e.g., design catalogs).

(28) Particular design guidelines for laying out of technical systems of all degrees of complexity, especially of machine elements.

FIGURE 9.5 List of information carriers for a design team.

(29) Particular form-giving guidelines for typical technical systems from various viewpoints, for example, manufacturing, strength, esthetics, ergonomics, etc.).

(30) Materials tables: designation according to standards, chemical analysis, physical property values, technological characteristics, heat treatments, weldability, application (e.g., temperature range), form of delivery (state, profile, sizes), supplier. Collected according to various viewpoints: mechanical, technological (heat treatment, welding), special uses (e.g., steam boilers, heat resistant steels, spring steels), climate suitability.

(31) Selection of materials in stock by the enterprise, types, profiles, dimensions; delivery conditions for material not in stock.

(32) Price lists for materials, either directly in local currency, or as relative numbers (e.g., VDI Guideline 2225).

(33) Weight tables for rapid calculation (estimation) of weight.

(34) Survey of the manufacturing possibilities of the enterprise, list of machine tools with parameters for machining, especially maximum and minimum size and precision capability; enterprise external capability for manufacture with rough delivery deadline capability.

(35) Recommended tolerances and fits, application, list of calibrated measuring tools (e.g., gap and plug gages).

(36) Measuring capability in the enterprise; best precision of measuring tools in manufacture, quality control, laboratories.

(37) Data collections from engineering sciences, standards, etc. (library).

(38) Sorts of representation, guidelines, enterprise conventions for completion of drawings.

(39) Principles for alterations and changes in manufacturing documentation, guidelines for procedures.

(40) Design log books, reports about individual design contracts.

(41) File card system for the important information carriers (books, papers, computer files) with brief details about the library, where they can be located; research reports for the relevant areas of the branch, various literature surveys (e.g., technical journals), copies of bibliographies from important branch-related books.

(42) Important journals directly applicable for the branch area (these should normally be held in the library), copies of articles and other documents.

(43) Listing of all purchased and circulated journals, evidence for readership (sign-off lists).

(44) Details to further data banks (connections, addresses, telephone numbers, contents, arrangement), extracts from the data banks for the branch, patent classes.

(45) Listing of specialist consultants available for partial areas, for example, stylists, patent lawyers, welding engineers, and so forth.

(46) Organization structure of the enterprise, relationships among sections and departments, especially with design.

FIGURE 9.5 Continued.

One derivation can be made from characteristic 3 — *recipients*. Completely different information systems are needed: (1) for *students* — as design-educational instruction, in which the design information is selected and ordered according to didactic and pedagogic concepts, and for general applicability; (2) for *teachers* or *researchers/scholars* — as scientific treatments with all reasoning, justifications, substantiations and foundations; and (3) for *practitioners* — for example, as design manuals in which prescriptive statements are contained, and the ordering occurs from pragmatic criteria (see Chapter 3).

A second derivation can be performed from characteristic 5 — *type of object*. The corresponding information for each TS-"sort" can be derived and concretized, for example, machine elements can be combined with EDS to form an SEDS (see Sections 7.1.1, 7.4 to 7.6).

Region of information		Disciplines (examples)
General sciences	(0)	Philosophy (theory of science, logic), psychology, ethics, art, politics, history, sociology, educational science, geography, biology (study of humans), botany, languages, economics, law
Basic sciences	(1)	Mathematics and geometry in all partial areas Representational geometry, descriptive geometry, constructions Physics in all partial areas (mechanics, dynamics, optics, acoustics, etc.) Chemistry (anorganic, organic) Cybernetics (systems theory, information theory)
Engineering sciences	(2)	Technical mechanics (statics, kinematics, dynamics) Solid mechanics (statics, dynamics, strength, fatigue, creep) Fluids (statics, dynamics, hydraulics) Gases (statics, dynamics, compressible flow, turbulence, fluidics) Electricity and magnetism, electronics Heat transfer, thermostatics, thermodynamics, chemical processes Materials (metallurgy, plastics, etc.) Enterprise economics and management Organization, planning Measurement, controls, instrumentation, transducers Automation, robotics, mechatronics Technical drawing, computer-aided representation Computer science and application Standardizing, unifying, typifying, intellectual property Process technology (working processes for technical systems)
Applied assisting Sciences	(3)	Technical esthetics Ergonomics Transportation, storage, packaging Working hygiene, safety, hazardous materials procedures, and so forth
Eng. design science	(4)	General questions, history, systematics of Engineering Design Science
Design branch Information	(5)	Design for properties and life cycle, including design for materials, modular and light—weight construction, lubrication
Theory of technical systems	(6)	General theory of technical systems Specialized theories of technical systems, For example machine tools, heat engines, transport machinery, water turbines, gearing, machine elements of all kinds
Design processes	(7)	Theory of design processes Design methodology and methods Management and leadership of the design process Working conditions and working means

(see also Figures I.11 and 12.6)

FIGURE 9.6   Regions of information for engineering designers.

9.1.4 INFORMATION ABOUT PROPERTIES — DESIGNERS' TECHNICAL (SPECIALIST) INFORMATION

This section should answer the question: "What information do we need in the operations of level H2 in Figures I.16, 2.5, and 2.7?" Figure 6.10 shows that all properties must interact with others, but specific directions are not shown. Different terms are used, for example, designers' specialist information (preferred term), domain information, discipline information, structured subject or branch information, "designerly" information, and so forth (see Figure 9.3).

Diagnosing and researching catastrophic failures is generally a way to encourage finding new information for engineering [466,467]. Experience is also increased by tracing fault occurrences in operating technical system (TS), such as gremlins, bugs, viruses, and so forth.

For instance, the resistance to load, the strength of a TS and its constructional parts, is usually a basic requirement for its ability to function, its "functionality," and its durability — an expression of *finality* (see Chapter 2 and Figure 2.3). Strength depends on the structure, form, and dimensions (sizes) of the TS and its constructional parts — an expression of *causality*. For designers to start designing and to achieve the necessary strength, they must have available information about quantitative and pseudo-quantitative relationships, and of heuristic advice and values of individual properties, that is, those that influence strength (see Section I.11.7). This is also a consequence and expression of the engineering principle discussed in Section 4.6, control audit, operation Op-P6.5. Strength is a property that must be anticipated by designers and that is actually achieved (realized) in the TS-life phases following designing (see Figure 6.14). All other properties must also be realized. Designers must possess, or have available, the appropriate information. Contents and form of this information are responsible for the technical and economic success of the TP(s)/TS(s), and for the *effectiveness* and time required for designing.

With respect to the contents of this information, we are concerned about possibilities for calculating, and about selecting appropriate forms and dimensions that exhibit a much less constrained set of relationships than are expressed in mathematical forms.

Information about establishing TS-form (form giving) and sizes (dimensioning) is needed, especially for constructional parts with respect to manufacturability and manufacturing costs. Such information for designers may be obtained by abstracting from practical shop-floor information (including "know-how") of operating a manufacturing technology, for example, for a particular method of manufacture: forms and shapes that can be manufactured, minimum sizes, obtainable ranges of tolerances and surface finish, manufacturing faults and causes for occurrence of scrap and need to rework, relationships to direct costs, and so forth.

The form and contents of this information should give clear recommendations to engineering designers about the needs expressed in relation to each of these manufacturing methods and about comparisons between manufacturing methods capable of delivering the same functional forms. In this sense the manufacturing department is an organization-internal "customer" of the designer's work. Figure 7.6 illustrates how this designers' specialist information may be abstracted from manufacturing technology, giving some design information regarding good form giving for castings. Similarly, this information should also provide guidelines about many other TS-properties, for example, appearance, ergonomics, and so forth. This set of disciplines is called "Design for X" ("DfX," compare [130]) and there should strictly be at least one such discipline for each class of TS-property indicated in Figures I.8 and 6.8–6.10.

Considering the tasks of design engineering from this perspective shows the complexity of designing as a process. Typical for establishing the individual design properties (e.g., dimensions, sizes) is that a partial determination must be derived from different dominating properties. For a car body shell made from sheet steel, the recommended thickness for strength will be different from that needed for TS-operational conditions "on the road," or for strength requirements during the manufacturing processes (deep drawing of sheet). Other (heuristic) recommendations arise with respect to thickness to give an adequate corrosion life and thickness for minimum cost.

Appropriate information for engineering designers (and for design work) for each of the TS-properties (external and internal, see Figures I.8 and 6.8–6.10) requires as precondition that a section on "designing to fulfill" that property (DfX) must exist in the designers' specialist information (see Figure 6.15 and Section I.11.9). Engineering designers normally develop such information by experience and observation, by obtaining a "feel" for the subject, and only rarely make retrievable records of these recommendations. The information in retrievable form should allow the gestation period that engineers need to "become good designers," typically 5 to 10 years, to be shortened. Only very few areas of designers' specialist information have been developed. Obtaining and deriving specific "DfX" information in complete and retrievable form on general, branch specific, and operational levels is needed.

9.1.5 Design for X — Information Structure

The information about how a property can be achieved by designing is gathered into a set of repositories called "DfX"; see Section I.11.9. As shown in Figure 6.15, such a repository should exist for each of the (classes of, and individual) external properties that relate to the life cycle of a TS, and for each of the operators in these life cycle processes.

Some of this information is available in the technical and scientific literature, usually in analytical expressions and data, and needs reformulating into information useful for designing.

The usual arrangement of the engineering sciences is determined by the phenomena under investigation, and their principles. In a typical thermodynamics book [519], the list of contents indicates concepts and definitions, work and heat, laws of thermodynamics, analyses that can be performed with these laws, the processes and cycles of compression and expansion and the resulting properties of the fluid, chemical reactions, phases and chemical equilibrium, and statistical and quantum mechanics. This arrangement of topics is suitable for learning thermodynamics, its theory and analytical application.

A more suitable arrangement for design engineering would be according to the effects that are achievable [286,457,540]. It would list the phenomena that are capable of achieving a particular effect (function), and their principles of action. A cross-reference of those phenomena that influence other properties would also be useful, for example, fluids with their pressures and temperatures investigated by thermodynamics must be contained (relationship to strength and mechanics of materials), fluid flow may need to be accommodated, and constructional parts (see Chapters 6 and 7) will be involved in performing the thermodynamic processes.

These insights may be used as basis for generating various checklists or catalogs that can be used by designers as part of their technical information and as methodical aids for designing. Some attempts to arrange information of "DfX" are discussed in Sections 9.1 to 9.3. They relate particularly to aspects of the quality of the list of requirements (see Section 4.3).

9.2 MASTERS

Information can be abstracted and externalized (see Section 10.4), and stored in the form of masters. Such stored information can evoke mental associations, and provide object information to support creativity. Many activities during designing can be associated with analyzing and synthesizing. In *analyzing*, an entity (a whole) is decomposed and explored. *Synthesizing* explores, selects, and unites the (often opposing or contradictory) units and moderates and overcomes any contradictions; it is not a direct inverse of analyzing.

For methodical and systematic designing, the description in the procedural model is applicable (Figures I.22 and 4.1), that is, goal-directed iterative and recursive synthesis from requirements to visualization, representation, and description of the complete TS — the carrier of the required properties (see Figures I.8 and 6.8–6.10). The phenomena and principles with which a function can be realized can thus be quickly found, and laws with which short-term and long-term behavior can be predicted (calculated) for a concrete case identified.

The existing experience information (object information) for design engineering is not, and almost cannot be, organized as precisely as the information in physics [98,556]. Some of this information is not available in a retrievable form, it is held in memories of designers. In addition this information is very extensive and depends on many elements. Therefore we must look for a different kind of resource; auxiliary materials as "masters," and especially in connection with a constructional structure as "form masters." A master can also be called an "idealization," an "archetype," a "graphical prototype," a "guiding image," a "pattern," an "exemplary scheme," and also a "model." Because of its special form and its particular content, the special term "master" is used here.

Masters are chosen for the more abstract structures of particular (recurrent) TS to present common and essential features and properties in concise ways. They are also chosen for special classes of form-giving zone in the constructional structure, where they should present the proven solutions, retrievably store the branch technical, heuristic, and empirical information. This should be done in a form that leaves the designers as much freedom (creativity) as possible and necessary for their task, and leads to generally valid solutions that they should imitate. A correctly chosen and well worked out master can deliver to the designers a basic conception (idea) that they can adjust and concretize to the conditions of their immediate case.

A procedural master for establishing sizes and other properties of journal bearings is shown in Figure 2.17. Some examples for several of the available TS-structures in Figure 2.18 should make the role of masters clear. A master for the details of a plain sliding (oil lubricated) bearing is shown in Figure 7.8. The solution of a gear box in Figure 4.6A, illustrates the role and execution of a master. Figure 4.6B shows several possible representations of masters, intentionally presented at different levels of abstraction and complexity. Part (A) of the figure shows the three most important forms of bearing arrangements in the style published by the manufacturers of rolling bearings. If the arrangement of a fixed and an axially movable bearing is chosen, the master of the fixed bearing as in part (B) (recommended as an organ structure) or at a more concrete level the one in part (C) can be used. For the choice of individual organs, masters for oil seals (part [D]), or axial locators and thrust fixtures (part [E]) are shown. If the choice is made according to the conditions of the case (mainly shaft diameter, rotational speed, and lubricants), a concrete bearing arrangement for the shaft emerges (see part [F]).

Many organization catalogs of products for OEM/COTS already show examples of how these products can and should be incorporated into a more complex product, so that they can fulfill their purpose. Comparisons among the offered products from different organizations are often difficult. Nevertheless, such examples can form the basis for constructional-structure masters for an organization using purchased constructional parts.

Some forms and other procedural instructions can also be regarded as masters, for example, a master for the arrangement of a morphological matrix is shown in Figure 4.4.

9.3   ENGINEERING DESIGN CATALOGS

The search for causes of the necessary effects (including suitable alternative principles and means) is supported by the natural and engineering sciences, and is profitably collected in *Engineering Design Catalogs*, collections and compilations of information made available to engineering designers, as useful and directly applicable as possible.

9.3.1   FORMS OF CAPTURED INFORMATION

The conventional arrangement of information is characterized by the titles of sciences and other disciplines: mathematics, physics, chemistry, history, fine arts, performing arts, and so forth. The engineering sciences follow this pattern (see Figure 12.6). Each area of information delimits its scope to consider an isolated phenomenon, usually by imposing some simplifying assumptions and axioms. It then categorizes that information in levels of hierarchical detail; see Section 12.3 [240], where possible described in mathematical/analytical models.

The main questions are: How can this phenomenon be described and categorized? How does this phenomenon behave (inputs, conditions, outputs, outcomes)? How can this phenomenon be modeled (in verbal, graphical, and symbolic/mathematical models)? How (and how well) can the behavior of this phenomenon be predicted? Prediction is not always possible, for example, where thinking of living things are involved (compare Section 11.1).

Information in a strictly categorized "dictionary" form is useful for design engineering, mainly in the analytical step Op-H3.3 "evaluate and decide"; see Section 2.4.1. The counterpart to a dictionary in this sense is a *thesaurus* (see Section 9.1.3). French [214] showed that this information can also help during conceptualizing to select among concepts and to establish some of the useful operating parameters for solution proposals.

For the basic operation Op-H3.2 "search for solutions," a different arrangement of information would be more useful. The main question is: With what phenomenon, principle or means can a desired outcome be achieved? This question needs a search across various disciplines (e.g., of science). Different phenomena and principles can deliver a particular (desired) outcome, more or less well, in some cases under different conditions. Presenting information in this way is the purpose of design catalogs.

9.3.2 PRINCIPLES OF DESIGN CATALOGS

Verein deutscher Ingenieure (VDI) issued a guideline for generating and presenting design catalogs [23]. Design catalogs are usually presented as tables, arranged according to methodical viewpoints, with information as complete as possible (within a given framework) and systematically categorized (see Figure 7.9). They permit targeted search for particular content.

Catalogs can be described by various characteristics [498], such as dimensionality, level of detail, predominant usage (steps of the design process, see Section I.12; method, Chapter 7; design operation, and basic operations, Chapter 4), and so forth. A distinction between design catalogs and masters (Section 9.2) is not always clear.

In their *arrangement*, most catalogs contain categorization, presentation, and extension or restriction parts for the information, and appendices if needed. These parts can be recognized by analogy in most scientific books: Categorization part = List of Contents; Presentation part = Main text, appendices, and so forth; Extension part = Index.

A *one-dimensional* catalog is arranged as a consecutive listing of the content items, one to each row. The categorization part is contained in the first (set of) column(s). The presentation part follows in the second (set of) column(s). An extension part may be attached in further columns. Figure 7.12 may be regarded as such a one-dimensional catalog.

A *two-dimensional* catalog has its categorization and extension parts both in the first (set of) column(s), and in the heading row(s). The presentation part is in the form of a matrix, each cell contains a presentation item. Further extension information may be given as transparent overlays to augment the information given in the cells. Figure 7.13 is an example.

A *three-dimensional* catalog can be formed by producing a set of two-dimensional catalogs with identical column and row headings, where the presentation of each separate (transparent) overlay sheet conforms to the different conditions for the third dimension.

Higher dimensionality may be possible by computer.

Concerning *detail*, a catalog may be set up according to stated *principles* [498]; it may provide a *survey*, giving the main classifications and sample presentation entries; it may give full *detail* of the classification and presentation entries, aiming to be comprehensive. In practice, all three levels of detail are useful for an iterative procedure: first take a "quick and dirty" survey of the situation [482], then go into just enough detail to explore the implications, and then go into detail only for those parts that are directly relevant (see Section 4.1).

Roth [498] divides catalogs into four types according to *content*.

Object catalogs are general, task independent; they collect basic object information that is generally applicable. Examples include: listings of physical phenomena and their (graphical and mathematical) models; properties (chemical, physical, and so forth) of materials; straight-line guidances.

Operations catalogs give instructions for procedures, they list steps and operations, conditions of use, and criteria for application. Typical are: rules for functions and function structures, for example, Figure 6.4; rules for variations of structures and forms, for example, Figure 2.14; operating instructions (see Section 2.5). Operations catalogs usually do not contain extension parts.

Solutions catalogs are task dependent, they try to present a complete selection of solutions, especially for the transitions from one TS-model to another. Many masters (see Section 9.2) may be regarded as (partial) solution catalogs.

Relationships catalogs show connections and relationships among functional features — they are closely related to TS-organs. Examples include: shaft-hub-connections; forms of impeller blade configuration for axial/radial compressors and turbines.

9.3.3 EXAMPLES

A typical solutions catalog may appear as in Figure 9.7. The first column lists different organs that can fulfill a function (e.g., transmission of torque between two shafts at various angles); the attached columns show parameter ranges and criteria for choice between the organ means. Figure 9.8 shows a catalog of principles based on electrical capacitance, with equations and examples of application. Many function structures can also be drawn as logic diagrams showing sequences of operations on materials, energy, and information (signals). Figure 9.9 shows the possible operations and their inverses, with appropriate symbols. Various ways of amplifying forces are shown in Figure 9.10, together with criteria for selection and sizing.

For a constructional structure, Figure 9.11 shows a selection of shaft-hub con-nections, showing the principles on which the catalogs are based, and a survey of five types of connection. The detail catalog (not illustrated) lists 24 variations on these 5 themes [498].

Several physical quantities (as functions) can be transformed. A matrix of possib-ilities of direct transformation is shown in Figure 9.12, for which the attached detail catalog lists 169 functional relationships of these transformations [178] (e.g., see Figures 9.8 or 9.10, compare also Section 7.5).

Various other forms of representation can be regarded as catalogs. Systematic variation of form is shown in Figures 2.15 and 8.7. Properties of humans in their relationships with TS is illustrated in Figure 6.1. Masters (see Section 9.2), can also function as catalogs. A procedural master for establishing dimensions of journal bearings appears in Figure 2.17, with appropriate design master shown in Figure 2.18.

The main compilations of engineering design catalogs have been published in German language [178,370,498]. Compilations for specialized fields in English include rotary piston devices [564], general mechanical engineering mechanisms [332,546], linkage mechanisms [381], analogies from nature [215], laws of physics, and their relationships suitable for research [268].

9.3.4 RELATIONSHIPS TO ORGANIZATION PRODUCT CATALOGS

Manufacturing organizations produce and publish advertising literature to describe the organization's products and their properties, including data, criteria and methods

Function: transmission of torque — shaft to shaft — Function > Organ > Parameter

Organ (Function-carrier); TS-"sorts" that can fulfill the function	Important properties (parameters and characteristics)							
	Possible shaft orientation	Maximum transmission ratio i_{max}	Maximum power P_{max} kW	Maximum circumferential velocity v_{max} m/s	Average efficiency η	Average life (hours of operation) t_L	Guide to relative cost	etc.
Straight-tooth spur gears	=	4 (7)	50.10^3	15 (100)	0.98	10^5		
Helical spur gears; angle drive helical gears	= l°	7 (10)	50.10^3	25 (150)	0.97	10^5	Typical values may be found for specific cases	
Worm gears	l°	4 (100)	50 (200)	15	0.4–0.7	10^3		
Friction wheel drives	=	7 (15)	100	15–20	0.96	2.10^3		
Chain drives	=	8 (10)	100 (3500)	40	0.9–0.97	$2-10.10^3$		
Flat belts and tapes	= (l°)	5 (20)	100 (3500)	35 (100)	0.88–0.95	Highly variable H		
Vee belts	= (V)	7 (10)	100 (1500)	60	0.96	Highly variable		
Rope drives	= (l°)	5						
Tooth (timing) belt drives	=							
etc.								

FIGURE 9.7 Solutions catalog — decision table: torque transmission.

**Solutions (principles) catalog:
transformation of energy or type of signal**

Cause ——— [▱] ——— Capacitance (change of)

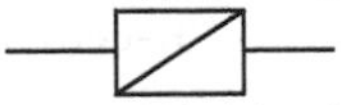

Function: measurement of displacement, Principle: electrical energy, Partial principle: capacitance

Cause	Physical source		Law	Literature	Example of application
01.11 Length, section area, volume	Capacitance (plate spacing)	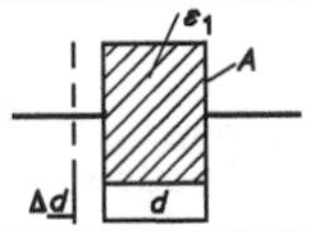	$\Delta C = C \dfrac{\Delta d}{d}$	[22], p. 150	Length measurement
	Capacitance (surface area)		$\Delta C = C \dfrac{\Delta A}{A}$	[22], p. 154	Variable tuning capacitor, angle measurement
	Movement of dielectric	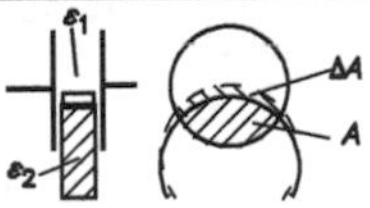	$\Delta C = C \dfrac{\Delta A}{A}(\varepsilon_1 - \varepsilon_2)$	[22], p. 154	Reader for punched card or tape
	Thickness of dielectric	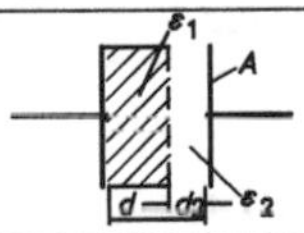	$\Delta C = \dfrac{C\, d_2\,(\varepsilon_1 - \varepsilon_2)}{d\varepsilon_2 - d_2(\varepsilon_1 - \varepsilon_2)}$	[22], p. 157	Thickness gauging
01.11 Length, section area, volume	Electrode gap	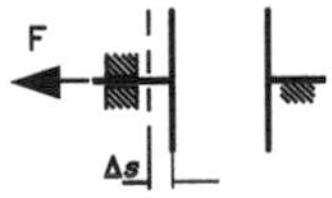	$\Delta C = \dfrac{2}{U^2} F \Delta s$	[40], p. 108	Force measurement
	Dielectric constant	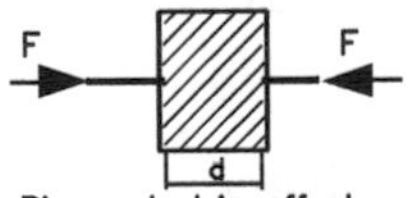 Piezo-electric effect	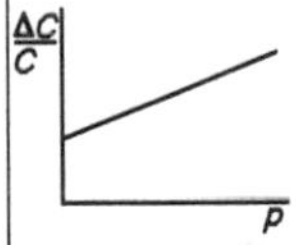	[8], p. 560	Pressure or strain measurement (transient)

Literature:
Koller, R. (1985) *Konstruktionslehre für den Maschinenbau* (2nd ed.), Berlin/Heidelberg: Springer-Verlag, p. 245

FIGURE 9.8 Design catalog: transformation of energy or signal.

(e.g., calculations) for selection, application, maintenance, and precautions for use of the product. They also show examples (masters) for typical applications of (OEM, COTS) product. Normally the product catalogs from a manufacturing organization do not contain comparisons between products from different organizations.

Wholesale and retail organizations produce catalogs that list ranges of products available through their agencies, usually with only a selection of the available data. Some comparisons among products are given, but only for the range of products that the organization can deliver.

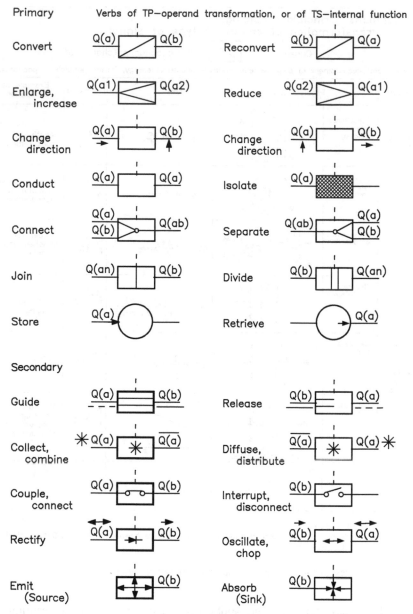

FIGURE 9.9   Basic physical operations — symbols for flowcharting.

Literature: *(see also Figures I.9, 7.12 and 7.13)*

Koller, R. (1985) *Konstruktionslehre für den Maschinenbau* (2nd ed.), Berlin/Heidelberg: Springer-Verlag, p. 37

General function	Configuration	Equation	Example of arrangement	Serial number	Amplification V	Stroke s	Friction effects	Main dimension l	No. of guides, 1 d.o.f.	Additional properties
	Categorization — Classifying factors	Presentation — Candidate solutions			Extension — Selection characteristics					
1	2	1	2		1	2	3	4	5	6
Force conversion system — organ structure	Wedge, inclined plane (screw)	$F_2=F_1 \cdot \cot(\alpha+2\rho)$	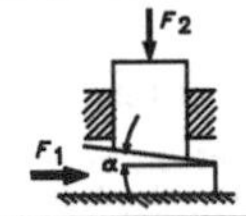	1	$V=\cot(\alpha+2\rho)$, $V_{max}\approx10$	$s_{2max}=(1/V)\cdot l$	Increased friction coeff. ρ reduces amplification V	$l=V \cdot s_{2max}$	3 sliding guides	Movement blocked when F_2 driving for $\tan\alpha<\rho$
	Connecting rod, push link	$F_2=F_1 \cdot \cot$	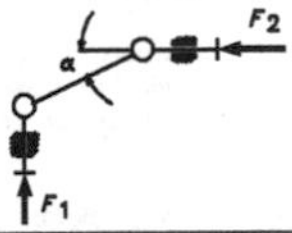	2	$V=\cot$, $V_{max}\to\infty$	$s_{2max}=\approx0.6\,l$	Small influence of friction coeff. because	$l=1.7s_{2max}$	2 sliding 2 rotating guides	Progressive force amplification
	Lever	$F_2=F_1 \cdot \dfrac{l_1}{l_2}$	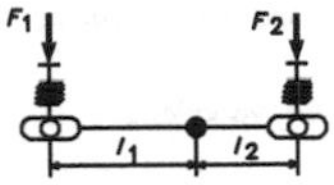	3	$V=l_1/l_2$, $V_{max}\to\infty$	s_{2max} Indefinite for wheel $\geq 2\,l_2$ for lever	of rotating joints	$l=2d$ for wheel, $l=l_1+l_2$ for lever	2 sliding 1 rotating guides	Transmission of unlimited motion is possible (wheel)
	Pulley hoist, block and tackle	$F_2=F_0+F_1$	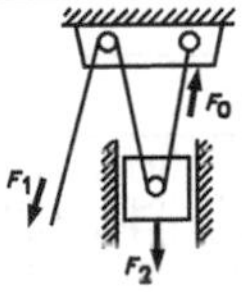		$V=2.n$, $n=$ number of lower pulleys, $V_{max}\approx8$	Depends on length of rope	Friction limits maximum amplification V	$l \geq s_{2max}$	1 sliding guide	Movement blocked when F_2 driving for
	Pressure in fluid	$F_2=F_1 \dfrac{A_2}{A_1}$	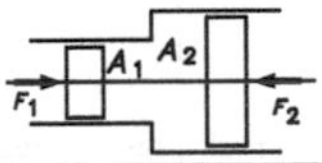		$V=\dfrac{A_2}{A_1}$		Almost no influence because of selection		2 sliding guides	

Literature
Roth, K. (1982) *Konstruieren mit Konstruktionskatalogen* (Designing with Design Catalogs),
Berlin/Heidelberg: Springer-Verlag

FIGURE 9.10   Design catalog — force amplification.

A suitable collection of information from such organization catalogs can help to compile more general design catalogs for use in designing.

9.4   STANDARDS, CODES OF PRACTICE, GUIDELINES

An engineering standard is a written document defining a requirements specification for a level of performance, a prescription of properties (including dimensions, quality) for technical products, for example, procedures, services, systems, constructional parts, materials, and so forth. A standard adopted for recurring use is a basis for implementing and certifying *quality assurance*, and reducing unnecessary variety in products and services (see [343]).

Application mostly implies voluntary acceptance of standards documents by an individual organization. Prescription can occur when a particular standard as agreed at an issue date is named as compulsory, and passed into law by a legislature. Compulsory prescription can occur after due process, from cases in courts of civil and criminal law, for example, aspects of product liability.

Principles for design catalog — shaft-hub-connections
Function: connect, means in principle

1	Purpose of the catalog	Survey and details of releasable mechanical connections between shafts and hubs, in five basic constructional forms: profiled shafts, use of a form element, interference fit, tensioning element, and pre-tensioned connection.
2	Design phase	Form-giving, constructional structure.
3	Usage	All connections between shafts or axles and hubs, for example for gearing, rollers, pulleys, levers, and so forth. The survey part allows general selection of the basic constructional form. The detail part increases the range of suggested solutions, and allows selection according to more detailed criteria.
4	Definition of primary term	A shaft-hub-connection is a connection between a nominal external cylindrical element and a nominal internal cylindrical element that is not movable in the radial direction, but that may or may not be movable in an axial direction.
5	Categorization	Sub-groups are formed by the type of surface closure: radial surface (form-closure) contact, tangential surface (friction-closure) contact, or combined (see Figure 7.14). A second sub-grouping occurs by direct or indirect action.
6	Variation	The detail catalog shows a total of 24 variations.

Survey catalog — shaft-hub-connections

Categorization		Presentation			Extension							Appendix
Type of surface closure	Type of force transmission	Formula	Name	Example of use	Transmitted moment	Moment dependancy	Axial force capability	Over-load behavior	Run-out adjustable	Hub axially movable	Hub rotatable	Comments
Normal (form-closure)	Direct	$M_t = \dfrac{dm}{2} A_{tot} \tau_{all}$	Profiled shaft		Large	Form factor	No	Fracture	Yes	Possible	Possible step-wise	
	Indirect	$M_t = \dfrac{dm}{2} A_p P_{all}$	Form element connection		Small		Possible			Possible		Simple assembly
Tangential (force-closure)	Direct	$M_t =$ $M_f =$ $\dfrac{dm}{2} F_f$ $\dfrac{dm}{2} \mu F_n$	Stressed fit		Large to small	Temperature, torque, axial force	Yes	Slip	Yes	Only if $F_a \geq F_f$	Continuous	
	Indirect		Stressing element	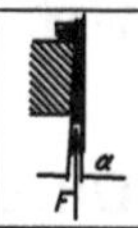	Mid				Possible			Low expense for manufacture and assembly
Mixed (form- and force-closure)	Indirect		Pre-stressed connection	Taper key in keyways	Small		Possible	Fracture	No	No	Possible	

Literature
Roth, K. (1982) *Konstruieren mit Konstruktionskatalogen* (Designing with Design Catalogs) 2nd ed., Vol. 2, Berlin/Heidelberg: Springer–Verlag, pp. 164—166

FIGURE 9.11 Design catalog — shaft-hub connections.

Input quantity → / Output quantity ↓	F 01	i 02	s 03	v 04	a 05	M 06	L 07	φ 08	ω 09	α 10	f 11	p 12	V 13	m 14	I 15	U 16	E 17	H 18	T 19	Q 20
Force F 01	13	1	15	10	1	1	1	3	3		1	1	1	5	2	2	1	1	2	2
Impulse i 02		1		1																
Length, displacement s 03	8		6	1		1					2	1		1		2			2	
Velocity v 04	2	1	2	3							1	1								
Acceleration a 05	1				1				1											
Moment (of force) M 06	1						1	1	2	1	1									
Rotational impulse L 07						1		1	1	1										
Angle φ 08						1														
Angular velocity ω 09			1			2	1													
Angular acceleration α 10						1			1											
Frequency f 11	1		1											1				1		
Pressure p 12	1		2	2									2					1		
Volume V 13											3									
Mass m 14															2	2				
Electrical current I 15	2		2						1		1	1			2	4				
Electrical voltage U 16	2		2								1	1			4	1			1	
Electrical field strength E 17																				
Magnetic field strength H 18																				
Temperature T 19	3																			
Electric charge Q 20	3		3								2				1					4

The entries in this matrix show the number of different ways in which the relationships (input to output transformations) of the physical quantities may be achieved. The listing in the catalog attached to the matrix in the literature shows the name, relationship and reference number, a diagram of the conditions, the relational equation(s), examples of use, and a literature reference. 169 relationships are listed in the catalog.

The reference numbers have the following form:

01.03–2

Output quantity F — Input quantity Displacement h — Serial number h

$F = f(h)$

(see also Figure 7.10)

Additional solutions may be found by chaining several of the displayed relationships to reach the desired output from the given input.

Example   Desired relationship:

$F = f(L)$

Generating a force from a rotational impulse;

Chain of effects:

(1) $M = f(L)$ Ref. No. 06.07–1

Generating a moment from a rotational impulse,

(2) $F = f(M)$ Ref. No. 01.06–1

Generating a force from a moment

Literature
Adapted from Ehrlenspiel, K. (1995) *Integrierte Produktentwicklung* (Integrated Product Development), München: Carl Hanser Verlag, 1995

FIGURE 9.12   Relationship matrix among physical quantities and laws.

Writing and using standards may be for reasons of reducing safety hazards, obtaining insurance coverage, protecting the environment, ensuring interchangeability and compatibility of constructional parts, long-term repair and replacement plans, independence from particular suppliers, reduction of excess variety, and outsourcing — (international) contract manufacture of parts. One source of information consists of fault reports of catastrophic failures, damage and personal injuries, which may lead to compulsory recall of a TS for modification, for example, grounding of aircraft, with instructions for modification site, or for return to the supplier or manufacturer.

Standards are written and accepted mainly by voluntary agreements (consensus) among representatives of suppliers, users and regulators (governments) to define and revise the scope and wording of standards documents. Usually this needs multiple voting, and so forth.

At times, an accepted market leader emerges for a TS-"sort" and dominates the market, for example, software operating systems for computers — CP/M, MS-Windows, UNIX, and so forth. The information provided by the leading organization is commonly called an *industry standard*.

9.4.1 Aspects of Organizations and Coordination

International standards are coordinated by International Organization for Standardization (Organisation Internationale de Normalization [ISO]) and International Electrical Commission (IEC), who are responsible for terminology, procedures, constructional parts, and materials.

Le Bureau International des Poids et Mesures (BIPM) with subgroupings of Bureau International de l'Heure (BIH), Le Comite International des Poids et Mesures (CIPM), Le Comite Consultatif des Unites (CCU), Le Comite Consultatif de Thermometrie et de Calorimetrie (CCTC), La Commission Internationale de l'Eclairage (CIE), and La Conference Generale des Poids et Mesures (CGPM), are responsible for Le Systeme International d'Unites (SI).

The North Atlantic treaty organization/Outreach and Technical Assistance Network (NATO/OTAN), advisory group for aerospace research and development (AGARD), the armed forces' Military Standards (MIL STD), and other similar alliances also issue standards with international validity.

At the next hierarchical level, national standardizing occurs in almost every country, usually by a single organization, which coordinates preparation and publication of standards, testing of products, constructional parts and services, management of certification programs, and is a dealership for standards from international organizations and from other countries.

National regulations, similar to standards, are issued by other organizations, for example, regulations about and registration of patents, copyrights, trade marks, registered industrial designs — "intellectual property" (see Section 11.3.2), and occupational safety and health for workplaces.

National codes of practice and guidelines for products, and codes of professional conduct and ethics, are issued by institutions of engineering in various branches. In most countries, these institutions are both learned societies (disseminating information), and regulators of the engineering professions. Some government and private special interest groups are responsible for particular usage of constructional parts or materials, for example, Concrete Users Association.

Testing and inspection for compliance with laws and standards (including certification tests and verification of measuring tools), and support for insurance, is performed by national research organizations, Lloyd's Register, Underwriters Laboratories, and so forth. Various national and local boards of trade, chambers of commerce, and public utilities sometimes publish rules and guidelines for special circumstances.

Organizations "in-house" make their own selections from the standards of preferred constructional parts and materials for use within the organization to restrict varieties (e.g., of fasteners) for inventory control, purchasing and cost control (see Figures 6.7 and 9.3). Offers for sale of equipment, constructional parts or materials by an organization, and rules for selection, are published in organization product catalogs.

Factual (object) information is available from ESDU (U.K.) [5], who collect and verify engineering data from many sources. National libraries provide reference services, online information retrieval from published sources, scientific numeric databases, and so forth, and are the highest resource level of library, interlibrary loans, and national library services.

9.4.2 EXAMPLES OF STANDARD SPECIFICATIONS

Boilers and pressure vessels illustrate needs for safety and inspection procedures regulated by a standard. The American National Standards Institute/American Society of Mechanical Engineers (ANSI/ASME) Boiler and Pressure Vessel Code [1] (in 25 loose-leaf volumes) covers general requirements for different size and duty ranges (including nuclear reactors), design (methods and features), fabrication, materials, welding, inspection and nondestructive examinations, qualification (of manufacturers, and of fabrication and user operators), care and operation, stress calculation methods (also with simplified formulae), corrosion allowances, tolerances (e.g., out-of-roundness), identification markings, and so forth.

An outline of some considerations is shown in Figure 9.13 [17,95].

NOTE: Lloyd's Register (originally "... of Shipping") was founded (1760) to publish lists of sailings and arrivals, established for inspection and classification of ships (1834), published regulations for inspection and classification of wooden ships (1836), and continued for iron and steel ships and their equipment. It is now an international organization for many types of equipment, for example, nuclear reactors and power plant. The "Dampfkessel-Untersuchungs-und Versicherungsgesellschaft auf Gegenseitigkeit" (Steam Boiler Investigation and Insurance Society for Mutual Protection) was founded in Austria (1872), and issued regulations for manufacture and use. In the United States, on March 10, 1905 in Brockton, Massachusetts, a boiler for steam pumps of a (horse drawn) fire appliance exploded, leaving 58 dead and 117 injured. The Commonwealth of Massachusetts ordered a boiler code to be written, *claimed by some to be "the first in the world."* This has been revised and enlarged to the ANSI/ASME Boiler and Pressure Vessel Code [1], and is now regarded as a definitive international standard.

Idealized vessel definitions and caution zones

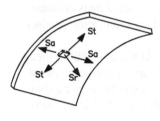

Caution (A):
Stress concentration
at head joint ——
Roark p. 445–512

Caution (B):
Flat plate as head ——
Roark p. 324–413

Caution (C)
External over-pressure
(po > pi) can cause
buckling ——
Roark p. 501

Theoretical stress conditions in shell

REMOTE FROM ENDS OF VESSEL:

Sa ... Axial stress, head ends active, longitudinal stress
—— would result in tension failure around circumference,
or shear failure.

St ... Tangential stress, shell active, hoop stress
—— would result in tension failure along body seam, or
shear failure

Sr ... Radial stress, pressure active thickness—dependent
—— effects of variation of stress from inside to
outside surfaces can be neglected if the ratio
$t/D < 0.05$ (1/20)

For equilibrium —— internal over-pressure equal to stress in material:

Axial loading: pressure on head end   =   stress in shell
$3.14159.D.D.(pi - po)/4$ = $3.14159.D.t.Sa$
assumed uniform over t: $Sa = (pi - po).D/4.t$ (tensile)

LONGITUDINAL loading: pressure on shell   =   stress in longitudinal seam
$D.L.(pi - po)$ = $2.L.t.St$
assumed uniform over t: $St = (pi - po).D/2.t$ (tensile)

MAX. SHEAR STRESS: $Ts = [St - (pi - po)]/2 = (pi - po).(D/4.t - 1)/2$

3–D Mohr's circle

Real deformation patterns

Cylindrical
expansion   Bulge

Bending

Bulge

pi – po

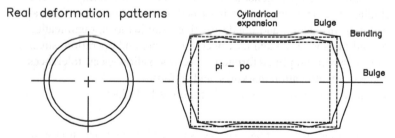

Ideal shape —— spherical (But difficult to manufacture)
Improved construction —— torispherical or elliptical heads, cylindrical shell

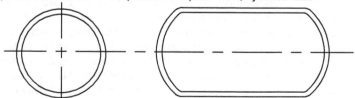

Literature
Young, W.C. and Budnyas, R. (2001) *Roark's Formulas for Stress and Strain* (7th ed),
New York: McGraw-Hill, http://www.UTS.com/TheUltimateReference

FIGURE 9.13   Thin-walled pressure vessels.

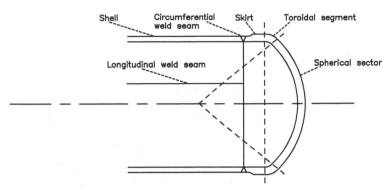

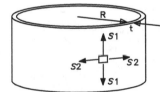

Shell   Circumferential   Skirt   Toroidal segment
         weld seam

Longitudinal weld seam   Spherical sector

TABLE 29 (excerpt) — Formulas for membrane stresses and deformations in thin‑walled
pressure vessels:
    v ... Poisson's ratio,   E ... Young's modulus of elasticity,
    psi ... Angular deflection at edge (positive when er increases with y),   phi ... Subtended angle,
    er ... Radial displacement (expansion),   ey ... Change in height dimension,
    S1 ... Meridional stress,   S2 ... Circumferential stress,
    R ... Radius of curvature of circumference,   R2 ... Principal radius of curvature,

Case 1: Cylindrical, uniform internal or external pressure, ($pi - po$), ends capped —
        conditions at points remote from the ends

$S1 = (pi - po).R \, / \, 2.t$
$S2 = (pi - po).R \, / \, t$
$er = (pi - po).R.R.(1 - v/2) \, / \, E.t$
$ey = (pi - po).R.y.(1 - v/2) \, / \, 2.E.t$
$psi = 0$

$R \, / \, t > 10$

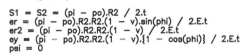

Case 3: Spherical, uniform internal or external pressure, (pi − po), edges supported
        tangentially

$S1 = S2 = (pi - po).R2 \, / \, 2.t$
$er = (pi - po).R2.R2.(1 - v).\sin(phi) \, / \, 2.E.t$
$er2 = (pi - po).R2.R2.(1 - v) \, / \, 2.E.t$
$ey = (pi - po).R2.R2.(1 - v).\{1 - \cos(phi)\} \, / \, 2.E.t$
$psi = 0$

$R2 \, / \, t > 10$

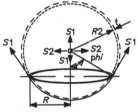

Case 5: Toroidal, uniform internal or external pressure, $(pi - po)$

$S1 = (pi - po).b.(r + a) \, / \, 2.t.r$
Max $S1 = (pi - po).(2.a - b) \, / \, 2.t.(a - b)$
        at point 0
$S2 = (pi - po).b \, / \, 2.t$
        throughout
$er = (pi - po).b.\{r - v.(r + a)\} \, / \, 2.E.t$

$b \, / \, t > 10$

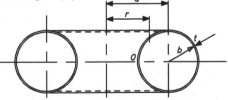

NOTES: at transitions, radial expansions and edge rotations should be as well matched as
      possible to reduce bending and shear stresses. Stresses should preferably be equal throughout
      shell and head, and welds should be remote from high−stress zones.

Literature:
Young, W.C. and Budnyas, R. (2001) *Roark's Formulas for Stress and Strain* (7th ed),
New York: McGraw−Hill, http://www.UTS.com/TheUltimateReference

FIGURE 9.13 Continued.

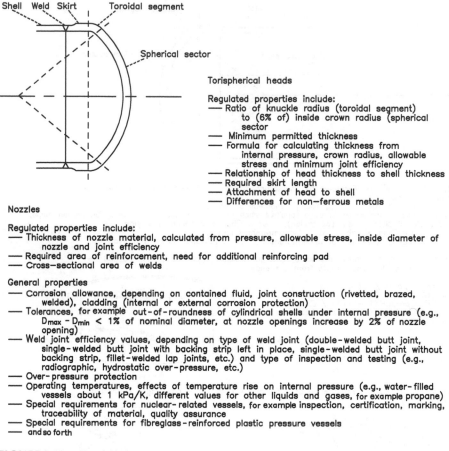

Shell Weld Skirt   Toroidal segment

Spherical sector

Nozzles

Torispherical heads

Regulated properties include:
— Ratio of knuckle radius (toroidal segment)
     to (6% of) inside crown radius (spherical
     sector
— Minimum permitted thickness
— Formula for calculating thickness from
     internal pressure, crown radius, allowable
     stress and minimum joint efficiency
— Relationship of head thickness to shell thickness
— Required skirt length
— Attachment of head to shell
— Differences for non-ferrous metals

Regulated properties include:
— Thickness of nozzle material, calculated from pressure, allowable stress, inside diameter of
     nozzle and joint efficiency
— Required area of reinforcement, need for additional reinforcing pad
— Cross-sectional area of welds

General properties
— Corrosion allowance, depending on contained fluid, joint construction (rivetted, brazed,
     welded), cladding (internal or external corrosion protection)
— Tolerances, for example out-of-roundness of cylindrical shells under internal pressure (e.g.,
     $D_{max} - D_{min} < 1\%$ of nominal diameter, at nozzle openings increase by 2% of nozzle
     opening)
— Weld joint efficiency values, depending on type of weld joint (double-welded butt joint,
     single-welded butt joint with backing strip left in place, single-welded butt joint without
     backing strip, fillet-welded lap joints, etc.) and type of inspection and testing (e.g.,
     radiographic, hydrostatic over-pressure, etc.)
— Over-pressure protection
— Operating temperatures, effects of temperature rise on internal pressure (e.g., water-filled
     vessels about 1 kPa/K, different values for other liquids and gases, for example propane)
— Special requirements for nuclear-related vessels, for example inspection, certification, marking,
     traceability of material, quality assurance
— Special requirements for fibreglass-reinforced plastic pressure vessels
— and so forth

FIGURE 9.13 Continued.

Rolling bearings (ball, roller, needle) illustrate needs for interchangeable constructional parts. For standardizing, a consistent terminology (nomenclature) is needed, and a full range of sizes, capacities, capabilities, and tolerances (dimensional and geometric), a *standard specification*. For application of bearings, designers need information about internal modes of action (principles and ways they operate, and limitations), and properties (load, speed, life, failures, and etc.). Much of this information is also standardized. For designing of bearing systems, a need exists for establishing actual operating forces and speeds, especially those imposed by operating conditions of the bearings, but including internal and external influences and redundancy of force transmission paths. This need extends to establishing influences of angular and axial deflection of shafts and housings from operating loads on bearing operation, internal load distribution among rolling elements, and influences of noncircular housing distortion under load.

Procedures of calculation to achieve adequate life of bearings are available from bearing catalogs issued by ball and roller bearing manufacturers, and for example, [596]. Considerations include freedom of motion in the main operating direction (e.g., rotation around centerline) and in auxiliary directions (axial for thermal expansion, angular deflection or pivotting of axis around a point on axis). Preloading of bearings can reduce (1) influences of clearances and (2) forces induced by bearing geometry, for example, angular contact, taper roller, and so forth.

Needs for lubrication, cleanliness, filtration, and cooling of the bearings and the lubricant are important design considerations. Available sizes, prices, delivery times, and so forth from likely suppliers and manufacturers are needed.

NOTE: Capacities, load characteristics, will be different from various manufacturers, but *testing methods* are standardized.

Correct handling, mounting and operating procedures and tools, and typical arrangements for mounting, retaining, positioning, sealing, and so forth, must be considered (see Figure 9.14).

Engineering designers should find out (for themselves) what the offered constructional parts (OEM, COTS) *can do* and how they operate, for example, by abstracting and analyzing, [133,188,342,521]. Asking others is not always best. Selling agents usually are directly interested in selling their products.

9.5 FORMS (PRO FORMA)

The term "pro forma" is understood to mean a previously prepared form, sheet, checklist or questionnaire, with spaces (categories, columns) for recording answers for prescribed questions. Though filling out pro forma appears routine (mechanical), it is of advantage if precise statements and completeness can be obtained quickly. Nevertheless, engineering designers, and other users of pro forma, need to understand the meanings of the contents.

Pro forma are often used for design aids, and they have clear relationships to other information carriers. A design catalog can be considered as a completed pro forma. A pro forma can be viewed as a draft for a master (e.g., Figure 2.18). Pro forma can also serve as communication, for example, belt and chain drives (Figure 9.15).

9.5.1 USE OF PRO FORMA IN METHODS

Some methods employ pro forma (see Chapter 8). The morphological matrix for establishing an organ structure serves as an example; see Figure 4.4 — functions are associated with classes of action principles and organs. Quality Assurance can use pro forma, for example, QFD [111,151,269], Figure 8.5, asks about requirements and properties, as perceived by (or for) customers, and shows their relationships to technical considerations. FME(C)A [554] prescribes a pro forma (Figure 9.16) with questions to analyze the possible (anticipated) faults or errors and their consequences.

Similarly a set of pro forma is suggested for using PABLA [378,537] — Program Analysis by Logical Approach, one of which is shown in Figure 9.17.

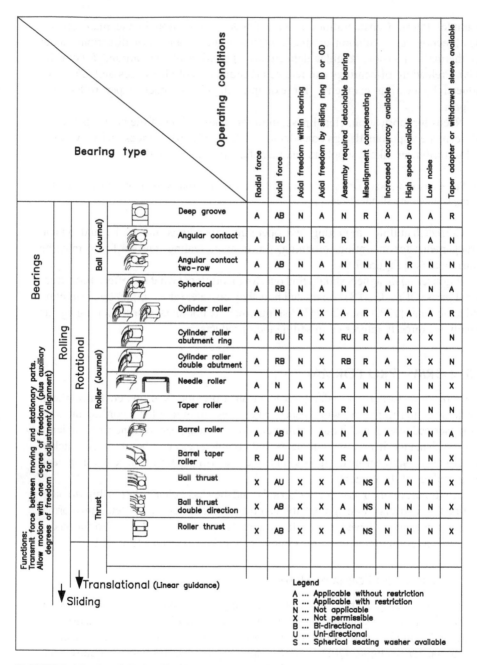

Bearing type				Operating conditions →	Radial force	Axial force	Axial freedom within bearing	Axial freedom by sliding ring ID or OD	Assembly required detachable bearing	Misalignment compensating	Increased accuracy available	High speed available	Low noise	Taper adapter or withdrawal sleeve available
Bearings	Rolling	Rotational	Ball (Journal)	Deep groove	A	AB	N	A	N	R	A	A	A	R
				Angular contact	A	RU	N	R	R	N	A	A	A	N
				Angular contact two-row	A	AB	N	A	N	N	N	R	N	N
				Spherical	A	RB	N	A	N	A	N	N	N	A
			Roller (Journal)	Cylinder roller	A	N	A	X	A	R	A	A	A	R
				Cylinder roller abutment ring	A	RU	R	X	RU	R	A	X	X	N
				Cylinder roller double abutment	A	RB	N	X	RB	R	A	X	X	N
				Needle roller	A	N	A	X	A	N	N	N	N	X
				Taper roller	A	AU	N	R	R	N	A	R	N	N
				Barrel roller	A	AB	N	A	N	A	A	N	N	A
				Barrel taper roller	R	AU	N	X	R	A	A	N	N	X
			Thrust	Ball thrust	X	AU	X	X	A	NS	A	N	N	X
				Ball thrust double direction	X	AB	X	X	A	NS	N	N	N	X
				Roller thrust	X	AB	X	X	A	NS	N	N	N	X

Functions:
Transmit force between moving and stationary parts.
Allow motion with one degree of freedom (plus auxiliary degrees of freedom for adjustment/alignment)

▼ Translational (Linear guidance)
▼ Sliding

Legend
A ... Applicable without restriction
R ... Applicable with restriction
N ... Not applicable
X ... Not permissible
B ... Bi-directional
U ... Uni-directional
S ... Spherical seating washer available

FIGURE 9.14   Bearing classifications and standards.

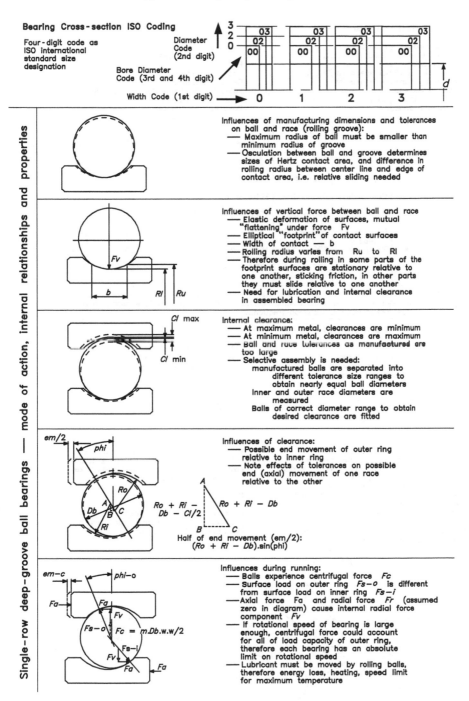

Bearing Cross-section ISO Coding

Four-digit code as ISO international standard size designation

Diameter Code (2nd digit)

Bore Diameter Code (3rd and 4th digit)

Width Code (1st digit)

Single-row deep-groove ball bearings — mode of action, internal relationships and properties

Influences of manufacturing dimensions and tolerances on ball and race (rolling groove):
— Maximum radius of ball must be smaller than minimum radius of groove
— Osculation between ball and groove determines sizes of Hertz contact area, and difference in rolling radius between center line and edge of contact area, i.e. relative sliding needed

Influences of vertical force between ball and race
— Elastic deformation of surfaces, mutual "flattening" under force Fv
— Elliptical "footprint" of contact surfaces
— Width of contact — b
— Rolling radius varies from Ru to Rl
— Therefore during rolling in some parts of the footprint surfaces are stationary relative to one another, sticking friction, in other parts they must slide relative to one another
— Need for lubrication and internal clearance in assembled bearing

Internal clearance:
— At maximum metal, clearances are minimum
— At minimum metal, clearances are maximum
— Ball and race tolerances as manufactured are too large
— Selective assembly is needed:
    manufactured balls are separated into different tolerance size ranges to obtain nearly equal ball diameters
    Inner and outer race diameters are measured
    Balls of correct diameter range to obtain desired clearance are fitted

Influences of clearance:
— Possible end movement of outer ring relative to inner ring
— Note effects of tolerances on possible end (axial) movement of one race relative to the other

$Ro + Ri - Db - Cl/2$

$Ro + Ri - Db$

Half of end movement (em/2):
$(Ro + Ri - Db).\sin(phi)$

Influences during running:
— Balls experience centrifugal force Fc
— Surface load on outer ring $Fs-o$ is different from surface load on inner ring $Fs-i$
— Axial force Fa and radial force Fr (assumed zero in diagram) cause internal radial force component Fv
— If rotational speed of bearing is large enough, centrifugal force could account for all of load capacity of outer ring, therefore each bearing has an absolute limit on rotational speed
— Lubricant must be moved by rolling balls, therefore energy loss, heating, speed limit for maximum temperature

$Fc = m.Db.w.w/2$

FIGURE 9.14 Continued.

For belt transmissions

Type and location of machine:
Power delivered by (electric motor,
    transmission, etc.):
Power (kW), rotational speed (hr/min):
Daily running duration, starts/hr:
Running roughness:
Diameter of small pulley,
    rotational speed:

Diameter of large pulley,
    rotational speed:
Pulley width:
Belt linear speed:
Centers distance:
Influence of chemicals, oil, humidity:
Running temperature:
Possible ordering quantity:

	Belt	σ_{Ult} N/mm^2	E N/mm^2	σ_{All} N/mm^2	v_{max} m/s	$b_{f\,max}$ 1/s
Textile	Cotton	300 — 700	350 — 700	3.9	40	30
	Rayon	420 — 620	3000 — 6000	4.5	40	30
	Polyamide	600 — 800	1000 — 2000	25	60	100
	Polyester	600 — 800	4000 — 8000	25	80	80
Composite	Polyamide tape	300 — 400	550 — 1000	25	80	100
	Polyamide cord	600 — 800	1500 — 2000	30	100	100
	Polyester cord	600 — 800	4000 — 8000	40	120	85

What you must know about belt materials: High specific load–carrying ability of belts made from synthetic fibers and plastics results in small cross–sections, higher permissible speeds, and higher bending vibration frequency

For chain transmissions

Power Input:

Machine, electric motor, internal
    combustion motor (number of
    cylinders), transmission, and so forth
Rotational speed:
Power:
Running roughness (smooth,
    cyclical, shock):
Number of starts per hour:

Chain drive:

Centers distance:
Tensioning provision (adjustable centers,
    tensioning wheel, tensioning slider, spring):
Desired or existing transmission ratio:
Shaft centres horizontal or vertical:
Protection against dust and dirt (shielded,
    encapsulated):
Lubrication (by hand, drop feed, oil bath,
    pressure lubrication):
External influences (heat, dust, humidity,
    temperature, fibers, etc.):

Power take-off:

Sort of driven machine:
Rotational speed:
Required power
    starting:
    normal operation:
    maximum:
Loading roughness (smooth,
    cyclical, shock):

Chain wheels:

Maximum permissible outside diameter:
Number of teeth (envisaged or executed):
Fixing to shaft (taper, taper sleeve, key,
    grub screw, etc.):

Chain:

Envisaged or executed chain:
Substitute for an existing transmission?
Maximum width of chain:

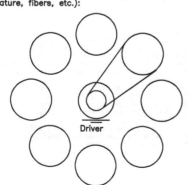

Driver

Position and direction
of rotation of the chain
transmission (enter onto
sketch)

FIGURE 9.15 Questionnaire for functionally determined properties of belt and chain transmissions.

Analysis of potential faults and failures, their effects and criticality

FMECA on System _____ FMECA on design layout/details _____ FMECA on manufacturing process *)

Assembly/Subassembly/Detail _____
Model type and year (name and number) _____
Planned release date _____

Responsible design or manufacturing section _____
Other impacted sections _____
Impacted supplier _____

Prepared by _____
Date (and revision) _____
Approved by _____

*) ... Delete items not referenced

System assembly process operation *) Name	System assembly process operation *) Function	Potential fault	Potential consequences of Fault	Potential causes of fault	Current state Control procedure	Occurrence A	Importance B	Detection C	Risk priority number (RPN) AxBxC	Recommended correction procedure	Responsibility (company or section)	Implemented measures (date of implementation)	Improved state Occurrence A	Importance B	Detection C	Risk priority number (RPN) AxBxC
Thread cutting	Cutting of screw threads in pre-drilled cover holes	Cut into water channel	Leakage of water and coolant; motor over-heating, valve damage	Wall thickness too small, core shifted in casting (wear on core box)	100 % inspection; water pressure test	3	10	5	150	Statistical process control	AuCast Ltd	Mean and range chart for dimn. with highest wear rates for cores (Feb 99)	3	10	5	50
Face milling	Finishing of sealing (gasket contact) surface	Sealing surface uneven	Leakage of combustion gases; rough uneven running	Pressure in holding jig not sufficient	5 % sample inspection of flatness in each batch	3	8	5	120	Clamping pressure monitoring	Technical Operations	Automatic force monitor, machine stops when pressure too low (Nov 98)	3	8	5	40
		Dirt and chips in jig			100% visual inspection of jig for cleanliness	2	8	5	80	Cleaning of jig before each item	Machine Operator	Automatic air pressure blast on jig (Dec 98)	2	8	5	40

Literature:
Vesely, W.E., Goldberg, F.F., Roberts, N.H. and Haasl D.F. (1981)
Fault Tree Handbook, NUREG-0492, Washington, DC: US Nuclear Regulation Commission

FIGURE 9.16 Failure mode, effects, and criticality analysis (FMECA).

Application	Influences	Available means
A1 — Case When should it be implemented? Special cases If something else happens	**A2 — Surroundings** Is it outdoors? Where can it be used? Is it hot or cold, dry or humid?	**A3 — Previous designs (layouts)** Can we use previous solutions? Are other solutions available? Coverage by patents?
A4 — Duration How long will it be used? Minimum usage period? Operational period?	**A5 — Safety/Security** What regulations and laws exist? Is it relatively dangerous/hazardous? Does it conserve energy? Are any occurrences known?	**A6 — Available plant** Which available plant can we use? Which available plant must we use?
A7 — Frequency How often? At what intervals?	**A8 — Supervision/control** Is it as good as before? Are good ideas available? Can something be adopted from other departments? Must basic properties be retained — noise, accuracy, etc.?	**A9 — Available assisting materials** Gases, electrical power, air, liquids, extracts, and so forth
A10 — Effects/Consequences After something else has happened? Before something else happens?	**A11 — Experiment and installation** Are tests needed? By whom? Where? How will it be installed? How will it be transported? Who will implement that?	**A12 — Experience** Who has done it so far? Who may possibly need to do it again? Who will do it this time?
A13 — Users Who will use it? What knowledge/skills do they have? Who is not allowed to use it? What training/education do they need?	**A14 — Time Plan** Absolute deadline for completion Is it routine? Does final completion depend on other facilities/people?	A15 —
A16 — Maintenance Should it be routinely performed? Should it work without maintenance for a period? Must maintenance demands be indicated automatically?	**A17 — Financing** What resources exist in-house? Who controls/audits them? How will they be controlled/audited?	A18 —
A19 — Acceptance by Personnel Will it be noisy? Will it vibrate? Will the operating personnel trust its safety, accuracy and reliability?	**A20 — Manufacture/production** Must it be brought to a particular place? What alterations are demanded for manufacture/production?	Professional title: Compiled by: Date of Issue:

Literature:
Latham, R.L. (1968) *A Guide to the Problem Analysis by Logical Approach System* (5 ed), UKAEA, AWRE Aldermaston
Terry, G.J. (1968) 'A Chart System to Help Designers' Chart. Mech. Eng. 15 No. 2, Feb., pp. 56—59

FIGURE 9.17 Problem analysis by logical approach (PABLA).

Gouvinhas [235] presents two pro forma for checking detail and assembly drawings with respect to machining and materials, and assemblability. It is apparent that the possibility of application of pro forma increases with the concreteness of the processed product "sort."

10 Technical Support for Design Engineering

Technical support (TS) includes all products that can assist an engineering designer's work, the operator "working means." These products assist in modeling TS(s) (see Figure 10.1), producing, storing, and retrieving drawings and other records, and providing support. Computers have become a major factor in design engineering. The information presented in this chapter is mainly situated in and around the "northeast" quadrant of Figures I.4, I.5, and 12.7.

10.1 COMPUTERS IN ENGINEERING DESIGN PROCESSES

Computers are currently a priority; finding a detailed explanation of the role of computers in design engineering is important and useful, to clearly show the relationships to the subject of this book. Engineering designers can design without computers. Even when they are designing with computers, designers must do some preparation work without computer assistance. Computers cannot design completely independently. To treat this problem as "automating design" is basically wrong. Some parts of designing may be automated, but generally computers are only tools to assist designing. If suitably used, computers are effective instruments that can help to solve many problems, contribute to improvements in TS(s), optimize quality, and improve and perfect the parameters of the design process. Computers have proved their worth in a wide range of the operations (see Figure 2.5): (1) For representation of the constructional structures and their constructional parts, using computer-aided design (CAD) (but see NOTE below). (2) For obtaining improved visualization, for example, by "virtual reality" 3-D imaging displayed with computers, by "rapid prototyping" in stereolithography using laser-cured plastics produced by computer-driven machines, or by "animating" images of a mechanism to display motion and allow checking for collisions and interferences. (3) For assessing the effects of dimensional and geometric manufacturing tolerances on assembly and motion. (4) For improving analysis and design calculations of constructional parts with respect to static phenomena, for example, stress analysis by finite elements, boundary elements, finite differences; optimization for single and multiple criteria; decision support. (5) For exploring dynamic phenomena, for example, kinematic and vibrational movement of mechanisms; computational fluid dynamics; chaos-theoretic instabilities. (6) For "what if" investigations about the consequences of proposed alterations to a system, for example, in its parameters, using DFMA and other programs. (7) For implementing changes and alterations in an existing system, and (more recently in some computer programs) propagating these changes to other constructional parts.

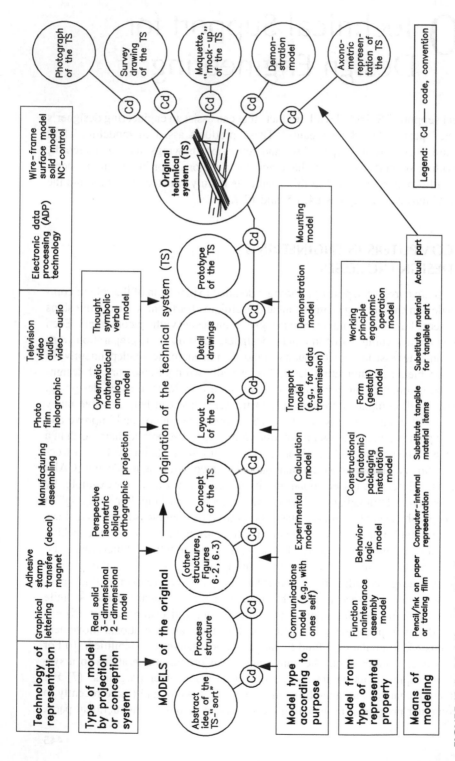

FIGURE 10.1   Modeling of technical systems.

(8) For text processing, word-processing. (9) For providing an information (data) base for engineering designers, setting up masters, catalogs, and pro forma. (10) For obtaining and verifying information, for example, via the Internet using browser programs. (11) For allowing a new arrangement of information, or directed search in existing organization schemes to obtain access to a particular element of information and knowledge from several locations, to cross-reference and call up by concepts, expected effects, functions, keywords, and so forth. (12) For transmitting data (e.g., drawings) using STEP and other protocols. (13) For attaching nongraphic properties to a computer-resident product model (capturing the "design intent"); and for other purposes.

NOTE: Acceptance by industry of early 2-D and 3-D CAD applications (due to their limitations) caused a drastic change in detail-design procedures. The earlier manual design method consisted of producing layout drawings, detailing, and then using an assembly drawing to check the details. CAD applications could not be used for layouts, most of them are still not suitable. Detail design of constructional parts tended to be allocated to different designers on their "own seat." Coordination among these specialists became difficult, and many errors resulted. The latest versions of some CAD applications are starting to allow "inheritance" of some properties from one constructional part to another, and automated check assembly (see Figure 10.2) [88,89].

The existing programs make it possible to perform more complex procedures, and process them faster than human capability allows, for example, for producing variants in the configuration and parametrization of parts in designing the constructional structure. Computers offer new possibilities for tasks that were too difficult without computers (see Figure 10.3), for example, strength calculations by finite element methods, or calculation of manufacturing costs during designing by means of the Herstell–Kosten–Berechnung (HKB) program [191,192,196,197] (compare Figure I.18). Integrating the information flow in organizations by computer integrated manufacturing (CIM) [539] shows possibilities.

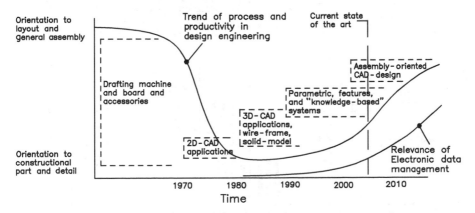

FIGURE 10.2 Progress of computer support of representation for design engineering. (Adapted from Burr, H., Vielhaber, M., Deubel, T., Weber, C., and Haasis, S., *J. Eng. Design*, Vol. 16, No. 4, 2005, pp. 385–398. With permission.)

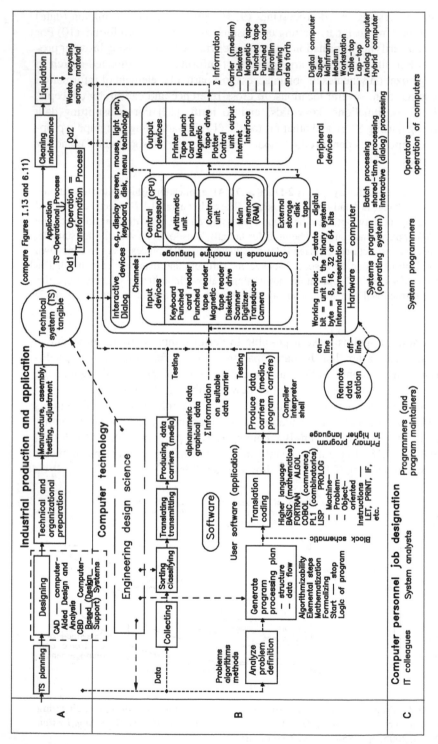

FIGURE 10.3   Engineering data processing/information technology—survey of CAD systems.

With introduction of computers, additional requirements arise, for example, engineering designers must at least be familiar with the rudiments of computer technology. They should be "computer literate"—be sufficiently familiar with operating systems and application programs to manage the local aspects of the computing facility, workstation or PC, including Internet connections. Design procedures and methods must be adapted to enable computer usage (see also Chapters 2 to 4). The organization must accept the presented form of computer data, this is currently the subject of intense activity in international standardizing. Coworkers must accept computers as useful devices, with rational and economic application.

Checking (verifying, auditing) of manufacturing information, drawings, and so forth, is a difficult task in the new computerized procedures. Procedures for checking for completeness and uniqueness of dimensioning, correctness to standards, and so forth, of manufacturing "drawings" that are only stored as computer data have not yet been adequately developed (see NOTE above).

Computers also present a range of dangers and risks. Among these is uncritical acceptance of computer data, and overvaluing of the results, because they lack transparency and can lead to wrong interpretations. Computers should not remain toys for a few enthusiasts, they must become proven and valued instruments for everyone.

At present, computers cannot be used in the whole design process as described by the procedural model (see Figures I.22 and 4.1). Their semantic "understanding" (capability of interpreting spoken or written words as instructions) is limited. There are currently no commercial computer programs that can help in the conceptualizing phases, stages, and steps of the procedural model, except the relatively trivial contribution from word processors and graphics programs as simple recording assistants (see case study 1.3). Computers cannot help in developing the structures of transformation processes, technological principles, technical processes or function structures, and the transition from one to another.

To a limited extent, design catalogs (see Chapter 9.3) have been transferred to computer-based operation, and developments in efficient usage have been reported [207].

Many organization catalogs of original equipment manufacturer/commercial off-the-shelf (OEM/COTS) products and related calculation methods have been implemented as computer tools, for example, rolling contact bearings and their calculations.

Developing and evaluating morphological matrices, recording and signaling incompatibilities between partial solutions, and helping to generate alternative concept proposals has been reported [77,578].

Analysis of proposed TS can be performed with computer applications, for example, [12]. In some of the design and analysis tasks at the constructional structure level, "critics" configured as resident expert systems can be implemented [139,180,532,533]. They monitor the progress of work and signal if they detect any potential difficulties—usually based on previously recognized limitations that have been directly programmed for recognition, or that are signaled and "interpreted" by computers from other indications (as "case-based reasoning"). They appear to be useful in the embodiment and detailing steps of the design process (constructional structure), from preliminary layout onward.

A preliminary layout is currently the most abstract point at which the graphic capabilities of computers can be effectively used. Some CAD 2-D and 3-D programs that are now available permit entry of a configuration (rough component structure), and will accept parametrization (entering and altering specific dimensions). They can correctly propagate data in all views, and among adjoining constructional parts. Features-based modeling permits entry of partial and incomplete constructional parts, for example, only active surfaces or manufacturing surfaces, and indicates the connections between them without specific dimensions. Extraction of features from solid model representations has been attempted, especially for manufacturing purposes.

Working from solid-model representations of concrete constructional structures, CAD programs allow generating tool paths for cutting tools, and extending these to full programs for running computer numerical control (CNC) machine tools to produce the constructional parts—computer-aided design/computer-aided manufacturing (CAD/CAM), for example [4]. Recent advances allow rapid prototyping, producing a solid constructional part in a plastic material by solid free-form manufacturing (stereolithography, fused deposition modeling, or similar proprietary methods), so that the represented forms can be handled by designers, and checked visually and by measurements.

Genetic programing introduces pseudo-random mutations into the search of a solution space to converge more rapidly to an optimum with less danger of "hanging up" in a suboptimal region of the search space [52,551,552].

To date, three computer applications have been developed using conceptual modeling as described in this book [275]. They have been under continual development since their first publication, and have found usage in various industrial organizations. Other programs are also available for these functions of engineering.

10.2 INFLUENCING COSTS—PROGRAM HKB

Influencing and controlling costs is currently one of the main tasks of organization management.

The task of reducing costs needs information, methods for effective procedures, and application of appropriate tools. Regarding considerations in *knowledge* and *methods*, the important items are (Figure 10.4, see Chapter 3): (1) knowledge about the product, especially the TS to be made; (2) knowledge about the manufacturing systems, including manufacturing technology and the available manufacturing equipment; (3) knowledge about the market regarding raw materials and semifinished goods; and (4) knowledge about cost calculation.

In general mechanical engineering, Figure 10.5 shows that a high proportion of product costs are *committed* by "design and development," with actual expenditure in the future, if the product is manufactured (see also Section I.12.1). The engineering designers establish all properties of technical system (see Figures I.8 and 6.8–6.10). Especially, they establish the sizes, shapes, and so forth, of all constructional parts, the internal properties of classes Pr10, Pr11, and Pr12 that are the causes for all the other properties, including their cost [304,314,315].

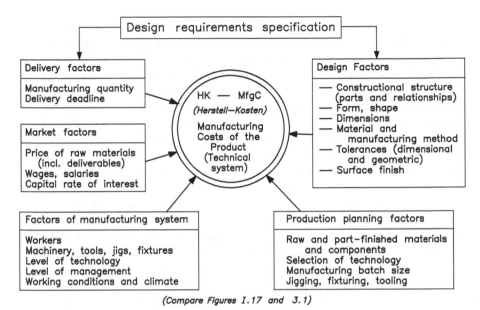

FIGURE 10.4 Range of influences on manufacturing costs. (Adapted by permission of Mirakon AG, Switzerland, http://www.mirakon.com.)

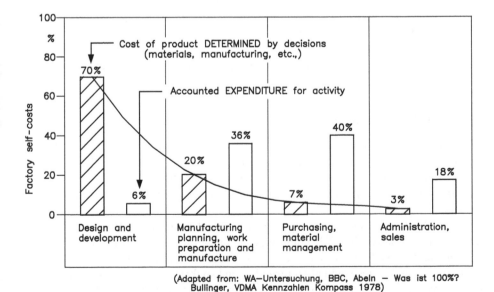

FIGURE 10.5 Influence of organization sections on manufacturing costs of product vs. accounted expenditures.

Costs can normally only be reliably determined (e.g., precalculated by a production planning department) when a *complete description* of the product is available [291], that is, all detail drawings (or their computer equivalents) are completed (see Figure I.18). Returning then to redesign the product is problematical because of the added costs and lost time. Precalculation of manufacturing costs as part of manufacturing process planning can at best deliver negative advice, for example, that a poor product concept (function and organ structure), a poor mode of construction, or poor form giving of subassemblies and constructional parts has been achieved, or that the cost targets have been exceeded and design work is wasted. The most sensible and understandable goal of recognizing costs *as early as possible* [179] is usually frustrated by a lack of suitable methods and data for the engineering designers.

Designers are traditionally oriented toward technical matters, and economic considerations are rejected because they are regarded as belonging to work preparation and manufacturing process planning. An experienced designer can usually make a qualitative judgment of costs (within limits, cheaper, or more expensive), but this is inadequate for early recognition of costs. Organizational methods cannot substitute for reliable cost data (e.g., see Chapter 8), total quality management (TQM) [10,108,125,549], quality function deployment (QFD) [75,111,151,259,269], Taguchi methods [391,468,496,530,531], and concurrent engineering.

In the near future the goal to be attained must be that products (TP and TS) are designed for their technical properties (e.g., strength, stiffness, durability), and for achieving a *cost goal*. Several potentially useful alternative solutions must be generated at each abstract level of designing, and these alternatives must be evaluated as objectively as possible with respect to properties, especially cost [304,314,450,451].

This section describes the software system HKB by Mirakon, Switzerland [191,192,196,197], which can present to engineering designers at their workplace the necessary cost data for a product in design layout, with proper regard to the available manufacturing facilities, production planning rules and cost structures for their particular organization.

10.2.1 THE HKB ARCHITECTURE

HKB (German "Herstell–Kosten–Berechnung"—manufacturing cost calculation, in English and German language) is a knowledge-based software system for calculating manufacturing costs of "components" (constructional parts) for "products" (TS) [14]. Its architecture is shown in Figure 10.6. HKB has been used in European industries since 1985.

Within HKB, a manufacturing system is modeled in a knowledge base. The organization data stored in HKB comprises the capabilities of the manufacturing machinery, tools, jigs, fixtures, cost rates, raw materials, semifinished parts, standard parts, cutting speeds, time calculation formulas, and so forth, and the manufacturing planning logic and rules. Much of this expert knowledge is supplied and must be maintained by the manufacturing planning personnel of the organization.

The input data for running the program is a description as represented in layout, detail, or assembly drawings of a tangible product—a constructional part, with geometry, features, and other elemental design properties; an assembly group;

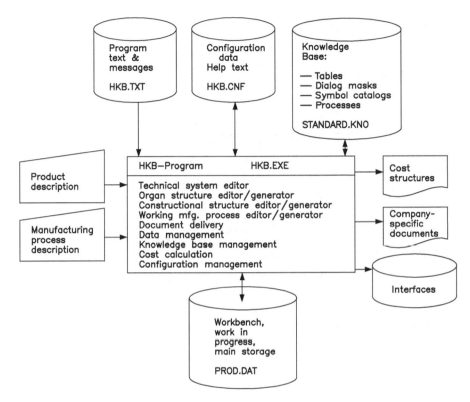

FIGURE 10.6   Architecture of HKB software system. (Adapted by permission of Mirakon AG, Switzerland, http://www.mirakon.com.)

a machine or plant. HKB can derive the manufacturing process, with times and costs for each manufacturing operation, and can structure the cost data in various ways. The cost-calculated items (products) are stored and managed in a database, which can also contain sample parts as basis for easy alterations and part families. The database can be searched to find appropriate parts by selecting suitable part properties as a filter.

10.2.2 HKB Methods

The software system HKB can be used in various departments of a typical manufacturing organization, especially design offices, production planning and work preparation, tendering, and purchasing. Each of the following methods and procedures for using HKB reaches higher into the processes shown in Figure 10.7.

The "Production Planner" Method: Production planners decide about the sequence of manufacturing operations needed to make a product. They know about machines, tools, jigs, and fixtures available to that organization. For the selected item to be calculated (product number, name, class, design data), HKB can derive the main processes and allow editing, calculate times and costs for each operation and material, and present this data in suitable form.

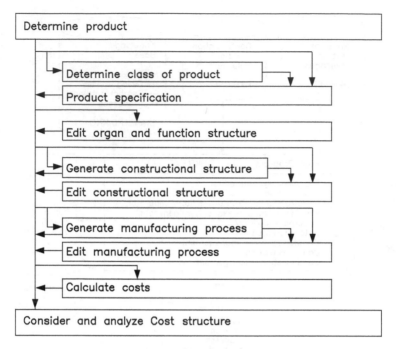

FIGURE 10.7 Process of using the HKB software system. (Adapted by permission of Mirakon AG, Switzerland, http://www.mirakon.com.)

The "Designer" Method: Engineering designers should be aware of the available manufacturing technologies, but they cannot know with what tools a product will be manufactured. They also do not know what manufacturing operations are needed and in what sequence they will be applied. Designers *do know fully* the constructional structure, they are the authors of that information. Designers can enter this constructional structure into HKB, by describing and editing all constructional parts, their features and their manufacturing properties, for example, the dimensions of features (surfaces) that need to be machined, the weight of a forging or casting, sizes of scantlings, and so forth. HKB can automatically generate a complete manufacturing process. This is a complex task for HKB—it must analyze the constructional structure, and build up a manufacturing process structure that is substantially different according to the logic contained in the data bank. This process will usually need further optimization or adaptation to available machinery—a task for the production planners (see the "production planner" method).

By entering several alternative constructional structures, in a "what if" study, designers can obtain reliable costs for the alternatives, to select the most suitable one. The designers, if they wish, can now edit the main process to make any needed alterations, for example, to investigate the effects of various variables such as batch size, to calculate the costs, and to present them in a suitable way.

The "Tendering" Method: This is the most demanding method for HKB cost calculation, because technical sales personnel cannot know the detailed constructional structure of a solution that has not yet been designed, and are therefore even further away from the operations that constitute the manufacturing process. They *do know* the technical functions that the customers demand of the product—the purposes and goals to be achieved by the product. These can be entered and edited as a function structure, and HKB asked to generate a constructional structure, as a parts list. This structure can be checked and edited manually, and the proposed manufacturing processes can be generated and costs calculated.

Comments: A product can be modeled by HKB in four structures (see also Figure 6.3): an organ structure; a constructical structure; a main process for manufacture; and a cost structure. All elements of these structures have mutual relationships with one another but usually not in one-to-one relationships (compare Figure 6.2). A constructional part "knows" from which functions and organs it was established and what manufacturing operations it has caused. A cost element "knows" where it belongs in the functional units, constructional parts, and manufacturing operations. The costs can be analyzed and categorized from three different viewpoints (see Figure 10.8). Each cost element in HKB depends on the function to be achieved, the feature of the constructional part and the method of manufacture of that feature.

HKB [191,192,196,197] presents a possibility of keeping a frequent score of cost targets during designing. Major cost centers and their reasons become transparent. Since engineering designers calculate costs, they improve their feel for costs, and learn to more effectively "design for cost." Education of young designers, production planners, and technical sales personnel can be more objective, efficient, comprehensive, and rapid. Know-how available within an organization can be structured

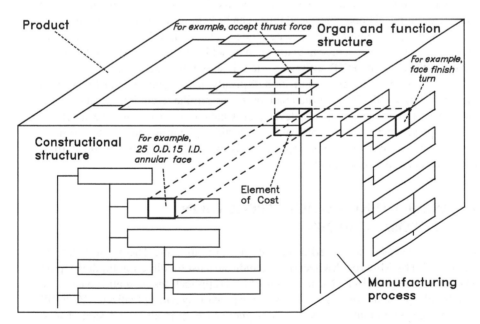

FIGURE 10.8 Principle for calculating costs. (Adapted by permission of Mirakon AG, Switzerland, http://www.mirakon.com.)

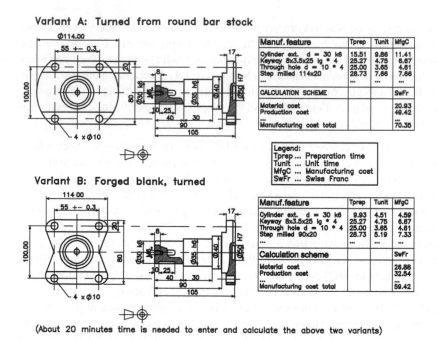

(About 20 minutes time is needed to enter and calculate the above two variants)

FIGURE 10.9   Cost calculation example: stub axle. (Adapted by permission of Mirakon AG, Switzerland, http://www.mirakon.com.)

within the production planning department, stored in a compact form, and used to influence design decisions. Short lead times can be kept for cost precalculation, generating a production plan and producing tenders, and design decisions. Objectivity and standardization in cost precalculation can be achieved. An example of the output to be expected from the HKB software system is shown in Figure 10.9.

Other programs for cost estimation typically use regression analyses from (commercially) available constructional parts, or similar techniques, and effectively work at the organ structure level (see Chapter 8). They assume a typical machine park, and do not enter the production planning phases to estimate the anticipated costs. One such application is available for free download from http://www.emachineshop.com [4].

10.3 ESTABLISHING LAYOUT—APPLICATION PROPOSAL CADOBS

This section describes an experimental attempt to realize the Theory of Technical Systems [304] by Mirakon, Switzerland, to indicate a useful direction for developing CAD procedures [193,194]. The main area of CAD applications currently concern the geometrical representation of TS that have been previously conceptualized. CADOBS is a software system that can assist the designer and help in all four design phases, task definition and clarification, conceptualizing, laying out, and detailing (compare Figure I.22).

The goal to support the designer in this typical task with respect to design time and quality, and to create a starting point for abstraction toward a general design system has been achieved and demonstrated by an example: the bearing arrangement of a shaft using rolling bearings.

An important condition for successful form giving is a division of a TS into smaller units, design zones and form-giving zones (see Section 6.4.4), which allow complex analyses and judgments from several points of view. Especially the complete working of the unit as part of the total system must be ensured. This requires an optimal selection of constructional parts and their action locations (organs) for all functions, including evoked functions—working with a complete function structure is essential. Elements of the organ structure must be available. All viewpoints that influence form giving must be considered, especially strength, stiffness, reliability, operability, transportability, manufacturability, and others (see Figures I.8, 6.8 to 6.10), without causing complicated interfaces with too many transferred parameters (see Chapter 3).

These demands can be fulfilled by means of a model that includes and describes an organism (see Figure 6.6), a *design group*. Such design groups will often also be sub-assemblies (manufacturing groups, assembly groups, and modules). Clear boundaries do not emerge with respect to the constructional parts, among other considerations.

The described approach is explained with an example. Consider a gear transmission as the whole of a TS. Its subdivision into *design groups* will yield: (1) Gear pairing or gearing system (gear ratio levels); an organism with the function "change rotational speed (number of revolutions per minute)." (2) Individual rotating shafts with complete bearing arrangements; an organism with the function "permit rotation of shafts" and "react gear-pair forces." (3) Housing: an organism with the function "provide connections and support." (4) Lubrication, cooling. The design process is demonstrated on the design group "rotating shaft."

10.3.1 DESIGN PROCESS ON THE COMPUTER

The design steps are more or less specific to each design group and are defined by the system administrator. The user can optionally use these steps or not, according to the personal working mode (see Section 8.1), and the procedural manner of the designer (Section 4.7).

Figure 10.10 shows a process for designing a shaft-bearing system. The order of working on the design steps is not prescribed; the designer can move iteratively and freely between the steps.

At some time during designing a shaft-bearing system, decisions must be made about the organism of the bearing arrangement (Figure 10.11). This decision causes the need to establish a crude constructional structure, and new functions such as: "axial fixing of the left bearing."

Shaft dimensioning is illustrated in Figure 10.12.

For each problem situation, the software offers masters. These can be parametrized and changed by the user, and inserted into existing design proposals. Figure 10.13 shows an example.

To properly support the pragmatic procedural steps, the computer-internal model of the object to be designed must be structured strictly according to the theory.

Design steps

1   Task definition
   1.1   Geometric arrangement of the shaft-bearing system
   1.2   Description of the environment
   1.3   Environment requirements
   1.4   Description of dynamics
      1.4.1   Normal state
      1.4.2   Critical state
   1.5   Functional requirements
   1.6   Ergonomic requirements for operation
   1.7   Production requirements
   1.8   Manufacturing cost targets
2   Shaft dimensioning
   2.1   Material choice
   2.2   Bearing diameters assumption
   2.3   Strength calculations
3   Conceptualization of the bearing system
   3.1   Organ structure
   3.2   Total arrangement
   3.3   Choice of kind of bearing
   3.4   Lubrication system
   3.5   Sealing system
   3.6   Protection system
4   Bearing calculation and choice of bearing types
5   Form-giving of the constructional structure
   5.1   Form (constructional) structure for bearing system—layout
   5.2   Form (constructional) structure of constructional part

FIGURE 10.10   Procedure in computer-aided design of a shaft-bearing system with rolling bearings. (Adapted by permission of Mirakon AG, Switzerland, http://www.mirakon.com.)

Only then can the program support the designer consistently. Concretely this means: (1) Clear separation between requirements, functions, organs, constructional parts (action locations), constructional relationships, and so forth, independent of when and how established. (2) Maintaining causal relationship among functions, organs, and construction parts in mappings that are usually not one-to-one. The program should not allow the designer to delete a constructional part or an action surface without consciously first deleting or changing the functional cause for this constructional part or feature. (3) Hierarchical structuring of the function, organ, and constructional structures. (4) Assignment of each element (function, organ, constructional part) to an abstract class where all properties are stored as knowledge elements. Figure 10.14 shows a computer-internal model of the structure of a design group.

10.3.2   CONFIGURATION OF CADOBS

Figure 10.15 shows the configuration and the data flow of CADOBS.

Two activities and functions in an organization are separated: (1) With respect to information, the system administrator (a design expert) prepares, changes, and completes the information database using personal knowledge, other experience and knowledge captured for designers (see Section 10.4), published literature or experiences that are supplied by the user through the experience file. This activity is supported by an information management program, as an integral part of CADOBS.

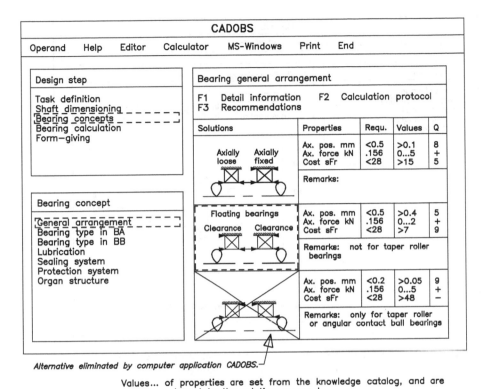

Values... of properties are set from the knowledge catalog, and are bound to the solution proposal
Requ. ... requirements are calculated by the application program
Q ... quality values are calculated from requirements and values
► ... direction of axial force capability

FIGURE 10.11 Computer screen snapshot—selection of bearing arrangement. (Adapted by permission of Mirakon AG, Switzerland, http://www.mirakon.com.)

(2) With respect to application, a designer designs a TS with the help of the program, and uses the knowledge stored in the program files by the system administrator.

Four groups of files have been established in CADOBS. (1) *Design data files* store the operand of the design process, the TS(s), in its current state—a bearing system, complete or incomplete. This file must allow data sets of variable length. (2) *Information files* store the reference tables that are generated and updated periodically by the system administrator. They are loaded from the central storage and interpreted for use by the program in the computer main memory. (3) *Experience files* collect information (experiences) originating from engineering practice and from applying the program. Each user can contribute to and profit from these files. The administrator systematizes and filters these experiences periodically and integrates them into the knowledge structure. (4) *Interface files* provide references for available CAD drawing systems, and for other programs such as cost calculation (HKB, see Section 10.2), FEM, DFMA, and so forth.

The design steps in the main menu of CADOBS are not firm elements of the program. They are determined by the administrator and entered in the program

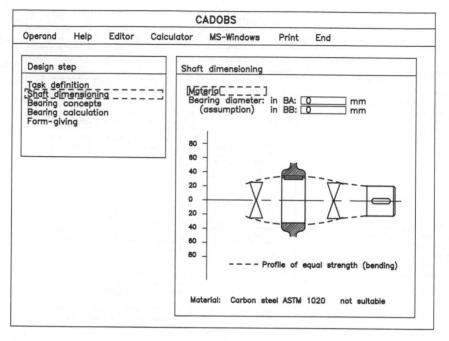

FIGURE 10.12   Computer screen snapshot—establishing shaft dimensions. (Adapted by permission of Mirakon AG, Switzerland, http://www.mirakon.com.)

usage table, by means of a meta-program language devised for CADOBS. With this language, other applications of the system should be realizable in different areas of engineering. The design activities are accessible at any time from function keys.

Several design variants should and can be pursued simultaneously. No sequence of design phases is enforced. The user can always move backwards and forwards in the main menu, and thus iteratively revise earlier decisions and requirements. Graphical freedom of form giving is maintained. In addition to the masters offered by the system, the user can call a simple drawing system as subroutine, and can thus at any time complete or change the drawing.

The structure of a digital model of a designed system (as operand of CADOBS) is dynamic, that is, the structure expands and shrinks with the insertion and deletion of data elements. It is hierarchical, without limit to the number of levels. An element can have several relationships at different hierarchy levels, and a relationship can influence several elements.

The information used in designing is stored in the form of tables, which contain data, and logical connections and programming instructions. They are arranged into four classes, general technical information (raw materials, form elements, influences, and so forth), general design information (functions, organs, constructional parts, connections, and so forth), information specific for a family of TS (requirements, organs, properties, and so forth, for example, of a rolling bearing system), and auxiliary tables (formulae, texts, figure codes, dialogue masks, organization data, and so forth).

Problem statement: Bearing position cover and seal, support bearing axially

Offered master solutions:

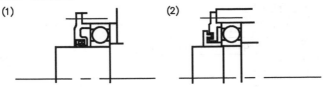

(1)   (2)

Parameters:
    For solution (1):   shaft diameter, bearing diameter, type of seal
    For solution (2):   shaft diameter, bearing diameter, labyrinth dimensions

Properties:
    For solution (1):   max. running speed = ...;   min. running accuracy = ...
    For solution (2):   max. running speed = ...;   min. running accuracy = ...

Concrete solution after selection of master, parametrization, alterations of
    individual constructional elements, and incorporation into the existing solution
    proposal

Solution proposal
following Master 1

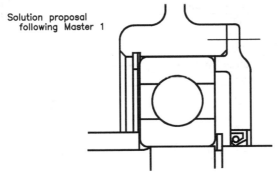

FIGURE 10.13   Example of a master. (Adapted by permission of Mirakon AG, Switzerland,
http://www.mirakon.com.)

10.3.3 RANGE OF PROBLEMS OF FORM GIVING

Form giving is the process of establishing the elemental design properties
(see Chapters 2 and 4). For constructional parts, form giving involves establishing the geometric form, dimensions, materials, tolerances, surface condition, kind
of production process, and so forth. For more complex TS, the arrangement of their
constructional parts must be added.

Predominantly this concerns the form of TS of low complexity, that is, constructional parts and assembly groups (subassemblies). The form of a "higher" TS is
synthesized from the forms of "lower" TS. Based on knowledge of form of the TS,
this view permits fulfilling the requirements for strength, costs, or other properties.

The task is to transform a TS-model from a sketch of principles (an organ
structure) into a TS-model with definitive form and other properties. The transition
from the organ structure to the constructional structure causes many considerations
(i.e., requirements and TS-functions), which should adequately secure the external
properties of the TS.

When laying out the software CADOBS, several additional requirements
were considered, for example, avoidance of generating impossible, unreal forms,

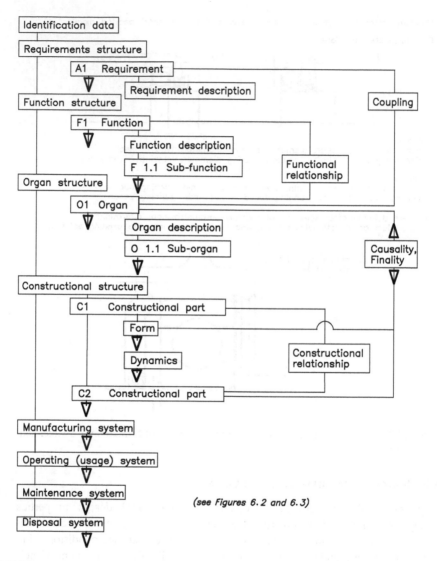

FIGURE 10.14 Computer-internal structure of a design group. (Adapted by permission of Mirakon AG, Switzerland, http://www.mirakon.com.)

structures, dimensions, and so forth. All necessary information and data for derivation of the manufacturing technology, strength calculations, costs, service life, and so forth, should be available, that is, an interface with HKB (Section 10.2).

Form is composed of individual form elements, whose structure is determined by a "skeleton"—an organ structure. A *skeleton*, derived from an organ structure, is a carrier of form elements, and is subdivided into segments and sectors. The sectors coincide with the boundaries of individual form elements (Figure 10.16). *Form elements* are elemental material basic bodies (especially concerning their surfaces, features), with a simple geometric form, differentiated into action and assisting

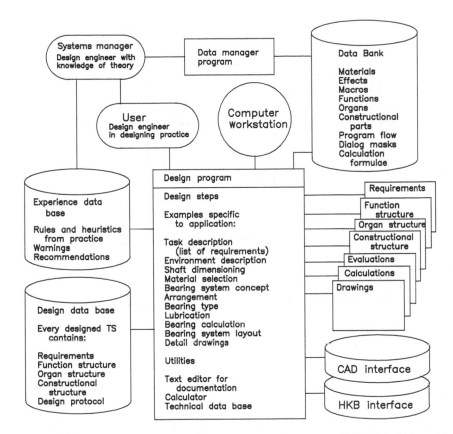

FIGURE 10.15 Architecture of the computer-aided design system CADOBS. (Adapted by permission of Mirakon AG, Switzerland, http://www.mirakon.com.)

elements. Action elements carry the action locations (partial organs); assisting elements connect the action elements into a material unit of the constructional part. *Action locations* mediate directly an input and output in the action pair (organ), and can be classified geometrically as action spaces (volumes), action surfaces, action lines, or action points. Action locations are realized by suitable form elements.

The form of a constructional part usually conforms to a form giving or a mode of construction principle, for example, aerodynamic form, constant thickness, filament wound lay-up molding. *Form giving* or *construction rules* deliver the knowledge for and about form-giving or modes of construction of a specific class of constructional parts or assembly groups, for example, regarding raw material, form, dimensions, production processes, and so forth. They thus fulfill a purpose, such as economy, possibility of production, capability for assembly, transportability, or appearance.

10.4 CAPTURING EXPERIENCE INFORMATION

This successful application program to capture information has been developed by Mirakon, Switzerland [195] within the KOMPASS Project, a collaboration between

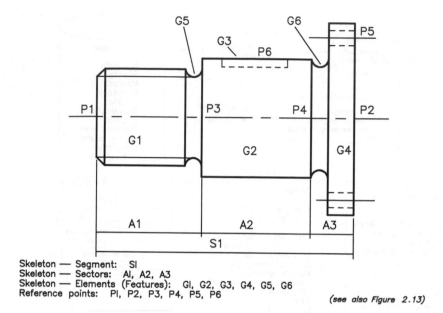

Skeleton — Segment: Sl
Skeleton — Sectors: Al, A2, A3
Skeleton — Elements (Features):   Gl, G2, G3, G4, G5, G6
Reference points: Pl, P2, P3, P4, P5, P6

(see also Figure 2.13)

FIGURE 10.16   Example of a form structure for a constructional part. (Adapted by permission of Mirakon AG, Switzerland, http://www.mirakon.com.)

ABB-Turbo-Systems, Eidgenössische Technische Hochschule (ETH) Zürich, and Mirakon.

Much of the engineering design information used by experienced designers in industry is held as tacit knowing, which can be lost to the organization when a designer retires or resigns, unless it can be captured. Several computer-based expert systems can capture information, mainly for diagnostics. The program application described here [195] is intended to capture design knowledge by relating a design situation (see Chapter 3) and its structure to the elements of the situation and appropriate masters (see Section 9.2).

The situation elements and masters formulate the collected information as a view of the actual range of situations. A *knowledge structure* can be defined for the inform-ation to be captured, preferably using TS-function structure, organ structure, and design properties as a basic pattern, for example, similar to Figure 10.14 (see also Figure 6.2).

The information must then be elicited and formatted, generally as lessons to be learned from particular situations that happened in previous design projects in which the experienced designers were involved; see the right-hand side of Figure 10.17. The information should be brought into a *system of statements* that is free of redundancy. Each item of information should be unique, and appear only in one place in the scheme. Limiting values for variables should appear as explicit statements. Each item of information must also be accompanied by a source statement, showing when and where the statement originated.

Each elicited item of information should be *categorized* by a thematic hier-archical ordering scheme based on practical and pragmatic considerations related to

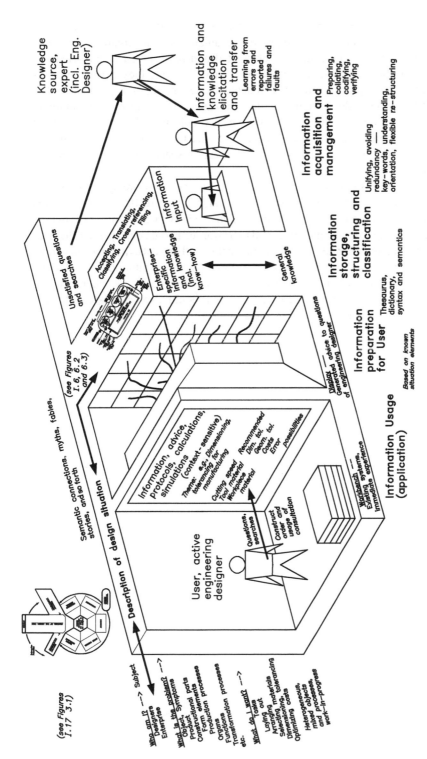

FIGURE 10.17 Capturing and arranging information in an organization. (Adapted by permission of Mirakon AG, Switzerland, http://www.mirakon.com.)

the knowledge structure; see the middle portions of Figure 10.17. The "organization" TrfS (see Figure I.7) can act as a "map" for organizing this information. The masters for each situation should also be abstracted and represented in the database. Some mutual influences of items of information will occur; they act as networking connections among the hierarchical statements, depending on the situation. The resulting database provides a view into the real situations that have been captured, and that represent the accumulated knowledge. It should include the machine park available to the organization, customer needs (e.g., see QFD, Chapter 8), competitor products (e.g., benchmarking, Chapter 8), organization-specific design conventions, and so forth.

The actual design situations (for a particular TS) as they arise are opportunities for using the captured information; see the left-hand side of Figure 10.17. The elements in each situation that are recognized by the designers allow them to inquire about the relevant information from the database, and to be warned about consequent influences of that information on other items. Designers can thus call on past experience in known situations, and can view the relevant masters to model their own situation. By adapting the situation elements and masters, they can also accommodate new situations. The added knowledge gained from these new situations should also be transferred to the database, either by the designers, or by the system administrator.

11 Human Resources, Management and Environmental Support for Design Engineering

The "Human System" as operator of the engineering design process can be described by a complex of interacting subjects, for example, human psychology, team work, organization management, and education for design engineering. The information presented is mainly situated in and around the "northeast" quadrant of Figures I.4, I.5, and 12.7.

11.1 PSYCHOLOGY OF THINKING

In designing, the engineering principle [425] states that: Engineering designers should produce their proposals only as accurately and completely as necessary, but also as coarsely, crudely and applicably as possible to achieve the necessary accuracy and completeness. This is the normal working mode of engineering designers, who mostly work on a project close to the deadline. This leaves little time to complete the project, search for alternatives, optimize, or reflect. A first idea is carried through until an acceptable solution is found, or the project is terminated—"satisficing" [506,507]. If a project is started when it is received, the subconscious mind can work on the problems, using incubation [563] (see Chapters 2 and 8). Systematic working demands early starting, and consistent steady working.

Engineering designers must take responsibility for proposals, but should not perform work beyond the state of confidence. Designing must be ended when the proposals can be accepted in the situation, optimal in principles, layout, embodiment and detail. The risk in this procedure must be accepted by the designers, with a realistic view of their capabilities.

Design engineering [425], consists of *anticipating* a possible change based on a future implementation of a TP(s)/TS(s). Designing depends on available information and theory about systems (see Chapters 5 to 7) and about designing (see Chapters 2 to 4). The products of design engineering are proposals. These cannot be *evaluated* as "true" or "false," or "probable" or "improbable"; they can only be evaluated and simulated as realizable or not, and valued better or worse than competing proposals.

Design engineering can only result in sufficiently complete and reliable information about the anticipated TP(s)/TS(s) if the designers can be sure to have considered all factors. Then, an anticipating proposal (and its documentation) for a designed TP(s)/TS(s) can be *evaluated as technically accomplishable*, if it can be confirmed with sufficient credibility and confidence that: (1) the TS(s) will fulfill the requirements under the circumstances of operation with sufficient reliability; (2) it is implementable or manufacturable under the given circumstances; (3) it complies sufficiently with the requirements of the manufacturing processes; and (4) all other requirements are fulfilled in acceptable ways to the user, customer, organization, legal and political authorities, the economy, culture, environment, and so forth.

Then also, an anticipating proposal (and its documentation) for implementing a TP(s) and manufacturing a TS(s) can be *evaluated as technically realizable*, if it can be confirmed with sufficient credibility and confidence: (1) that it can be implemented and manufactured under the given circumstances; (2) that the proposed sequence of implementing and manufacturing operations as specified will fulfill the required purposes of the TP(s)/TS(s) with sufficient reliability; and (3) that the requirements of the field are acceptably fulfilled.

To verify the accomplishability and realizability, the proposals must be tested in a design audit, by experiments, simulations, models, samples, and prototypes of the complete system and of suitable parts. Proposals should be confirmed before release for manufacture or implementation.

The engineering principle [425] must be tempered by a human trend to over-estimate one's own capabilities and knowledge—over-confidence (compare Section 11.1.8). Over-confidence seems to be prevalent when defining design tasks—designers (even though their preparation is not adequate) frequently think they understand the problem.

Three kinds of action modes exist [427]: (1) *Normal operation* (intuitive, second nature procedure) runs activities from the subconscious in a learned and experienced way, at low mental energy, giving an impression of competence [161,456,458] (see Section I.1). If difficulties arise, the action departs from the normal, and higher energy is needed. (2) *Risk operation* uses the available experiences (and methods) together with partially conscious rational and more formalized methods, in an unplanned trial and error behavior, which can occasionally be very effective. (3) *Safety* or *rational operation* needs conscious planning for systematic and methodical work, with conscious processing of a plan, because competence is in question, but this mode must be learned before attempting to use it.

The proportion of systematic, methodical work should be increased, especially for team consultations. This methods-conscious mode of working, and appropriately documented results, should be demanded by higher management.

Normal, routine, operation is mainly preferred and carried out by an individual. Risk operation tends to demand team activity, the task becomes nonroutine, consultations can and should take place — "bouncing ideas off one another," obtaining information and advice from experts, reaching a consensus on possibilities and preferred actions, and so forth. Consultations are best if the participants are of approximately equal experience or status, or if there is a large gap in experience from questioner to consultant. Personal contact tends to be quicker at lower mental energy than obtaining information from (written) records [32,425].

Nonroutine situations often produce critical situations in a design process [208,209,211,212], for example, during: (1) defining the task, analysis and decisions about goals; (2) searching for and collecting information; (3) searching for solutions; (4) analyzing proposed solutions; (5) deciding about solutions; (6) managing disturbances and conflicts, individual or team.

Designing in engineering has technical, economic, human, sociological, and psychological dimensions (see Figure I.15), which needs aspects, context, and consequences of psychology.

11.1.1 MEMORY AND THINKING OPERATIONS

If the aim is to alter the "results" of "behaviors" of human beings, and their *mind sets* or *mental maps*, the most effective way is to change beliefs and truths, attitudes and opinions. Trying to alter behaviors seems to produce only temporary changes.

In a simplified view we can distinguish between working (short-term) and long-term memory. Working memory is restricted to 7 ± 2 "thought chunks" or less for intellectual processing [101,416–420]. Each thought chunk (and its information content) can be simple or complicated, and details or extended connections need further thought chunks. Three items, and three relationships among them constitute six chunks. If mental capacity is exceeded, something is lost and the outcome may be failure [438]. Externalizing thoughts in sketches, and mentally interacting with them, is thus important (see Sections I.7.1, 6.10, and 12.1.8). A change in levels of abstraction or detail is relatively easy, but transferring chunks from one level to another is difficult—overview of a broader situation can hardly be maintained while considering detail.

Working memory can hold a seven-digit number in mind long enough to recite three or four digits backwards, or hold a ten-digit number in mind [117, p. 200]. Working memory is retained for about a maximum of 25 s, parts of the contents are continually refreshed or deleted, and parts can be transferred into long-term memory. Transfer into long-term memory needs "rehearsal," by reciting, reformulating, repeating in working memory, repeated reacting, and reviewing a conversation during pauses. Any organized procedure, whether physical or mental, systematic, methodical, stereotype, or guided by prejudices, is internalized into the subconscious by learning, repetition, and practice [153]. Predominantly those parts of the procedure that are useful for the given (practiced) task are absorbed. The transferred contents are normally incorporated and structured or restructured in a person's idiosyncratic way, into learned and experienced "tacit" knowing. "Knowing" (a process) can only be performed by the mind.

After internalizing, a person does not need the formal instructions, can forget them, and even forget that the instructions (and methods) are being used. The activity will progress "naturally," intuitively, at low mental energy. Mental structures of different people are probably similar. Recall from long-term memory presents some difficulty.

NOTE: Damasio [117, p. 227] states: "... memories are not stored in facsimile fashion and must undergo a complex process of reconstruction during retrieval, ... events may not be fully reconstructed, may be reconstructed in ways that differ from the original, or may never again see the light of consciousness."

Working memory, and transfer between working and long-term memory, can be aided by external representations, for example, sketching, note taking [41]. Necessary or useful modes of operation in thinking and acting include (see Sections I.11.1, I.12.6, 2.2.2.1, and 2.4):

1. An *iterative* mode—a task is repeated (systematically, intuitively, or mixed), each time with better understanding and knowledge about the circumstances and proposed solutions, and thus a preferred solution is approached.
2. A *recursive* (decomposing) mode—a task is decomposed into smaller parts, each part task treated by itself (but at least under partial consideration of other parts), and the resulting partial solutions are combined.
3. An *interactive* mode—one or several thought chunks are captured and considered in sketches or other notes, for example, on a computer screen— the interplay between working memory and the activity help to expand the thoughts, adding completeness and precision.
4. A *searching and selecting* (problem solving) mode—initially several solution principles are proposed and processed to a certain maturity, and only then a selection is made.
5. An *abstracting and concretizing* mode—although the goal is a concrete TP(s) and TS(s), occasional work of abstracting and on different levels of abstraction can help.
6. A *sequential* mode—a (partial) problem is treated one step at a time, a reductionist way.
7. A *simultaneous*, concurrent, parallel mode (usually only possible if performed in a team) — several (partial) problems or steps are treated at the same time, holistically.

These modes of operation can and should be utilized in continuous interplay, adapted to the problem. Neither the path of the solution process, nor the solution preferred by a certain examiner can be predicted, both can be guided by consciously applying suitable theories and methods (see Figure 12.10). Higher management may demand recorded evidence about the product against possible law suits for product liability, after the pattern of a systematic and methodical procedure. Any results obtained in an "intuitive" way (in normal operation) must then be brought into the method retrospectively, which can also serve as control and audit.

11.1.2 EMOTIONS, HUMAN INTELLECTUAL DEVELOPMENT

Emotions [116,233] produce a natural quick reaction that is acquired through genetic transfer and lifelong experience—nature and nurture. Learning takes place by repeated programming of the brain cells and their synapses [117,175,224,232]. Conscious actions (rationality) and motivation are much slower—reaction times of 0.1 s are common for trained activities, and are longer when deliberate thinking and deciding is needed. About 90% of normal reactions are emotional and 10% under conscious control.

11.1.3 INTELLIGENCE AND PERCEPTION

Intelligence may be defined as the ability to manipulate information to generate meaning and knowledge [117, p. 198]. Guilford [242,243] discovered that the mental abilities can be presented on three axes, each with some subgroups (see Figure 11.1).

Cognitive thinking uses "intellectual operations" on suitable "intellectual content" (information), to obtain "products of thinking." An ability is composed of an element of each of the three axes, 120 combinations. Each person can master only a limited fraction of these abilities. Operations of the design process need only some of these intellectual abilities, for example, in the basic operations (see Chapters 2 and 4), "perception and memory" relate to clarifying the task, Op-H3.1; "convergent

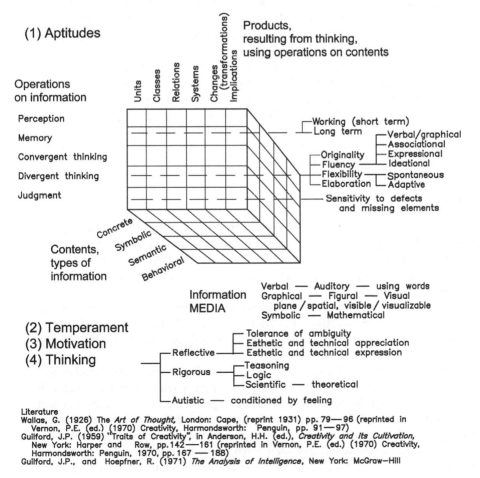

Literature
Wallas, G. (1926) The *Art of Thought*, London: Cape, (reprint 1931) pp. 79—96 (reprinted in Vernon, P.E. (ed.) (1970) Creativity, Harmondsworth: Penguin, pp. 91—97)
Guilford, J.P. (1959) "Traits of Creativity", in Anderson, H.H. (ed.), *Creativity and its Cultivation*, New York: Harper and Row, pp.142—161 (reprinted in Vernon, P.E. (ed.) (1970) Creativity, Harmondsworth: Penguin, 1970, pp. 167 — 188)
Guilford, J.P., and Hoepfner, R. (1971) *The Analysis of Intelligence*, New York: McGraw-Hill

FIGURE 11.1   Guilford's model of the structure of human intellect.

production" occurs in evaluating, Op-H3.3; "divergent production" is searching for solutions, Op-H3.2; and "judgment" is involved in evaluating and deciding, Op-H3.3. The most important ability is divergent production (see also [563])—creative thinking, at an appropriate level of abstraction, and constrained by the design situation as perceived. This ability is subdivided into four subgroups as shown in Figure 11.1. Guilford also proposed elements of the ways of thinking: temperament, coordinated with psychological type (see Section 11.1.5); motivation; and thinking in its manifestations.

Perception is a part of interrogating the world around us by using our sensory capabilities and deals with the observable reality. Mental models are structural analogies of the world, usually incomplete and selective, which explain only the necessary parts, are often incorrect, and are adjusted and improved by recursion and iteration. They are not in 1:1 correspondence with the reality.

Apperception [502] receives nonperceivable kinds of information. Examples of apperceptive information are: the possible, infinity, force, stress, mass, speed, friction, magnetic flux, symmetry, and so forth, which cannot be observed directly, can only be inferred using learned abstractions, and are then conceptually added to the mental representations. A similar process occurs in *situatedness* [96,227] (see Section 12.3.1).

Nevala [438] suggests that human thinking proceeds in four stages, starting from a self-consistent apperceptive mental representation. Thinking then passes through an automatic reconstructing process, a decoding of inconsistencies of the apperception. Reflection, exploration, and divergent thinking involves coping with these inconsistencies. This leads to constructing a consistent and integrated mental representation, synthesizing and convergent thinking, by integrating the content elements and the relevant organizational information.

11.1.4 MOTIVATION

Some theories describing *personal* motivation seem to fit together [165] (see Figure 11.2). In a hierarchy of needs [402,403], depending on the current situation, a person can exist at one hierarchical plane. As danger threatens, persons drop at least a plane. The uppermost plane of self-actualization/self-fulfillment can be expected only for short periods.

Some offerings work as incentives (motivators) if they promise an improvement over the current conditions [264–266]. Offer of an available asset (hygiene elements) does not motivate; withdrawing the asset acts as a disturbance, depressing the Maslow motivation.

Porter and Lawler [471] postulated a circuit of "effort"—which can stimulate further efforts, and is a condition for preserving a position in a Maslow plane. A break in the circuit can result in dropping to a lower Maslow plane. Parallel to (and independent of) Maslow, Vickers [555] proposed and investigated five levels of communication, which agree fairly well with the existence levels in the Maslow hierarchy. Motivation, and trust, can quickly be reduced or suppressed; it needs only a threat of redundancy (loss of job). Construction and advancement of motivation and trust can only proceed slowly.

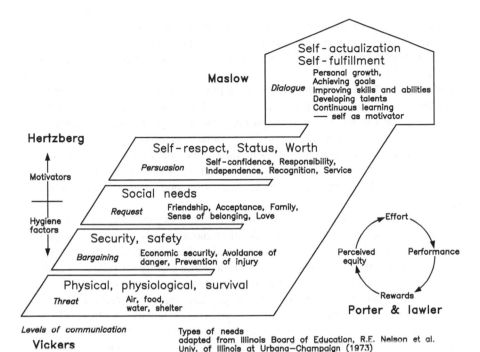

Types of needs
adapted from Illinois Board of Education, R.F. Nelson et al.
Univ. of Illinois at Urbana—Champaign (1973)

This diagram indicates five levels of human needs identified by Maslow, together with objects and values that characterize these need levels.

Related to the needs are five typical levels of communication identified by Vickers.

Hertzberg found that if a person is operating at a particular level of needs (e.g., Social Needs)
— Any offered factors below that level act as hygiene factors — abscence of the factor leads to regression to a lower level of need, presence of the factor does not provide additional motivation;
— Any offered factors at or above that level act as motivators.

Porter and Lawler identified a cycle of effects that act in relation to human motivation
— Normally, effort leads to performance, performance leads to rewards, received rewards lead to a perception of equity, and on the basis of perceived equity further effort results
— If this cycle is broken at any point, the other effects tend to be diminished.

Literature
Maslow, A.H. (1943) "A Theory of Human Motivation", Psychol. Rev. 50, p. 370—396
Maslow, A.H. (1954) *Motivation and Personality*, New York: Harper & Row
Hertzberg, F. (1966) *Work and the Nature of Man*, Harcourt Brace & World
Hertzberg, F. (1968) "One more time: how do you motivate employees?", Harvard Business Review 46 1968, pp. 53—62
Hertzberg, F., Mausner, B., and Snyderman, B. (1959) *The Motivation to Work*, New York: Wiley
Porter, L.W., and Lawler, E.E. (1974) *Behavior in Organizatons*, New York: McGraw-Hill
Vickers, Sir G. (1967) *Towards a Sociology of Management*, London: Chapman & Hall

FIGURE 11.2 Human needs.

11.1.5 PSYCHOLOGICAL TYPES

The Greek physician and philosopher Galen (about 130 to 200) divided persons into four types: choleric/hot-tempered, optimistic, phlegmatic/calm and melancholic/mournful. Immanuel Kant (1724–1804) declared choleric and melancholic as high in emotionality, optimistic and phlegmatic as low in emotionality, choleric and optimistic as changeable (extrovert), and phlegmatic and melancholic as unchangeable (introvert). Wilhelm Wundt (1832–1920) [586] united these attempts (see Figure 11.3A).

From his own observations in psychological practice (without controlled experiments) Carl Gustav Jung [338] proposed a categorization as a hypothesis: (1) the *nature* of a person can be: (a) rational/judging, showing preference for planning and organizing, having things settled; or (b) irrational/perceiving, preferring flexibility and spontaneity, keeping options open; and (2) the *attitude* within each nature can be: (a) extrovert (E) implying extrinsic motivation, and focus on people and things; or (b) introvert (I) implying intrinsic motivation, and focus on thoughts and concepts.

Attitude is subdivided by applying four functions: (1) thinking (T), objectively analyzing causes and consequences, deciding based mainly on logic—yields a differentiation from ET to IT—allocated to rational/judging natures; (2) feeling (F), subjectively evaluating, being people-centered, deciding mainly on values—yields a differentiation from EF to IF—allocated to rational/judging natures; (3) sensing (S), working bottom–up from specific to general, driven by facts and data, oriented to details and to the here-and-now—yields a differentiation from ES to IS—allocated to irrational/perceiving natures; and (4) intuiting (N), working top–down from general to specific, driven by concepts and meanings, oriented to theory and speculation, and to the future—yields a differentiation from EN to IN—allocated to irrational/perceiving natures.

NOTE: "N" is used for intuiting because "I" is already used for introversion.

The combinations of attitudes and functions have influences on communication (see Figure I.5), with oneself and with other persons, and consequences for personal and interpersonal outlooks, attitudes, encounters and reactions to the communication, but not for the quality (character, content, goodness) of the communication itself.

The Myers-Briggs Types Indicator (MBTI), modified from Jung's scheme, is investigated by questionnaires. It interprets four "dimensions" and 16 combinations of types (see Figure 11.3B). A small tendency in one direction is interpreted as the resulting designation and is assumed to be unchanging (invariant) during a person's life.

The Berkeley personality profile declares five styles [254]: expressive style, interpersonal action style, working style, emotional style, and intellectual style. A questionnaire with 35 questions (7 for each style) was developed, and includes instructions for calculating the scale values and entering them on charts, giving comprehensive explanations about the results.

The Personal Empowerment through Type (PET)-diagram [106,353] (see Figure 11.3C), is an accurate version of Jung's theory. Eight groups are shown, with continuous scales between E and I on each of four "dimensions" (eight manifestations) of the functions. A numerical measure of the manifestations is obtained from a questionnaire, with scores out of 30. The highest score shows the main function and the lowest the subordinate function. These functions can be influenced by

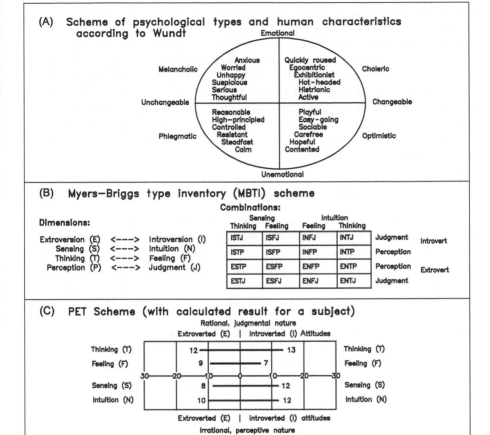

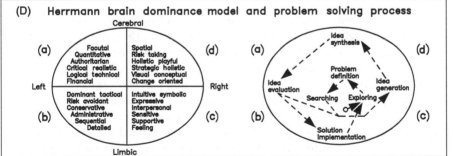

Literature
Wundt, W. (1903) *Grundzüge der physiologischen Psychologie* (Basic Concepts of Physiological Psychology), Bd. 3, Leipzig
Jung, C.G. (1923) *Psychological Types*, London: Routledge; and Princeton, NJ: Princeton Univ. Press, 1971
Myers, I.B., & McCaulley, M.H. (1985) *Manual: A guide to the development and use of the Myers-Briggs Type Indicator* (2nd ed.), Palo Alto, CA: Consulting Psychologist Press
Cranton, P., & Knoop, R. (1995) "Assessing Jung's Psychological Types: The PET Type Check", *Psychology Monographs*, Vol. 121, No. 2, p. 249—273
Herrmann, N. (1990) *The Creative Brain*, Lake Lure, NC: Brain Books

FIGURE 11.3 Models of psychological types.

the individual person, the largest improvement can be in the main function, and the subordinate function can almost not be changed.

Influences of types were investigated (by PET) regarding communication, conflict resolution, stress, kind of problem solving, decision making, team work, preferred modes of operation, learning and teaching styles, leadership style, and management (relationships between employer and employee) [102–106,353–364]. For instance, conflict can be constructive or destructive, and is rarely "zero-sum"—a "win" for one is a "loss" for the other. The literature [354] suggests some mechanisms, procedures, and methods for conflict resolution: *ignoring*; *forcing*; *referring*; *debating*; *opposing*; and *problem solving* (compare Figure 2.7).

11.1.6 RIGHT AND LEFT BRAIN HEMISPHERES

The right brain hemisphere in a human is mainly responsible for the actions of the left body side and vice versa. A productive connection (*corpus callosum*) coordinates between the brain hemispheres. Language centers and analytical processes are located in the left brain hemisphere, working serialistically/sequentially. Figural processing and creative processes are more developed in the right brain hemisphere, working holistically. These abilities cohere in part with the psychological type, but also with the other parts of thinking [117].

Combining emotions (Section 11.1.2) with the capabilities of the brain hemispheres, Herrmann [263] proposed a brain dominance model (Figure 11.3D), with steps and associated mindsets of problem solving (compare Section 2.4.1). The resulting model fits fairly well with the Wundt scheme of psychological types (Figure 11.3A).

11.1.7 CREATIVITY

Ideas that are apparently generated spontaneously, and in problem solving and designing, reflect creativity, for which persons must have [153]: (1) an adequate *knowledge of objects and principles* (see Chapters 5 to 7), including *tacit (intuitive) knowledge*, possessed by or available to an individual or team; (2) a *knowledge of processes*, especially about design and problem-solving processes (see Chapters 2 to 4); (3) adequate *judgment*, a sense of what is reasonable to expect under the circumstances; (4) an *open-minded attitude*, the *essence of creativity*, a willingness to accept ideas and suggestions (generated by self or others) and to associate them with other knowledge; (5) sufficient *motivation* (refer Section 11.1.4), including self-motivation and externally induced motivation; and a *sense of care and attention* for excellence; (6) ability to *communicate* to make the generated proposals visible, to present the proposals in useful forms (compare Figure I.5); (7) an appropriate level of *stress*, too much or too little stress can reduce creativity; and (8) recognition and *ownership* of *the existence of a problem* (may be a part of "motivation").

There is no guarantee that a person will be creative, it depends on the situation (Chapter 3).

Rhodes [483] identified four regions that influence creativity—person, process, product and press (or place)—the four strands operate functionally in unity; "press" (or place) defines the climate in which people operate. The overlap with the listing of creativity is obvious.

Creativity, generating novel ideas, occurs as a result of a natural tension between intellectual and intuitive mental modes [442]. The *intellectual mode* ("left-brain"—Section 11.1.6) recognizes a problem and can analyze its nature. Dissatisfaction can then arise in the *intuitive mode* ("right-brain") to solve the problem. For creativity, oscillatory *tension* must exist between the intellectual and the intuitive mental modes.

Wallas [563] showed that problem solving occurs in four stages: (1) preparation, (2) incubation, (3) illumination, and (4) verification (compare Chapter 8). A problem should be studied; the subconscious should be allowed to work in incubation; an idea will emerge, sometimes as a flash of insight, but not under the control of the solver; the idea must then be verified.

11.1.8 Thinking Errors

Thinking errors, misunderstandings, fallacies, and failures, were explored by Dörner [131] and others. Errors occur by not recognizing changed circumstances, a firm conviction of one's own correctness, omitting inquiries about information, and so forth. Errors occur in a team, for instance by "group thinking" and by an uncritical "follow-my-leader" agreement with the dominant personality.

Errors can occur through linearization of intermeshed, complex, nontransparent, dynamic circumstances, false classification of processes as *causal*, instead of as *emergent* (compare Section 12.2.1). Humans rely too much on intuition, internal voices, and emotions (Müller's [425] normal, routine operation).

Because of complexity, many features and variables should be considered simultaneously. Dynamics raise questions of stability; unstable oscillation behavior can occur. Goals should be set and clarified before an action is started [189]. Dörner [131] and Frankenberger [211] describe consequences of some defective goal definitions.

From cognitive psychology [41], Reason [480] describes: (1) *slips and lapses* as errors leading to incorrect actions when correct actions were intended; (2) *mistakes* result from incorrect judgments and assumptions, lead to actions that execute as intended but result in failure, are more subtle, complex, less understood than slips, and thus present a greater danger.

"Critical situations" (see Figure 11.4 and Section 11.2) relate to errors [153].

11.1.9 Acceptance of Change

Acceptance for methods was explored by Müller (see [153,456], and Section 11.5.11) and depends on many psychological elements. People rarely depart from experiences and learned (intuitive) methods, the "unknown" displays dangers and risks. The difficulty and time needed for learning an approach leads initially to reduced performance, and the later increase may not be obvious (see Section I.4.1). Learning methods under stress is poor motivation to good performance. Methods need more mental energy, but can expand the solution field. Methodology does not restrict creativity, only negative attitudes to method can negatively influence creativity.

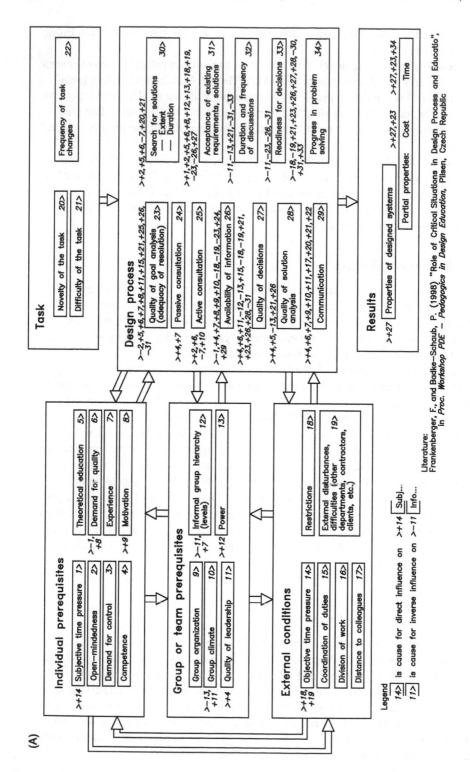

(A)

Task

| Novelty of the task | 20> |
| Difficulty of the task | 21> |

| Frequency of task changes | 22> |

Design process

>−2,+5,+6,+7,+8,+11,+15,+21,+25,+26, −31

| Quality of goal analysis (adequacy of resolution) | 23> |

>+4,+7

| Passive consultation | 24> |

>+2,+6, −7,+10

| Active consultation | 25> |

>−1,+4,+7,+8,+9,+10,−18,−19,−23,+24, +29

| Availability of information | 26> |

>+4,+6,+11,−12,−13,+15,−18,−19,+21, +23,+26,+28,−31

| Quality of decisions | 27> |

>+4,+5,−13,+21,+26

| Quality of solution analysis | 28> |

>+4,+6,+7,+9,+10,+11,+17,+20,+21,+22

| Communication | 29> |

>+2,+5,+6,−7,+20,+21

| Search for solutions | 30> |
| — Extent |
| — Duration |

>+1,+2,+5,+6,+8,+12,+13,+18,+19, −23,−26,+27

| Acceptance of existing requirements, solutions | 31> |

>−11,−13,+21,−31,−33

| Duration and frequency of discussions | 32> |

>−11,−23,−26,−31

| Readiness for decisions | 33> |

>−18,−19,+21,+23,+26,+27,+28,−30, +31,+33

| Progress in problem solving | 34> |

Individual prerequisites

>+14

| Subjective time pressure | 1> |

>−1, +8

| Open−mindedness | 2> |

| Demand for control | 3> |

>+9

| Competence | 4> |

Theoretical education	5>
Demand for quality	6>
Experience	7>
Motivation	8>

Group or team prerequisites

>−13, +11

| Group organization | 9> |

>−11, +7

| Group climate | 10> |

>+4

| Quality of leadership | 11> |

| Informal group hierarchy (levels) | 12> |
| Power | 13> |

External conditions

>+18, +19

| Objective time pressure | 14> |

Coordination of duties	15>
Division of work	16>
Distance to colleagues	17>

| Restrictions | 18> |

| External disturbances, difficulties (other departments, contractors, clients, etc.) | 19> |

Results

>+27

| Properties of designed systems | >+27,+23 | >+27,+23,+34 |
| Partial properties: | Cost | Time |

Legend

| 14> | is cause for direct influence on | >+14 | Subj... Info... |
| 11> | is cause for inverse influence on | >−11 | |

Literature:
Frankenberger, F., and Badke−Schaub, P. (1998) "Role of Critical Situations in Design Process and Education", in *Proc. Workshop PDE − Pedagogics in Design Education*, Pilsen, Czech Republic

Mechanisms leading to successful and deficient decisions

(B) Example mechanism leading to decisions for successful proposals

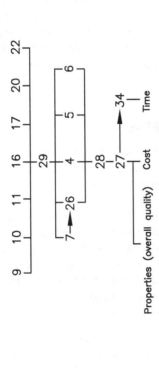

(C) Example mechanism leading to decisions for deficient proposals

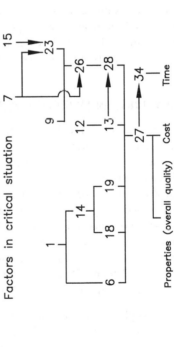

Literature
Frankenberger, F., and Badke–Schaub, P. (1998) "Role of Critical Situations in Design Process and Education", in *Proc. Workshop PDE — Pedagogics in Design Education*, Plisen, Czech Republic

FIGURE 11.4 Factors of critical situations in design engineering.

11.1.10 ETHICS

Products should conform to accepted standards of ethics, and to the laws and standards of the country or region [567], for direct usage of the product, and for consequential short-term and long-term influences, for example, on the natural, physical, cultural, and economic environment [460]. Ethical dilemmas, to be resolved by judgment and appropriate behavior, occur with products and procedures, where the rules do not deliver clear guidelines.

11.1.11 CLOSURE

These points give a brief survey about human thinking. Only a holistic consideration of many models and their probable relationships can facilitate this survey. Advice for an individual must be to proceed in a systematic and methodical way if possible, or at least to check and examine the results of an intuitive procedure using the methodology. Some psychological types (especially the intuitive ones) find working with methods uncomfortable, but they can learn—and when methods are no longer conscious, these individuals are again comfortable.

11.2 TEAM WORK IN DESIGN ENGINEERING

Literature on teamwork [114,346,347,399,461,489,511,587] deals with management procedures and functions. Significant work in designing is performed in teams, but each team member has individual responsibilities, and routine tasks must still be performed between team meetings.

Drucker [136] distinguishes three kinds of team. These types cannot easily be mixed; a change of operating procedure and management structure within an organization is difficult. An actor can change from one kind of team to another, but most easily into the neighboring team type. Concurrent engineering (see Chapter 8) demands a type 2 (or type 3) team. Total quality management (TQM), and flexible manufacture demands a type 3 team. Design engineering usually needs a type 3 team for conceptualizing and embodying, and type 2 or 3 for detailing.

Type 1: participants play *on* a team, not *as* a team; typified, for example, by baseball, or a hospital operating theater. Actors hold a fixed position, expertise and responsibility, and train for repetitive tasks. Substituting for each other is not possible or desirable. Coordination is by prior planning, allowing a multilevel hierarchy of command, and a loose organization of individuals responsible for pieces of the solution [529].

Type 2: participants play *as* a team, typified by, for example, volleyball, a symphony orchestra, a hospital emergency ward. Actors occupy a relatively fixed position, expertise and responsibility, are trained for a task, but coordinate their tasks flexibly with those of others. This requires a coach or conductor who carries the responsibility and gives directives. "Integration" of knowledge is needed among the team members, to reach a common understanding in terms of [373]: (1) "know-what"—definitions and facts, team members use the same language; (2) "know-how"—personal expertise and actions available to the team, and transferring information among team members; (3) "know-why"—knowledge of causes and influences, understanding of a system; (4) "know-where"—means to coordinate the efforts of team members. Inter-personal relationships must be developed, and leadership roles defined (see Section 11.1.2 and [233]), giving a flat management hierarchy of maximum 2 or 3 levels.

Type 3: participants play as equals *in* a team, typified by, for example, a jazz combo, a doubles tennis match. Such teams have a maximum of seven to nine people. Actors occupy preferred positions, but can and do cover for each other; they flexibly adjust to the strengths and weaknesses of others, almost as a conditioned reflex. Leadership changes from time to time according to the needs of the team. Very rapid reaction to a situation is possible.

Tacit knowledge (see Section 11.1.7) is essential, but communication skills and social competence are also needed. Effective teamwork needs effort, dedication, and development in a structured way [382]. Behaviors associated with effectiveness are [459]: (1) *Collective decision making* — Effective teams discuss decisions to reach consensus. Ineffective teams are ruled by a strong team member. (2) *Collaboration and interchangeability*—On an effective team, members help one another, even when the task is outside the member's discipline and expertise. Ineffective teams have members working independently. (3) *Appreciation of conflicts and differences*—Effective teams expect conflicts, resolve them openly, and use them to explore alternatives and improve their decisions. Ineffective teams tend to preserve surface agreement, avoiding conflict. (4) *Balanced participation*—On effective teams, members balance the demands of the team against other responsibilities, compensate for situations where team members reduce their efforts. Ineffective teams have one or two members who do most of the work. (5) *Focus*—Effective teams focus on important goals, and pace themselves to achieve them. When an effective team lags, every member helps to get back onto the schedule. Ineffective teams either spend too much time on early tasks and find that they must rush to meet the deadline, or they spend too little time on the early tasks and find that correcting the deficiencies takes up too much of the remaining time. (6) *Open communication* — Members of effective teams inform each other about the events that might influence the work of the team, are open about the progress (or lack it) in their work. Communication is open, spontaneous, and nonthreatening in the team meetings. (7) *Mutual support*—On effective teams, the members support one another, and actively show their appreciation for the efforts, ideas, constructive criticisms and help. (8) *Team spirit*—Members of effective teams show loyalty. Members of ineffective teams use the team as a place to work, or as a hindrance to achieving their own goals.

Formal techniques for conflict resolution are known [354] (see Section 11.1.5).

When a team is collected, the members usually experience four phases of operation [333,534], often with recursions to earlier phases: (1) *forming*—team rules are established, consciously or intuitively; (2) *storming*—first conflicts appear; some individuals start to oppose the work of the team; (3) *norming*—the team settles down to the common task and begins serious work; and (4) *performing*—the team works together toward the goal.

Engineering designers, who perform *routine* work as individuals, at times reach decision points—*critical situations*—that should be discussed in team meetings. Efficiency and effectiveness of design work in industry are determined by technical problems and several nontechnical factors [208–212]: (1) *individual prerequisites*: domain specific knowledge, experience, style of problem solving, open-mindedness (e.g., for new measures, ideas, suggestions), and so forth; (2) *group or team prerequisites*: informal or formal hierarchy, conflict resolving capability, group cohesiveness, style of communication, group climate, and so forth; (3) *external conditions*: working environment, organization situation, management style, and so forth; (4) *task*: complexity, novelty, difficulty of designing, and so forth; (5) *design process*: (a) *problem solving*—see Figure 2.7; (b) *TS progress*—see Figures I.17 and 4.1; and (c) *external influences*, disturbances, conflicts, task changes, and so forth; and (d) *results of designing*: quality, external properties of the TP(s)/TS(s).

Various factors influence the critical situations, and the relationships influence the results of designing (see Figure 11.4A). Some factors have direct relationships; an increase in one factor causes an increase in another, for example, factors 14 and 1. Other factors have an inverse relationship; an increase in one factor causes a decrease in another, for example, factors 11 and 12.

Successful decisions about a design solution resulted from factors shown in Figure 11.4B. Figure 11.4C, shows factors that led to deficient decisions and solutions, the main loss was in total quality, and in the partial property of time.

Teams should be composed of a mixture of psychological types (Section 11.1.5), bringing different experiences, object, and process knowledge together. Leadership within the team can change according to the situation, personalities, and expertise of the participants. Interpersonal relationships within teams demand training in the work (and conflict resolution) in the team.

Leadership in a team is a controlling and regulating function in a social system [55] (see Sections I.12.2 and 12.2.2). Social systems consist of partial systems that pursue their own objectives while contributing to the common goals. Three requirements for leadership are: (1) *Content-related*—activities needed for tasks and to solve problems: (a) formulation of goals and decisions to clarify the goals and tasks; (b) searching, discussing, and selecting solutions as a supervising function; and (c) diagnosing sources of failures and developing alternatives. (2) *Process-related*—deals with the design task, and directs activities toward structuring and coordinating people and processes: (a) scheduling (time related) activities that coordinate personnel, equipment, information, and so forth; (b) personnel, and procedures of monitoring and controlling; and (c) allocation of resources, including financing, staffing, and materials. (3) *Relationship-oriented*—to pursue the goals with or against the support of others and ensure motivation of participants: (a) interpersonal strategies to influence others who pursue their own and organization goals; (b) processes of detecting, analyzing and solving conflicts and opposing positions; and (c) coaching, motivating and supporting, rewarding performance, and building team identification for organization goals.

Humans tend to seek power, and make themselves indispensable to an organization, for example, by hoarding experience information [90]. Team members should show inter-personal tolerance, and be open to suggestions and ideas.

11.3 MANAGEMENT OF DESIGN ENGINEERING

Management has two overlapping and supporting functions related to design engineering [249]. One concerns the product range of an organization, and managing "the designs"; the other relates to managing the design process. Both require knowledge of the organization and its products, and the cultural, political, economic and legal situation in the target region for the product (see Chapter 3). Some psychological types may be more suitable for these functions [359–361] (see Section 11.1.5).

11.3.1 GENERAL ASPECTS OF DESIGN MANAGEMENT

Managing "the designs" of offered products (technical process [TP] and technical system [TS]) should take account of market needs. This design management should plan and implement a product, supply chain, production and distribution, servicing and customer relations, and so forth [8,22,30,44,136,271,272,478]. *Product planning* and IPD (see Section 11.4) should maintain a succession of products. Aspects of efficiency, flexibility, and innovation need to be considered [368].

A requirement specification should feed the customers' real needs into the engineering design process. Various methods (see Chapter 8) are available, for example, in TQM [108,125,323,325,326,339–341,549], and within TQM, quality function deployment (QFD) [75,111,151,259,269] provides explicit consideration of trade-offs. Cooperation of designing with manufacturing should ensure suitable trade-offs and synergies. Planning and supervising the tasks with respect to times and resources can be helped by network diagraming, critical path method (CPM), and program evaluation review technique (PERT) [246], which assume that the processes are predictable. The design structure matrix [83,184,185,330,525] can assist in task scheduling and some predictions.

Decisions made during designing commits about 60 to 80% of the expenditure for the product, if it is manufactured (see Figure 10.5). Design decisions also determine part of manufacturing and other "downstream" life cycle processes and costs, engineering designers make *dispositions* [47]. Early consultation should influence the relevant design decisions.

An organization that wishes to implement *just in time* (JIT) delivery or *quick demand ordering* of constructional parts to supply the manufacturing and assembly operations must brief and supervise any outside suppliers in the supply chain.

Under *concurrent engineering* (see Figures 2.10 and 2.11), the most important details are *frozen* during the dimensional layout, and the manufacturing processes, jigs and tooling are also largely established. This underlines the need for thorough checking and verification of the layouts and details, and traceability of all design decisions [526]. Any later changes made to manufacturing drawings incur large costs. Prototype testing, the task of a development department, should avoid modifications of the TS(s) where possible.

Value engineering [222,261,414,424], cost–benefit analysis [217,379], failure mode, effects, and criticality analysis (FMECA), design for manufacture and assembly (DFMA), and QFD (among others, see Chapter 8) can usefully be applied as review or auditing techniques during the layout stages. Using Taguchi methods [391,468,496,530,531], experiments can be performed to investigate variability in some parameters of constructional structures and the elemental design properties, and of manufacturing that will influence the performance of the product.

Completion of the detail and assembly drawings (or their equivalent) would be the last opportunity to stop the project before manufacturing commitments are made. Sales and marketing can now prepare for product launch, with adequate knowledge about TS-properties, including costs. Additional documentation for assembly, testing, adjusting, operating, maintaining, repairing, and so forth, can be prepared.

11.3.2 INTELLECTUAL PROPERTY

Management should protect the intellectual property of the organization. In most countries, laws, regulations, and relevant organizations exist to provide procedures for this protection. In general, protection includes establishing ownership of the property, and maintaining records of the nature and contents. If a violation of the property rights occurs, the owner can take legal action against the violator, usually in civil courts. These summaries are incomplete; information is available from government, for example, Department of Consumer and Corporate Affairs in Canada.

11.3.2.1 Copyright

The law regulates copying any written work, for example, the contents of books, papers, drawings, photographs, and so forth. The authors own the copyright, unless they were employed to create the work. Copyright is automatic in Canada, and extends to 50 years from the death of the author; registration is needed in the United States and many other countries.

11.3.2.2 Industrial Design

An industrial design (noun) is an original shape, pattern, configuration, or ornamentation applied to an article of manufacture that was made by an industrial process, and is judged only by the eye. The industrial design must be registered within 1 year of publication. Protection in other countries requires separate application to each country.

11.3.2.3 Trade Mark

A word, symbol, logo, picture or a combination of these that distinguishes products of an organization from those of others in the market place, a trade mark, may be registered. Protection in other countries requires separate application to each country.

A *certification mark* is a type of trade mark that is used to distinguish goods or services that meet a defined standard, for example, the CSA or UL marks on electrical equipment, the ISO 9000:2000 mark to show approval of a quality management system, ISO 14000:1995 mark to show environmental conformity [93,113,328], and so forth.

If a word in a trade mark becomes common usage for a generic product, the right to the trade mark may be withdrawn—for example, in England, "doing the hoovering"—vacuum cleaning—caused the Hoover Company to loose that part of its trade mark rights.

11.3.2.4 Patent

Traditionally, a patent gives the right to exclude others from making, using, or selling an invention within the country. Any new and useful process, machine, composition of matter, or improvement in these items, which shows inventive ingenuity, may be patentable. A patent may be granted if the invention has not been described in any publication anywhere in the world before the application is filed. A patent is valid in Canada for up to 20 years. Protection in other countries requires separate application to each country.

A *patent specification* is a legal document that describes a particular invention. Its language is important, especially what is interpreted as being described. A patent attorney can advise. Possibilities exist for avoiding or evading a patent (without actually infringing it) by devising something similar that is not quite described in the patent. Whether this is a valid solution can only be decided in a civil court action. The courts of law may remove the right to protection [255]: (1) for unreasonable or unexcused delay in the assertion of a claim for infringement (filing a suit), and prejudice to the defendant (the alleged infringer) resulting from the delay, typically a maximum of 6 years; or (2) affirmative action by the patent holder, which lead the infringer to believe that the patent holder has abandoned the claim, for example, threatening to take legal action and then delaying the actual filing of a suit.

11.3.2.5 Other Protection

Intellectual property may be protected by keeping it as a *trade secret*, under common law. Such protection is not effective if an item can be "reverse engineered." Trade secrets are often protected by marking "confidential" or "secret," and limiting the number of people with access.

11.3.3 MANAGEMENT OF THE ENGINEERING DESIGN PROCESS

This duty is to manage the process and progress of design work [301]. It includes organizing the design office, the procedures in designing, the design personnel and other members of the design team, and the resulting documentation. Traceability [526] and audit of all design decisions, and verification of all design data are important.

Routine work in design engineering is predictable in its requirements for time and resources. As innovation and novelty increase, unforeseen iterations and recursions, and a need to search for information and candidate solutions, make the design duration become more unpredictable.

Various influences of the design situation and management styles are shown in Figure 11.5. The behavior grid relates mainly to general management and production.

The expended costs of designing rise progressively as the state of the TP(s)/TS(s) becomes more concrete. Conceptual designing is relatively low in costs. A large proportion of the design cost is expended in producing detail drawings, and the decisions account for a large proportion of the potential manufacturing costs of the product. Skills, knowledge, and ability are essential, especially at the detail design level,

(A) Behavior grid of management styles

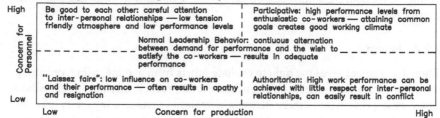

(B) Design situations and contingency

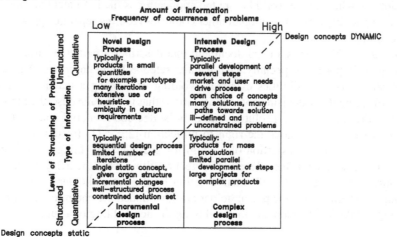

(C) Design management procedures

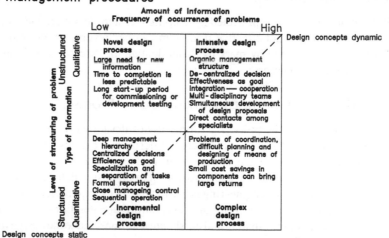

Literature
Volk, H. (1984) "Führen nach dem Verhaltensgitter" (Leading According to the Behavior Grid), *Technische Rundschau*
    Vol 51 and WDK 11— *Führung im Konstruktionsprozess* —— *Reading* (Leadership in the Engineering Design Process
    — Reading), Zürich: Heurista, p. 71–73
Slusher, E.A., Ebert, R.J., and Ragsdell, K.M. (1989) "Contingency Management of Engineering Design", Paper C377/010
    in *Proc. of the Inst. of Mech. Eng. ICED 89 Harrogate*, London: I.Mech.E., pp. 65–75
Perrow, C. (1967) "A Framework for the Comparative Analysis of Organizations", *American Sociological Review*,
    Vol. 32, p. 194–208
Daft, R.E., & Lengel, R.H. (1986) "Organizational Information Requirements, Media Richness and Structural Design",
    Management Science, Vol. 32, No. 5, pp. 554–571

FIGURE 11.5   Situation of design problem type and management.

because a minor change of geometry, or omission of a radius and transition of surfaces, can lead to early failure of the TS(s), for example, by material fatigue, or unnecessary increase of committed manufacturing costs, for example, for special processes or tooling.

The problems are worth the attention of professional engineers at this level of concretization, and consultation with other specializations in the organization should be active.

Management makes many decisions about any design process, including auditing and giving clearance (permissions, releases) for further work at the various stage-gates (see Figures I.17 and 4.1), the "evaluate–improve–check" blocks. Information about the state of work in designing a TP(s)/TS(s) must be made available to the management team, preferably by a *product champion* who is knowledgeable and enthusiastic, and has sufficient authority.

Management also creates time pressures, which may result in a need to "crash" a project, or hold a "charette"—as a deadline approaches, time becomes limited, resulting in overtime and overly rapid progress, to the detriment of care and innovation.

Assessment from criteria of optimal quality of the TS is basically correct, but often too late because it occurs after the design work, and possibly manufacture, have been completed. The quality of the causes for these criteria contributes to achieving the aims. Such causes can be derived from management (leadership), and formulated as: (1) construction and constitution of an organizational structure; (2) establishing the aims and methods of progress, controls and checks; (3) acquisition, documentation, and communication of information, including feedback from downstream activities; (4) leadership and motivation of personnel (see Section 11.1); (5) organization and application of technical means; (6) level of specialist knowledge in designing (see Chapter 9); and (7) organization and methodology of leadership activities.

Design engineering requires a resilient style of participative management, depending on the design situation. Management styles, atmosphere and contingencies influence the status, motivation and dedication of design personnel, and their perceived rewards and equity; see Figure 11.2 and [289,293,301]. A decisive role in managing design engineering is played by considering the operators and their optimal use. Design managers should have broad design knowledge, to provide expert control and advice.

11.4   INTEGRATED PRODUCT DEVELOPMENT

Many products have little or no engineering content, will be produced in larger quantities, and their primary property is "appeal to customers" (see Sections I.7.1 and 6.11.10). For these products, integrated product development (IPD) [44,178] emphasizes industrial design [204], and then, if needed, involves design engineering; the tasks overlap (see Figure 7.10).

IPD provides guidelines for forming and operating design teams, integrating the needs and inputs to designing from organization functions (sales and advertising, manufacturing, maintenance, and service), and where possible representing potential customers. Methods for IPD have been introduced into industry [267]. Many products from IPD processes in a particular organization have been conceived as variants based on a common platform [374]. The variants are often interchangeable *modules* with different capabilities for functioning. The interfaces between modules and platform are therefore of prime importance. The theories and methods presented in this book should be applicable to most instances of IPD.

11.5 INTRODUCTION OF SYSTEMATIC AND METHODICAL DESIGNING INTO PRACTICE

Introducing any new methods, procedures, knowledge, and so forth is difficult (see Section 11.1). People can be expected to resist change, unless they have motivation and incentive. The recommendations in this book, expanded from Engineering Design Science (EDS) [315], provide guidelines about the changes, and their introduction into practical design processes in industry and the more idealized processes for education.

11.5.1 CURRENT SITUATION OF ENGINEERING DESIGNERS AND DESIGN PROCESSES

In current engineering practice, designers usually work intuitively, based on their education and past experiences. This results in most of the published descriptions of design engineering (see the NOTE in Section I.11.1). They learn from "sitting by Nellie" [239,240], in an apprenticeship mode, so well that the operations run subconsciously and intuitively. Designers (like other people) are not fully conscious of their actions (see Section 11.1) (rationality vs. emotions). Work takes place under time pressure, externally imposed deadlines, and starts when the deadline approaches (compare Section 11.1) (subconscious and incubation). Work is not fully traceable and design audits cannot reveal the reasons for decisions nor the considered alternatives, which causes problems if a liability case is lodged.

In their education, designers have obtained little of the collected procedural knowledge about designing. They may have heard about design methodologies. They generally and falsely believe that creativity is inborn, and that methods will reduce their creativity.

They may be guided by publications that describe new methods, which: (1) are general, and not immediately and directly applicable for their area of work; (2) assume that they would have to learn it all before they could apply any of it; (3) present the knowledge in a heterogeneous form without a supporting theory; or (4) are commercially advertised as answer to all problems.

Management of the organization demands a high quality of product, but does not demand installing new methodical tools, except for computer programs and management techniques advertised by commercial vendors.

Designers and managers have no proof that a new method offers a better product, a larger sales volume, or more effective treatment of design problems, and proof is difficult to establish. They fear complex explanations, often delivered without supporting theory or credible examples.

Educators have little experience of design engineering in industry, and lack a general basis of theory, for example, as presented by EDS [315].

The theories and methods presented in this book were developed in reaction to the defects of design engineering found in practice, especially since the 1960s. Various "industry-best practice" methods have been introduced (see Chapter 8), but their impact on design engineering has been small.

The defects include insufficient attention to the definition and understanding of the design task, its contents, and form. Organization-internal task definition (design specification) does not exist. In most situations, the first solution is adopted and adapted until it is just acceptable, "satisficing" [167,506,507]. No alternatives or variants are proposed. An optimal solution is hardly to be expected under these conditions. Where alternative solutions are proposed and discussed, they exist mostly in the constructional structure. Important and innovative solution possibilities are available from the task definition, transformation process structure, technologies structure, function structure, and organ structure, that is, from conceptualizing, but are not used.

Properties of the TP(s)/TS(s) are not systematically examined during designing. An optimal quality of product is hardly reached, and the potential is not explored. Reliable methods and technical information for establishing and controlling the individual properties of the product are missing. Avoidable changes are often necessary in the manufactured product, in order to make the product acceptable. The design process is not traceable. ISO 9000:2000 [10] prescribes that design work and its documentation should be transparent, but does not show how.

The design process is not accomplished as planned, making checking, controlling and managing the design process difficult. Alteration of plans should be allowed, but should not derail the design process. Inertia, as failure to depart from existing solutions, prevents product innovation.

Formal reports about the design process are hardly ever produced, or are prepared with insufficient arguments and explanation of the individual decisions. Possibly a very subjective report is presented. When a product failure occurs, the causes cannot be clearly determined. Learning from failures is then difficult, and training of novice designers is inadequate. Ability and readiness for cooperation, mutual support and supply of suitable information is also lacking.

These listed defects should be seen as features of the present status of engineering design work. This book shows some possible solutions, and presents argument to consider ways to achieve improvement, rationalization and completion, to overcome deficiencies of the existing situation. This book shows some possible solutions.

11.5.1.1 Conditions for Improvement

The starting point for a completion to improve designing is the procedural model of design engineering (Figures I.22 and 4.1). This design process (see Chapters 2 and 4) transforms the input information about requirements into a description of the product ready to be manufactured or implemented. The quality of the results of the engineering design process depends on: (1) the quality of the contract, as given by customers and management; (2) the quality of the technology of the design process; and

(3) the quality of the operators—designers, working means, environment, information, and management.

Generally, the engineering designers must master individual activities to the best of their abilities, and must obtain maximum usage from the design potential, especially from technical means and information. They must also be concerned about their mental hygiene.

The system of knowledge about design engineering presented in this book, must be adapted to the design situation (see Chapter 3), within the organization, and adapted and concretized for a certain engineering specialty, and a particular organization. This also depends on a time frame, a local, national and world economy, and a country. Most of the elements will remain constant during a longer time span (Figures I.17 and 3.1), especially the environment (FE) and organization factors (FO) (see Chapter 3).

The knowledge and models that have been concretized for the constant elements are usable for the specific conditions of each contract, and can create the requirements for a planned completion of the design process. This planning must be flexible to allow adjustment for any unforeseen circumstances that may arise during design engineering, especially if the requirement for novelty of the product is high (innovation).

Designers normally design only for one "sort" of TP(s) and TS(s), the organization's product range. This "sort" of TS(s) has typical functions, properties, level of complexity, degree of difficulty, and novelty, and the function structure and organ structure are usually unaltered.

A design department forms one element of the organization. The degree of innovation of the TS(s) is clear; this stabilizes the structure of the organization, creates traditions, and influences the realization of contracts, and the cooperation with other sections or departments, both within the organization and with outside agencies.

Designers form an element of a particular design department and team, even if acting as self-employed consultant designers. This makes the conditions for specialization and cooperation possible, including the internal system, working conditions, cooperation with others, and so forth (see Section 11.2). Designers are bound by employment conditions or contracts to the technical structure of the organization, and its policies and directions.

A designer is an individual, who has certain personal characteristics (see Section 11.1), relationships to motivation, the work, creativity, capability for cooperation and team functioning (Section 11.2), and a level and orientation of education, expertise, and experience.

Each contract, task, or project has specific traits that influence the design process, for example, the deadline may result in time pressure, with a large influence on the development of a solution.

11.5.2 COMPLETION OF THE DESIGN PROCESS

The quality of the design system (Figure 2.1) depends mainly on the quality of: (1) the contract; (2) the technology of the engineering design process; and (3) the operators and their effects (see Section 11.5.1.1). In Figure 11.6 these three areas are listed

	What? area to be completed	Goal, aim	HOw? with what? measures	Starting material
Od operand	Assigned design problem, design brief Order, contract List of requirements, design specification	Complete, clear, well formulated, quantified, realizable, substantiated, and attractive design problem corresponding to the ideas of the customers (design brief as assigned, and designers' working design specification)	— Methodology of processing — Masters or forms for given tasks in individual branches (as concretely as possible) — Computer processing	Theory of technical systems — Life cycle, Figures I.14 and 6.11 — Classes of properties, Figures I.15, 6.8, 6.9 and 6.10 — Design tactics

Tg technology of designing	Behavior in the design situation
	
	Strategy
	
	Tactics
		Optimal behavior to be attained in important, standardized design situations, for individual designers and/or a design team (i.e., its members)
		
		Optimal procedural plan established for each order, contract or design task
		
		Optimal methods selected for standardized design activities, optimal selection of methods for unusual design activities (menu, tool-kit)
			— Establishing "standardized design situations"
			— Instructions for conduct (guidelines) for individual situations
			— Establishing "classes of orders, contracts or tasks"
			— Master procedural plans for individual classes
			— Establishing "standard activities"
			— Master tactics for activities
			— Positioning of 'design methodology'
				Chapter 3
				   Design Situation
				Section 9.2
				   masters
				Procedural model, Figures I.17 and 4.1
				Chapter 6.6
				   Properties
				Chapter 1
				   Case Examples
				Chapter 8
				   Methods
				Chapter 2
				   Design Operations
				Section 2.4.1
				   Basic Operations

Ot operators of the design process (designing)	Designer	Qualified designer with abilities: flexibility, life-long learning, responsibility, team work, mastery of suitable technical knowledge, understands designing, etc.	— Appropriate job description and capability profile — Careful selection for employment — Motivating — Study, continuing education, training — Leadership in work — Teamwork	Figure 2.6 WDK 11 Leadership [Hubka 1985b] WDK 21 Education [Eder 1992a] WDK 24 Creativity [Eder 1996a]
	Technical working means	Available: optimal hardware and optimal individual software maximum know-how maximum usage, application, employment (technical and time)	— Investigation of fitness and utilization of existing technical means. Aims for application — Better utilization — organization — Acquisition of means — Organization of individual processes (specific software)	Chapter 10 Computers

	Technical information
	— Branch object and
	— Design process information
		Quick and reliable finding of the necessary technical information, which corresponds in content and form to the needs of the designer (object information and design process information)	— Construction of information systems
			— Ensuring the currency and accuracy through flow and feedback of information from production, usage (application), disposal, and so forth
				Chapter 9
				   Information Systems
				Section 6.6
				   Properties
				Chapter 8 Methods
				Section 9.5 Forms
				Section 11.1
				   Psychology

	Design leadership, management	Create a powerful, flexible design potential in a design system	— Optimal organization — Optimal cooperation with other organization departments — Introduction of methodical and systematic designing — Securing of operators: quality of the persons, technical means, information, leadership (incl. self-control) and work conditions	WDK 11 Leadership [Hubka 1985b]
	Working conditions, environment	Achieving optimal working conditions	— Securing of suitable physical and psychological working conditions	

FIGURE 11.6 Completion of the engineering design process.

with appropriate subdivision in the first column. In the second column the goal is formulated: "What is to be completed?" The third column shows ways and resources for achieving quality of the areas, and measures to arrive at the goals. The last column contains starting materials, with references to the literature that deals either with this problem or with further details and references. It is necessary to consider concrete situations to obtain an effective set of instruments for designers.

The recommended procedure is: (1) Determine the defects in design engineering by analyzing completed contracts. Engineering designers (and managers) must be convinced that an improvement is needed, at least in effectiveness of designing and in record-keeping. (2) Establish the areas to be completed and improvement goals, task division, priorities, and deadlines. (3) Process the partial materials according to the instructions contained in the sections and supplements of this book. (4) Gradual transfer into the engineering practice.

With a "systems view," some improvement can be reached by individual measures — and gradual introduction of these measures is recommended. An eventual complete application will show the important influences, when the interrelationships among individual measures become effective.

11.5.3 COMPLETION OF THE TASK DESCRIPTION (ORGANIZATION-INTERNAL DESIGN CONTRACT)

Part of the contract to deliver a TP(s) and TS(s) to the customer is the contract to *design* the system, usually internal to the organization, as the input to the design process (see Figures 2.1 and 4.1). The initial problem is to define the requirements (including needs, constraints) for the future TP(s) and TS(s). These requirements should describe the conception of the future TP(s) and TS(s), preferably of expected performance, as a *design specification* (see Section 4.3.1).

These requirements determine the future properties of the TP(s)/TS(s) to be designed (see Chapters 5 and 6). Relevant areas of the engineering design theory are: (1) origination and existence—Figures I.13 and 6.14; (2) classes of properties—Figures I.8 and 6.8; (3) relationships among classes of properties—Figures 6.9 and 6.10; and (4) establishing and determining the properties of technical systems (extent of feedback loops)—Figure I.18.

As shown by the feedback loop in the procedural model (Figures I.17 and 4.1), it is necessary to consult with the external or internal customer, stakeholder, or client, when establishing the design specification, and in subsequent stages of design development, which can help to ensure improvement in the quality of the contract or order.

If the client, customer, or other stakeholder is outside the organization, that is, through the sales office and distribution section, and has a relatively clear idea of the desired product ("made to measure or custom manufacture," Figure 6.18, lower cycle), then the engineering designers should strive to meet the customers' requirements as given. Advising and questioning of the contract by the engineering designers should be accepted, negotiated, and regarded positively.

For an organization-internal client (e.g., management, product planning, sales), for example, for series or mass production ("design for the market," Figure 6.18, upper cycle), the designers or any representatives of the design process can and should be involved in negotiating the product plans. This is also valid for team work (see Section 11.2), and for *simultaneous* or *concurrent* engineering (see Chapter 8). A similar stage is present as a basic operation, Op-H3.1 (Figure 2.7), some practical knowledge is included in Section 2.4.

11.5.3.1 Status in Engineering Practice

Engineering practice shows that the task definition as given to the designers in most cases is poor, incomplete, unordered, not quantified, often even unrealizable, containing opposing and contradictory requirements, or too restrictive in its references to existing real or process systems.

This design specification should neither be confused with the *terms of reference* attached to an employment position, which describes the generic duties of a person, nor with a contract from a customer, which tends to emphasize the financial and legal contract obligations.

11.5.3.2 Goals for Completion

The goal is a suitable quality of the task definition, the design specification should be complete, clear, current, and realizable, with time deadlines and justifications (see Section 4.3.1).

This should be a firm duty, yet there must be a possibility of questioning the contract. There is always a certain freedom of interpretation, ambiguity of language, but this must be supported with consultation, avoiding unfounded assumptions. A balance must be struck between statements that are too restrictive to allow a wide-ranging search for solutions, and statements that allow too much freedom for creativity that may give latitude for unsuitable solutions.

11.5.3.3 Improvement Measures: Tactics and Technology

A quality design specification complies with the list in Section 4.3.1. Additional methods could be needed for items of completeness, quantification, and currency.

Chapter 8 contains many known methods and indicates their fitness for individual phases, stages and steps of the design process (see Figures 8.1 and 8.2). The classes of properties (Figures I.8 and 6.8–6.10), can be used as a classification for the list of requirements.

In a certain work area, that is, in a certain "sort" of TS, the answers to these questions are expected to be the same or they will be very closely related. This repetition can be used to generate a "requirements list pro forma" (see Section 9.5 and Figure 2.18), which can reduce the design specification to a simpler completion of a form.

Where the assortment of products to be designed is not completely uniform, a requirements master (see Section 9.2) can be assembled to show the pertinent similarity of classes of TS.

Designers must obtain a good understanding of the task—some effort is necessary in the task formulation. Reading a design specification can contribute to this understanding, but usually cannot fully achieve it. The designer must argue actively with the task. The task definition must also continuously be examined during the subsequent engineering design process, questions referred back to the customers, and its formulation kept up to the latest state, because the understanding of the engineering designers grows continuously during the search for solutions.

11.5.4 COMPLETIONS OF THE TECHNOLOGY OF DESIGNING

For the range of problems of systematic and methodical procedure in design engineering, the possible design transformations can in principle be arranged into four classes (see Figure I.20), and for each into strategy and tactics. The fundamental procedure (Figure 4.1) is composed of individual operations that solve the partial tasks with the help of methods or working principles of designing (see Chapter 8). This procedure is valid for each degree of complexity (see Figure 6.5).

These procedures are neither compulsory, nor linear. *Iterative* working, including reflective reviewing and improving, and *recursive* working is almost always essential (see Chapter 2). This implies concurrent working on several different parts of the problem at differing levels of abstraction. Good record-keeping is essential in the recommended procedures.

These processes for novel designing (Figure 2.10), and redesigning (Figure 2.11), with their record-keeping, are transparent, comply with ISO 9000:2000 [10], and are suitable for team work and for detecting the causes of possible errors. During the process, characteristic design and critical situations emerge, some of which are often repeated, and this can be used for rationalizing the design process.

11.5.4.1 Status in Engineering Practice

Systematic and methodical designing has not yet found the position in engineering practice justified by its rationalizing potential. Objective and subjective reasons can be found (see Section 11.5.1). Changes in technology are a difficult area, since changes of personal work styles (procedural manner, see Sections 4.7 and 8.1) can push staff members toward denial.

11.5.4.2 Goals of Completion in Design Technologies

The goal is a transparent design process of type D, or at least C (Figure I.20), that is, application of systematic and methodical designing as technology, or at least methods-aware designing. This means creating conditions in which: (1) an optimal design procedure is found for each contract; (2) an optimal method is found for each operation in standard situations; (3) an optimal behavior is found by and for designers, and used for standard situations in designing; and (4) methods-awareness can guide designers into behavior that can also deal with nonstandard situations that may even extend the designers' range of expertise [425].

11.5.4.3 Means for Completion in Design Technologies

A new quality of the design process can be reached by assembling a plan that respects the situations of the organization. This plan will take a longer time span to implement, and will take different forms in individual organizations.

The strategy for introducing systematic methods should define partial goals to adapt the methods for selected areas, suitable elements of the task, and of the organization. The areas determine which can be detached from the whole, and reintegrated after improvement. An organization, or a designer, can begin perfecting the individual operations, by selecting an area that promises quick and easy success, so that staff members are motivated to accept the newly introduced methodology. Support from higher management for introducing new methods is important.

The basic operations (Section 2.4.1), can successfully be used, for example, by recommending them for a designer's procedure in an important design situation. The organization should establish a special staff position for introducing new methods, for example, a design methodologist, who takes over the preparation and consultation regarding professionalism, competence, and capacity.

11.5.5 COMPLETION OF THE OPERATOR "ENGINEERING DESIGNER"

The most important factor in designing is without doubt the operator "designer." The qualifications and competencies of designers [155,456] play a decisive role (see Section I.1). These can be described by the personal characteristics of designers: (1) What can they do—where can they work with respect to specific branch capabilities? (2) What do they want to do—what needs, professional value systems, personal goals and motivation do they have? and (3) What are their personal values — moral and ethical profile, and attitude to work?

The qualifications of designers include the aggregation of knowledge, abilities, competencies and experiences in the area of a TP-"sort," the relevant object information, and other abilities (see Figures 2.6 and 6.1A). Pedagogics has found that persons who have a high degree of general education also achieve branch (professional) qualification more easily. A complete picture of qualifications emerges only from demonstrated performance, that is, long-time results of work, in hindsight. The branch-related qualification of the designer expands through additional elements of team work (see Section 11.2).

11.5.5.1 Status in Engineering Practice

Although the situation is different in individual countries and organizations, design offices and teams, the status in engineering practice is generally unsatisfactory. Discussions about completion of this status (e.g., [61,189]) always show a lack of capable engineering designers. Analysis of the deficiencies in design practice and their causes (see Section 11.5.1), leads to the engineering designers, their qualifications and competencies. Top-level management should assume responsibility for the initiative toward this completion.

11.5.5.2 Goals for Completion of the Operator "Engineering Designer"

The professional profile of the engineering designer should approach that of an ideal designer (see Figures 2.6A and B). Each designer should aim to complete the needed knowledge for the design function (Figure 2.6B). The designers' development should be adapted to the changing conditions. Keeping up to date in the developing information area is an essential duty, and includes the working (design) methods. Necessary requirements are human characteristics: frankness, openness, flexibility, motivation, responsibility; a high degree of general and specialized education; ability to work in a team; and ability to continuously develop and educate. Preconditions for this are certain personal traits: character, temperament, attitude and outlook, needs and interests; some of these are dynamic, many are constant or develop slowly during life.

11.5.5.3 Means for Completion of the Operator "Engineering Designer"

Means to increase the qualifications include: (1) continuing education for completion of object and design process information; (2) personal study for the completion of knowing, subject to availability of suitable literature; and (3) "on-the-job instruction" for acquiring experiences and skills, obtained either directly from the executed contract, or from special projects (education parallel with practical work) in an appropriate climate.

The personal characteristics, which play such an important role in design engineering, include responsibility, care and attention, motivation, contact possibilities, and others. Several possibilities, experiences and specialists exist to cover these aspects.

Acquisition of knowledge in EDS is one of the goals for expanding the professional qualifications, design process information as explained in this book, and its basis and orientation. A further need is to ensure translation into the practical application.

Individual study, for example, distance learning, demands a strong personal engagement and is effective when the learner is technically guided. "On-the-job instruction" delivers known and unrecognized parts of information, which are accepted together with the acquisition of traditional practices, and usually with few explanations or references to a theory base.

11.5.5.4 Means for Stimulating Motivation

Motives to induce a certain conduct and behavior result mainly from external stimuli, which are rated by designers from the point of view of their own needs. Motivation (see Section 11.1.4), and its origins cannot be fully determined, because accidental inspirations of affective, emotional or intellectual origin play a role. Among the important stimuli are material (salaries, rewards) and psychological (praise, career prospects); see also Section 11.1.

11.5.6 COMPLETION OF THE OPERATOR "TECHNICAL MEANS"

The operator "technical means" include tools and equipment (see Section 2.2.4.2). The evolutions of the computer promise a strong increase of this operator's importance (see Chapter 10). A variety of technical means can be used in different activities.

A high level has been reached in communications, where contacts among designers and with remote partners are easy, and deliver new conditions for the flow of information, for example, interactive work over the internet, teleconferencing, exchange of graphical data (drawings, photographs).

The design process consists of operations at various levels of complexity and abstraction (see Figures I.21, 2.5, and 2.7). Some of these operations can be automated, in others the computer can assist as coworker of the designers. This means that both the knowledge about the TP(s)/TS(s) to be designed, and the information about the design process, must be arranged from the beginning, complete and well classified, to consider using the computer.

The computer works productively, rapidly, reliably, and practically without errors—however, many of the errors that do occur can be difficult to detect; others result in loss of work, data, and so forth. Mainly for the basic operations of problem solving, an intelligent application of the computer can increase both the quality of the results and the productivity of design work.

11.5.6.1 Status in Engineering Practice

The present status of the application of technical means in practice can only be roughly described, according to some criteria. From the point of view of hardware and software, the quantitative situation is relatively good. The problem is whether the existing systems, central computers, personal computer (PC) or workstation, are advantageous for designing. Software is missing mainly as individual programs for the specific design work of the engineering designers in an organization. Much of the software is written by programmers who are not fully aware of the needs of engineering designers — they rely on the adaptability of humans to situations.

Computer applications are often implemented as "island solutions" with little possibility of transferring data and information from one application program to another. Information about TP and TS, and about the design process as basis for computer work is sparse, or does not exist.

11.5.6.2 Goals for Completion of the Operator "Technical Means"

In the engineering design process, as in all other processes, optimal technical means should be selected for the situation in the organization, that is, hardware in general, and software for computers. The further goal is the intense application of these technical means. This is difficult because of rapid developments, especially of computer hardware and software.

11.5.6.3 Means for Completion of the Operator "Technical Means"

The computer depends on the necessary object information and design process information. Material and financial means, for example, investment funds, must be available. The completions can be executed in this area at two levels. The first is to improve the usage of existing arrangements, that is, through better exploitation of their abilities, by means of specific software, and better usage in time. The second is through acquisition of new technical means that are advantageous for the specific organization conditions. The decision about which arrangement is profitable for the organization should be based on the whole system of elements as shown in Figure 11.6.

11.5.7 COMPLETION OF THE OPERATOR "INFORMATION FOR DESIGNERS"

Both the tacit, internalized knowledge (knowing) of the designer from education and experience, and the externally stored information, are brought together under the term "branch information" (see Chapter 9). The properties, status, classification and working with information are treated among the working principles in Chapter 8.

Designers need and use large quantities of branch information, and it must be of good quality, and procured quickly and reliably. Especially for critical information, designers should not rely on their own memories, such information should be verified from external sources. We can rate the quality of information according to the criteria stated in Section 9.1.1. Engineering designers must possess "their own" information system, in part internalized in their individual memory. Internalized information should be captured and verified into an external information system (see Chapter 10) for one possibility. Archiving the information, and keeping the holdings up to date, is another problem in the information system.

11.5.7.1 Status in Engineering Practice

This important area is still characterized by an information system for engineering designers that consists of a handbook [54,198,199,225,238,407,463,513,594], some text books brought from the engineering designer's educational experience, a few catalogs collected from exhibitions and suppliers, some standards, and gathered experiences from the practice. Experiences have remained in the designer's memory, and are refreshed from case to case by notices in a log or note book (see also Chapters 8 and 10). Engineering designers, as part of their continuing education, should seek and be given better access to newly developed knowledge, object information as input for a particular task, and engineering design process.

11.5.7.2 Goals for Completion of the Operator "Information for Designers"

Designers must always record good quality information according to the criteria in Chapter 9, to obtain it efficiently, in the shortest possible time, and to participate in developing and actualizing the information in an information system. Initiative, motivation, and support from the top levels of management are needed for designers to become aware of the newest insights. The upper levels of organization should therefore know about these sources.

11.5.7.3 Means for Completion of the Operator "Information for Designers"

Developing a special information system on any plane and in accordance with the situation of the organization, and its combination with the individual systems of designers, is an obvious demand. Only a member of the organization can obtain, organize, and submit the information about the organization's products in their life cycle phases.

Decisive for the construction of a lasting information system is EDS, as shown in Chapter 9. An information system is conditional on the cooperation of designers and staff members, thereby discovering a vast knowledge potential. For example, it should be possible for an engineering designer to generate a morphological matrix (see Chapter 8 and Figure 4.4) for the internal and transboundary functions of their TS-"sort," as part of the information system.

11.5.8 COMPLETION OF THE OPERATOR "MANAGEMENT OF THE DESIGN PROCESS"

Managing in relation to the design process is discussed in Section 11.3. Managing any process, including organizing, leadership, staffing, goal setting, resourcing, financing, and so forth, implies aiming for effectiveness, precision of goals, and economy. Managing has the character of leadership, one way must be selected from several variants, and it has the character of a regulation and control system, a predetermined goal is to be reached and disturbing factors and conflicts must be eliminated or reduced. Leadership must plan, execute or ensure (delegate) execution of the task, and enforce discipline. Without direction (self-direction or externally prepared direction) the engineering designers cannot reach their goals.

The competence of the design manager should contain qualifications for design engineering and for management. This depends on many conditions, for example, the size and structure of the design section and the quality of the team or project leaders, and so forth. It is necessary that the design manager understands the work of the engineering designers (see also [249,301]).

11.5.8.1 Status in Engineering Practice

With regard to the conventional design work (Figure I.20B), management has not been confronted with a systematic and methodical model of the design process in its whole breadth. For that reason, no relevant experiences can be reported, nor completion goals established.

11.5.8.2 Goals for Completion of the Operator "Management of Design Processes"

The goal here is to develop a new, well functioning design system, which would be flexible and would create a high design potential; see design situation (Chapter 3 and Figure 2.5).

11.5.8.3 Means for Completion of the Operator
"Management of Design Processes"

Measures in connection with the goals for the other operators of the design process include: (1) construction, composition, and organization of the execution system, that is, designers, their working means and environment; (2) determination of the goals and methods for progress, checks, verifications, and controls; (3) acquisition, documentation and communication of information, within and outside the organization; (4) leadership and motivation of the personnel (working climate); (5) organization and application of technical means; (6) level of special knowledge in the design section, and in the coordinated team; and (7) general organization and methodology of management actions, capabilities for planning, decision making, execution, and examination.

To realize the individual goals, management must decide on a strategy that corresponds to the situation of the organization and the determined priorities.

11.5.9 COMPLETION OF THE OPERATOR "ENVIRONMENT
INFLUENCES"—"WORKING CONDITIONS"

In comparison with the other operators, environment has relatively minor importance. Improvement in this area is quickly realizable, at low costs and with surprisingly good results.

Environment influences can be divided into macrosituations ("design situation," Chapter 3) and microsituations. The microsituation depends directly on each staff member in the organization, the set of conditions under which designing proceeds can be divided into physical and social–psychological parts, including motivation and creativity.

11.5.9.1 Status in Engineering Practice

Engineering designers will usually state that their working space could be improved in its size and placement, equipment, arrangement (distribution), embellishment (aesthetics), physical condition (light, noise, colors, climate, temperature, humidity, and hygiene), and its psychical–social atmosphere: relationships between workers, mutual understanding or intolerance and tension.

11.5.9.2 Goals for Completion of the Operator
"Environment Influences"

The goal to be considered is satisfaction of the design workers. This relates to influences on motivation (see Section 11.1), and opinions will not fully agree about individual problems.

11.5.9.3 Means for Completion of the Operator
"Environment Influences"

The questions are directed to design management; the problems belong to their competence. Much depends on the working organization, leadership, evaluation, reward structure, promotion possibilities, encouragement and support, formation of structures of teams, responsibility, and so forth.

11.5.10 REMARKS TO THE RESULTING CONCRETIZATION OF THE PROCEDURES

The goals and measures for individual operators of the design process are summarized in Figure 11.6, and reveal several impulses, for each designer, and for others in the organization. Quantifying these influences shows especially the dominant role of designers (see Figure I.23).

A recent study of "potential for success" of design methodology [63] is based on the opinions of 32 designers in different kinds of design processes. The conclusions agree basically with the insights presented in this book.

Concretization of procedures of designing, for example, based on Figures I.22 and 4.1, shows that the function structure is likely to be the same for all TS of a TS-"sort," the organ structure also remains identical, and some parts of the con-structional structure differ at most in dimensions (sizes). The procedure is simplified, but solving an engineering design problem with this procedure can still be complex.

For introducing systematic and methodical designing into the practice of the engineering designer, proceeding in at least two (or better in three) stages is recommended: (1) Process the relevant models and information (of transformation processes, technologies, TS, TS-properties, and so forth) from the recommendations in this book. The immediate purpose is to concretize them for the level of the TS-"sort," in suitably small substages. Because some of the SEDS (Chapter 7 and Section 12.5) that pursue the same goal are beginning to be built, this stage will be partly elimi-nated in the future. (2) Concretize and complete the practically constant organizational elements and the important design elements (see Chapter 3). (3) Respect the special conditions of the particular contract, and the available design potential that will solve the task (see Chapter 3).

Stages 1 and 2 can be judged as preparation. Their results for constant elements, developed over a longer time period, can be a starting point for the third stage, and transfer can be considerably shortened, and engineering design work can be continuously improved, as recommended in TQM [108,125,549] and ISO 9000:2000 [10] (see Chapter 8).

In the development of this system, constructional parts have a particular posi-tion (see Chapter 7). Engineering designer will recognize this area as "machine elements" (see Section 7.5). During study and application, designers will discover fundamental differences in the view of these elements, and in their assortment and selection (see Figures 7.12 and 7.13).

11.5.11 ACCEPTANCE BARRIERS FOR SYSTEMATIC PROCEDURES

Acceptance Barriers exist for systematic procedures (compare Sections I.4.1 and 11.1.9). It may be surprising that industry has not yet accepted the newer methods (including EDS), and generally does not even know about them [155,218, 220,426,427,456,510]. The term EDS is generally misunderstood. Yet the methods that industry does accept and use (e.g., TQM, QFD, Taguchi, see Chapter 8) are claimed as "industry best practice," and industry wants academia to accept these methods. An explanation for this delay in accepting "foreign" results (in both directions) is needed. Unfamiliar terminology, outside ones own experience, and use of familiar words in a different context make the transition more difficult.

Even the individual "industry-best-practice" methods are each used in only a small fraction of industry. Methods tend to be more useful for clarifying problems, and for conceptualizing solution proposals, in which active creativity may be essential. For embodying and elaborating, creativity is less essential, and more experience information is required, although many innovations can be implemented at this level.

Members of an organization must have "ownership" of the method, and adapting and championing a method is a difficult task for which time is usually not available. Unless a visible success is attained in the first few attempts at usage, the method is likely to fall into disuse almost immediately. Champions emerge when economically or politically powerful bodies ask for action: top-level executives, selling organizations (e.g., the "industry-best-practice" methods), standards ISO 9000:2000 [9,10] and ISO 14000:1995 [11], or by setting requirements.

When engineering designers meet a novel problem outside their immediate experience, more formal procedures and methods are needed. Such methods must usually be known in advance of the need to use them. Learning "on the job" is difficult, unless it is supported by management.

A good understanding of the method and its underlying theory is important for theory-based methods. Then the procedure that is prescribed or recommended for the method makes more sense, therefore producing less stress, and a better direction toward the goals.

One to two human generations are needed for a new insight (a change in the disciplinary matrix, a paradigm shift) to be generally accepted [376,377]. There is always resistance to change and to accept good points from other schools of investigation, especially if the expected leap is large [426,427] (see Sections I.4.1 and 11.1.9).

When a method is well known to the designer, it can at best be run from the subconscious, and the users can then even deny that they are using the method. It is necessary for engineering designers to learn methodology during their engineering education. Then the methods are familiar enough to apply, even if there is resistance from a supervisor. The beneficial results of teaching design methodology have been demonstrated [132,216,245,457], after 25 years of teaching, and after some graduates entered industry as engineering designers.

11.6 EDUCATION FOR DESIGN ENGINEERING

NOTE: Isaac Watts stated some insights into education [570], for example, as quoted in [183, p. 26, 27, and 38].

In most of continental Europe, higher education in engineering concentrates on "laying out" as a fundamental activity of a designer [156,157,160,161,170,171, 317,320]. Therefore, an engineering graduate as a rule masters individual branch directions and disciplines like mechanical or electrical engineering relatively well. Because designing is now involved with hybrid technical system and manufactures (e.g., mechatronics, nanotechnology, lean manufacturing), a new situation emerges. Engineers educated in a particular branch of engineering must also study other branches and disciplines, at least to obtain awareness and knowing, so that they can understand cooperating colleagues in a team.

In the English-speaking world, the focus (but in greater depth) is on engineering scientific analysis. Synthesis is not a simple inversion of analysis. Designers must also acquire the expertise of the particular industry sector in which they find employment. They must obtain information to achieve the "new properties" of the TP(s)/TS(s) such as ergonomics, esthetics, economics, marketing, and so forth, to understand causes and consequences. The special mixture of properties is again specific to TS-"sorts."

The aim of engineering education (see also [163]) should be to achieve *competency* of graduates in analyzing and synthesizing TP(s) and TS(s) [456,458] (see Section I.1).

A curriculum and teaching plan, and its execution in an educational process, should achieve the educational goals in a preplanned way through the choice of the educational material, and the teaching regulations. It should therefore define the subject matter, its volume, scope and detail, and its sequence. It should define relationships among the topics [165], and demonstrate these to students. Pedagogics [155,163], teaching/learning theory and strategy, and didactics, teaching/learning tactics and methods, show suitable kinds of instruction, "how to teach?," and recommend teaching media, "with what to teach?" Discussions about compatibility of the curriculum and the freedom of instruction, part of academic freedom, are needed.

11.6.1 PEDAGOGICS

Pedagogics as a theory base should provide a reason and guidelines for the contents (object information) and the methods (process information). The overall content for learning is "design engineering"; see for example, [145,147,150,159,160, 163,170,171,290,310].

"Learning" is used (1) as a *noun*, the sum of the acquired (internalized and categorized) information directly accessible to a learner, but also (2) as a *verb*, the process of acquiring information (compare Sections 12.2.1 and 2.2). Learning should be supported by a body of theory and by experience from practice.

Object information is usually examinable by relatively objective tests. Procedural (mental process) information tends to be learned and its use tends to become automatic, subconscious—as soon as it has been sufficiently well absorbed, the user is no longer aware of its use.

11.6.2 DESIGNING

The usual first difficulty in starting to design is where and how to begin, overcoming a natural fear of reaching into the unknown. Designing needs a variety of information, some of which must be internalized and understood, but much must be retrieved from external sources, interpreted and understood. Designing is also difficult because it needs an appropriate openness of mind, a capability to search for alternative solutions, creativity [153,276,315]. Design engineering is also a social activity, it influences the structure and operation of society (see Figure I.15), and takes place within a social structure of a "designing and manufacturing team," usually an industry, an organization. Education of engineering designers, should take these considerations into account—it is consequently more difficult than education for the analytical engineering sciences. The basic phenomena must be understood by developing a "feel" for them, but the human, social, economic, environmental, and other fields must also be understood. There are many opportunities for designers, if their capabilities are suitable (see Figure 11.13). Designers need time to become familiar with a TS-"sort" and the procedures of their industry branch, and organization.

11.6.3 Problems of Engineering Design Education

The subject in this enquiry is (1) the TP(s) and TS(s), (2) the design processes and designers, and (3) the educational systems, including the teachers, learners and organization.

Reducing the time, from the reported 10 years, for an engineering designer to reach adequate competency in designing is an important aim. In order to achieve the competencies [456,458] (see Section I.1), the aspects and parts of designing must be defined, including object and design process information, which (1) can be taught and learned from formal teaching, and from the experiences of experts; (2) can be learned from experience with designing, (a) with prior explanation of context, (b) with simultaneous explanation, (c) with *post hoc* explanation, or (d) without explanation; (3) can *only* be learned with personal experience (is practically unexplainable); or (4) cannot be learned, because it depends on genetic, social and cultural background (diffusion and infusion), environment, motivation, current situation, and so forth.

EDS [287,304,315], and design methodology [305,314] can provide a basis for designing and for education in designing [150]. Theories and methods of education, pedagogics and didactics need to be explored in the context of the teaching system, teaching and learning processes [170].

11.6.4 Teaching System, Teaching and Learning Process

A model of the teaching system can be derived from the transformation system (Figures I.6 and 5.1). The demands on design education can be shown with the model of the teaching and learning process (see Figure 11.7). Humans possess some abilities for design engineering, and should be brought from "lacking knowledge of designing" into a state of adequate knowledge.

As operators of this transformation, the teachers work with their technical means and their knowledge, in a certain environment, coordinated by the management of the teaching process. The "technology" for these transformation processes resides in pedagogics (as strategy) and didactics (as tactics), therefore the organization of the teaching/learning process.

Students are the operand of the teaching and learning process, their peculiar feature is their *active and reactive* role (distinct from the reactive role of most operands

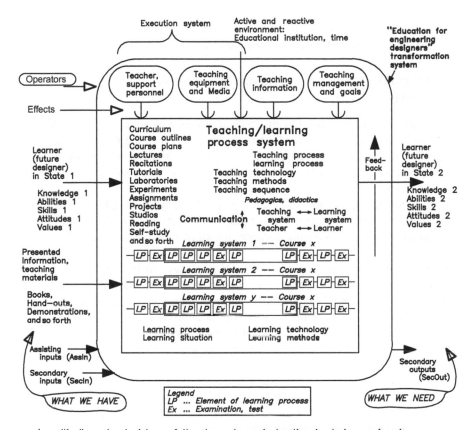

FIGURE 11.7 Model of the teaching and learning process.

in transformations), as living, thinking, and feeling humans with prior knowledge and experience. Their learning processes proceed during their studies of the subject matter, for example, object and process knowledge of designing, parallel to the teaching processes, or even outside them. Students possess different levels of knowledge, abilities, skills, attitudes, and values—although all have much in common. Many facets of an individual's behavior are gained through engaging in various social situations and through social interaction, of which teaching/learning in the classroom is one. People learn about things that are not explicitly taught, and many of those things are learned through simply being involved in situations and cultural activities. For that reason, pedagogics regards the learner as a "learning system," a psychostructure.

Teachers in group instruction, with their own psychostructure, are presented with several different learning systems, psychostructures.

11.6.5 Educational Theories—Theories of Teaching and Learning

Comenius (Jan Amos Komensky, Czech, 1592–1670), an educational reformer who revolutionized teaching methods for languages, is regarded as the founding father of pedagogics, the science of teaching. He recognized the need for repetition in learning, by seeing/hearing/using and "reflecting" on the same subject matter in different situations.

Important influencing factors in education were defined by Frank [206], following P. Weimann. They are collected into the six pedagogic variables shown in Figure 11.8 as the partial or contributing systems, compare *topika* of Aristotle in Figure 8.7.

1. Educational results—*why, for what?*—(1) purposes, goals, objectives of teaching/learning; (2) expected outcomes in knowledge, abilities, skills, attitudes, and values; (3) applicability, forms of measurement of outcomes. During learning, parts of the presented information will not be included in the learner's mental structure, and some knowledge, abilities, skills, attitudes, and values must be "unlearned" as obsolete, and to allow new learning.

2. Psychostructure—*who?*—(1) student, customer, user, operator, individual, team, organization, society—individual differences, academic abilities, prior knowledge and preparation, motives and incentives; (2) teacher, instructor, tutor, role model, and so forth.

3. Subject matter—*what?*—(1) nature, contents, arrangement of the presented learning materials and tasks, planned structure and content of the curriculum, and academic program; (2) form of presentation of the learning materials and tasks, as perceived by the learner—for example, a formal curriculum with courses, supervised practice, and apprenticeship.

4. Social structure—*where?*—environment, space, educational management—situation or environmental variables for learners, and for teachers, which may be subtle and complex in their effects on learning.

5. Media (learning and teaching means)—*with what?*—with what means, objects, tools, systems—equipment used by teachers and learners, including books, chalkboard, computers, and so forth.

6. Teaching method (teaching technology)—*how, when?*—with what procedures, processes, methods, strategies, tactics—timing, sequence, repetition—nature and quality of instruction, conditions of practice, guidance, modes of presentation (e.g., hands-on activities, reflection, and so forth), order of presentation ("from particular to general" and "from general to particular").

These pedagogic variables are correlated with the constituents of the teaching and learning system (see Figures 11.7 and 11.8). This model of learning can also

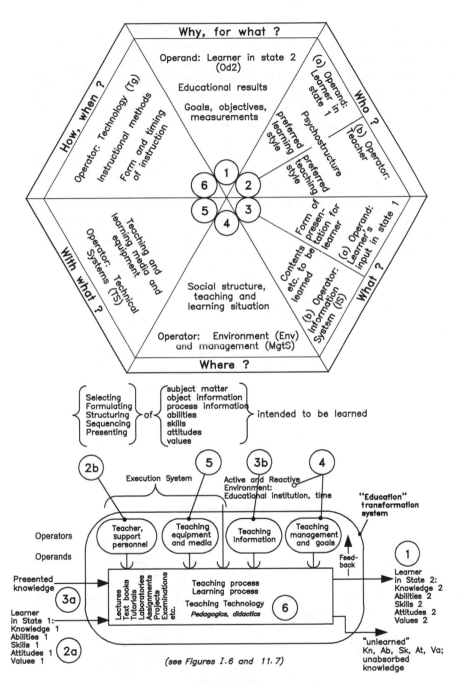

FIGURE 11.8 Teaching and learning system and variables.

Literature
Frank, H. (1969) *Kybernetische Grundlagen der Pädagogik* (Cybernetic fundamentals for pedagogics), Baden-Baden: Agis—Verlag
Taylor, F.W. (1919) *Principles of Scientific Management*, New York: Harper

be interpreted for autonomous learning, for example, from a text book or computer-delivered material. The "teacher" {(2) in point 2} prepares the subject matter in advance, based on the means and media (5) and teaching information {(2) in point 3} available—management (4) enters in the production process for these teaching materials. The technology (6) is then that the learner {(1) in point 2} searches out the available material {(1) in point 3}, studies it, reflects on it, and critically incorporates some relevant parts into his/her mental structures. In collaborative or cooperative learning, one learner at times plays the role of a teacher.

General considerations of pedagogics and educational theories, and of didactics as the tactics of teaching and learning, were presented in [156,157,170,171]. References included: structures of learning from Feldman and Paulsen [190]; skills, strategies, and events of learning as defined by Gagné [223]; descriptions of learning styles according to Kolb [367] (see Figure 11.9, and compare to the models of Wundt and Herrmann, Figure 11.3 [171]); learning patterns according to Perry [464,465]; influences of psychology and psychological type; taxonomies of educational objectives by Bloom [68] and Krathwohl [371]; Guilford's model of human intellect [242,243] (Figure 11.1); cognitive styles according to Messick [413]; constructivism [446]; learning styles from Reichman and Grasha [481]; cooperative learning as propounded by Smith [516]; formative evaluation and feedback to learners, and summative evaluation; roles of teachers and instructors according to McKeachie [408] and first-order principles by Boice [70]. Other concerns are about age-relationships—for example, it is now known that neurological development ends approximately at age 18. Influences of psychological type (of the teacher and of the learner) on education is the subject of [102–106], and is likely to influence performance in design engineering.

"We know much about stimulating and guiding learning, and need not wait for final or conclusive answers from experimental educational research" [547].

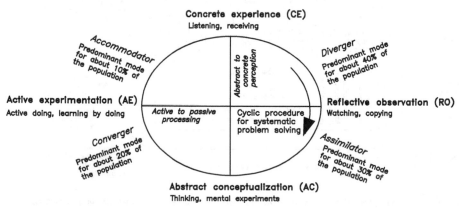

Literature
Kolb, D.A. (1984) *Experiential Learning: Experience as the Source of Learning and Development,* Englewood Cliffs, NJ: Prentiuce-Hall

FIGURE 11.9 Kolb's model of thinking styles for learning.

Tests and examinations at the beginning of an educational experience inquire about state Od1, tests at intermediate stages allow remediation (formative evaluations), and at the end determine state Od2 of the learner's knowledge, ability, skill, attitudes, and values (summative evaluation).

11.6.5.1   Goal of the Teaching/Learning System

For students, accomplishing state Od2, adequate knowledge and capability of designing, can be set as educational goal. This state must be clearly and accurately defined with reference to engineering design abilities. Each teaching process is a part of the total upbringing and these wider goals are also essential. Effectiveness as goal puts further demands on all process factors.

Design education is contained in the curriculum, and its methods must be adapted to that curriculum. The decisions about the curriculum—study goals and study plans—are the responsibility of the management of the teaching/learning processes.

The goals of the teaching/learning processes, especially for high education, can be characterized as "make learning possible," that is, transmitting of teaching contents, but also providing favorable conditions (see operator "environment") for the learning systems, the students [223].

11.6.5.2   The Learning Person (The Learning System) — Operand of the Process

The input to the teaching/learning processes depends in part on the desired output, that is, the goal of the teaching/learning processes. Prerequisite knowledge and personal, psychological, and biological characteristics, possessed by a candidate for admission into the teaching/learning system should be defined.

11.6.5.3   Effects of the Teacher—Operator 1

The teaching contents are set when establishing the curriculum. The selection of teaching content in the instructional situation is a didactic task of the teacher. EDS recognizes two kinds of teaching/learning content [150] (compare Figures I.4, I.5, and 1.6): (1) information about the object of designing, the TP(s)/TS(s), and (2) information about the design process, including methods, abilities, skills, values, and attitudes of engineering designers.

Design engineering puts particular demands on the teaching/learning process, and especially on the teacher. As shown in Figure 11.10, in order to teach effectively, instructors for design engineering should be familiar with current thinking about design, the "west" hemisphere, and with educational theories and methods [170,171], the "east" hemisphere.

The information about objects includes the general appreciation of TP and TS, their constituents, life cycles, properties, development in time, and so forth, (see Chapters 5 to 7). A more specialized kind of information about objects includes the branches or sectors of industry, and the academic disciplines—TS-"sorts" (e.g., cranes, pumps, turbines, domestic appliances)—which should also feature in

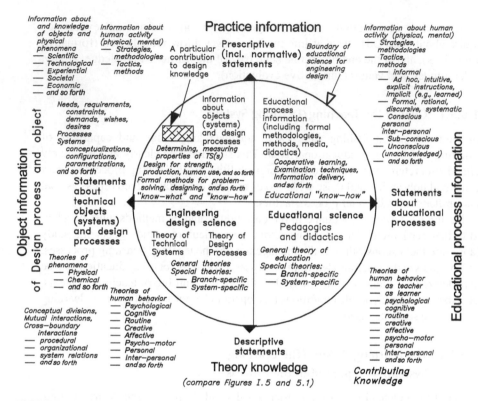

FIGURE 11.10   Model (map) of educational information for design engineering.

engineering design education. This should be in a matrix relationship to the engineering sciences, which deal with the isolated phenomena and principles, and their analysis. The TS-"sorts" and the engineering sciences should be treated from related points of view. The aspect in common is the design activity, which needs separate treatment, but which should be included in the science instruction and in design branches on a uniform basis. Only then will students fully understand designing.

Suitable teaching methods, and instruction methodology for an educational course or lesson in typical situations are important elements for successful education, and consequently one of the problems of pedagogics. An educational procedure can and should be developed from first principles. Design teachers currently either come to technical universities as proven engineering designers from industrial practice, or they have a postgraduate degree from a university without engineering practice, and probably neither have any education in pedagogical matters.

The decision about the kind of teaching presentation implies certain methods and roles of the teacher, especially concerning the communication of the teacher with students. Individual instruction involves two-way communication—teaching system with learning system. Group or parallel instruction also involves communication among the participating learning systems (students) — collaborative learning.

It is necessary to explore the roles of lectures, recitations, practice problems, and projects, and of individual, (Whimbey) pair, and team (collaborative) work, for presentation and acquisition of object knowledge and design process/methods information. Experiential and project-based learning are necessary, but not sufficient.

There is a need in teaching and learning to provide a theoretical explanation (by lecture, or by assigned reading); demonstration (e.g., by sample case studies; see Chapter 1 and [308]); problems and projects on relatively simple design tasks involving conceptualizing, laying out and detailing (preferably with supervision by an experienced and knowledgeable engineering designer who understands the theory); and more comprehensive design projects with progressively less supervision. This should be distributed in all levels of the curriculum, with cross-referencing from the engineering science and humanities courses. Nevertheless, various teaching presentations have different impact on students (see Figure 11.11).

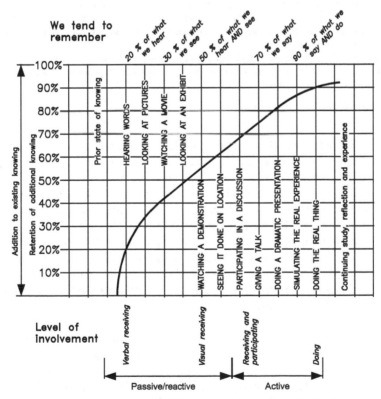

Typical activities in which students are involved and their relationship to the effectiveness of the learning experience measured by percentage of retention of the learned object and process knowledge.

A piece of wisdom credited to Confucius says:
Tell me and I will forget
Show me and I will remember
Involve me and I will understand
Take one step back and I will act.
These statements are best interpreted in combination:
Do all four and I will become competent.

Data from Cone of Learning, developed and revised by Bruce Hyland from material by Edgar Dale, University of Texas at Austin, 1987

FIGURE 11.11   Experience and learning.

1. The principle of TOTALITY demands that for every detail of the subject matter (contents for learning) the context relative to the totality of subjects becomes active. Every subject should be observed and developed from all aspects, should be supported by prior experience (of staff and student), and should indicate the scope of new tasks.
2. The principle of ACTIVITY should bring the students (the learners) into an internalized relationship with the subject matter, such that they are actively confronted by their task with their whole inner unity.
3. The principle of INDIVIDUALITY requires that the educational experience is fitted to the individuality of each student (learner), and should consider the uniqueness in time and situation of the active participation of the student.
4. The principle of CURRENCY is predicated on the fact that the formative action on the student (learner) depends on the current educational situation. The sitation must be considered in light of the subject matter (learning content and form of presentation) that appeals to the student, and under what subjective conditions that student approaches the situation.
5. The principle of RELEVANCE is satisfied when all details are brought into a conclusive and sensible sequence that reveals the truth of the subject matter. A logical closure can also be postulated that tries to achive a measure of completeness of the subject matter to provide a surveyable orienation and order.
6. The ACQUISITION OF KNOWLEDGE should develop the ability to approach new tasks, and to independently find ways of solving (open- and closed-ended) problems in a planned (methodical, systemaric) fashion.

FIGURE 11.12   Principles of education.

Even if no reliable method is available for design education, some educational principles are known, which can make possible the construction of methods (e.g., see Figure 11.12). The principle of repetition (Comenius, Section 11.5.5) is important.

For learning design engineering [455], (1) "what has been heard is not yet understood," (2) "what is understood does not yet give the ability to act," and (3) "being able to act does not mean being able to act optimally." Successful learning of object information and of design process information (including design methodology) requires education. This is more than transmitting information, and more than exposing students to design projects. Use of methods, and understanding their theories, must be exercised and practiced. Successful use requires experience of use, and capability to select the appropriate methods.

Regarding the sequencing of individual items of educational content be presented, the principles "from the known to the unknown," "from the abstract to the concrete" and "from the concrete to the abstract" are applicable. The interplay of the analytical (engineering sciences) subject matter, the humanities, and the engineering design process information need to be considered, including a possible and preferable attempt at "just-in-time" delivery of information, for example, [474].

11.6.5.4   Effects of the Teaching Means (Media)—Operator 2

Technical means of various kinds (e.g., media) can help the teacher to transmit the teaching contents, object, and process information, more clearly, vividly, attractively, and effectively. A new situation concerning technical means emerges if, for instance, skills for operating certain instruments is prescribed as educational goal, for example, engineering drawing with instruments or computers, calculating with computers, experimenting, simulating, and so forth. The possibilities offered by technical means are not always used to advantage, and some chances are neglected.

11.6.5.5 Effects of the Environment—Operator 3

Two classes of environment influences should be distinguished for the teaching system: (1) the direct environment of the teaching/learning processes, for example, a classroom in which the process is executed in material, psychological, and organizational aspects—this "atmosphere" is important for teaching success, learning; and (2) the conditions and influences of the macroenvironment—country, organization, and economic systems, including laws for the teaching/learning situation, and social or financial influences (student homes, scholarships, fees, loans, and school budgets).

11.6.5.6 Effects of Teaching Information—Operator 4

This operator represents the influence of pedagogics to accomplish the desired educational transformation [170,171]—instruction methods, didactics, pedagogical psychology, social pedagogy, and so forth. The information system also contains the subject matter to be instructed—what is to be learned, a part of the operand of teaching, together with the student in state Od1. Figure 11.13 shows a TS-life cycle, indicating the employment opportunities for educated and trained engineers, and interactions of transformations and designers with society.

The subject of teaching and learning for design engineering, especially for the activities of designing, is very complicated. The theories and the methods for both object areas (teaching and learning, and designing in engineering) and their relationships need to be explored, illuminated, and clarified. The goal is to develop a rational curriculum and set of teaching methods to educate potential designers, so that they can become effective as designers in the shortest possible time.

11.6.5.7 Effects of Teaching Organization
(Management)—Operator 5

The functions of management of the teaching/learning system only indirectly influence pedagogics and didactics. They include establishing the goals of teaching/learning through the curriculum and organization, including the choice of the operators (teachers, technical means, teaching conditions) and their realization, financial security, and operative leadership of the process. Pedagogics should be an important aspect in the decisions of leadership. The choice of teachers should consider their teaching capability, as well as their expert knowledge. Most engineering persons have not graduated from pedagogical education; it is important that part of continuing education should include pedagogics and didactics [171].

11.6.6 Curriculum Structure—Role for TTS and EDS

A proposed curriculum for design engineering is shown in Figure 11.14 [308], based on using EDS [315] as a central theme.

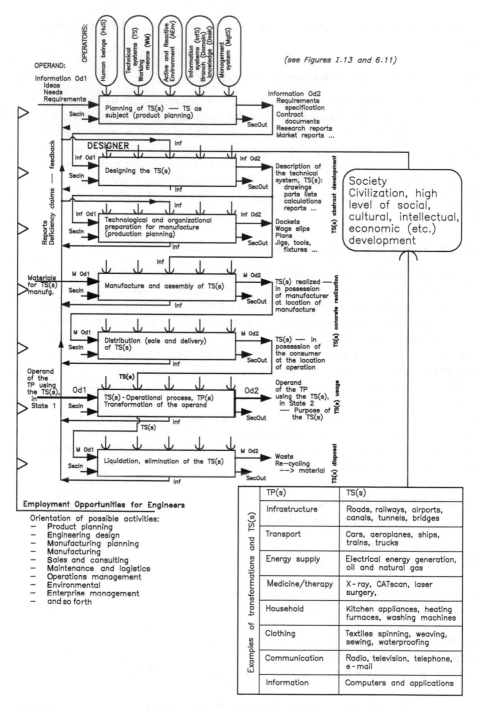

FIGURE 11.13   Engineering and designing in societal context.

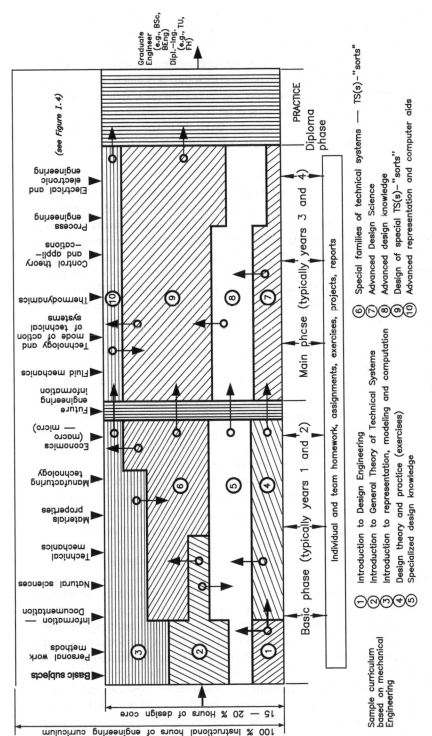

FIGURE 11.14 Model of curriculum for engineering and engineering design education.

The educational goals should include acquiring a design capability, a capability for engineering synthesis, a deep understanding of *technology* and *engineering*, their societal context, and the relationships among subject topics [162,165].

The range of information, knowledge, and data needed (at least at the level of awareness) by engineering designers to perform their duties is much larger and more interconnected than for almost any other profession (see Figure I.3). Interconnections among theories in this information should provide benefits to engineering practitioners for designing, diagnosing, analyzing, or other task [167]. The range of knowledge, relationships, and typology of design engineering [164] should be a major factor in engineering education.

Following an overview as introduction (item 1 in Figure 11.14), the continuous process of teaching/learning of design engineering (i.e., learning to design) is built on four pillars: (1) EDS—to transmit the general knowledge and skills of designing TP(s)/TS(s), items 2, 4, and 7—theoretical and practice-related object and design processes information. (2) Special TS-"sorts"—object-related information about important TS-"sorts" used as examples, and abilities needed for designing them, particularly items 6 and 9: (a) elemental engineering systems (see Figures 7.12 and 7.13), and the relevant theories to describe their behaviors; (b) exemplary TS-"sorts" with pedagogic value for design teaching (e.g., cranes, turbo-machinery, machine tools); and (c) application-oriented TS-"sorts," reflecting the specialty of the engineering school, item 9. (3) modeling—extended qualification for representation, experimentation, communication, information storage, and visualization, particularly with computers, items 3 and 10. (4) special information, see also Figure I.4—information, including data and knowledge needed to realize various properties of TP/TS, particularly items 5 and 8: (a) manufacturability of TS—design for manufacture, information about manufacturing methods, representing the TS (and its constructional parts) to be manufactured with full information about "design intents," acceptable tolerancing, and so forth; (b) suitability for transportation—design for transportation; (c) reliability of TS—design for reliability; (d) economy and marketability—design for economy and marketability; and (e) other properties (see Figures I.8 and 6.8 to 6.10).

The model indicates the relationships with arrows that show the material and temporal precedence coordination, for example, instruction in the "special theory of technical systems" (item 6) presupposes knowledge and understanding of the general theory (2). "Special knowledge" (5), builds on "introduction to TTS" (2). "Advanced design knowledge" (8) is conditional on the "basic design knowledge" (5).

Information and knowledge elements can be useful for some other courses. For instance, system theory, system engineering can act as prior knowledge in general and for the theory of technical systems, because both TP and TS are classes of systems. Fundamental mathematical, geometric, physical, chemical, and biological information are general inputs that must appear in each engineering curriculum; see [16], but also the methods of iterative approach to establishing the desired properties. Extended physical and chemical information depend on the specialized training, for example, some topics from mechanics. General technology can build on the theory of transformation processes (see Figures I.6 and 5.1), and can clarify the overview and technological classes, and the importance of technology in society. Basics of management should bring out the principles and forms of leadership/direction of the econo-manufacturing systems as special area for the student. Management of the design process itself, of the documentation of the captured "design intent," teamwork, social and cultural structure [162], psychology, motivation, and so forth is also necessary. For good understanding of TS-properties (see Figures I.8 and 6.8 to 6.10), the curriculum should offer other knowledge areas, for example, human factors, ergonomics, aesthetics, ecology, and so forth—properties class Pr5. The knowledge of these sciences offers the basic knowledge for the area of "Design for X." Basics of economics (macro and micro, property class Pr9) should be present early in the curriculum, so that students from the beginning include the economic system in their considerations. To understand the engineering working methods, especially of systematic designing, a fundamental knowledge of the psychology of thinking would be of advantage. In order to solve a problem in a concrete situation, the engineer must obtain and master the fundamental sociological, political, and legal concepts—especially important are patent laws, ethics, and personal responsibility.

All engineering activities can benefit from the holistic consideration of this book, and the "outputs" in all other subjects in the engineering curriculums gain in meaning.

It is possible and desirable to introduce the suggested changes in successive portions. The full effect of justifying other courses and subjects in an engineering curriculum, and giving guidance on systematic designing will only be realized once the whole scheme has been adopted.

Supervised project work [504], including demonstrations and self-directed project work is essential for learning to design, providing that it is coupled with the theoretical studies from this book. Self-study, using currently available media, means and facilities for communication is not sufficiently applied, and has not reached the stage of maturity.

For learners, topics to design education include: preferred and alternative learning styles; fitting any presented material into the learners' mental models or changing them; doing by learning and learning by doing; and motivation.

For teachers: they have two professions, engineer and teacher; being a role model for learners; needing to learn, interpret and organize information, including experience to formulate the presentations to learners, as providers of information, but not the only source; acting as tutors and mentors by giving sympathetic help; motivation (self and of learners); acting as judges and assessors (as friend or enemy?) in formative evaluation by giving feedback to learners, and in summative evaluations as end-of-course grading assessments.

It is at times important to change focus during designing. Four points of focus were mentioned: the problem (needed during problem definition—design specification), the developing solution (with alternatives, during conceptualizing, laying out and detailing), ideas (searching for solutions, in product development searching for potential products that can deliver a marketing advantage), and the line of thought (the engineering design process, and its structure).

11.6.7 Closure

The subject of engineering design education is extremely complex, and cannot easily be resolved. Curriculum changes cannot be made rapidly. If any change is made, a transition period must be allowed, that is, from the start of the changed course to the end of the graduating year for that cohort of students, 1 to 5 years, depending on the length of the degree program. Universities then require a period of stability, typically 5 to 10 years. This contrasts with the rapid change of knowledge in some fields of engineering. Wholesale change is neither possible nor desirable in the short term, but change is urgently needed.

Part V

Meta-Knowledge Related to Design Engineering

Information exists at several hierarchical levels, from the most abstract and general to the specific. The science about design engineering is more abstract than the relevant activities, it is meta-knowledge, which is outlined in this chapter. The information is mainly situated in and around the "south" hemisphere of Figures I.4, I.5, and 12.7.

12 Engineering Design Science

Outlining the systematic foundation for this book, based on Engineering Design Science (EDS) [315], is the goal. Two related themes can lead to rationalizing of designing: (1) the processes of design engineering and (2) the development of the TP(s) and TS(s). Both themes concern two aspects of information, the theory related and the practice related.

12.1 INFORMATION, KNOWLEDGE, DATA, AND SO FORTH

A first discussion may be found in the Introduction, Section I.8. Information exists in two forms, (1) as *records*, and (2) in *thought*. Information is usually transmitted as *signals*.

12.1.1 QUALITY OF INFORMATION

Correct and reliable information, at the right time, and in a usable form is currently one of the problems (see also Section 9.1.1). Today's situation is practically the same for design engineering, for research, and for management, and can be characterized by: (1) information and information media are over abundant and redundant; (2) only a part of the information is usable; (3) the life of some information is short; (4) terminology is not sufficiently unified or agreed; (5) abstracts and summaries of documents do not represent the contents well enough; (6) too little time is available to absorb, process, and understand information; (7) obtaining and processing information is costly; and (8) the user has unsatisfactory knowledge about documentation systems. Some important aspects of working methods in this area should be indicated.

12.1.2 PROPERTIES OF TECHNICAL INFORMATION

"Technical information" (including specialist design information) implies (1) recorded and documented information, knowledge and insights, and (2) remembered knowing, understanding, experience, and intelligence of an individual or a team. This information ranges from the laws of natural sciences to the data concerning production and other life cycle operations of technical process (TP) and technical system (TS). The *quality of information* (compare Section 12.1.1) depends mainly on: (a) correctness of contents; (b) intelligibility; (c) uniqueness, clarity, accuracy;

521

(d) availability at required time; (e) verifiability, source references, independent checking; (f) completeness, no sections out of context; (g) form; (h) type of information carrier/medium; (i) author and publisher of the information, reputation; (j) accessibility, ease of obtaining information; and (k) potential user group.

12.1.3 State of Information

The state of information is decisive for its applicability and possibilities of use, distinguishing

1. *Unfixed information*, in human memory, hardly accessible, likely to be incomplete, with little possibility of verification, may be recalled and usually incompletely returned.
2. *Fixed information*, recorded in an information carrier (medium, document), arranged in a well-defined way, generally accessible and unerasable.

Fixed information is only systematically accessible if it is properly documented and cataloged.

12.1.4 Classification System for Information

Documentation procedures should characterize the contents, for example, by cataloging the information carrier into classes of a filing and retrieval system. Classification systems are described in Section 9.1.3. These elements are not isolated; the sum of elements and relationships show the structure of the system of knowledge in general.

12.1.5 Working with Information

Two *strategies of working* with information can be distinguished.

1. *Systematically and methodically collecting* the information, when studying, reading periodicals, or visiting a technical exhibition.
2. *Problem-oriented searching*, when particular information is needed to solve a problem, for which engineering designers must usually: (1) question their own knowledge (knowing); (2) question their fellow workers—probably most used; (3) search in their own library; (4) search in other data banks or systems for information carriers, for example, reports such as [5,7,345]; (5) obtain a search from experts in an information service; (6) obtain a search from computer-resident files; and others.

All engineering designers should have sufficient *relevant knowledge* about data storage and retrieval (see Section 8.2).

It is useful for engineering designers and a design team to build their own information [282], study books and periodicals in their own area, and maintain awareness in other areas.

12.1.6 INFORMATION, TRANSMISSION, COMMUNICATION

A first discussion may be found in Section I.8.4. Any composite message usually requires a sequencing of messages. Communication may be active (sending) or reactive (receiving)—transmitting mediates between sender and receiver (see Figure I.12). Nonverbal communication contains body language, gestures, and implications.

Depending on the relative *locations* of the sender and the receiver, communication may be: immediate and direct; interactive and computer aided; and intermediate and indirect.

NOTE: Items of information may be classified from various viewpoints and manifestations [146]. The boundaries between classes are indistinct; the manifestations can reveal the context of information.

12.1.7 DESIGNING AND INFORMATION

Designing ranges from artistic to technical. Design engineering is a subset of designing and of engineering [145,314]; see Section I.10. The TS(s), and its TP(s), may range from simple to complex (see Figure 6.5). TP(s) and TS(s) may range from being novel, to being assembled in a known way from selected constructional parts (see Section 6.11).

TP and TS may involve a variety of information, from a single discipline (e.g., principle or branch of science), or from many. TP(s) and TS(s) may range from being easy, to being difficult to design. The design process itself may take on various forms with respect to information support (see Figure I.20). To be effective, an engineering designer requires information and knowing, and practice in design engineering—experience. Important elements are the following

1. *Knowing about TP and object systems*, TP and TS, in general, and in particular forms, awareness of and ability to find information; the knowledge exists as: (1) theory of TS [304], TTS—general for technical object and process systems, and specialized (branch specific); (2) engineering sciences of individual properties, scientific/mathematical methods for analysis of existing or proposed TP and TS, knowledge about natural phenomena; (3) other information about various properties, for example, the influences of neglecting some features or phenomena; (4) information about how properties can be achieved [314]: (a) designing for needed processes that transform operands (see Chapter 5); (b) designing for individual properties of TS (see Chapters 6 and 7); (c) designing to include all necessary constituents of processes and systems (see Chapters 5 to 7); (d) designing for life cycle of a product (see Chapter 6); (e) state of the art in the field.
2. *Knowing what to do*, and how to achieve results (see Chapters 2 to 4): (f) theory of design processes [287], and TP and TS models; (g) methodology related to the theory [314], *strategic* approach and system of methods to rationally help designers without restricting their freedom of action and creativity, with examples of usage [305] (see Chapter 1); (h) a set

of *tactics*, techniques, procedures, and methods for individual operations of designing [111,335], for example, problem solving, evaluating, creativity, and so forth (see Chapter 8).

3. *Knowing the conditions and tools* that can effectively and efficiently support design engineering, working means, for example, sketch pads, drafting equipment, computers, and so forth.

12.1.8 INFORMATION, INTERNALIZED, TACIT

A first discussion may be found in Section I.8.8. The attention span of humans, and their working (short-term) memory capacity, is limited [101,416–419] (compare Section 11.1). A need for sketching and modeling exists to enable communicating with oneself, and for team discussions (see Section 11.2). Comments by one participant often trigger mental associations for another. Knowing is not neutral and value-free [446] it is individually constructed and shared by direct and indirect *dialog* among people, depends on the context, and concerns the TP, the TS, and the design process.

12.1.9 REASONING AND INFERENCE

Reasoning is falsely identified with discursive deductive inference [174]. Discursive reasoning is a step-by-step process, which also appears with nondeductive reasoning. Discursiveness is apparently opposed to intuition, and is equated with rationality, but this excludes intuition from rationality. Reasoning in science and technology contains and requires rationality and intuition to proceed; intuition is indispensable.

Rational inference may use categorical syllogisms or propositional syllogisms; only propositional syllogisms are considered, especially implication. Best known are the *modus ponendo ponens* and the *modus tollendo tollens*; only the first is considered.

Let **p** and **q** be propositions, from which it is known that **p** implies **q**. The syllogism then reads

Premiss   **p** → **q** (always if **p** then **q**)
Premiss   **p** (**p** is true)
Conclusion   **q** (**q** is true)

This inference is a form of *deduction*. To know that "always" applies, every occurrence of the phenomenon would need to be observed—an impossibility.
The inference

Premiss   **p** → **q**
Premiss   **q**
Conclusion   **p**

is forbidden by deductive logic, yet it occurs continually in daily life, in medicine (inferring from symptoms to the disease), in jurisdiction (inferring from clues to the criminal), in astronomy (inferring from the cosmic background radiation to the big bang), and so forth.

The inference of *innoduction*

Premiss	**q**
Conclusion 1	**p**
Conclusion 2	**p → q**

seems unbelievable, yet it is the pattern for innovative designing, and presupposes intuition.

The inference of *induction* is the process of constructing scientific laws and theories. The premisses are several observations; the conclusion is a general statement as an implication:

Premiss 1	**p1 → q1**
Premiss 2	**p2 → q2**
Premiss n	**pn → qn**
Conclusion	**p → q** (always if **p** then **q**)

In strict deductive logic, inferring from the particular to the general is forbidden, yet the development of science depends on it. In daily life, management, and design engineering, induction is normal, which implies that *applications* of science must be heuristic [365].

Science and daily life also depend on *abduction* and *reduction*, simplifying the situation so that it may be better understood, and can be synthesized into a more *holistic* view.

One of the tasks of EDS is to explore the reasoning patterns needed by designers, how to use them and how to safeguard them from errors. This safeguarding function is in part performed by engineering design methodology, and the engineering design methods.

12.2 SYSTEM

A first discussion may be found in Section I.6. Defining a terminology is not popular, choosing suitable technical terms is difficult, and reading definitions is usually resisted. This step is unavoidable to reduce misunderstandings. Criticisms come easily from people who have chosen a different term to express a similar content. This can be avoided if the opinions are confronted and discussed, and unity can be reached. A tradition exists of intuitively accepting and retaining some terms, or coining new terms from analogies, almost without questioning, and without exact definition. For example, the terms "machine" and "engine" have differing content— for example, "engine" can be a motor for automotive machinery, a device for regular scribed scale rulings, a search program for computer information retrieval, or other items.

Some terms are only used in one scientific field, for example, buckling resistance. Other terms are used in normal language, and have imprecise or multiple interpretations. In a technical or scientific field, they are either used with a defined content, for example, system, tension, or have different meaning, for example, sleeve, bush, tooth, ram. The difficulties are increased by different meanings in other languages.

The selection of technical terms for this book was based on the following principles.

1. Use of existing expressions in their conventional meaning, but more precisely defined.
2. Connections with general and engineering sciences, including mathematics.
3. Use of an international set of words to help attain international understanding.
4. Consistent use of the same word for the same meaning.

Other literature has helped in the search for terminology and definitions [251,292,493,498]. Various terms are accompanied by symbols or abbreviations that assist formalization, and as a short-hand notation, see the list of symbols. In Section 12.2.1, the terms regarding "systems" are placed in the sequence in which they are needed for defining other terms of design knowledge, for example, "set" helps to define "system." Some attempts have been made to formalize "design" [493,588–591]. In this book, stringing two or more nouns together is avoided, because ambiguity may result from complex nouns. The combinations of nouns listed here are deliberate exceptions, they are intended as composite nouns, and could be hyphenated. Sentences are structured such that any activity is referred to by a verb, for example, designing.

12.2.1 FORMAL BASIC DEFINITIONS

Some formal definitions were published in [304, pp. 241–261].

Set: A *set* is a collection of objects, thoughts or opinions (its elements) included in a conceptual whole, delimited by an agreed boundary. The relationship of interest is "belonging."

System (S): A *system* is a finite set of *elements* collected into a whole under well-defined rules, whereby *relationships* exist among the elements, and to the system's environment.

The terms "element" and "system" are *relative*. An element can be regarded as a system, and a system can be regarded as an element within a larger system, that is, systems are hierarchical, at different levels of complexity (see Chapter 6). Each system can therefore be studied at different *levels of discrimination*, which depend on the specific field of engineering (or science, sociology, history, art, etc.), on the abilities of our sensing organs, and on the technical means available to assist them. Other terms are always connected with the term "system," for example, purpose, behavior, structure, environment, input, output, property, and state (see Chapter 6).

System—purpose: Every artificial system serves a purpose, described with reference to a goal system, for example, an intended purpose. A *goal system* is a set of goals and their relationships. Goals may be hierarchically related to one another, as "super goals" and "sub- (or partial) goals."

System—behavior: *Behavior* is an inherent property of a system, successively attained states caused by internal or external, short- or long-term influences. Goal systems or systems of terms (terminologies) have no behavior. The purposeful behavior of an object system is termed its *purpose function* or teleological function [493] (see Figure 6.4). "Behavior" can be applied in a broader sense than "function," which is connected with a desired action.

System—structure (Str): A structure is the union of the set of elements and the set of their relationships to one another and to the environment of the system. Elements may be conceptual, abstract, or tangible. Relationships may be deliberate or accidental, desired or undesired, and definite or undefined. "Structure" is usually applied to the internal arrangement, configuration, conformation, constitution, or construction of a system. Structure and behavior (where applicable) are the most important *properties* of a system (see also Chapter 6).

System—relationship between behavior and structure: The behavior of a closed system with a defined structure acting or reacting to given circumstances is determined by its structure(s). The behavior of a system can be predicted only within (statistical) limits due to uncertainty, and to how well the circumstances can be assumed and modeled. Conversely, a behavior *does not* determine a unique structure. The same behavior can usually be achieved and realized by several different structures.

System—environment (Env): The *environment* of a system **S** is everything that is not included in that system. Practical considerations restrict the environment to all systems (external to **S**) that have at least one element: (1) its output must be simultaneously an *input* to an element of the system **S**, or (2) that receives as its input an *output* from an element of the system **S**. This "close" environment is termed the *active and reactive environment* (AEnv) of the system **S**. The "remote" environment is termed the *general environment* (GEnv), and contains the geosphere, atmosphere, biosphere (including humans), technosphere, and astrosphere.

System—input (In), output (Out): The inputs and outputs (quantities) are connections between a system and its environment, that is, across the system boundary. Inputs and outputs are couplings, desired and undesired, and so forth, which consist of M, E, I. An *input* represents the external relationship from the environment to the system, entering at a *receptor*. The input can be an effect exerted on the system, a coupling, or the state of a property of a system; primary (main), assisting or secondary; and supporting, neutral or disturbing for the system. The totality of inputs is termed the total input and is represented by an aggregate of individual inputs. An *output* represents the external relationship from the system to the environment, presented at an *effector*. The output can be an effect delivered by the system, a coupling, or the state of a property of a system; primary (main) or secondary; and beneficial, neutral or detrimental (usable or waste) to the environment. The output of a system is the set of those outputs of all elements that do not constitute inputs to other elements of that system. A particular case of output delivers a suitably transformed effect to an input to the same system, as an external *feedback*.

System—properties (Pr): A *property* is any characteristic, attribute, or quality that is possessed by and describes or characterizes an arbitrary object, a system, and each of its elements and relationships (see Chapter 6). These are an integral and inherent part of the system, and define it more precisely. Each property can appear in various states, differences between systems occur in the states of particular properties. Properties exhibit various manifestations (qualitative properties), and some have values and units of measurement (quantitative properties). No existing object is without properties, that fact would be its property.

System—evaluation: A combination of the states of a selected number of properties and their manifestations and measures characterizes a system as accurately as possible, for the purpose of *evaluating* it (see Chapter 6). Partial evaluations consider a limited range of states or properties, and results are termed *partial values*. These may be aggregated in a suitable way to a *total value*, as a basis for judging the quality of that system.

System—state: The total of the states of all properties of a system at a given time is termed the *state* of the system, which can be regarded as an aggregate of the individual states of properties (see Chapters 2 and 6). Two systems can be equivalent; any change in measures between the two is termed a *difference*. A difference may exist between two (static) states of a system, or during the (dynamic) transition from one state to another, when the difference can occur differentially (if the resulting transition is continuous) or in discrete steps.

System—model: The generalized model of a system is shown in Figure 12.1.

System—types: Some of the important classes of criteria are

1. By the position of the system in a hierarchy: super system, system, partial or subsystem.
2. By connections to the environment: open, with defined environment, and at least one input or output; or closed, with no connections to the environment.

NOTE: This is different from the usual mathematical modeling of the dynamics of a system: an open-loop system has feedback by human operations and a closed-loop system has internal feedback.

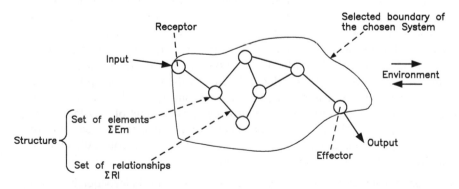

FIGURE 12.1 Model of a system.

3. By the system's changes of state: dynamic, its state is variable with time; or static, its state is constant in time (a relative statement with respect to the observed properties).
4. By the system's anticipated behavior: *deterministic*, a statement about the behavior depends only on the states of the system, which is *causal*, a clearly defined cause results in a predictable (but not certain) consequence, and a clear end to the process; or *stochastic* (probabilistic), only prediction of the variability of behavior is possible, the system is *emergent*, the processes are continuous and usually equilibrium seeking.

NOTE: In any system both types can be simultaneously active; usually if the macroprocess is causal, the microprocess could well be emergent.

5. By concreteness of the elements: concrete, the elements are tangible objects or operations; or abstract, the elements are intangible.
6. By its origin: natural or artificial.
7. By the dependencies of its outputs: combinatorial, the outputs depend only on its inputs; or sequential, the outputs depend on its inputs and other quantities, synergics may exist.
8. By the complexity of the system's structure: extremely complex, for example, a brain, a country's economy; very complex, for example a fully automatic factory, an organization's economy; complex, for example, a passenger car, a city library; and simple, for example, a family library, a screw.
9. By the types of elements: object (object system), elements are abstract or tangible things; or process (process system), elements are operations and activities.

System—types of task: Three types of tasks may be asked with respect to systems:

Analysis: given a structure, find its behavior and other properties.
Synthesis: given the desired behavior and other requirements, find a structure that satisfies the behavior and requirements—usually one of several possible structures.
Black box problem, Identification: given a system, of which the structure is unknown or only partly known, find its behavior (and its inputs and outputs), and possibly its structure.

Analysis is a one-to-one transformation, and is in some ways an inverse of synthesis. Synthesis goes far beyond an inverse of analysis, it is almost always a transformation with alternative means and arrangements, a one-to-many (or few-to-many) transformation. Synthesizing is the more difficult kind of action. A generally held conviction wrongly claims that all synthesis is "creative" and "intuitive"—yet many methods can help in synthesis (see Chapter 8). The term "creative synthesis" should be used only for new and previously unknown results of synthesis, for example, radical patents, or a synergistic formation.

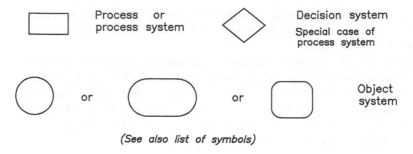

(See also list of symbols)

FIGURE 12.2   Graphical symbols for two types of system.

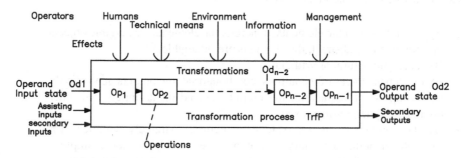

FIGURE 12.3   Model of a transformation process.

System—symbolic representation: A system is represented by shapes. Different symbols are chosen for process and object systems (see Figure 12.2) and the list of symbols.

Process (P): "Process" refers to a procedure, a transformation, an activity, a happening, a change in something during a period of time; see ISO 9000:2000 [9], article 3.4.1.

Changes continuously occur in nature, for example, the evolution of living matter, plants, animals, and so forth. Objects that appear to be stable (a mountain) suffer change by erosion, weathering, aging, and so forth.

Humans organize various *artificial* processes to achieve the changes that they think are necessary or desirable; *transformations*, consisting of *operations* (see Section I.9.1).

Operand (Od) is used for objects that are changed, animate and inanimate materials, energy and information (M, L, E, I), including data, commands, understandings, and so forth. An operation within a transformation is the consequence of an *effect* interacting with an operand. Each effect is based on physical, chemical, biological and other phenomena, and is described by a (fixed or flexible) rule, prescription, recipe, principle, regulation, algorithm, or *technology*. A set of *operators* contribute to exerting the active and reactive effects necessary for performing the transformations (see Chapter 6).

The model of a transformation process is shown in Figures 12.3, I.6, 5.1, and 5.4. The effects of the operators consist of M, E, I. Further details about transformation systems, and about processes may be found in Chapters 5 and 6.

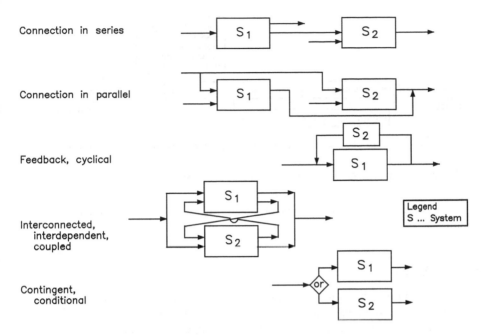

FIGURE 12.4   Types of couplings between systems.

NOTE: In mathematics, an operand, the input state, is transformed into a result, the output state, by applying various operators, including equals, plus, minus, multiply by, divide by, differentiate, integrate, sin, cos, tan, and many more.

Relationship (R): A relationship is the real or meaningful dependence or interaction among two or more objects or phenomena of abstract or concrete kind. Relationships connect individual elements into a system. Several types exist, for example, coupling types are shown in Figure 12.4, and coupling can be with respect to M, E, I.

Regulating, controlling: "Regulating" and "controlling" are two *types of process.*

1. *Regulating,* where one or more magnitudes, as input values or set points, influence a magnitude as output value or controlled variable, using the governing rules of the system. It is characteristic for a causal chain to produce results.
2. *Controlling* (see Figure 12.5) is a process where the output magnitude is monitored or sampled, and compared to a desired magnitude, a standard, or a set point. Depending on the result of the comparison, the input magnitude is influenced to cause a subsequent more favorable comparison of the output values. This results in a control loop closed by feedback, either by a human operator (open-loop control), or by a TS (self-acting, closed-loop or automatic control). The system may be stable, critical, or unstable.

NOTE: Firing a brand-new sports rifle for the first time at a target requires regulating; there is no previous reference to predict the properties and behavior of the system. Firing a second

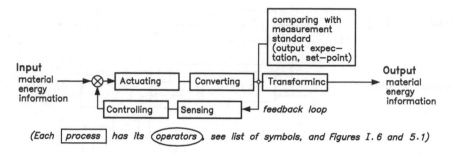

(Each | process | *has its* (operators), *see list of symbols, and Figures I. 6 and 5.1)*

FIGURE 12.5   General control system (transformation process).

shot, and adjusting the rifle sights, are control actions. Regulating is usually subsumed into controlling.

12.3 SCIENCE

A first discussion may be found in Sections I.1 and I.9.2. Both a *reductionist* view and a *holistic* understanding of a system are needed for the full implications of the system to be discovered. Equally, both synthesis and application of information must be accompanied by analysis.

NOTE: The maturity of a science about a body of knowledge (e.g., knowledge about designing) may range through: (1) description of phenomena, natural history phase; (2) categorization in terms of apparently significant concepts, taxonomy, Carolus Linnaeus (1735); (3) ordered categorization whose pattern may be deemed a model, the evolutionary taxonomy or periodic table phase; (4) isolation and test of phenomena, with implied reproducibility by independent observers; and (5) quantification, classical physics phase in which mathematical relationships are formulated [240]. Even the first is acknowledged as "science," for example, biology. Those areas of knowledge that reach level (5) still need the other levels. Many areas of science cannot reach level (5). EDS cannot even reach level (4) (see Figure I.5), because humans are idiosyncratic and have their own decision powers. Some knowledge used for design engineering exists at high scientific levels and is used for analysis, but choices and values are human.

Each *science* is a system of ordered knowledge about a bounded region, which includes theories and hypotheses about the region and its behaviors. Science and scientific information are often regarded as "value free," yet the values of the researchers and their funding agencies and employers are reflected in the resulting information [446]. The boundaries between sciences are fluid, ill defined, with overlaps, and their maturity will differ.

12.3.1 Science Hierarchy

Depending on the subject, sciences can be arranged into a hierarchy (see Figure 12.6), where the more specific levels should inherit the features of the next more general level.

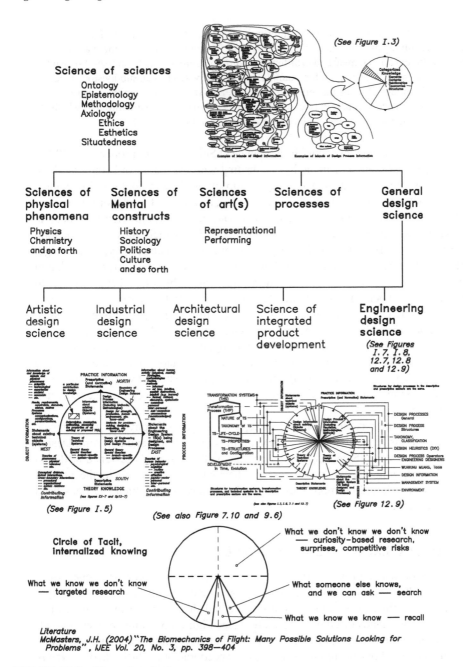

FIGURE 12.6 Hierarchy of sciences.

NOTE: Sciences can be characterized by different criteria that are used to hierarchically sub-divide the science, compare Figure 12.6. The natural sciences divide into biology, physics, chemistry, and so on. Biology has subdivisions of botany and zoology, and zoology has subdivisions of ichthyology, herpetology, and so forth. The hierarchical arrangement is not unique, a different order of applying the criteria leads to a different hierarchy. For Newtonian mechanics, applying the criteria "static" and "dynamic" before or after fluid, thermal, and solid leads to a different arrangement and to different relationships among the subclasses. Some interdisciplinary regions also develop, for example, biochemistry. Each such subdivision "inherits" a substantial amount of its characteristics, structure, and knowledge from the superior level. The relevant science at an appropriate level provides the theoretical structure into which the available information can be categorized (see Figures 12.9 and 7.1).

The most general level should be a Science of Sciences, with properties of [279]: (1) *ontology*, that branch of metaphysics dealing with the nature of being, the theoretical philosophy of being and knowing, a philosophy of the mind; (2) *epistemology*, the theory of the method or grounds of knowledge, its origins, foundations, limits and validity, including the degree of acceptance; (3) *methodology*, the science of method, relationships among subject, theory, and method [351], which allows classification and prescriptive formulation of the body of methods used in an activity; (4) *axiology*, the theory of values, advocating value judgments, which includes: (a) *ethics*, relating to morals, treating of moral questions, moral correctness, honor, and conformance to standards of good behavior and (b) *aesthetics*, appealing to the senses, conscious and subconscious perception and apperception (see Section 11.1.3), cognitive processes; and (5) *situatedness* [96,227], interactions among three "realities": (a) an external world; (b) an interpreted world obtained by interrogating, perceiving, and interpreting the external world to formulate concepts of existence; and (c) an expected world obtained by hypothesizing and designing an anticipated external world, formulating goals, and taking action to realize that world based on a theory of the expected outcomes [351].

Sciences at the next level should include art in general, physical phenomena, mental constructs (e.g., history, sociology, politics), processes, and a general Design Science [279]. Design Science(s) are concerned with "designs" (designed artifacts) and with designing (design processes) as their subject. Attached to the general Design Science should be more specific Design Sciences, of which EDS has been developed [315].

12.4   ENGINEERING DESIGN SCIENCE

A *system of knowledge* for and about design engineering has been developed since about 1960 [281], by observing, researching, and theorizing based on design experiences. Before this, information about objects and their engineering sciences had been codified as islands of object information (see Figure I.3). Some process knowledge has been published as individual methods, but hardly integrated into a scheme. The information about designing was largely anecdotal, with little theory. Newly developed knowledge was needed to perform the transition shown in Figure I.3, a purpose of EDS [315]. This science about designing of TP(s) and TS(s) should present: (1) a conception of relationships between the processes of design engineering and the objects being designed, especially causal connections; (2) a system of laws or principles to define a paradigm of action that supports intuitive and emotional procedures—including a procedural model (see Chapter 4); (3) a sphere of speculative thought or doctrine; (4) abstract information, as contrasted to practice; and (5) relationships among subject, theory, and method (see Figure I.2), as one of the guiding premises of EDS.

This system of knowledge, EDS, should be of general utility, particularly if the design problems are situated in the more abstract levels of modeling, the phase of conceptualizing for designing novel TP(s) and TS(s). A systematic and consistent procedure without theoretical discontinuities, jumps or intuitive leaps, is best for finding an optimal solution in minimum time, with a reasonably high probability of success.

Intuition has its place among the designers' tools (see Section 12.1.9). Yet systematic work wins out unless the problem remains at the concrete and routine level, from the dimensional layout into details. At these concrete levels, the trained human mind tends to be quicker if left to itself, using intuitive working modes, and learned object and design process knowledge, without the need to follow a prescribed procedure. However, the human mind is unreliable (see Section 11.1), and therefore a systematic procedure based on EDS, and a preferably independent check (design audit, systematic reflection) must be implemented, both on the process of design engineering and on the results. Proven experiences are valuable and not to be ignored. They will need to be reorganized to effectively help engineering designers, and it is necessary to close any gaps that may emerge. This implies an enrichment of knowledge, which itself brings new ideas, and inspires designers.

12.4.1 STRUCTURE OF ENGINEERING DESIGN SCIENCE

Perceived knowledge of branch-related objects and of design processes is of differing quality and exhibits varying possibilities for systematizing, indicated in Figure I.3.

NOTE: In EDS, the information is systematized using a morphology (see Section 4.4.2 and Chapter 8). Characteristics of statements of design science, and their manifestations:

1. Methodological category of statement: (1) primarily descriptive: theorizing, not just a narrative; (2) primarily practice related: (voluntary) prescriptive, and (compulsory, obligatory) normative.
2. Empirical support for statement: (1) prescientific (practice—experiences); (2) scientific: singular "understanding"; and (3) scientific: inductive, deductive, innoductive, abductive—statistical.
3. Recipient of statement (typically): (1) novice, student; (2) practitioner, engineering designer; and (3) teacher, researcher. This is better regarded as a continuum, not discrete manifestations.
4. Aspects of designing: (1) the tangible TS and its TP, the engineering product as it exists; (2) the process of design engineering, and object or process-related information to assist designing.
5. Range of objects as subject of statement: (1) universal: all artificial real and process systems; (2) tangible TS and TP; and (3) branch-specific reference objects of engineering (see Chapter 6): machine systems, building systems, and so forth, constructional parts (see Chapter 7).
6. Experience and status of author: (1) branch or discipline of the authors; (2) position in organization; and (3) primary activity: practice, research, education.

7. Declared aims of researcher: (1) "automation" of (all or part of) a process of design engineering; (2) better empirical foundation for a methodology or method; (3) others.
8. Other classes.

These classes indicate that many SEDS can and should be formulated, in addition to this general EDS (see Chapter 7). Two viewpoints, items 1 and 4, in this morphology seem to be important [303,304,307,309,315], as shown in Figure 12.7, compare Figures I.4 and I.5.

The aspect of designing, item 4, is represented in the "west" by the *operand* of designing—the TS and TP as it exists, the "as is" state, and in the "east" by the *design process*, including any TP- or TS-related heuristics, the "as should be" state. The resulting hemispheres are shown in Figure 12.7 (I.4 and I.5) as subsets of EDS. A second division, the methodological category, item 1, distinguishes *descriptive* statements (theoretical—"south"), from *prescriptive* (practical/advisory—"north") and *normative* statements (practical/compulsory/obligatory, regulative). The contents of the two "southern" quadrants are clear from the descriptions in Figure I.4. The "northeast" quadrant contains methods and heuristics based directly on EDS; other methods and heuristics reside around this quadrant. The "northwest" quadrant contains typical classes of properties and other TS-related information derived from EDS applicable to a particular TS-"sort."

EDS intends to provide a classification framework for information, and shows the relationships of the included information with other areas (see Figure I.5). The descriptive (theoretical) knowledge is presented in this book as a set of interrelated models. Some relationships are indicated in the figures, and illustrated in Figure 12.8.

The descriptive and prescriptive information are best structured in the same way. The respective structures of information in the "north" and "south" quadrants within each hemisphere can be identical (see Figure 12.9). The terminology may be different in the related quadrants. For example, the theory of properties is the descriptive (theoretical) basis for the prescriptive information of "Design for X" ("DfX"), and provides a structure for this information.

With respect to the quality of statements attached to the prescriptive regions (the "north" hemisphere), the statements should be derived from, or at least coordinated with, the descriptive regions (the "south" hemisphere). For TP and TS, and for design processes, the triad "subject–theory–method" should be used, compare Klaus [351] in Section I.1.

NOTE: If the information about a *subject* has been well abstracted and codified, that is, the *theory* is well formalized, even at the lowest level [240] (see NOTE in Section 12.3), *and* the prescription of a *method* is derived from the theory, then the method qualifies as part of the science. Other methods, and their adaptations or applications, should be located outside the boundary of the science. This book contains mostly mixed descriptive and prescriptive statements.

A misinterpretation of *theory* occurs in the opinions of theoreticians who claim that a phenomenon "is governed by theory." On the contrary, every theory is governed by the phenomenon [152]; see also the NOTE in Section I.5. A more important

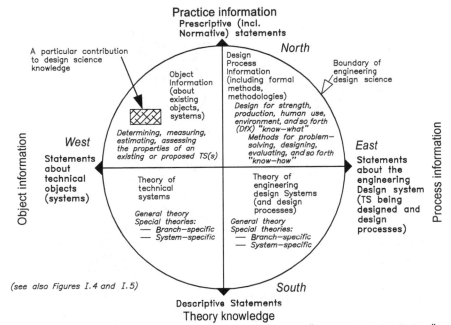

Practice information
Prescriptive (incl. Normative) statements

(see also Figures I.4 and I.5)

Descriptive Statements
Theory knowledge

— Descriptive (theoretical) statements about technical systems: The "Theory of Technical Systems" describes, explains, establishes and substantiates the structures, their elements, properties (and quality, modes of action, functions, etc.) of existing technical systems.
— Prescriptive statements about technical systems: The statements about actual manifestations and values of properties of existing or proposed technical systems.
— Descriptive statements about the design process: The "Theory of Design Processes" describes, explains, establishes and substantiates the elements, properties, sequences and effects (actions, results, successes) abstracted from actually observed (including intuitive) and possible (including systematic and methodical) design processes in their socio–technical context, including all company–related, operational, organizational and leadership aspects.
— Prescriptive (voluntary and normative) statements about designing of technical system: The "application and use–related design information" (domain information in its narrower sense) contains the know–how regarding ways of satisfying (realizing, achieving, synthesizing) the functions in a future technical system, i.e. the knowledge about the ways and means in which technical products can and must be laid out and detailed in concrete forms to fulfill the required functions. These must be verified by suitable analysis and audit. The statements find their theoretical foundation in the engineering sciences, in the collected heuristic information from practical experience, and by a further step in the pure sciences which provide the connections to the descriptive statements about technical systems.
and Prescriptive (voluntary and normative) statements about the design process: Design process information contains indications about all factors of the design process, and their technologies. Of particular importance is the "methodology of designing" (design methodology, technology of designing) which shows ways (methods, procedures, strategies, tactics) for successfully performing and managing design processes in an industrial context. This contains in its widest sense all the formal and ideal means of assistance (including methods) which designers can use in their tasks of analyzing, thinking out (conceptualizing, inventing), representing (modeling), and evaluating (including calculating) designed systems (products and processes, etc.

Literature:
Hubka, V. and Schregenberger, J.W. (1989) "Eine Ordnung Konstruktionswissenschaftlicher Aussagen" (An Arrangement for Classifying Design–Scientific Statements), VDI–Z, vol. 131 no. 3.
Hubka, V. and Schregenberger, J.W. (1988) "Eine neue Systematik konstruktionswissenschaftlicher Aussagen — Ihre Struktur und Funktion" (A New Systematic Order for Design-Scientific Statements — its Structure and Function), in V. Hubka, J. Baratossy & U. Pighini (eds), WDK 16: Proceedings of ICED 88, Budapest: GTE u. Zürich: Heurista, Vol. 1, p. 103–117
Hubka, V. and Schregenberger, J.W. (1987) "Paths Towards Design Science", in W.E. Eder (ed), WDK 13: Proceedings of the 1987 International Conference on Engineering Design, New York: ASME, pp. 3.14
Hubka, V and Ropohl, G. (1986) "Was ist ein technisches System? Zur Grundlegung der Konstruktions wissenschaft" (What is a Technical System? To the Foundation of Design Science), VDI–Z vol. 128, no. 22, pp. 864–874
Stegmüller, W. (1973) Wissenschaftliche Erklärung und Begründung (Scientific Explanation and Substantiation), Berlin: Springer
Schregenberger, J.W. (1986) "Erfolgreicher konstruieren — aber wie?" (Designing More Successfully — But How?), Schweizer Maschinenmarkt, no. 23, pp. 46–49.

FIGURE 12.7 Model (map) of engineering design science.

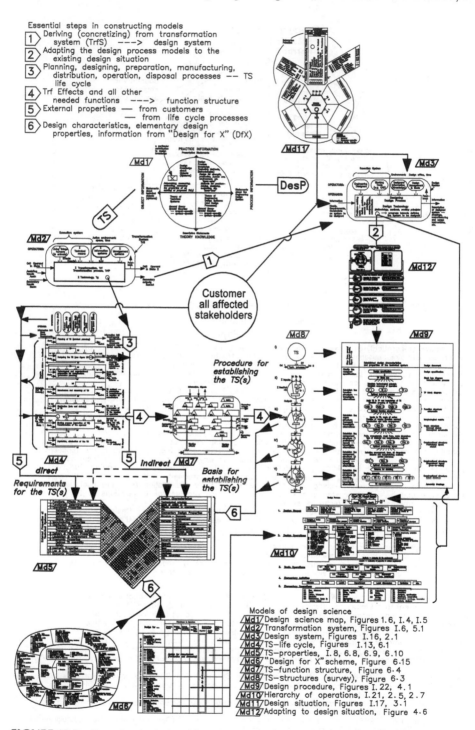

Essential steps in constructing models

1 > Deriving (concretizing) from transformation system (TrfS) ---> design system
2 > Adapting the design process models to the existing design situation
3 > Planning, designing, preparation, manufacturing, distribution, operation, disposal processes -- TS life cycle
4 > Trf Effects and all other needed functions ---> function structure
5 > External properties — from customers — from life cycle processes
6 > Design characteristics, elementary design properties, information from "Design for X" (DfX)

Customer all affected stakeholders

Procedure for establishing the TS(s)

Basis for establishing the TS(s)

direct
Requirements for the TS(s)

Indirect

Models of design science
Md1 / Design science map, Figures 1.6, I.4, I.5
Md2 / Transformation system, Figures I.6, 5.1
Md3 / Design system, Figures I.16, 2.1
Md4 / TS—life cycle, Figures I.13, 6.1
Md5 / TS—properties, I.8, 6.8, 6.9, 6.10
Md6 / "Design for X" scheme, Figure 6-15
Md7 / TS—function structure, Figure 6-4
Md8 / TS—structures (survey), Figure 6-3
Md9 / Design procedure, Figures I.22, 4.1
Md10 / Hierarchy of operations, I.21, 2.5, 2.7
Md11 / Design situation, Figures I.17, 3.1
Md12 / Adapting to design situation, Figure 4.6

FIGURE 12.8   Procedural relationships among models of engineering design science.

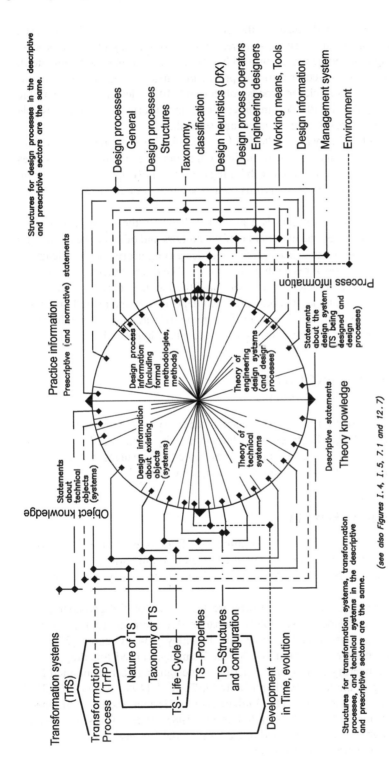

(see also Figures 1.4, 1.5, 7.1 and 12.7)

FIGURE 12.9 Engineering design science—structure.

misinterpretation occurs with respect to the term *practice*. Every practice is concerned with two aspects (see Figure I.2): (1) a *subject* of concern—a tangible and a process object (including a phenomenon about which a theory exists or may be formulated); and (2) a *method* (formalized or informal procedure, strictly or adaptively applied) (a) for operating the tangible or process object and (b) of designing the tangible or process object, that is, the design procedures, and information about possible configurations, structures, limiting values, and so forth.

A similar falsehood consists of "technology transfer," if it implies only transfer from research into industry, that is, from theory into practice. Transfer always goes both ways, even though some "high-tech" things spectacularly appear to come from pure theory and research. "Practice" is thus seen to be a fuzzy conglomerate of subject, theory, and method.

The theory developed for EDS is a grounded theory in the sense proposed by Glaser [231]. It is based on the experience of the authors and others in designing TP and TS, and in observations and discussions about design engineering with colleagues, compare the four methods of research in Section I.1. The development of theory and analysis can never be completed. Our theory, and the methods derived from it, are well grounded, logical, and applicable, but much depended on the mental processes of the participants (compare Section 11.1). This involves "serendipity, unanticipated, anomalous, and strategic finding (that) gives rise to a new hypothesis" [412].

12.5 SPECIALIZED ENGINEERING DESIGN SCIENCES (SEDS)

The EDS outlined in Section 12.4 is as abstract as necessary to deliver a general theory and system of knowledge. Each region of human activity would need to adapt this general EDS for its own conditions and circumstances, the changes will depend on the design situations that arise in an organization; see Chapter 7.

12.6 HISTORICAL SURVEY

The historical developments about designing and EDS (see chapter 3 of [315]) incorporate existing information, and show diverse efforts in various places. These represent reactions to perceived situations, and efforts to clarify and rationalize designing.

Noteworthy among the earlier efforts are the works of Wörgerbauer [585], Kesselring [350], and Rodenacker [487]. Separate developments took place with Hansen [251,253], Dietrych [295], Jones and Thornley [334], Gregory [239,241], and others. The "bottleneck design" (1965) in Germany, caused founding of several chairs of engineering design, and experienced design engineers appointed from industry, resulting in books by Pahl and Beitz [457], Roth [498], Koller [369,370], Ehrlenspiel [178,179], and others. The Guidelines of the German Society of Engineers (e.g., VDI-Richtlinie 2221 [24,25]) started around 1985. These efforts were aimed at generating pragmatically formulated procedural models for design processes, and providing supporting structures for information.

Several reports were generated in England, among which were those by Feilden [189], Moulton [423], Corfield [100], Finniston [200], and Lickley [385], after which some research centers for engineering design were founded at universities. The Design Methods Group existed in the United States since about 1960, but engineering design was actively supported only from 1985 onwards with the NSF Initiative [15]. The emphasis in the English-speaking countries has been mainly on creativity, and on computer support of designing.

Works by Eekels [490,492], Andreasen [44–46], Yoshikawa [588–591], Hongo [274], Cross [109–111], Pugh [475,476,478], and Ullman [548] feature design engineering, theory, methods, and similar topics. The considerations have more recently expanded to include topics of general management [571], product management [8,22], safety and product liability [30], integrated product development and innovation [44,178,271,272], and ecological and environmental issues [7,62,160,186,237,244,582].

Glossary

This glossary presents selected definitions. The word appears in bold at its first appearance in the text.

abstracting   Simplifying and representing a concrete reality, such that a (relatively small) set of essential features (properties, structures, elements, relationships) is emphasized.

accuracy   In conformance to a standard, for example, quality of conformance with a detail or assembly drawing; within permitted *tolerance* range. See also *precision*.

apperception   [502] Receives nonperceivable kinds of information. Examples of apperceptive information are: the possible, infinity, force, stress, mass, speed, friction, magnetic flux, symmetry, and so forth, which cannot be observed directly, can only be inferred using learned abstractions, and are then conceptually added to the mental representations.

artifact   Any tangible or perceived object that is a result of human activity or interpretation.

behavior   Characterized by successive *states*, including manifestations and values of the properties, of a *system* in response to its environment and to received stimuli (M, E, I).

causality   An essential assumption to engineering: it is necessary to rely on cause-consequence chains to *design* or control our environment. In causality, the starting *state* (analytically) determines the outcome, an idea often wrongly assumed to apply to *teleology*. The opposite concept, *finality*, is equally important for engineering, especially to explain design engineering. A goal (especially of designing) is usually conditional on a purpose characterized by a desirable end situation. The concept of *finality* describes the finishing *state* to be reached by synthesis (see Figures I.19 and 2.3). Progress is manifested by change from (current) goals toward a transformed *state*, similar to the colloquial "means to achieve ends," and at times "ends to justify means." Goals, needs, and *means* are interdependent and occur with hierarchical relationships.

cause   An initiator of a change. The colloquial combination is "cause and effect," but we have chosen "effect" as the active *output* of a technical system, and we couple "cause and consequence."

concretizing   The process of making an idea or concept more concrete, working toward *realizing* it—the opposite direction to *abstracting*.

configure   Establish an arrangement of organs (OrgStr) and constructional parts (CStr).

consequence   See *cause*.

controlling   Guiding a *system* to achieve a "set point." We distinguish "Regulating" (setting or varying the target *state*, the "set point") and "controlling," generally considered almost interchangeable, so that both functions may be recognized in analogy to some societal tasks, such as setting the standards (e.g., legislating) and complying with them (e.g., enforcing the law).

customer   *Organization*-internal or *organization*-external *stakeholder* and other interested parties, especially the recipient of a *product* (goods and services).

danger   The nature of a potential adverse event. See also *risk* and *hazard*.

descriptive   Information can be descriptive (theory, *knowledge*), or *prescriptive* (advisory, including *heuristic*) and normative (compulsory, established by a standard or law, regulative); see Chapter 12. The English language uses "descriptive" as narrative, anecdote, or essay with little *theory*. "Descriptive" in EDS [315] implies declarative and theoretical concepts, abstracted from and supported by narrative and empirical results.

design   Typical dictionary definitions [2,13] read:

> *design* (n.) (1) mental plan, scheme of attack; (2) purpose, end in view, adaptation of means to ends; (3) preliminary sketch for picture, plan of building, machine, and so forth; arrangement of lines, drawing, and so forth, decorating or distinguishing a thing (delineation, pattern; act of making these); and (4) established form of a product; general idea, construction from parts.
>
> *design* (vb.) (1) set (thing) apart *for* person, destine (person, thing) *for* a service; (2) contrive, plan; purpose, intend; (3) make preliminary sketch of (picture), draw plan of (future building, etc.); and (4) be a designer.
>
> *designer* (n.) in verbal senses; draughtsman who makes plans for manufacture; person who prepares designs (n.) for clothing, stage productions, and so forth.
>
> *designing* (adj.) in verbal senses; crafty, artful, scheming; hence *designingly*.

"The design" (n.) usually refers to the appearance of an existing object or pattern, for example, on wallpaper, or of a car. "Design" (vb.) is an activity, mainly performed by human beings. It is ambiguous when speaking, writing, or hearing about "the design of a gear box," is the intent (a) the resulting pattern, that is, an actual (or proposed) TS "gear box?" or (b) the activity of designing?

We strictly differentiate between (1) a process system and tangible object system that is to be designed and (2) the activity of designing (the design process system in particular). We use "design" connected with an adjective, for example, "engineering design," meaning the process of designing in engineering, but we prefer "design engineering," or as an adjective, for example, "design process." We thus hope to avoid ambiguity between the process of designing (one of many processes) and the products (goods and services, tangible and process systems) being designed.

The literature offers over a hundred definitions (compare [315]), many connected with the aims of designing, others describe partial contents of design activities, or human qualities to enable designing. Many industries simply use the term "engineering" a product, considering its *operational* processes, and its manufacturing, realization, and implementation process.

effect The active *output* of an operator (e.g., a TS) that acts directly on an operand, or indirectly through another operator, and consists of material, energy, and information (M, E, I).

effectiveness "Doing the right things" [136]; the extent to which planned activities are realized and planned results achieved (ISO 9000:2000 [9], article 3.2.14).

efficiency "Doing things right" [136]; relationship between the results achieved and the resources used (ISO 9000:2000 [9], article 3.2.14).

elemental See *partial*.

establish The process of generating a preferred solution, and eliminating, or deferring from consideration, those solutions that are not considered appropriate or *optimal*.

external to Applied to anything that is topologically outside the currently defined boundary of a TP or TS. See *topology*.

finality See *causality*.

form Operands in TP may exist as inanimate materials, living things, energy, and information (M, E, I, L). Inputs/outputs of TS may exist as inanimate materials, energy, and information (M, E, I). Materials (in a solid state) can have "shape," but all can exist in different forms. "Form" is adopted as the more general term. "Shape" and "form" are often used colloquially as synonyms. An intuitive difference is

> *Shape* (as a noun, with related verbs to shape, to mold) is an inclusive term for the designation of geometric classes, for example, lines, planes and surfaces, solid bodies; the manifestation of "shape" for bodies is for example, cylinder, cone, block.
>
> *Form* (including build; character; configuration; architecture; visual, geometric, connective, volumetric arrangement; juxtapositions) is the total perception of a formation, implying consistency (gas, liquid, solid), and where applicable also the shape and dimensions.

function May refer (1) to the way a *technical system* works or (2) to what it does. We distinguish the internal actions (how it is capable of operating, its mode of action, resulting from its action chains and internal action processes) from its intended actions (what it does or is intended to do, its capabilities for action, its function, its exerted effects anticipated to result from the externally established purpose functions it is intended to perform). Each function should be formulated as a combination of a verb (or a verb phrase) with a noun (or noun phrase).

functionality The general and particular capability of a process system and an object system to do something, especially the variety of capabilities.

group Inanimate or animate collection of objects. See also *team*.

hazard Property of a *system* posing a *danger* of causing an adverse event. See *risk* and *danger*.

heuristic From the Greek *heuristiko* (to find, to discover [2,13]); "arguments and methods of demonstration that are persuasive rather than logically compelling" [365, p. 60]. "Heuristic reasoning is reasoning not regarded as final and strict but as provisional and plausible only, whose purpose is to discover the solution to the present problem. ... We shall attain complete certainty when we shall have obtained the complete solution, but before obtaining certainty we must often be satisfied with a more or less plausible guess" [470]. For applications, certainty is unattainable, "all is heuristic" [365, p. 111].

hypothesis Supposition made as a basis for reasoning, without assumption of truth, or as starting point for further investigation of known facts (see *theory*); groundless assumption; an assumption made especially to test its logical or empirical consequences [2].

implementing The process of making a process *system* into an operational reality.

input Any *system* has primary, and secondary inputs and outputs (M, E, I) that cross the boundary of the system, sometimes called throughputs, which are subjected to a change of *state*.

knowledge (1) Knowing, familiarity gained by experience (of person, thing, fact); (2) person's range of information; (3) theoretical or practical understanding (of subject, language, etc.); (4) the sum of what is known (every branch of knowledge); (5) (Philosophy) certain understanding, opposite to opinion [2]. (NOTE: Implies that knowledge is mostly tacit, internalized.)

main The decisive part required for the *system* to function. The main transformation process describes the transformation of an operand as a *consequence* of the main effects delivered by the operators. It is always accompanied by assisting processes, and so forth. The TrfP is caused by operating the TS, or by the TS being operated, for example, by a human, other TS, or other operator.

manifestation Most properties can exist (be manifested) in various distinctive forms, measures, values, and states. A "color" can exist as blue, red, and so forth and in various intensities. We have chosen "manifestation" for the possible discrete ways in which a general property can be embodied; the word "form" has been reserved for a specific use.

means Anything that can be used to make progress toward a goal, "means to an end." We have tried to avoid using the word "means" to imply "meaning/understanding."

methodical (design engineering). The *heuristic* use of established and newly developed methods in an engineering *design* process, including "industry best practice," strategic or tactical, formalized and intuitive. See also *systematic*.

modular Consisting of modules, interchangeable standardized parts or units of construction with common interfaces that can perform similar functions.

operational   A TS in a *state* capable of operating, or being operated, capable of delivering effects to an operand, whether that operand is present or not, provided that suitable *inputs* are delivered to the TS.

operational process   Synonym of a TP as executed by a TS when operating or being operated.

optimal   Implies a near optimum for a particular range of conditions, including the perceptions of *stakeholders*. "The optimum" can hardly ever be achieved in real and complex circumstances.

organ   An active or reactive connection between constructional parts, or across the TS-boundary, at action locations. We distinguish organs from constructional parts, which are connections between organs.

Organs are analogs of the physiological interpretation in biology, which describes what is done by the living *organism* in that region. The physiologically important part of the heart is the inner surface (helped by valves) that draws blood (as operand) from the veins, and delivers it to the arteries. A skeletal joint that separates two bones and allows movement between them is an organ. The capability of activity characterizes the organ. Constructional parts are analogs of the anatomical interpretation in biology. Each bone, ligament, or muscle is an anatomic part.

In some cases, a single unit acts as both a constructional part and an organ. The liver and the spleen are cases where the operations are not clearly separable from the constructional parts.

organism   A coordinated collection of *organs* that can fulfill several functions, or a higher function. The heart–lung combination may be regarded as an oxygenating organism, but a whole animal is also called an organism.

organization   Includes company, corporation, enterprise, firm, institution, association, society, federation, council, government assembly, administration, and so forth (ISO 9000:2000 [9], article 3.3.1).

output   See *input*.

outside   See *external to*.

parametrize   Establish a transition from an *organ* structure or preliminary layout to a definitive layout and details by establishing dimensions.

partial   A partial transformation, partial operation, partial *function*, partial *system*, partial process, partial *organ*, and so forth, constitutes a part of a more complex transformation, operation, *function*, *system*, process, or *organ*. A partial transformation etc. may or may not be able to exist by itself. At the simplest possible level, it is an *elemental* transformation etc. A consistent expression of these subdivisions, using adjectives "partial" and "elemental."

precision   Closeness to the standard set for that quantity, see also *accuracy* and *tolerance*.

prescriptive   See *descriptive*.

product Any *output* from an *organization* or part of an *organization* intended to be received by a *customer*, *stakeholder*, or other *organization*-internal or *organization*-external party; this *output* should be economically useful, including "services, software, hardware, and processed material". See ISO 9000:2000 [9], article 3.4.2. (but see also Section I.1).

realizing The process of making an idea, concept, or *system* into an operational reality.

regulating See *controlling*.

risk Probability that an adverse event will occur, see also *danger* and *hazard*.

science Defined as [2]: (1) *knowledge*; (2) systematic and formulated *knowledge*; (3) branch of *knowledge* (especially one that can be conducted on scientific principles), organized body of the *knowledge* that has been accumulated on a subject [13] (from Latin "scientia"—having knowledge); (3) a branch of study concerned with observation and classification of facts and especially with the establishment of verifiable general laws; (4) accumulated systematized *knowledge* especially when it relates to the physical world. (NOTE: This definition implies that *knowledge* is mostly recorded, documented. See *knowledge*.)

sequencing An arrangement of elements that influence other elements in some way, arranged in series, parallel, cyclic, interconnected, recursive, and so forth, compare Figure 12.4.

"Sort" TS-"sort" indicates which branch of engineering organization is responsible for this TS.

stakeholder Person with interest, usually monetary, in a process and its outcomes.

state The way that a TP or TS exists at any one time, including its properties, their manifestations, and their measures, values and scales of measurement. "Condition" is used to make statements about imposed circumstances that are essential to the existence of a TP or TS, especially limits. "Constraints" are a form of condition.

statement Expression in words, thing stated, formal account of facts [2].

state of the art The best available *technology* in a specific field at any one time; knowing, "know-how" and "know-what" available to a person is the *personal state of the art* [365].

synergy Combined influence that exceeds the sum of the individual influences. The *behavior* of a *system* is an aggregate of behaviors of subsystems, plus synergy, the behaviors that arise only through the interactions of the subsystems.

system Consists of elements (of various kinds), and their mutual relationships (internal and cross-boundary), within a defined boundary, connected to an active and reactive environment (the surroundings of the boundary). The boundary is selected according to the contents of the collection and its purpose. It is useful to understand systems theory, that is, the concepts, contents, and *behavior*, but not necessarily as a mathematical model. The study of systems has progressed through cybernetics, systems behavior, control theory, artificial intelligence, and so forth.

systematic (design engineering). The heuristic-strategic use of a *theory*—EDS [315]—to guide the *design* process, see also *methodical*.

team *Group* of humans that interact and work together for a common goal.

technical system (TS) Man-made tangible object *system* that can be used to deliver the effects to perform one or more operation(s) in a transformation process. See Chapters 6 and 7.

technology With a definite or indefinite article, a (or the) technology is the specific way of delivering an *effect* (M, E, I) directly to an operand, interpreted as the means, active and reactive effects, and knowledge needed to perform manufacturing or other transforming operations; and the answer to "how, with what effects is a transformation achieved?" Effects produced (e.g., by a TS) are exerted by an operator, and are received by an operand.

Without a definite or indefinite article (as a generic term), technology is the general manifestation of technological *means* and their significance for humans and society, and is increasingly used to denote any application of information to enable designing and producing a TP and TS. "Technology" in this general sense covers "the science of practical or industrial arts" [2], and the experimental (craft-based) "know-how" that is outside *science*. Technology is then usually replaced by "the technological sphere" to convey the functions of a device, and the functions of the *means* for manufacturing that device, including commerce, transportation, management, and so forth, to realize and distribute it.

teleology The view that any developments (transformations, changes) are caused by the purpose that is served by them. In our view there is no implication of predestination, that is, that causes must lead to unalterable outcomes in the future. The path to the causes and purposes can be traced back in time, but not in an infinite regression to an ultimate *cause*. The path forward is not determined. We are concerned with the useful aspects of teleology—the linking of causes and consequences. It is not necessary that all events (as consequences) must have identifiable causes, just those that we wish to influence in a predictable way. We intend to influence some aspects of engineering, particularly the processes of designing and teaching/learning. For this purpose, it is necessary to adopt a teleological approach to those aspects that can be influenced, and therefore the two concepts of *causality* and *finality* are needed by engineers.

term A collective and abstracting expression in words that is to be consistently used.

terminus technicus A defined technical expression as normal language for this book.

theory Supposition or system of ideas explaining something, especially one based on general principles independent of the facts, phenomena, and so forth, to be explained; opposite to *hypothesis*; speculative (especially fanciful) view; the sphere of abstract knowledge or speculative thought; expositions of the principles of a *science* and so forth; collection of propositions to illustrate principles of a subject [2]: (1) the general principles drawn from any body of facts (as in *science*); (2) a plausible or scientifically acceptable general principle offered to explain observed facts; (3) *hypothesis*, guess; and (4) abstract thought [13].

tolerance   Degree of refinement in measurement and so forth, for example, actual variation of a particular dimension on one constructional part (related to surface finish), actual variation of that dimension on successive parts, and achieved standard deviation (related to "six sigma" limits and Taguchi experimentation recommendations). See also *accuracy*.

topology   Study of geometrical properties and spatial relations that are unaffected by continuous change of shape or size of figures [2]. Consider a "donut" or "bagel," a geometrical torus formed by rotating a circle around an axis that does not intersect the circle. In normal language, this body "has a hole inside it," but where the hole starts is undefined on the continuous surface. Starting from any point on the surface, any other point may be reached without departing from the surface. All of the space around the body is "external to" the torus, including the hole. All of the space filled by the (cooked) dough is "internal to" the torus, and is covered by the surface—which is the boundary between "internal" and "external." The relationship of "internal" and "external" remains unchanged independent of the way the torus is stretched or distorted. If the torus is compressed such that the axis intersects the circle, the "hole in the donut" disappears, but the relationship of "internal" and "external" remains unchanged.

A tube or pipe may be formed by distorting the circle of a torus into an elongated figure of two (parallel) lines with half-circles connecting their ends. "Internal to" the pipe is the material, for example, copper, "external to" the pipe is (1) the conventional outside space and (2) the inside space that can contain a different material (fluid or granular solid), for example, as operand of a flow process. The pipe experiences the pressure difference from inside to outside, and reacts with internal stress and strain, but the pipe "does not know" about the flow rate. The water valve, Section 1.2, can operate without water and without being connected to pipes. If water is present as operand, the valve will react by containing the water and exerting counterpressure as its main *effect*.

Material and living operands in a transformation *process* must therefore be considered "external to" the TS under consideration—the arbitrary boundaries of the TS can and should be drawn in a suitable way. A similar situation can be assumed by analogy regarding energy and information. The gearbox in Figure 4.5 may rotate without transmitting energy. When the gear box experiences static or rotational torque on the input and output shafts (applied and resisted from "external to" the TS "gearbox"), it reacts by stress and strain "internal to" the constructional parts, and by force reactions between them and to ground. The gearbox "does not know" whether any energy is transmitted as operand, its effects are the reactions to the conditions applied to the TS. The example of an operational amplifier in the NOTE in Section 4.4.2 shows an equivalent situation for signals, that is, information as operand. Energy and information are "carried by" the TS, the TS acts on or reacts to the presence of the operand of the TP.

References

[1] *ANSI/ASME Boiler and Pressure Vessel Code*, New York: ASME, 1992.

[2] *The Concise Oxford Dictionary* (7th edn.), Oxford: Oxford University Press, 1984, http://www.askoxford.com.

[3] DARPA, U.S. Defense Advanced Research Projects Agency, http://www.darpa.mil.

[4] eMachineShop, free download of software, 2004, http://emachineshop.com.

[5] ESDU, Engineering Science Data Unit, http://www.ihs.com/engineering/edu-design-guidelines.html.

[6] Geomate ToleranceCalc 4.0, http://www.inventbetter.com.

[7] Granta Design Limited, *Cambridge Engineering Selector CES for Windows*, Cambridge, UK: Granta Design Ltd, 2006, http://www.grantadesign.com.

[8] *Instructions for Consumer Products*, Department of Trade and Industry, London: HMSO, 1988.

[9] *ISO 9000:2000 Quality Management Systems—Fundamentals and Vocabulary*, Geneva: ISO, 2000, http://www.iso.ch.

[10] *ISO 9000:2000 Quality Management Systems—Requirements*, Geneva: ISO, 2000.

[11] *ISO 14001:1995 Environmental Management System*, Geneva: ISO, 1995.

[12] KISSsoft (Rel04/2004), *Calculation Programs for Machine Design*, KISSsoft AG, Frauwis 1, CH-8634 Hombrechtikon, Switzerland, 2004, http://www.KISSsoft.ch.

[13] *The Merriam-Webster Dictionary*, New York: Pocket Books, 1974, http://www.m-w.com/cgi-bin/dictionary.

[14] Mirakon, *Das Mirakon System*, Fabrik am Rotbach, Bühler, Appenzell AR, Switzerland, 2006, http://www.mirakon.com.

[15] National Science Foundation: Program Announcement, *Design Theory and Methodology*, No. OMB 3145-0058 12/31/85, 1985.

[16] Report on Evaluation of Engineering Education (L.E. Grinter, chairman), *J. Eng. Educ.*, September 1955, pp. 25–60.

[17] *Simplified Methods in Pressure Vessel Analysis*, ASME/CSME Montreal Pressure Vessel and Piping Conference, New York: ASME, 1978.

[18] Standard for Integration Definition for Function Modeling (IDEF0), *Draft Federal Information Processing Standards*, Publication FIPS PUB 183 December 1993, Gaithersburg, MD: National Institute of Standards and Technology, 1993.

[19] VDI Richtlinie 2223:1969, *Begriffe und Bezeichnungen im Konstruktionsbereich* (Terminology and Designations in the Design Engineering Region), Düsseldorf: VDI, 1969.

[20] VDI Richtlinie 2225:1975, Blatt 1 and 2, *Technisch-wissenschaftliches Konstruieren* (Technical-Scientific Designing), Düsseldorf: VDI-Verlag, 1975.

[21] VDI Richtlinie 2222:1977 (1), *Konstruktionsmethodik: Konzipieren technischer Produkte* (Design Methodology: Conceptualizing Technical Products), Düsseldorf: VDI, 1977.

[22] VDI Richtlinie 2220:1980, *Produktplanung—Ablauf, Begriffe und Organisation* (Product Planning—Procedure, Terminology and Organization), Düsseldorf: VDI, 1980.

[23] VDI Richtlinie 2222:1982 (2), *Konstruktionsmethodik: Erstellung und Anwendung von Konstruktionskatalogen* (Design Methodology: Generation and Use of Design Catalogs), Düsseldorf: VDI, 1982.

[24] VDI Richtlinie 2221:1985, *Methodik zum Entwickeln und Konstruieren technischer Systeme und Produkte* (Methodology for Developing and Designing Technical Systems and Products), Düsseldorf: VDI, 1985.

[25] VDI Guideline 2221, *Systematic Approach to the Design of Technical Systems and Products*, Düsseldorf: VDI (ed. K.M. Wallace), 1987.

[26] VDI Richtlinie 3780:2000, *Technikbewertung—Begriffe und Grundlage* (Technology Evaluation—Terminology and Fundamentals), Düsseldorf: VDI, 2000.

[27] VDI Richtlinie 2206:2004, *Konstruktionsmethodik für mechatronische Systeme* (Design Methodology for Mechatronic Systems), Düsseldorf: VDI, 2004.

[28] WDK 18: *Proceedings of the Institution of Mechanical Engineers, International Conference on Engineering Design*, ICED 89 Harrogate, London: IMechE, 1981.

[29] WDK 28: *Proc. International Conference on Engineering Design*, ICED01 Glasgow, August 21–23, 2001, London: IMechE, 2001.

[30] Abbott, H., *Safer by Design: The Management of Product Design Under Strict Liability*, London: Design Council, 1988.

[31] Adams, J.L., *Conceptual Blockbusting*, Reading, MA: Addison, Wesley (3rd edn) 1986 (1st edn., 1974).

[32] Ahmed, S. and Wallace, Identifying and Supporting the Knowledge Needs of Novice Designers within the Aerospace Industry, *J. Eng. Design*, Vol. 15, No. 5, 2004, pp. 475–492.

[33] Albers, A. and Birkhofer, H., *Teaching Machine Elements at Universities—Towards Internationally Harmonized Concepts in View of Industrial Needs*, in [387], Vol. 2, 1999, pp. 1043–1046.

[34] Albers, A., Matthiesen, S. and Ohmer, M., An Innovative New Basic Model in Design Methodology for Analysis and Synthesis of Technical Systems, in *DS 31—Proc. ICED 2003 Stockholm*, paper 1228 on CD-ROM, 2003.

[35] Albers, A., Burkhardt, N. and Ohmer, M., Principles for Design on the Abstract Level of the Contact & Channel Model, in *Proc. TMCS 2004 Lausanne*, 2004, pp. 87–94.

[36] Albrecht, K. and Bradford, L., *The Service Advantage: How to Identify and Fulfill Customer Needs*, Homewood, IL: Dow Jones-Irwin, 1989.

[37] Albrecht, K., *The Only Thing That Matters: Bringing the Power of the Customer into the Center of Your Business*, New York: HarperCollins, 1972.

[38] Alexander, E.R., Design in the Decision-Making Process, *Policy Sci.*, Vol. 14, 1982, pp. 279–292.

[39] Altschuller, G.S., *Erfindungen—(K)ein Problem in Russland (Inventing—(Not) A Problem in Russia)*, Berlin: Verlag Tribüne, 1973, translated from a much earlier Russian-language publication.

[40] Altschuller, G.S., *Creativity as an Exact Science: Theory of the Solution of Inventive Problems* (2nd edn.), New York: Gordon & Breach, 1987.

[41] Anderson, J.R., *Cognitive Psychology and Its Implications*, New York: Freeman, 1995.

[42] Andreasen, M.M., Darstellungsmöglichkeiten beim Konzipieren (Representation Possibilities in Conceptualizing), *Schw. Masch.*, Bd. 78, H. 4 and 6, 1978.

[43] Andreasen, M.M., *Syntesemetoder på Systemgrundlag—Bidrag til en Konstruktionsteori* (Synthesis Methods Based on Systematic Approach—Contributions to a Design

Theory), Thesis, Department of Machine Design, Lund Institute of Technology, Sweden, 1980.

[44] Andreasen, M.M. and Hein, L., *Integrated Product Development*, London: IFS Publ., and Berlin/Heidelberg: Springer-Verlag, 1987.

[45] Andreasen, M.M. and Ahm, T., *Flexible Assembly Systems*, London: IFS Publ., 1988.

[46] Andreasen, M.M., Kähler, S., Lund, T. and Swift, K.G., *Design for Assembly* (2nd edn.), London: IFS Publ., and Berlin/Heidelberg: Springer-Verlag, 1988.

[47] Andreasen, M.M. and Olesen, J., The Concept of Dispositions, *J. Eng. Design*, Vol. 1, No. 1, 1990, pp. 17–36.

[48] Andreasen, M.M., Wognum, P.M. and McAloone, T., Design Typology and Design Organization, in *Proc. Design 2002*, University of Zagreb, 2002, pp. 1–6.

[49] Aristotle, *Organon Section Topika*, 384–322 BCE.

[50] Aristotle, *Rhetoric*, 384–322 BCE.

[51] Ashby, M.F., On the Engineering Properties of Materials, *Acta Metall.*, Vol. 37, No. 5, 1989, pp. 1273–1293.

[52] Ashley, S., Engineous Explores the Design Space, *Mech. Eng.*, February 1992, pp. 49–52.

[53] Aurisicchio, M., Bracewell, R. and Wallace, K., Characterizing the Information Requests of Engineering Designers, in *Proc. International Design Conference— Design 2006*, Dubrovnik, 2006, pp. 1057–1064.

[54] Avallone, E.A. and Baumeister, T. III, *Mark's Standard Handbook for Mechanical Engineers* (10th edn.), New York: McGraw-Hill, 1996 (also on CD-ROM interactive).

[55] Badke-Schaub, P. and Stempfle, J., Analyzing Leadership Activities in Design: How do Leaders Manage Different Types of Requirements, in *Proc. International Design Conference—Design 2004*, Dubrovnik, 2004 (on CD-ROM).

[56] Barnes, J. (ed.), *The Complete Works of Aristotle* (The revised Oxford translation), Princeton, NJ: Princeton University Press, 1985.

[57] Barnes, J. (ed.), *The Cambridge Companion to Aristotle*, Cambridge: University Press, UK, 1995.

[58] Barnett, R.L. and Brickman, D.B., Safety Hierarchy, *J. Safety Res.*, Vol. 17, No. 2, 1986, pp. 49–55.

[59] Barwell, F.T., *Lubrication of Bearings*, London: Butterworths, 1956.

[60] Beitz, W., Übersicht über Konstruktionsmethoden (Survey of Design Methods), *Konstruktion*, Jg. 24, H. 2, pp. 68–72 and 3, pp. 109–114, 1972.

[61] Beitz, W., Engpass Konstruieren—Situation in der Bundesrepublik Deutschland (Bottleneck Designing—Situation in the Federal Republic of Germany), in [298], 1983, pp. 7–15.

[62] Berendt, S., Jasch, Chr., Peneda, M.C. and Weenen, H. van, *Life Cycle Design: A Manual for Small- and Medium-Sized Enterprises*, Berlin: Springer-Verlag, 1997.

[63] Birkhofer, H., Design Practice and Design Education—What Shall We Teach and Learn in Engineering Design, in [491], 1993, pp. 1746–1755.

[64] Birkhofer, H., The Power of Machine Elements in Engineering Design—A Concept of a Systematic, Hypermedia-Assisted Revision of Machine Elements, in [485], 1997, Vol. 3, pp. 679–684.

[65] Birkhofer, H., Mechatronic Machine Elements—The Contribution of Design Science to the Efficiency and Effectiveness of Teaching and Learning, International Workshop EED—Education for Engineering Design, November 23–24, 2000, State Scientific Library, Pilsen, Czech Republic, 2000.

[66] Birmingham, R., Cleland, G., Driver, R. and Maffin, D., *Understanding Engineering Design*, London: Prentice-Hall, 1997.

[67] Blanchard, B., *Systems Engineering Management*, Hoboken, NJ: Wiley, 2004.

[68] Bloom, B.S. (ed.) *Taxonomy of Educational Objectives: Cognitive Domain*, New York: McKay, 1956.

[69] Bodily, S.E., *Modern Decision Making: A Guide to Modelling with Decision Support Systems*, New York: McGraw-Hill, 1985.

[70] Boice, R., *First-Order Principles for College Teachers*, Bolton, MA: Anker Publications, 1996.

[71] Booker, P.J., *Principles and Precedents in Engineering Design*, London: Institution of Engineering Designers, 1962.

[72] Boothroyd, G. and Dewhurst, P., *Product Design for Assembly*, Warfield, RI: Boothroyd Dewhurst Inc., 1987.

[73] Boothroyd, G., Dewhurst, P. and Knight, W., *Product Design for Manufacture and Assembly*, New York: Marcel Dekker, 1991.

[74] Boothroyd, G., *Assembly Automation and Product Design*, New York: Marcel Dekker, 1991.

[75] Bossert, J.L., *Quality Function Deployment*, New York: Marcel Dekker, 1990.

[76] Brad, R.G., Straub, C.P. and Prober, R. (eds.), *CRC Handbook of Environmental Control, Vol. 1—Air Pollution*, Boca Raton, FL: CRC Press, 1972, pp. 448–451 and 550–555.

[77] Braha, D., *Satisfying Moments in Synthesis* (Preliminary Version), unpublished 2006.

[78] Braun, T. and Lindemann, U., Method Adaptation—A Way to Improve Methodical Product Development, in *Proc. International Design Conference—Design 2004*, Dubrovnik, 2004, pp. 977–982.

[79] Breiing, A., Methoden gibt es viele (Methods are Multitude), *Schw. Masch.*, No. 9, 1990, pp. 44–47, and [490], 1990, pp. 92–94.

[80] Breiing, A., Prozessualer Ablauf des Verfahrens (Process Development of Methods), *Schw. Masch.*, No. 12, 1990, pp. 52–57, and [490], 1990, pp. 95–98.

[81] Breiing, A., Neue Gesichtspunkte zur Bewertung technischer Systeme (New Viewpoints to Evaluation of Technical Systems), in *Proc. EVAD 1992—Evaluation and Decision in Design*, Praha, Mai 20–21, 1992, Zürich: Heurista, 1992.

[82] Breiing, A. and Fleming, M., *Theorie und Methoden des Konstruierens*, Berlin: Springer-Verlag, 1993.

[83] Browning, T.R., Applying the Design Structure Matrix to System Decomposition and Intregration Problems: A Review and New Directions *IEEE Trans. Eng. Manag.*, Vol. 48, No. 3, 2001, pp. 292–306.

[84] Bruch, C., Handling Undesired Functions During Conceptual Design—A State- and State-Transition-Based Approach, in *Proc. International Design Conference—Design 2004*, Dubrovnik, 2004, pp. 827–832.

[85] Bucciarelli, L.L., *Designing Engineers*, Cambridge, MA: MIT Press, 1994.

[86] Buergin, C., Integrated Innovation Capability, in *Proc. International Design Conference—Design 2006*, Dubrovnik, 2006, pp. 455–462.

[87] Burkhardt, R., Volltreffer mit Methode—Target Costing, *Top Bus.*, Vol. 2, 1994, pp. 94–99.

[88] Burr, H., Deubel, T., Vielhaber, M., Haasis, S. and Webar, C., Challenges for CAx and EDM in an International Automotive Company, in [205], pp. 309–310 executive summary, (on CD-ROM) 2003.

[89] Burr, H., Vielhaber, M., Deubel, T., Weber, C. and Haasis, S., CAx/Engineering Data Management Integration: Enabler for Methodical Benefits in the Design Process, *J. Eng. Design*, Vol. 16, No. 4, 2005, pp. 385–398.

[90] Buss, D., Manipulation in Close Relationships: Five Personality Factors in Interactional Context, *Journal of Personality*, Vol. 60, 1992, pp. 477–499.

[91] Bussmann, K.F., *Industrielles Rechnungswesen*, Stuttgart: Pöschel, 1973.

[92] Buzan, T., Head Strong, Mind Map Books, 2001, http://mind-map.com, http://visual-mind.com, http://smartdraw.com.

[93] Cascio, J. (ed.), *The ISO 14000 Handbook*, Milwaukee, WI: ASQ Quality Press, 1996.

[94] Christianson, L.L. and Rohrbach, R.P., *Design in Agricultural Engineering*, St Joseph, MI: American Society of Agricultural Engineers, 1989.

[95] Chuse, R., *Unfired Pressure Vessels: The ASME Code Simplified*, New York: F.W. Dodge, 1990.

[96] Clancy, W., *Situated Cognition*, Cambridge: University Press, UK, 1997.

[97] Colombo, G., De Angelis, F. and Formentini, L., A Mixed Virutal Reality—Haptic Approach to Carry Out Ergonomic Tests on Virtual Control Boards, in *Proc. TMCE 2006 Ljubljana*, 2006, pp. 349–356.

[98] Constant, E.W., II, *The Origins of the Turbojet Revolution, Johns Hopkins Studies in the History of Technology*, Baltimore, MD: Johns Hopkins U.P., 1980.

[99] Corbitt, R.A. (ed.), *Standard Handbook of Environmental Engineering*, New York: McGraw-Hill, 1990, pp. 4.42–4.49.

[100] Corfield, K.G., *Product Design*, London: NEDC, 1979.

[101] Cowan, N., The Magical Number 4 in Short-Term Memory: A Reconsideration of Mental Storage Capacity, *Behav. and Brain Sci.*, Vol. 24, No. 1, 2001, http://www.bbsonline.org/documents/a/00/00/04/46/index.html.

[102] Cranton, P., *Teaching Style and Type*, Sneedsville, TN: Psychological Types Press, Publ. 25, 1994.

[103] Cranton, P., *Learning Style and Type*, Sneedsville, TN: Psychological Types Press, Publ. 26, 1995.

[104] Cranton, P., *Transformative Learning and Type*, Sneedsville, TN: Psychological Types Press, Publ. 27, 1995.

[105] Cranton, P., *Teaching, Learning and Type*, Sneedsville, TN: Psychological Types Press, Publ. 28, 1995.

[106] Cranton, P. and Knoop, R., Assessing Jung's Psychological Types: The PET Type Check, *Psych. Monog.*, Vol. 121, No. 2, 1995, pp. 249–273.

[107] Crawford, R.P., *The Technique of Creative Thinking*, Burlington, VT: Fraser, 1954 (reprint 1964).

[108] Crosby, P.B., *Quality is Free: The Art of Making Certain*, New York: McGraw-Hill, 1979.

[109] Cross, N. (ed.), *Developments in Design Methodology*, Chichester: Wiley, 1984.

[110] Cross, N., Kees, D. and Roozenburg, N., *Research in Design Thinking*, Delft: University Press, 1992.

[111] Cross, N., *Engineering Design Methods* (Strategies for Product Design, 2nd edn.), London: Wiley, 1994.

[112] Csa yi, V., *Evolúciós rendszerek*, Budapest: Gondolat Könyvkiadó, 1988.

[113] Culley, W.C., *Environmental and Quality System Integration*, Boca Raton, FL: Lewis Publishers, 1988.

[114] Cushman, D.P. and King, S.S., *Continuously Improving an Organization's Performance: High-Speed Management*, Albany, NY: SUNY Press, 1997.

[115] Daft, R.E. and Lengel, R.H., Organizational Information Requirements, Media Richness and Structural Design, *Mgt. Sci.*, Vol. 32, No. 5, 1986, pp. 554–571.

[116] Damasio, A.R., *Descartes' Error: Emotion, Reason and the Human Brain*, New York: Avon, 1994.

[117] Damasio, A., *The Feeling of What Happens: Body and Emotion in the Making of Consciousness*, London: Heinemann, 1999.

[118] Dawes, R.M., *Rational Choice in an Uncertain World*, San Diego, CA: Harcourt Brace Jovanovich, 1988.

[119] DeBono, E., *The Use of Lateral Thinking*, Harmondsworth: Penguin, 1971.

[120] DeBono, E., *PO: Beyond Yes and No*, Harmondsworth: Penguin, 1973.

[121] DeBono, E., *Eureka! Illustrated History of Inventions from the Wheel to the Computer*, London: Thames & H., 1979.

[122] DeBono, E., *Tactics: The Art and Science of Success*, London: Collins, 1985.

[123] DeBono, E., *The Six Thinking Hats*, Little, Brown, 1985.

[124] Delbecq, A.L. Van de Ven, A.H. and Gustafson, D.H., *Group Techniques for Program Planning*, Glenview, IL: Scott, Foresman, 1975.

[125] Deming, W.E., *Out of the Crisis*, Cambridge, MA: MIT, 1986.

[126] Deming, W.E., *New Economics for Industry, Government and Education*, Cambridge, MA: MIT Center for Advanced Engineering Study, 1993.

[127] Descartes, R., *A Discourse on Method* (transl. J. Veitch), London: Dent Books, 1912.

[128] Dieter, G.G., *Engineering Design*, New York: McGraw-Hill, 1991.

[129] Dixon, J.R., *Design Engineering: Inventiveness, Analysis and Decision Making*, New York: McGraw-Hill, 1966.

[130] Dixon, J.R., Engineering Design Science: The State of Education, *Mech. Eng.*, Vol. 113, No. 2, 1991, pp. 64–67.

[131] Dörner, D., *Die Logik des Mißlingens: Strategisches Denken in komplexen Situationen* (The Logic of Failure: Strategic Thinking in Complex Situations), Reinbek bei Hamburg: Rowohlt Verlag, 1989; and The Logic of Failure, New York: Metropolitan Books, 1996.

[132] Dörner, D., Ehrlenspiel, K., Eisentraut, R. and Günther, J., Empirical Investigation of Representations in Conceptual and Embodiment Design, in [316], 1995, pp. 631–637.

[133] Doughtie, V.L. and Vallance, A., *Design of Machine Members*, New York: McGraw-Hill, 1964.

[134] Dreibholz, D., Ordnungsschemata bei der Suche von Lösungen, *Konstruktion*, Vol. 57, 1975, pp. 233–239.

[135] Drucker, P., *Post-Capitalist Society*, New York: Harper Business, 1993.

[136] Drucker, P.F., *The Effective Executive*, New York: Harper & Row, 1964.

[137] Dutch, R.A. (ed.), *Roget's Thesaurus of English Words and Phrases*, Harmondsworth: Penguin, 1966.

[138] Dylla, N., *Denk- und Handlungsabläufe beim Konstruieren* (Thinking and Action Progress in Designing), München: Hanser, 1991.

[139] Dym, C.L., Expert Systems: New Approaches to Computer-Aided Engineering, *Eng. with Comp.*, Vol. 1, No. 1, 1985, pp. 9–25.

[140] Eder, W.E. and Gosling, W., *Mechanical Systems Design, Commonwealth and International Library*, #308, Oxford: Pergamon Press, 1965.

[141] Eder, W.E., The Design Method—Definitions and Methodologies, in S.A. Gregory (ed.), *The Design Method*, London: Butterworths, 1966, pp. 19–31.

[142] Eder, W.E., Design Research—Technologies and Varieties of Design, in S.A. Gregory (ed.), *The Design Method*, London: Butterworths, 1966, pp. 311–315.

[143] Eder, W.E. (ed.), *WDK13—Proc. ICED 87*, Boston, MA and New York: ASME, 1987.

[144] Eder, W.E., Engineering Design—Practice and Theory, New York: ASME, Paper 86-DET-113, Design Engineering Technical Conference, 1986.

[145] Eder, W.E. and Hubka, V., Knowledge Structure for Engineering Design and Education, in *People Make the Difference, Proc. ASEE 1989 Annual Conference*, Lincoln, Nebraska, 25–29 June 1989, pp. 778–780.

[146] Eder, W.E., Knowledge Engineering—An Extended Definition, in G. Akhras and P.R. Roberge (eds.), *Proc. Symposium/Workshop Applications of Expert Systems in DND, RMC*, March 1989, pp. 11–20.

[147] Eder, W.E., Hubka, V., Melezinek, A. and Hosnedl, S., *WDK 21—ED—Engineering Design Education—Ausbildung der Konstrukteure—Reading*, Zürich: Heurista, 1992.

[148] Eder, W.E., Vergleich einiger bestehenden Bewertungsmethoden in der Anwendung (Comparisons of Some Existing Evaluation Methods in their Application), in *Proc. EVAD 1992—Evaluation and Decision in Design, Praha*, May 20–21, 1992, Zürich: Heurista, 1992, pp. 50–57.

[149] Eder, W.E., Bekannte Methodiken in den U.S.A. und Kanada (Design Methods in the U.S.A. and Canada), *Konstruktion*, Springer-Verlag, 46 Heft 5, Mai 1994, pp. 190–194.

[150] Eder, W.E., Developments in Education for Engineering Design: Some Results of 15 Years of WDK Activity in the Context of Design Research, *J. Eng. Design*, Vol. 5, No. 2, 1994, pp. 135–144.

[151] Eder, W.E., Methode QFD—Bindeglied zwischen Produktplanung und Konstruktion (Method QFD—Connective Link between Product Planning and Design), *Konstruktion*, Springer-Verlag, 47 Heft 1/2, February 1995, pp. 1–9.

[152] Eder, W.E., Engineering Design—Art, Science and Relationships, *Des. Stud.*, Vol. 16, 1995, pp. 117–127.

[153] Eder, W.E. (ed.), *WDK 24—EDC—Engineering Design and Creativity—Proceedings of the Workshop EDC, Pilsen*, Czech Republic, November 1995, Zürich: Heurista, 1996.

[154] Eder, W.E., Benchmarking—Bedeutung für Konstruktionswissenschaft (Benchmarking—Importance for Design Science), WDK Workshop Alpina 96, March 1996.

[155] Eder, W.E., Design Modeling—A Design Science Approach (and Why Does Industry Not Use It?), *J. Eng. Design*, Vol. 9, No. 4, 1998, pp. 355–371.

[156] Eder, W.E., Einstein Got It Wrong (for once)—Some Consequences for Problem-Based Learning, in *Proc ASEE Annual Conference, Educational Research and Methods Division*, Seattle, WA, June 28 to July 1, 1998, Washington, DC: ASEE, (CD-ROM), 1998.

[157] Eder, W.E. (ed.), *Proc. International Workshop PDE—Pedagogics in Design Education*, November 19–20, 1998, State Scientific Library, Pilsen, Czech Republic, 1998.

[158] Eder, W.E., Survey of Pedagogics Applicable to Design Education, An English-Language Viewpoint, International Workshop PDE—Pedagogics in Design Education, November 19–20, 1998, State Scientific Library, Pilsen, Czech Republic, 1998.

[159] Eder, W.E., Pedagogics and Didactics for Education in Engineering Design, International Workshop EED—Education for Engineering Design, November 23–24, 2000, State Scientific Library, Pilsen, Czech Republic, 2000.

[160] Eder, W.E., Designing and Life Cycle Engineering—A Systematic Approach to Designing, *Proc. Inst. Mech. Eng.* (UK), Part B, *J. Eng. Manuf.*, Vol. 215, No. B5, 2001, pp. 657–672.

[161] Eder, W.E. and Hubka, V., Curriculum, Pedagogics, and Didactics for Design Education, in [29], Vol. 4, 2001, pp. 285–292.

[162] Eder, W.E., Social, Cultural and Economic Awareness for Engineers, in *CSME Forum 2002, Queen's University*, Kingston, ON, abstract p. 130, on CD-ROM, 2002.

[163] Eder, W.E., Education for Engineering and Designing, in *Proc. International Design Conference—Design 2002*, Dubrovnik, May 14–17, 2002, on CD-ROM, 2002.

[164] Eder, W.E., A Typology of Designs and Designing, in *DS 31—Proc. ICED 03 Stockholm*, pp. 251–252 (Executive Summary), full paper number 1004 on CD-ROM, Glasgow: The Design Society, 2003.

[165] Eder, W.E., Integration of Theories to Assist Practice, in *Proc. IDMME 2004— 5th International Conference on Integrated Design and Manufacturing in Mechanical Engineering*, Bath, UK, April 5–7, 2004, on CD-ROM.

[166] Eder, W.E. and Hosnedl, S., Information—A Taxonomy and Interpretation, in *Proc. International Design Conference—Design 2004*, Dubrovnik, May 18–21, 2004, pp. 169–176.

[167] Eder, W.E., Reflections About Reflective Practice, in *Proc. International Design Conference—Design 2004*, Dubrovnik, May 18–21, 2004, pp. 177–182.

[168] Eder, W.E., Machine Elements—Integration Of Some Proposals, in *Proc. AEDS 2004 Workshop, The Design Society—AEDS-SIG*, November 11–12, 2004, Pilsen, Czech Republic, 2004, on CD-ROM.

[169] Eder, W.E., Design Sciences—An Overview, in *Proc. AEDS 2004 Workshop, The Design Society—AEDS-SIG*, November 11–12, 2004, Pilsen, Czech Republic; and *Proc. PhD 2004, 2nd International PhD Conference on Mechanical Engineering*, November 8–10, 2004, Srni, Czech Republic, both on CD-ROM.

[170] Eder, W.E., Survey of Pedagogics Applicable to Design Education: An English-Language Viewpoint, *Int. J. Eng. Edu.*, Vol. 21, No. 3, 2005, pp. 480–501.

[171] Eder, W.E. and Hubka, V., Curriculum, Pedagogics and Didactics of Design Education, *J. Eng. Design*, Vol. 16, No. 1, 2005, pp. 45–61.

[172] Eder, W.E. and Weber, C., Comparisons of Design Theories in *Proc. AEDS 2006 Workshop*, October 27–28, 2006, Pilsen, Czech Republic, 2006, on CD-ROM.

[173] Eekels, J., The Engineer as Designer and as a Morally Responsible Individual, *J. Eng. Design*, Vol. 5, No. 1, 1994, pp. 7–23.

[174] Eekels, J., On the Fundamentals of Engineering Design Science: The Geography of Engineering Design Science. *J. Eng. Design*, Vol. 11, No. 4, 2000, pp. 377–397; Vol. 12, No. 3, 2001, pp. 255–281.

[175] Egan, K., *The Educated Mind: How Cognitive Tools Shape Our Understanding*, Chicago, IL: University of Chicago Press, 1997.

[176] Ehrlenspiel, K., Vorschläge zur Integration von methodischem Konstruieren und Maschinenelemente (Suggestions for Integration of Methodical Designing and Machine Elements) in *Proc. Workshop Methodisches Konstruieren und Maschinenelemente*, Milano: Politecnico, pp. 1–13, 1984.

[177] Ehrlenspiel, K. and John, T., Inventing by Design Methodology, in [143], Vol. 1, 1987, pp. 29–37.

[178] Ehrlenspiel, K., *Integrierte Produktentwicklung—Methoden für Prozeßorganisation, Produkterstellung und Konstruktion* (Integrated Product Development—Methods for Process Organization, Product Realization and Design), München: Carl Hanser Verlag, 1995.

[179] Ehrlenspiel, K., Kiewert, A. and Lindemann U, *Kostengünstiges Entwickeln und Konstruieren* (Cost Favorable Developing and Designing), Berlin: Springer-Verlag, 2005, 5th ed. (1st edn., 1985).

[180] Eisenberger, M., The Mechanical Strength Critic in CASE, Report EDRC 12–36–90, Pittsburgh, PA: The Engineering Design Research Center, Carnegie-Mellon University, 1990.

[181] Ellis, G.W., Rudnitsky, A. and Silverstein, B., Using Concept Maps to Enhance Understanding in Engineering Education, *Int. J. Eng. Edu.*, Vol. 20, No. 6, 2004, pp. 1012–1021.

[182] Eloranta, K.T., Hilliaho, E. and Riitahuhta, A., Radical Innovation: A Quest for Conceptual Creativity, in *Proc. TMCE 2004*, Lausanne, 2004, pp. 221–231.

[183] Entwistle, N. and Hanley, M., Personality, Cognitive Style and Students' Learning Strategies, *Higher Education Bulletin*, Vol. 6, No. 1 (Winter 1977–78), Institute for Post-Compulsory Education University of Lancaster, pp. 23–43; and *2nd Congress of the European Association for Research and Development in Higher Education*, Louvain-la-Neuve, August 30 to September 3, 1976; and *Proc. Congress Instructional Design in Higher Education: Innovations in Curricula and Teaching* (ed. A. Bonboir, 2 vols.), Louvain-la-Neuve, 1977, pp. 105–124.

[184] Eppinger, S.D., Whitney, D.E., Smith, R.P. and Gebala, D.A., Organizing the Tasks in Complex Design Projects, in *ASME Conference on Design Theory and Methodology*, New York: ASME, 1990, pp. 39–46.

[185] Eppinger, S.D., Module Based Approaches to Managing Concurrent Engineering, in [312], 1991, pp. 171–176.

[186] Ernzer, M. and Birkhofer, H., Life Cycle Design for Companies—Scaling Life Cycle Design Methods to the Individual Needs of a Company, in [205], 2003, pp. 393–394.

[187] Eversheim, W., Schuh, G. and Caesar, C., Variantenvielfalt in der Serienproduktion (Variant Abundance in Series Production), *VDI-Z*, Vol. 130, No. 12, 1988, pp. 45–49.

[188] Faires, V.M., *Design of Machine Elements* (4th edn.), New York: Macmillan, 1965.

[189] Feilden, G.B.R., *Engineering Design, Report of Royal Commission*, London: HMSO, 1963.

[190] Feldman, K.A. and Paulsen, M.B., *Teaching and Learning in the College Classroom*, Needham Heights, MA: Simon & Schuster (Ginn Press), 1994.

[191] Ferreirinha, P., Meier, P. and Hubka, V., Kalkulationssystem für Neuentwicklungen (Cost Calculation System for New Developments), *Schw. Masch.*, Vol. 82, No. 9, 1982.

[192] Ferreirinha, P., Herstellkostenberechnung von Maschinenteilen in der Entwurfsphase mit dem HKB Programm (Manufacturing Cost Calculation of Machine Parts in the Layout Phase with the HKB Program), in V. Hubka (ed.), *WDK 12: Theory and Practice of Engineering Design in International Comparison, Proc. ICED 85 Hamburg*, Zürich: Heurista, 1985, pp. 461–467.

[193] Ferreirinha, P., Konstruktionswissenschaft und Rechnerunterstützte Modellierung des Konstruktionsprozesses (Design Science and Computer Supported Modeling of the Design Process), in [143], 1987, pp. 455–462.

[194] Ferreirinha, P., Hubka, V., Andreasen, M.M. and Rosenberg, R., Rechnerunterstütztes Entwerfen und Detaillieren von Konstruktionsgruppen (Computer Supported Layout and Detailing of Design Groups), in [306], Vol. 2, 1988, pp. 105–116.

[195] Ferreirinha, P., Oral presentation about an *Industry Application*, given at the Workshop WDK/MeKoME, Rigi Kaltbad, March 1988.

[196] Ferreirinha, P., *Rechnerunterstützte Vorkalkulation im Maschinenbau für Konstrukteure und Arbeitsvorbereiter mit HKB* (Computer Assisted Pre-Calculation in Mechanical Engineering for Designers and Production Planners with HKB), in [311], 1990, pp. 1346–1353.

[197] Ferreirinha, P., Hubka, V. and Eder, W.E., Early Cost Calculation—Reliable Calculation, not just Estimation, in P.J. Guichelaar (ed.), *Design for Manufacturability 1993 Conference Proc*, New York: ASME, DE, Vol. 52, 1993, pp. 97–104.

[198] Fink, D.G., *Electronics Engineers' Handbook* (4th edn.), New York: McGraw-Hill, 1996.

[199] Fink, D.G. and Beaty, H.W., *Standard Handbook for Electrical Engineers* (13th edn.), New York: McGraw-Hill, 1993.

[200] Finniston, Sir M.F., *Engineering Our Future: Committee of Inquiry Report into the Engineering Profession*, Cmnd. 7794, London: HMSO, 1980.

[201] Fisher, R.A., *The Design of Experiments* (6th edn.), London: Oliver & Boyd, 1951.

[202] Fisher, R.A. and Yates, F., *Statistical Tables for Biological, Agricultural, and Medical Research*, Edinburgh: Oliver & Boyd, 1957.

[203] Flursheim, C., *Engineering Design Interfaces*, London: Design Council, 1977.

[204] Flurscheim, C.H., *Industrial Design in Engineering: A Marriage of Techniques*, London: The Design Council and Berlin/Heidelberg: Springer-Verlag, 1983.

[205] Folkeson, A., Gralén, K., Novell, M. and Sellgren, U. (eds.), *DS 31—Proc. ICED 2003 Stockholm*, Glasgow: The Design Society, 2003.

[206] Frank, H., *Kybernetische Grundlagen der Pädagogik* (Cybernetic Fundamentals for Pedagogics), Baden-Baden: Agis-Verlag, 1969.

[207] Franke, H-J., Löffler, S. and Deimel, M., Increasing the Efficiency of Design Catalogues by Using Modern Data Processing Technologies, in *Proc. International Design Conference—Design 2004*, Dubrovnik, 2004, pp. 853–858.

[208] Frankenberger, E., Badke-Schaub, P. and Birkhofer, H., Factors Influencing Design Work: Empirical Investigations of Teamwork in Engineering Design Practice, in [485], 1997, pp. 387–392.

[209] Frankenberger, E. and Badke-Schaub, P., Integration of Group, Individual and External Influences in the Design Process, in [210], 1997.

[210] Frankenberger, E., Badke-Schaub, P. and Birkhofer, H. (eds.), *Designers: The Key to Successful Product Development*, Berlin/Heidelberg: Springer-Verlag, 1997.

[211] Frankenberger, E., *Arbeitsteilige Produktentwicklung* (Product Development with Task Distribution), Fortschrittsberichte VDI Reihe 1, No. 291, Düsseldorf: VDI, 1997.

[212] Frankenberger, E. and Badke-Schaub, P., Role of Critical Situations in Design Processes and Education, in *Proc. WDK International Workshop PDE—Pedagogics in Design Education*, Pilsen: West Bohemia University, 1998 (on CD-ROM).

[213] Freeman, P., *Lubrication and Friction*, London: Pitman, 1962.

[214] French, M.J., *Conceptual Design for Engineers* (2nd edn., revised edition of Engineering Design: The Conceptual Stage), London: Design Council and Berlin: Springer-Verlag, 1985.

[215] French, M.J., *Invention and Evolution; Design in Nature and Engineering*, Cambridge: University Press, UK, 1988.

[216] Fricke, G. and Pahl, G., Zusammenhang zwischen Personenbedingtem Vorgehen und Lösungsgüte (Relationship between Personally Conditioned Procedure and Quality of Solution), in [312], 1991, pp. 331–341.

[217] Frost, M., *Value for Money—The Techniques of Cost–Benefit Analysis*, Aldershot: Gower, 1971.

[218] Frost, R.B., Why Does Industry Ignore Design Science?, *J. Eng. Design*, Vol. 10, No. 4, 1999, pp. 301–304.

[219] Frost, R.B., A Suggested Taxonomy for Engineering Design Problems, *J. Eng. Design*, Vol. 5, No. 4, 1994, pp. 399–410.

[220] Fulcher, A.J. and Hills, P., Towards a Strategic Framework for Design Research, *J. Eng. Design*, Vol. 7, No. 2, 1996, pp. 183–194.

[221] Gadamer, H.-G., *Truth and Method* (2nd edn.), London: Sheed and Ward, 1985.

[222] Gage, W.L., *Value Analysis*, London: McGraw-Hill, 1967.

[223] Gagné, R.M., *The Conditions of Learning* (3rd edn.), New York: Holt Rinehart & Winston, 1977.

[224] Gallagher, W., *Just the Way You Are: How Heredity and Experience Create the Individual*, New York: Random House, 1996.

[225] Gaylord, E.H. and Gaylord, C.N., *Structural Engineering Handbook* (2nd edn.), New York: McGraw-Hill, 1979.

[226] Gerö, J.S. and Kannengiesser, U., The Situated Function–Behaviour–Structure Framework, *Des. Stu.*, Vol. 25, No. 4, 2003, pp. 373–391.

[227] Gerö, J.S., Situated Design Computing: Introduction and Implications, in *Proc. International Design Conference—Design 2004*, Dubrovnik, May 18–21, 2004, pp. 27–36, on CD-ROM.

[228] Giampà, F., Muzzapappa, M. and Rizzuti, S., Design by Function: A Methodology to Support Designer Creativity, in *Proc. International Design Conference—Design 2004*, Dubrovnik, 2004, pp. 225–232.

[229] Giapoulis, A., Schlüter, A., Ehrlenspiel, K. and Günther, J., Effizientes Konstruieren durch Generierendes und Korrigierendes Vorgehen, in [316], 1995, pp. 477–483.

[230] Gill, R., Gordon, S., Moore, J. and Barbera, C., The Role of Knowledge Structures in Problem Solving, in *Proc. ASEE Annual Conference 1988*, Washington: ASEE, 1988, pp. 583–590.

[231] Glaser, B.L. and Strauss, A.L., *The Discovery of Grounded Theory: Strategies for Qualitative Research*, Chicago, IL: Aldine Publ., 1967.

[232] Glynn, I., *An Anatomy of Thought: The Origin and Machinery of the Mind*, London: Weidenfeld & Nicholson, 1999.

[233] Goleman, D., *Emotional Intelligence*, New York: Bantam Books, 1995.

[234] Gordon, W.J.J., *Synectics: The Development of Creative Capacity*, New York: Harper & Row, and Collier, 1961.

[235] Gouvinhas, R.P. and Corbett J., A Design Method Developed for Production Machinery Companies, in [29], 2001, pp. 67–74.

[236] Grabowski, H., Lossack, R.-S. and Bruch, C., Requirements Development in Product Design—A State- and State Transition-Based Approach, in *Proc. TMCE 2004 Lausanne*, 2004, pp. 1087–1088, and on CD-ROM.

[237] Graedel, T.E., *Streamlined Life Cycle Assessment*, Upper Saddle River, NJ: Prentice-Hall, 1998.

[238] Green, R.E. (ed.), *Machinery's Handbook* (25th edn.), New York: Industrial Press, 1996.

[239] Gregory, S.A. (ed.), *The Design Method*, London: Butterworths, 1966.

[240] Gregory, S.A., Design Science, in [239], 1966, pp. 323–330.

[241] Gregory, S.A., *Creativity and Innovation in Engineering*, London: Butterworths, 1972.

[242] Guilford, J.P., Traits of Creativity, in H.H. Anderson (ed.), *Creativity and Its Cultivation*, New York: Harper & Row, 1959, pp. 142–161 (Reprinted in P.E. Vernon (ed.), *Creativity*, Harmondsworth: Penguin, 1970, pp. 167–188).

[243] Guilford, J.P. and Hoepfner, R., *The Analysis of Intelligence*, New York: McGraw-Hill, 1971.

[244] Guine, J.G. (ed.), *Handbook on Life Cycle Assessment*, Dordrecht: Kluwer Academic, 2002.

[245] Günther, J. and Ehrlenspiel, K., How Do Designers from Practice Design?, in [210], 1997, pp. 85–97.

[246] Halcomb, J., *Project Manager's PERT/CPM Handbook*, Sunnyvale, CA: Halcomb Assoc., 1966.

[247] Hales, C., Designer as Chameleon, *Des. Stu.*, Vol. 6, No. 2, 1985, pp. 111–114.

[248] Hales, C., *Analysis of the Engineering Design Process in an Industrial Context* (2nd edn.), Winetka, IL: Gants Hill Publ., 1991 (1st edn. 1987).

[249] Hales, C., *Managing Engineering Design*, Harlow, Essex: Longman Scientific & Technical, 1993.

[250] Hall, A.D., *A Methodology for Systems Engineering*, Princeton, NJ: Van Nostrand, 1962.

[251] Hansen, F., *Konstruktionssystematik* (Design Systematics, 2nd edn.), Berlin: VEB Verl. Technik, 1966.

[252] Hansen, F., *Adjustment of Precision Mechanisms*, London: Iliffe, 1970.

[253] Hansen, F., *Konstruktionswissenschaft—Grundlagen und Methoden* (Design Science—Basis and Methods), München: Carl Hanser, 1974.

[254] Harary, K. and Donahue, E., *Who Do You Think You Are? The Berkeley Personality Profile*, San Francisco, CA: HarperSanFrancisco, 1994.

[255] Harrington, C.L., To Retain Rights, Patent Holders Must Act Swiftly, *Mech. Eng.* (ASME), Vol. 114, No. 7, July 1992, p. 77.

[256] Harris, J.S., New Product Profile Chart, *Chem. Eng. News*, Vol. 39, 1961, pp. 110–118, and [490], 1990, pp. 65–72.

[257] Hatchuel, A. and Weil, B., A New Approach of Innovative Design: An Introduction to C-K Theory, in [205], 2003, pp. 109–110, executive summary (on CD-ROM).

[258] Hatchuel, A., LeMasson, P. and Weil, B., The Design of Science-Based Products: An Interpretation and Modelling with C-K Theory, in *Proc. International Design Conference—Design 2006*, Dubrovnik, 2006, pp. 33–44.

[259] Hauser, J.R. and Clausing, D., The House of Quality, *Harvard Bus. Rev.*, May–June, 1988, pp. 63–73.

[260] Heinrich, W., Eine systematische Betrachtung der konstruktiven Entwicklung technischer Erzeugnisse, *Maschinenbautechnik*, No. 5, 1973 to No. 12, 1976.

[261] Heller, E.D., *Value Management: Value Engineering and Cost Reduction*, Reading, MA: Addison-Wesley, 1971.

[262] Hellfritz, H. (ed.), *Innovation via Galeriemethode* (Innovation via the Art Gallery Method), Königstein/Taunus, 1978.

[263] Herrmann, N., *The Creative Brain*, Lake Lure, NC: Brain Books, 1990.

[264] Hertzberg, F., Mausner, B. and Snyderman, B., *The Motivation to Work*, New York: Wiley, 1959.

[265] Hertzberg, F., *Work and the Nature of Man*, Harcourt Brace & World, 1966.

[266] Hertzberg, F., One More Time: How Do You Motivate Employees?, *Harvard Bus. Rev.*, Vol. 46, 1968, pp. 53–62.

[267] Heßling, T. and Lindemann, U., Introduction of the Integrated Product Policy in Small and Medium-Sized Enterprises, in *Proc. International Design Conference—Design 2004*, Dubrovnik, 2004, pp. 1057–1062.

[268] Hix, C.F., Jr. and Alley, R.P., *Physical Laws and Effects*, New York: Wiley, 1958.

[269] Hofmeister, K., Quality Function Deployment: Market Success Through Customer-Driven Products, in E. Graf and I.S. Saguay (eds.), *Food Products Development from Concept to the Marketplace*, New York: Van Nostrand Reinhold, 1990.

[270] Hofstadter, D.R., *Gödel, Escher, Bach: An Eternal Golden Braid* (20th Anniversary Edition), New York: Basic Books, 1999.

[271] Hollins, B. and Pugh, S., *Successful Product Design*, London: Butterworths, 1989.

[272] Holt, K., *Product Innovation*, London: Newnes-Butterworths, 1977.

[273] Holt, K., Brainstorming—From Classics to Electronics, in [153], 1996, pp. 113–118.

[274] Hongo, K., Revitalization of Machine Elements in Machine Design Education, in [28], 1989, pp. 1139–1146.

[275] Hosnedl, S., Higher Effectiveness in the Creation and Applicability of CAD Software, in *Proc. 11th Polish Conference on Theory of Machines and Mechanisms*, Zakopane, Polsko: Politechnika Šlaska, 1987, pp. 69–82.

[276] Hosnedl, S. Comments to Creativity in Design Education, in [154], 1996, pp. 153–155.

[277] Hosnedl, S., Borusikova, I. and Wilhelm, W., TQM Methods from the Point of View of Design Science, in [485], Vol. 1, 1997, pp. 391–394.

[278] Hosnedl, S. and Eder, W.E., Engineering Design Science—Advances in Definitions, in *Proc. AEDS—Workshop, Pilsen, Czech Republic*, 2005, on CD-ROM.

[279] Horvath, I., On Methodical Characteristics of Engineering Design Research, in *Proc. International Design Conference—Design 2004*, Dubrovnik, May 18–21, 2004, on CD-ROM.

[280] Howe, A.E., Cohen, P.R., Dixon, J.R. and Simmons, M.K., Dominic: A Domain-Independent Program for Mechanical Design, *Int. J. Artif. Intell. Eng.*, Vol. 1, No. 1, 1986.

[281] Hubka, V., Der grundlegende Algorithmus für die Lösung von Konstruktionsaufgaben (Fundamental Algorithm for the Solution of Design Problems) in XII. Internationales Wissenschaftliches Kolloquium der Technischen Hochschule Ilmenau, Heft 12 Sektion L–Konstruktion, Ilmenau: T.H.I., 1967, pp. 69–74.

[282] Hubka, V., Der Konstrukteur als Informationsverbraucher, *Schw. Masch.*, Bd. 73, H. 38, 40 and 42, 1973.

[283] Hubka, V., *Theorie der Maschinensysteme*, Berlin: Springer-Verlag, 1974.

[284] Hubka, V., Bewerten und Entscheiden beim Konstruieren (Evaluating and Deciding during Designing), *Schw. Masch.*, Bd. 74, H. 20, pp. 45–49, H. 22, pp. 41–46, H. 24, pp. 53–56, 1974, and [490], 1990, pp. 13–26.

[285] Hubka, V., Intuition und Konstruktionsgefuhl, *Schw. Masch.*, Bd. 75, H. 50, 1975.

[286] Hubka, V., Klärung der Aufgabestellung, *Schw. Masch.*, Bd. 76, No. 3 and 13, 1976.

[287] Hubka, V., *Theorie der Konstruktionsprozesse* (Theory of Design Processes), Berlin: Springer-Verlag, 1976.

[288] Hubka, V., Darstellen und Modellieren beim Konstruieren (Representing and Modeling in Designing), *Schw. Masch.*, Bd. 76, H. 33, 35 and 37, 1976.

[289] Hubka, V., Arbeitsbedingungen beim Konstruieren (Working Conditions in Design Engineering), *Schw. Masch.*, Bd. 77, H. 16, pp. 56–59, H. 18, pp. 34–37, H. 20, pp. 57–58, H. 22, pp. 32–35, H. 24, 32–33, 1977.

[290] Hubka, V., *Konstruktionsunterricht an Technischen Hochschulen* (Design Education in Universities), Konstanz: Leuchtturm Verlag, 1978.

[291] Hubka, V., Konstruieren verlangt ein neues Wirtschaftlichkeitsdenken (Designing Demands New Economic Thinking), *Management Zeitschrift*, Vol. 49, No. 4, 1979.

[292] Hubka, V., Andreasen, M.M., Eder, W.E., Pighini, U., Schlesinger, A. and Wyss, M. *WDK 3: Fachbegriffe der wissenschaftlichen Konstruktionslehre in 6 Sprachen* (Terminology of the science of design engineering in 6 languages), Zürich: Heurista, 1980.

[293] Hubka, V., Planen der Konstruktionsarbeit (Planning of Design Work), *Schw. Masch.*, Bd. 80, H. 50 and 53, 1980.

[294] Hubka, V. (ed.), *WDK 5: Konstruktionsmethoden in Übersicht, Proc. ICED 81 Rome*, Vol. 1 of *Proc. ICED 81 Rome*, Milano: tecniche nuove, 1981.

[295] Hubka, V. (ed.), *WDK 9: Dietrych zum Konstruieren* (Dietrych about Designing), Zürich: Heurista, 1982.

[296] Hubka, V., Systematische Heuristik (Systematic Heuristics), *Schw. Masch.*, Bd. 83, H. 38, 1983, pp. 33–35.

[297] Hubka, V. (ed.), *WDK 8: Konstruktionsmethoden—Reading*, Zürich: Heurista, 1982.

[298] Hubka, V. and Andreasen, M.M. (eds.), *WDK 10: CAD, Design Methods, Konstruktionsmethoden: Proc. ICED 83 Copenhagen* (2 vols.), Zürich: Heurista, 1983.

[299] Hubka, V., *Theorie Technischer Systeme* (Revised from Theorie der Maschinensysteme 1974, 2nd edn.), Berlin: Springer-Verlag, 1984.

[300] Hubka, V. and Ferreirinha, P., Rechnergestützte Kalkulation im Maschinenbau (Computer Assisted Cost Calculation in Mechanical Engineering), *Schw. Masch.*, Bd. 85, H. 6, 8 and 10, 1985.

[301] Hubka, V. (ed.), *WDK 11: Führung im Konstruktionsprozess—Reading* (Leadership in the Design Process), Zürich: Heurista, 1985.

[302] Hubka, V. and Ropohl, G., Was ist ein technisches System? Zur Grundlegung der Konstruktionswissenschaft (What is a Technical System? To the Foundation of Design Science), *VDI-Z*, Vol. 128, No. 22, 1986, pp. 864–874.

[303] Hubka, V. and Schregenberger, J.W., Paths Towards Design Science, in [143], 1987, pp. 3–14.

[304] Hubka, V. and Eder, W.E., *Theory of Technical Systems: A Total Concept Theory for Engineering Design*, New York: Springer-Verlag, 1988 (completely revised translation of [299], 1984).

[305] Hubka, V., Andreasen, M.M. and Eder, W.E., *Practical Studies in Systematic Design*, London: Butterworths, 1988 (English edition of WDK 4–Fallbeispiele, Zürich: Heurista, 1981 and 1983).

[306] Hubka, V., Baratossy, J. and Pighini, U. (eds.), *WDK 16: Proc. ICED 88*, Budapest: GTE and Zürich: Heurista, 1988.

[307] Hubka, V. and Schregenberger, J.W., *Eine neue Systematik konstruktionswissenschaftlicher Aussagen—Ihre Struktur und Funktion* (A New Systematic Order for Design-Scientific Statements—Its Structure and Function), in [306], Vol. 1, 1988, pp. 103–117.

[308] Hubka, V., A Curriculum Model—Applying the Theory of Technical Systems, in W.E. Eder and G. Kardos (guest eds.), *Special Issue on Education for Engineering Design*, *Int. J. Appl. Eng. Ed.*, Vol. 4, No. 3, 1988, pp. 185–192.

[309] Hubka, V. and Schregenberger, J.W., Eine Ordnung Konstruktionswissenschaftlicher Aussagen (An Arrangement for Classifying Design-Scientific Statements), *VDI-Z*, Vol. 131, No. 3, 1989, pp. 33–36.

[310] Hubka, V. and Eder, W.E., Design Education and Design Science, in *Proc. S. Neaman International Workshop on The Place of Design in the Engineering School*, Haifa, Israel: S. Neaman Press, 1989, pp. 31–55.

[311] Hubka, V. and Kostelic, A., *WDK 19: Proc. 1990 International Conference on Engineering Design*, ICED 90 Dubrovnik, Zürich: Heurista and Judeko, 1990.

[312] Hubka, V. (ed.), *WDK 20: Proc. ICED 91 Zürich*, Zürich: Heurista, 1991.

[313] Hubka, V., *Lagebericht Konstruktionswissenschaft* (Situation Report Design Science), in [312], 1991, pp. 761–768.

[314] Hubka, V. and W.E. Eder, *Engineering Design*, Zürich: Heurista, 1992 (2nd edn., Hubka, V., *Principles of Engineering Design*, London: Butterworth Scientific, 1982, transl. and ed. W.E. Eder from Hubka, V., *WDK 1—Allgemeines Vorgehensmodell des Konstruierens* (General Procedural Model of Designing), Zürich, Heurista, 1980).

[315] Hubka, V. and Eder, W.E., *Design Science: Introduction to the Needs, Scope and Organization of Engineering Design Knowledge*, London: Springer-Verlag, 1996, http://deseng.ryerson.ca/DesignScience/ (completely revised edition of Hubka, V. and Eder, W.E., *Einführung in die Konstruktionswissenschaft* (Introduction to Design Science), Berlin, Springer-Verlag, 1992).

[316] Hubka, V. (ed.), *WDK 23—Proc. International Conference on Engineering Design—ICED 95 Praha*, Zürich, Heurista, 1995.

[317] Hubka, V., Design Education in an Engineering Curriculum, International Workshop EED—Education for Engineering Design, November 23–24, 2000, State Scientific Library, Pilsen, Czech Republic, 2000.

[318] Hubka, V. and Eder, W.E., Theory of Technical Systems and Engineering Design Synthesis, in A. Chkrabarti (ed.), *Engineering Design Synthesis*, London: Springer-Verlag, 2001, pp. 49–66.

[319] Hubka, V. and Eder, W.E., Functions Revisited, in [29], Vol. 1, IMechE Paper C586/102, 2001, pp. 69–76.

[320] Hubka, V. and Eder, W.E., Pedagogics of Engineering Design Education, *Int. J. Appl. Eng. Edu.*, Vol. 19, No. 6, 2003, pp. 798–809.

[321] Hundal, M.S., *Systematic Mechanical Designing: A Cost and Management Perspective*, New York: ASME Press, 1977.

[322] Hundal, M.S., Engineering Management for Rapid Product Development, in [491], 1993, pp. 588–595.

[323] Imai, M., *Kaizen*, New York: Random House, 1986.

[324] Isaksson, J., Mapping Haptic Properties in Product Development Work, in *Proc. International Design Conference—Design 2004*, Dubrovnik, 2004, pp. 1069–1074.

[325] Ishikawa, K., *Guide to Quality Control*, Tokyo: Asian Productivity Press, 1982.

[326] Ishikawa, K., *What is Total Quality Control? The Japanese Way*, Englewood Cliffs, NJ: Prentice-Hall, 1985.

[327] Ivashkov, M. (unpubl.) Establishing Priority of TRIZ Inventive Principles in Early Design, TU Eindhoven, Building and Architecture Adanpad 5, 2005.

[328] Jackson, S.L., *The ISO 14001 Implementation Guide*, New York: Wiley, 1997.

[329] Janhager, J. and Hagman, L.A., Approaches for the Identification of Users and their Relations to the Product, in *Proc. TMCE 2004 Lausanne*, pp. 345–354, and on CD-ROM, 2004.

[330] Jarratt, T., Eckert, C. and Clarkson P.J., The Benefits of Predicting Change in Complex Products: Application Areas of a DSM-Based Prediction Tool, in *Proc. International Design Conference—Design 2004*, Dubrovnik, 2004, pp. 303–308.

[331] Jarret, J. and Clarkson P.J., The Surge-Stagnate Model for Complex Design, *J. Eng. Design*, 13, September 2002, pp. 189–196.

[332] Jensen, P.W., *Classical and Modern Mechanisms for Engineers and Inventors*, New York: Marcel Dekker, 1991.

[333] Johnson, S.D., Siriya, C., Yoon, S.W., Berrett, J.V. and LaFleur, J., The Development and Group Processes of Virtual Learning Teams, *Comput. Educ.*, Vol. 39, No. 4, 2002.

[334] Jones, J.Ch. and Thornley, D.G., *Conference on Design Methods*, Oxford: Pergamon Press, 1963.

[335] Jones, J.Ch., *Design Methods: Seeds of Human Futures* (2nd edn.), New York: Wiley, 1980.

[336] Jones, J.C., Article: 'If I Were You,' *Design Research Newsletter*, No. 61, April–June 1998, p. 3.

[337] Julier, G., *The Culture of Design*, London: Sage Publications, 2000.

[338] Jung, C.G., *Psychologische Typen*, Zürich, 1921; and *Psychological Types*, London: Routledge, 1923, and Princeton, NJ: Princeton University Press, 1971.

[339] Juran, J.M. and Gryna, F.M., Jr., *Quality Planning and Analysis: From Product Development Through Use* (2nd edn.), New York: McGraw-Hill, 1980.

[340] Juran, J.M., *Management of Quality* (4th edn.), New York: J.M. Juran, 1981.

[341] Juran, J.M., *Juran on Planning for Quality*, New York: Free Press, 1987.

[342] Juvinall, R.C. and Marshek, K.M., *Fundamentals of Machine Component Design* (3rd edn.), New York: Wiley, 2000.

[343] Kahn, W.A. and Raouf, A., *Standards for Engineering Design and Manufacture* Dekker Mechanical Engineering, Vol. 196, New York: Marcel dekker, 2005.

[344] Kato, T, Ikeyama, T. and Matsuoka, Y., Basic Study on Classification Scheme for Robust Design Methods, in *Proc. 1st International Conference on Design Engineering and Science*, Vienna, Austria, October 28–31, 2005, pp. 37–42.

[345] Katz, R.H., *Information Management for Engineering Design Applications*, New York: Springer-Verlag, 1985.

[346] Katzenbach, J.R. and Smith, D.K., *The Wisdom of Teams: Creating the High-Performance Organization*, Boston, MA: Harvard Business School Press, 1993.

[347] Katzenbach, J.R., *Teams at the Top: Unleashing the Potential of Both Teams and Individual Leaders*, Boston, MA: Harvard Business School Press, 1998.

[348] Kazakçi, A.O. and Tsoukias, A., Extending the C-K Theory: A Theoretical Background for Personal Design Assistants, *J. Eng. Design*, Vol. 16, No. 4, August 2005, pp. 399–411.

[349] Kesselring, R., *Bewertung von Konstruktionen* (Evaluation of Designs), Düsseldorf, VDI-Verlag, 1951.

[350] Kesselring, F., *Technische Kompositionslehre* (Study of Technical Composition), Berlin/Heidelberg: Springer-Verlag, 1954.

[351] Klaus, G., *Kybernetik in philosophischer Sicht* (Cybernetics in Philosophical View, 4th edn.), Berlin: Dietz Verlag, 1965.

[352] Klaus, G., *Wörterbuch der Kybernetik* (Dictionary of Cybernetics), Frankfurt: Fischer, 1969.

[353] Knoop, R. and Cranton, P., *Jung's Psychological Types*, Sneedsville, TN: Psychological Types Press, Publ. 34, 1995.

[354] Knoop, R., *Conflict Resolution and Type*, Sneedsville, TN: Psychological Types Press, Publ. 20, 1995.

[355] Knoop, R., *Problem Solving and Type*, Sneedsville, TN: Psychological Types Press, Publ. 22, 1995.

[356] Knoop, R., *Team Work and Type*, Sneedsville, TN: Psychological Types Press, Publ. 23, 1995.

[357] Knoop, R., *Leadership and Type*, Sneedsville, TN: Psychological Types Press, Publ. 29, 1995.

[358] Knoop, R., *Charismatic and Transformational Leadership and Type*, Sneedsville, TN: Psychological Types Press, Publ. 30, 1995.

[359] Knoop, R., *Management and Type*, Sneedsville, TN: Psychological Types Press, Publ. 31, 1995.

[360] Knoop, R. and Cranton, P., *Manager-Staff Relations and Type*, Sneedsville, TN: Psychological Types Press, Publ. 32, 1995.

[361] Knoop, R., *Decision-Making Styles and Type*, Sneedsville, TN: Psychological Types Press, Publ. 33, 1995.

[362] Knoop, R., *Communication and Type*, Sneedsville, TN: Psychological Types Press, Publ. 19, 1995.

[363] Knoop, R., *Stress and Type*, Sneedsville, TN: Psychological Types Press, Publ. 21, 1996.

[364] Knoop, R. and Cameron, P., *Work Preferences and Type*, Sneedsville, TN: Psychological Types Press, Publ. 24, 1996.

[365] Koen, B.V., *Discussion of the Method: Conducting the Engineer's Approach to Problem Solving*, New York: Oxford University Press, 2003.

[366] Koji, Y., Kitamura, Y. and Mizoguchi, R., Towards Modeling Design Rational of Supplementary Functions in Conceptual Design, in *Proc. TMCE 2004 Lausanne*, 2004, pp. 117–130, and on CD-ROM.

[367] Kolb, D.A., *Experiential Learning: Experience as the Source of Learning and Development*, Englewood Cliffs, NJ: Prentice-Hall, 1984.

[368] Kollenburg, P.A.M. van and Bakker, R.M., Competence Related Education of Engineers, in *Proc. TMCE 2006 Ljubljana*, 2006, pp. 1021–1028.

[369] Koller, R., *Konstruktionsmethode für Maschinen-, Geräte- und Apparatebau* (Design Method for Machine, Device and Apparatus Construction), Berlin/Heidelberg: Springer-Verlag, 1979.

[370] Koller, R., *Konstruktionslehre für den Maschinenbau* (Study of Designing for Mechanical Engineering, 2nd edn.), Berlin/Heidelberg: Springer-Verlag, 1985.

[371] Krathwohl, D.R., Bloom, B.S. and Masia, B.B. (ed.) *Taxonomy of Educational Objectives: Affective Domain*, New York: McKay, 1965.

[372] Krchl, H., Erfolgreiche Produkte durch Value Management, in [312], 1991, pp. 246–253.

[373] Kremer, D. and Birnzeisler, B., Improving the Efficiency and Innovation Capability of Collaborative Engineering: The Knowledge Integration Training for Teams (KITT), in *Proc. TMCE 2004 Lausanne*, 2004, pp. 935–943, and on CD-ROM.

[374] Kristjansson, A.H., Jensen, T. and Hildre, H.P., The Term Platform in the Context of a Product Developing Company, in *Proc. International Design Conference—Design 2004*, Dubrovnik, 2004, pp. 325–330.

[375] Kuate, G., Choulier, D., Deniaud, S. and Michel, F., Protocol Analysis for Computer-Aided Design: A Model of Design Activities, in *Proc. TMCE 2006 Ljubljana*, 2006, pp. 693–704.

[376] Kuhn, T.S., *The Structure of Scientific Revolutions* (2nd edn.), Chicago, IL: University of Chicago Press, 1970.

[377] Kuhn, T.S., *The Essential Tension: Selected Studies in Scientific Tradition and Change*, Chicago, IL: University of Chicago Press, 1977.

[378] Latham, R.L., *A Guide to the Problem Analysis by Logical Approach System* (5th edn.), Aldermaston: UKAEA, AWRE, 1968.

[379] Layard, R. (ed.), *Cost–Benefit Analysis*, Harmondsworth: Penguin, 1972.

[380] Leech, D.J., *Management of Engineering Design*, London: Wiley, 1972.

[381] Lent, D., *Analysis and Design of Mechanisms* (2nd edn.), Englewood Cliffs, NJ: Prentice-Hall, 1970.

[382] Lewis, P., Aldridge, D. and Swamidass, M., Assessing Teaming Skills Acquisition on Undergraduate Project Teams, *J. Eng. Edu.* (ASEE), Vol. 87, No. 2, April 1998, pp. 149–155.

[383] Leyer, A., *Machine Design*, Glasgow: Blackie, 1974.

[384] Li, M., Stokes, C.A., French, M.J. and Widden, M.B., Function-Costing: Recent Developments, in [491], 1993, pp. 1123–1129.

[385] Lickley, R.L., *Report of the Engineering Design Working Party*, London: SERC, 1983.

[386] Linde, H. and Hill, B., *Erfolgreiches Erfinden—Widerspruchsorientierte Entwicklungsmethodik* (Successful Inventing—Contradiction-Oriented Development Method), Darmstadt: Hoppenstedt, 1991.

[387] Lindeman, U., Birkhofer, H., Meerkamm, and Vajna, S. (eds.), *WDK 26—Proc. ICED 99 Munich*, Technische Universität München, 1999.

[388] Linn, C.E., *Market Dynamics: The Development of Value in Branded Goods, Systems and Services* (2nd edn.), Stockholm: Meta Management AB, 1996.

[389] Linn, C.E., *Brand Dynamics*, Norcross, GA: Institute for Brand Leadership, 1998.

[390] Liu, D.H.F. and Liptak, B.G. (eds.), *Environmental Engineers' Handbook* (2nd edn.), Boca Raton, FL: Lewis Publishers, 1996, pp. 353–361.

[391] Lochner, R.H. and Matar, J.E., *Designing for Quality: An Introduction to the Best of Taguchi and Western Methods of Statistical Experimental Design*, White Plains, NY: Quality Resources, 1990.

[392] López-Mesa, B. and Thompson, G., On the Significance of Cognitive Style and the Selection of Appropriate Design Methods, *J. Eng. Design*, Vol. 17, No. 4, August 2006, pp. 371–386.

[393] Lossack, R.-S., Foundations for a Universal Design Theory—A Design Process Model, in *Proc. Int. Conf. 'The Sciences of Design—The Scientific Challenge for the 21st Century'*, INSA-Lyon, France, March 15–16, 2002.

[394] Lossack, R.-S., Foundations for a Domain Independent Design Theory, in *Annals of 2002 Int. CIRP Design Seminar*, Hong Kong, May 16–18, 2002.

[395] Louridas, P., Design as Bricolage: Anthropology Meets Design Thinking, *Des. Stu.*, Vol. 20, No. 6, November 1999, pp. 517–535.

[396] Lowka, D., *Über Entscheidungen im Konstruktionsprozeß* (About Decisions in the Design Process), Thesis D 17, T.H. Darmstadt, 1976.

[397] Luckman, J., An Approach to the Management of Design, *Oper. Res. Quarterly*, Vol. 18, No. 4, 1967.

[398] Lumsdaine, D., *Creative Problem Solving: Thinking Skills for a Changing World*, New York: McGraw-Hill, 1995.

[399] Manz, C.C., Mansco, D., Neck, C.P., and Manz, K.P., *For Team Members Only: Making Your Workplace Team Productive and Hassle-Free*, New York: AMACOM, 1997.

[400] Marca, D.A. and McGowan, L., *SADT—Structured Analysis and Design Technique*, New York: McGraw-Hill, 1988.

[401] Marples, D.L., The Decisions of Engineering Design, London: Institution of Engineering Designers, 1990; and *IRE Trans. Eng. Manag.*, EM–8 1961, pp. 55–71.

[402] Maslow, A.H., A Theory of Human Motivation, *Psychol. Rev.*, Vol. 50, 1943, pp. 370–396.

[403] Maslow, A.H., *Motivation and Personality*, New York: Harper & Row, 1954.

[404] Matchett, E.D. and Briggs, A.H., Practical Design Based on Method (Fundamental Design Method), in [239], 1966, pp. 183–200.

[405] Matousek, R., *Engineering Design: A Systematic Approach*, London: Blackie, 1963; Original German edition: *Konstruktionslehre des allgemeinen Maschinenbaues*, Berlin/Heidelberg: Springer-Verlag, 1957.

[406] Mayer, H., Stark, H.L. and Ford, R., One Foot in Jail: Mitigating the Influence of Errors on the Outcome of Design Processes for Industrial Plant, in *Proc. International Design Conference—Design 2004*, Dubrovnik, pp. 389–400, 2004.

[407] Maynard, H.B., *Industrial Engineering Handbook* (3rd edn.), New York: McGraw-Hill, 1971.

[408] McKeachie, W.J., *Teaching Tips, A Guidebook for the Beginning College Teacher* (8th edn.), Lexington, MA: Heath, 1986.

[409] McMahon, C.A., Cooke, J.A. and Coleman, P., A Classification of Errors in Design, in [485], 1997, Vol. 3, pp. 119–124.

[410] Meißner, M., Meyer-Eschenbach, A. and Blessing, L., Adapting a Design Process to a New Set of Standards—A Case Study from the Railway Industry, in *Proc. International Design Conference—Design 2004*, Dubrovnik, 2004, pp. 401–408.

[411] Mekid, S., Design Strategy for Precision Engineering: Second-Order Phenomena, *J. Eng. Design*, Vol. 16, February 2005, pp. 63–74.

[412] Merton, R.K., *Social Theory and Social Structure*, Glencoe, IL: Free Press, from Chapter III, paraphrased in Glaser, B.L. and Strauss, A.L., *The Discovery of Grounded Theory: Strategies for Qualitative Research*, Chicago, IL: Aldine Publishing, 1967, p. 2.

[413] Messick, S., *Individuality in Learning: Implications of Cognitive Styles and Creativity for Human Development*, San Francisco, CA: Jossey-Bass, 1976.

[414] Miles, L.D., *Techniques of Value Analysis and Engineering* (2nd edn.), New York: McGraw-Hill, 1972.

[415] Miller, D.W. and Starr, M.K., *The Structure of Human Decisions*, Englewood Cliffs, NJ: Prentice-Hall, 1967.

[416] Miller, G.A., The Magical Number Seven, Plus or Minus Two: Some Limits on Our Capacity for Processing Information, *Psychol. Rev.*, Vol. 63, 1956, pp. 81–97.

[417] Miller, G.A., Information and Memory, *Sci. Am.*, Vol. 195, 1956, pp. 42–46.

[418] Miller, G.A., Human Memory and the Storage of Information, *IRE Trans. Inf. Theory*, Vol. IT-2, No. 3, 1956, pp. 128–137.

[419] Miller, G.A., Galanter, E. and Pribram, K., *Plans and the Structure of Behavior*, New York: Holt, Rinehart & Winston, 1960.

[420] Miller, G.A., Assessment of Psychotechnology, *Am. Psychol.*, Vol. 25, No. 11, 1970, pp. 911–1001.

[421] Morrison, D., *Engineering Design: The Choice of Favourable Systems*, London: McGraw-Hill, 1968.

[422] Mortensen, N.H., Function Concepts for Machine Parts—Contribution to a Part Design Theory, in [387], Vol. 2, 1999, pp. 841–846.

[423] Moulton, A.E. (ed.), *Engineering Design Education*, London: The Design Council, 1976.

[424] Mudge, A.E., *Value Engineering*, New York: McGraw-Hill, 1971.

[425] Müller, J., *Arbeitsmethoden der Technikwissenschaften—Systematik, Heuristik, Kreativität* (Working Methods of Engineering Sciences, systematics, heuristics, creativity), Berlin: Springer-Verlag, 1990.

[426] Müller, J., Akzeptanzbarrieren als berechtigte und ernst zu nehmende Notwehr kreativer Konstrukteure—nicht immer nur böser Wille, Denkträgheit oder alter Zopf (Acceptance Barriers as Justified and Serious Defence Reaction of Creative Designers—Not Always Ill Will, Thinking Inertia or Old Hat), in [312], 1991, pp. 769–776; and Acceptance Barriers as Justified and Serious Defence Reaction of Creative Designers—Not Always Only Ill Will, Thinking Inertia or Old Hat' in [153], 1996, pp. 79–84.

[427] Müller, J., Akzeptanzprobleme in der Industrie, ihre Ursachen und Wege zu ihrer Überwindung (Acceptance Problems in Industry, Their Causes, and Ways to Overcome Them), in G. Pahl (ed.), *Psychologische und pädagogische Fragen beim methodischen Konstruieren* (Psychological and Pedagogic Questions in Systematic Designing), Ladenburger Diskurs, Köln: Verlag TÜV Rheinland, 1994, pp. 247–266.

[428] Murrell, K.F.H., Data on Human Performance for Engineering Designers, *Engineering*, Vol. 184, 1957, in five parts: Part I, August 16, pp. 194–198; Part II, August 23, pp. 247–249; Part III, September 6, pp. 308–310; Part IV, September 13, pp. 344–347; Part V, October 4, pp. 438–440.

[429] Muther, R., *Systematic Layout Planning (SLP)*, Boston, MA: Industrial Education Institute, 1961.

[430] Myers, I.B. and McCaulley, M.H., *Manual: A Guide to the Development and Use of the Myers–Briggs Type Indicator* (2nd edn.), Palo Alto, CA: Consulting Psychologist Press, 1985.

[431] Nadler, G., *Work Systems Design: The IDEALS Concept*, Homewood: Irwin, 1967.

[432] Nadler, G. and Hibino, S., *Breakthrough Thinking: Why We Must Change the Way We Solve Problems, and the Seven Principles to Achieve This*, Rocklin, CA: Prima, 1987.

[433] Nagel, E., *The Structure of Science*, London: Routledge, p. 8, 1961.

[434] Naumann, F.K., *Failure Analysis: Case Histories and Methodology*, Metals Park, Ohio, OH: American Society for Metals, 1983.

[435] Nevala, K., The Knowledge Representations in Paper Machine Design, an essay for the Connet-Course Ckt 221C *Mental and Other Representations*, at the Virutal University, Finland, 2003, http://www.connet.edu.helsinki.fi/FLE/fle_users/nevala

[436] Nevala, K., Constructive Engineering Thinking: Embodying Social Desires, in *Proc. Psykocenter Symposium on Social and Cultural Dimensions of Technological Development*, November 3–4, 2003, University of Jyväskylaä Agora Center, 2003, pp. 158–168.

[437] Nevala, K. and Karhunen, J., Content-Based Design Engineering Thought Research at Oulu University, in S. Hosnedl (ed.), *Proc. AEDS Workshop 2004, Pilsen*, November 11–12, 2004, on CD-ROM.

[438] Nevala, K., *Content-Based Design Engineering Thinking*, Academic Dissertation, University of Jyväskalä, Finland, Jyväskalä: University Printing House, 2005, http://cc.oulu.fi/~nevala (includes [436,437,439,501,502]).

[439] Nevala, K., Mechanical Engineering Way of Thinking in a Large Organization. A Case Study in Paper Machine Industry, in S. Hosnedl (ed.), *Proc. AEDS Workshop 2005*, Pilsen, November 3–4, 2005, on CD-ROM.

[440] Nevala, K., Saariluoma, S. and Karvinen, M., Design Engineering Process from Content-Based Point of View, in *Proc. International Design Conference—Design 2006*, Dubrovnik, 2006, pp. 1543–1550.

[441] Newell, A. and Simon, H., *Human Problem Solving*, Englewood Cliffs, NJ: Prentice-Hall, 1974.

[442] Nezel, I., *Allgemeine Didaktik der Erwachsenenbildung* (General Didactics of Adult Education), Bern: Haupt, 1992.

[443] Norris, K.W., The Morphological Approach to Engineering Design, in Jones, J.Ch. and Thornley, D.G., *Conference on Design Methods*, Oxford: Pergamon Press 1963, pp. 115–140.

[444] Novak, J.D. and Goodwin, D.B., *Learning How to Learn*, New York: Cambridge University Press, 1984.

[445] Novak J.D., The Theory Underlying Concept Maps and How to Construct Them, 2000, http://cmap.coginst.uwf.edu/info/

[446] Novodvorsky, I., Constructing a Deeper Understanding, *The Phys. Teach.*, Vol. 35, 1997, pp. 242–245.

[447] Oglesby, S. Jr. and Nichols, G.B., *Electrostatic Precipitation*, New York: Marcel Dekker, 1978.

[448] Osborn, F.E., *Applied Imagination—Principles and Procedure of Creative Thinking*, New York: Scribner's, 1953.

[449] Ott, H., *Vorlesungen: Maschinenelemente* (Lecture Notes: Machine Elements), Zürich: ETH, 1980.

[450] Ott, H. and Hubka, V., Vorausberechnung der Herstellkosten von Maschinenteilen beim Entwerfen (Pre-Calculation of Manufacturing Costs of Machine Parts during Layout), *Schw. Masch.*, Vol. 83, No. 31, 1983.

[451] Ott, H. and Hubka, V., Berechnung der Herstellkosten von Schweissteilen (Calculation of Manufacturing Costs of Welded Parts), *Schw. Masch.*, Vol. 86, No. 7, 1986.

[452] Otto, K. and Wood, K., *Product Design*, Prentice-Hall, 2000.

[453] Ottosson, S., *When Time Matters*, plenary presentation at TMCE 2004 Lausanne, obtained from the author, 2004.

[454] Pahl, G., Klären der Aufgabestellung, *Konstruktion*, Vol. 24, No. 1, pp. 30–33, 1972.

[455] Pahl, G., Notwendigkeit und Grenzen der Konstruktionsmethodik (Necessity and Limitations of Design Methodology, in [311], 1990, pp. 15–30.

[456] Pahl, G., Ergebnisse der Diskussion (Results of the Discussions), in G. Pahl (ed.), *Psychologische und Pädagogische Fragen beim methodischen Konstruieren: Ergebnisse des Ladenburger Diskurses vom Mai 1992 bis Oktober 1993* (Psychological and Pedagogic Questions in Systematic Designing: Results of a Discourse at Ladenburger from May 1992 to October 1993), Köln: Verlag TÜV Rheinland, 1994, pp. 1–37.

[457] Pahl, G., Beitz, W., Feldhusen, J. and Grote, H-K., *Engineering Design* (3rd edn.), London: Springer-Verlag, 2007 (1st edn. 1984) (ed. and transl. K. Wallace and L. Blessing), translated from 2003 (5th edn.), of Pahl, G. and Beitz, W., Feldhusen, J. and Grote, H.-K. *Konstruktionslehre, Methoden und Anwendungen* (7th edn.), Berlin/Heidelberg: Springer-Verlag, 2007 (1st edn. 1977).

[458] Pahl, G., Transfer Ability as Educational Goal—Results of a Discourse at Ladenburg, in [153], 1996, pp. 133–138; and [316], 1995, pp. 247–252.

[459] Panitz, B., Briefings: Team Players, *ASEE PRISM*, December, 1997, p. 9.

[460] Papanek, V., *Design for the Real World*, Totonto: Bantam, 1972.

[461] Parker, G.M., *Cross-Functional Teams*, San Francisco, CA: Jossey-Bass, 1994.

[462] Perrow, C., A Framework for the Comparative Analysis of Organizations, *Am. Sociol. Rev.*, Vol. 32, 1967, pp. 194–208.

[463] Perry, R.H. and Green, D.W., *Perry's Chemical Engineers' Handbook* (7th edn.), New York: McGraw-Hill, 1997.

[464] Perry, W.G., *Forms of Intellectual and Ethical Development in the College Years: A Scheme*, New York: Holt, Rinehart, 1970.

[465] Perry, W.G., Jr., Cognitive and Ethical Growth: The Making of Meaning, in A. Chickering, and Assoc. (eds.), *The Modern American College*, San Francisco, CA: Jossey-Bass, 1981.

[466] Petroski, H., *To Engineer Is Human*, New York: St. Martin Press, 1985.

[467] Petroski, H., *The Pencil: A History of Design and Circumstance*, New York: Knopf, 1989.

[468] Phadke, M.S., *Quality Engineering Using Robust Design*, Englewood Cliffs, NJ: Prentice-Hall, 1989.

[469] Pinder, A., Model of Models, *Design Research* (Newsletter of the Design Research Society), No. 16, April, 1983.

[470] Polya, G., *How to Solve It*, Princeton, NJ: Princeton U.P., 1945.

[471] Porter, L.W. and Lawler, E.E., *Behavior in Organizatons*, New York: McGraw-Hill, 1974.

[472] Poser, H., Wissenschaft und Lehre—Wertfrei? Max Weber und die Ingenieurwissenschaften (Science and Education—Value-Free? Max Weber and the Engineering Sciences), in A. Melezinek (ed.), *Unique and Excellent—Proc. 29th Internationales Symposium Ingenieurpädagogik 2000*, Alsbach/Bergstraße: Leuchtturm-Verlag, 2000, pp. 47–54.

[473] Poser, H., Computergestütztes Konstruieren in philosophischer Perspektive (Computer Aided Design in Philosophical Perspective), in G. Banse and K. Friedrich (eds.), *Konstruieren zwischen Kunst und Wissenschaft* (Designing Between Art and Science), Berlin: rainer bohn verlag (edition sigma), 2000, pp. 275–287.

[474] Proulx, D., Brouillette, M., Charron, F. and Nicolas, J., A New Competency-Based Program for Mechanical Engineers, in *Proc. CSME Forum 1998*, Toronto, 1989.

[475] Pugh, S., Concept Selection—A Method that Works, in [294], 1981, pp. 497–506, and [490], 1990, pp. 73–82.

[476] Pugh, S., *Specification Phase, Curriculum for Design: Preparation Material for Design Teaching*, Loughborough: SEED Publishers, 1986.

[477] Pugh, S., Organising for Design in Relation to Dynamic/Static Product Concepts, in [28], 1989, pp. 313–334.

[478] Pugh, S., *Total Design: Integrated Methods for Successful Product Engineering*, Wokingham and Reading, MA: Addison-Wesley, 1991.

[479] Quirk, G.C., *Logic Design Factors*, Report No. R61 P006, Ordnance Department, Defense Electronics Division, General Electric, 1961.

[480] Reason, J.T., *Human Error*, New York: Cambridge University Press, 1990.

[481] Reichmann, S.W. and Grasha, A.F., A Rational Approach to Developing and Assessing the Construct Validity of a Student Learning Style Scales Instrument, *J. Psychol.*, Vol. 87, 1974, pp. 213–223.

[482] Rheinfrank, J.J., ASEE-DEED BULLETIN, Vol. 14, No. 1, 1989, p. 12.

[483] Rhodes, M., An Analysis of Creativity, *Phi Beta Kappan*, Vol. 42, 1961, pp. 305–310.

[484] Ridley, M., *Genome*, New York: Perennial (HarperCollins), 2000.

[485] Riitahuhta, A. (ed.), *WDK 25—Proc. ICED 97*, Tampere: Tampere University, 1997.

[486] Rockstroh, W., *Die technologische Betriebsprojektierung* (Technological Project-Engineering for Enterprises), Berlin: VEB Verlag, 1977.

[487] Rodenacker, W.G., *Methodisches Konstruieren* (Methodical Design, 4th edn.), Berlin/Heidelberg: Springer-Verlag, 1991 (1st edn., 1970).

[488] Rolfe, S.T. and Barsom, J.M., *Fracture and Fatigue Control in Structures*, Englewood Cliffs, NJ: Prentice-Hall, 1977.

[489] Romig, D.A., *Breakthrough Teamwork: Outstanding Results Using Structured Teamwork*, Chicago, IL: Irwin Professional, 1996.

[490] Roozenburg, N. and Eekels, J. (eds.), *WDK 17: EVAD—Evaluation and Decision in Design—Reading*, Zürich: Heurista, 1990.

[491] Roozenburg, N. (ed.), *WDK 22: Proc. ICED 93 Den Haag*, Zürich: Heurista, 1993.

[492] Roozenburg, N.F.M. and Eekels, J., *Product Design: Fundamentals and Methods*, Chichester: Wiley, 1995.

[493] Ropohl, G., *Eine Systemtheorie der Technik*, München: Carl Hanser, 1979.

[494] Ropohl, G., Die Wertproblematik in der Technik (Value Problems in Technology), in [490], 1990, pp. 162–182.

[495] Roseman, M.A. and Gerö, J.S., Purpose and Function in Design: From the Socio-Cultural to the Techno-Physical, *Des. Stu.*, Vol. 19, No. 2, 1989, pp. 161–186.

[496] Ross, P.R., *Taguchi Techniques for Quality Engineering*, New York: McGraw-Hill, 1988.

[497] Roth, K., Franke, H.J. and Simonek, R., Algorithmisches Auswahlverfahren zu Konstruktion mit Katalogen (Algorithmic Selection Procedure for Designing with Design Catalogs), *Feinwerktechnik*, Jg. 75, No. 8, 1975.

[498] Roth, K., *Konstruieren mit Konstruktionskatalogen* (Designing with Design Catalogs, 2nd edn., 2 vols.), Berlin/Heidelberg: Springer-Verlag, 1995 (1st edn., 1982).

[499] Ruiz, C. and Koenigsberger, F., *Design for Strength and Production*, London: Macmillan, 1970.

[500] Rutz, A., *Konstruieren als gedanklicher Prozeß*, München: Technical University, Thesis, 1995.

[501] Saariluoma, P, Nevala, K. and Karvinen, M., Content-Based Design Analysis, in J.S. Gero and N. Bonnardel (eds.), *Studying Designers '05, Key Center of Design Computing and Cognition*, University of Sydney, Sydney, Australia, 2005, pp. 213–228.

[502] Saariluoma, P., Nevala, K. and Karvinen, M., The Modes of Design Engineering Thinking, in *Computational and Cognitive Models of Creative Design, hi'05*, Heron Island, Australia, December 10–14, 2005.

[503] Salustri, F.A. and Parmar, J., Product Design Schematics: Structured Digramming for Requirements Engineering, in *Proc. International Design Conference—Design 2004*, Dubrovnik, 2004, pp. 1453–1460.

[504] Samuel, A., *Make and Test Projects in Engineering Design*, London: Springer-Verlag, 2006.

[505] Schmidt-Kretschmer, M. and Blessing, L., Strategic Aspects of Design Methodologies: Understood or Underrated?' in *Proc. International Design Conference—Design 2006*, Dubrovnik, 2006, pp. 125–130.

[506] Schön, D.A., *The Reflective Practitioner: How Professionals Think in Action*, New York: Basic Books, 1983.

[507] Schön, D.A., *Educating the Reflective Practitioner: Towards a New Design for Teaching and Learning in the Professions*, San Francisco, CA: Jossey-Bass, 1987.

[508] Schregenberger, J.W., Erfolgreicher konstruieren—aber wie? (Designing More Successfully—But How?), *Schw. Masch.*, Vol. 86, No. 23, 1986, pp. 46–49.

[509] Seidenschwarz, W., *Target Costing Marktorientiertes Zielkostenmanagement*, München: Vahlen, 1993.

[510] Sheldon, D., *Does Industry Understand and Adopt Design Science and Tools*, unpublished paper presented at ICED 97 Tampere [485], 1997.

[511] Sherriton, J. and Stern, J.L., *Corporate Culture, Team Culture: Removing the Hidden Barriers to Team Success*, New York: AMACOM, 1997.

[512] Shewhart, W.A., *Economic Control of Quality of Manufactured Product*, New York: van Nostrand, 1931.

[513] Shigley, J.E. and Mischke, C.R., *Standard Handbook of Machine Design* (2nd edn.), New York: McGraw-Hill, 1996.

[514] Shinno, H., Yoshioka, H., Marpaung, S. and Hachiga, S., Quantitative SWOT Analysis on Global Competitiveness of Machine Tool Industry, *J. Eng. Design*, Vol. 17, No. 3, June 2006, pp. 251–258.

[515] Slusher, E.A., Ebert, R.J. and Ragsdell, K.M., Contingency Management of Engineering Design, in [28]. Paper IMechE C377/010, 1989, pp. 65–75.

[516] Smith, K.A., The Craft of Teaching Cooperative Learning: An Active Learning Strategy, in *Proc. Frontiers in Education Conference 1989*, Binghamton, New York, October 15–17, 1989, pp. 188–193.

[517] Smith, J. and Clarkson, P.J., Design Concept Modelling to Improve Reliability, *J. Eng. Design*, Vol. 16, No. 5, October 2005, pp. 473–492.

[518] Smithers, T., On Knowledge Level Theories of Design Process, unpublished, extended version of paper in Gerö, J.S. and Sudweeks, F., *Artificial Intelligence in Design '96*, Academic Press, 1999.

[519] Sonntag, R.E. and Van Wylen, G.J., *Introduction to Thermodynamics: Classical and Statistical*, New York: Wiley, 1971.

[520] Spooner, A. (ed.), *The Oxford Minireference Thesaurus*, Oxford: Clarendon Press, 1992.

[521] Spotts, M.F., *Design of Machine Elements* (6th edn.), Englewood Cliffs, NJ: Prentice-Hall, 1985.

[522] Starfield, A.M., Smith, K.A. and Bleloch, A.L., *How to Model It: Problem Solving for the Computer Age*, New York: McGraw-Hill, 1990.

[523] Starr, M.K., *Product Design and Decision Theory*, Englewood Cliffs, NJ: Prentice-Hall, 1963.

[524] Stegmüller, W., *Wissenschaftliche Erklärung und Begründung* (Scientific Explanation and Substantiation), Berlin: Springer, 1973.

[525] Steward, D.V.,*Using the Design Structure Method*, Washington, DC: NSF Report, 1993.

[526] Storga, M., Traceability in Product Development, in *Proc. International Design Conference—Design 2004*, Dubrovnik, 2004, pp. 911–918.

[527] Strasser, C. and Grösel, B., A Landscape of Methods—A Practical Approach to Support Method Use in Industry, in *Proc. International Design Conference—Design 2004*, Dubrovnik, 2004, pp. 1167–1172.

[528] Suh, N.P., *Principles of Design*, New York: Oxford University Press, 1989.

[529] Swan, B.R. et al., A Preliminary Analysis of Factors Affecting Engineering Design Team Performance, in *Proc. 1994 ASEE Annual Conference*, Washington, DC: ASEE, 1994, pp. 2572–2589.

[530] Taguchi, G., *Introduction to Quality Engineering: Designing Quality into Products and Processes*, White Plains, NY: Quality Resources (Asian Productivity Organisation), 1986.

[531] Taguchi, G., *Taguchi on Robust Technology Development: Bringing Quality Engineering Upstream*, New York: ASME Press, 1993.

[532] Talukdar, S., Sapossnek, M., Hou, L., Woodbury, R., Sedas, S. and Saigal, S., *Autonomous Critics*, Report EDRC 18-13-90, Pittsburgh, PA: The Engineering Design Research Center, Carnegie-Mellon University, 1990.

[533] Talukdar, S. and Christie, R., *An Extended Framework for Security Assessment*, Report EDRC 18-15-90, Pittsburgh, PA: The Engineering Design Research Center, Carnegie-Mellon University, 1990.

[534] Tavcar, J., Benedidic, J., Duhovnik, J. and Zavbi, R., Creativity and Efficiency in Virtual Product Development Teams, in *Proc. TMCE 2004 Lausanne*, 2004, pp. 425–434, and on CD-ROM.

[535] Taylor, F.W., *Principles of Scientific Management*, New York: Harper, 1919.

[536] Terninko, J., Zusman, A. and Zlotin, B., *Systematic Innovation: An Introduction to TRIZ* (Theory of Inventive Problem Solving), Boca Raton, FL: St. Lucia Press, 1998.

[537] Terry, G.J., A Chart System to Help Designers, *Chart. Mech. Eng.*, Vol. 15, No. 2, 1968, pp. 56–59.

[538] Thevenot, H.J. and Simpson, T.W., Commonality Indices for Product Family Design: A Detailed Comparison, *J. Eng. Design*, Vol. 17, No. 2, April 2006, pp. 99–119.

[539] Thomson, V. and Graefe, U., *CIM—A Manufacturing Paradigm*, National Research Council Canada, Division of Mechanical Engineering Report DM-6, NRC No. 26198, 1968.

[540] Tjalve, E., Formulierung der Konstruktionsziele, *Schw. Masch.*, Vol. 77, No. 36, 1977.

[541] Tjalve, E., Andreasen, M.M. and Schmidt, F.F., *Engineering Graphic Modelling*, London: Butterworths, 1979.

[542] Tjalve, E., *A Short Course in Industrial Design*, London: Newnes-Butterworths, 1979.

[543] Tomiyama, T., A Design Process Model that Unifies General Design Theory and Empirical Findings', in *Proc. 1995 ASME Design Engineering Technical Conference*, Vol. 2, 1995, pp. 329–340.

[544] Tsourikov, V., Inventive Machine: Second Generation, *AI and Soc.*, Vol. 7, No. 1, 1993, pp. 62–78.

[545] Tuomaala, J., Creative Engineering Design—Summary of a Book, in [153], 1996, pp. 23–33.

[546] Tuttle, S.B., *Mechanisms for Engineering Design*, New York: Wiley, 1967.

[547] Tyler, L.E., *Individuality: Human Possibilities and Personal Choice in the Psychological Development of Men and Women*, San Francisco, CA: Jossey-Bass, 1978.

[548] Ullman, D.G., *The Mechanical Design Process*, (3rd edn) New York: McGraw-Hill, 2003 (1st edn., 1992).

[549] Underwood, A., *TQM Handbook on Total Quality Management* (Canadian Aerospace Industry), Ottawa: Industry Canada, 1993.

[550] Upton N. and Yates, I., Putting Design Research to Work, in [29], Vol. 4, 2001, pp. 51–58.

[551] Vajna, S., Clement, S., Jordan, A and Bercsey, T., The Autogenic Design Theory: An Evolutionary View of the Design Process, *J. Eng. Design*, Vol. 16, No. 4, 2005, pp. 423–440.

[552] Vajna, S., Edelmann-Nusser, J., Kittel, K. and Jordan, A., Optimization of a Bow Riser Using the Autogenic Design Theory, in *Proc. TMCE 2006 Ljubljana*, 2006, pp. 593–601.

[553] Vance, D. and Eynon, J., On the Requirements of Knowledge Transfer Using Information Systems: A Schema Whereby Such Transfer is Enhanced, in E. Hoadley and Benbaast (eds.), *American Conference on Information Systems*, MD: Association of Information Systems, 1998, pp. 632–634.

[554] Vesely, W.E., Goldberg, F.F., Roberts, N.H. and Haasl D.F., *Fault Tree Handbook*, NUREG-0492, Washington, DC: US Nuclear Regulation Commission, 1981.

[555] Vickers, Sir G., *Towards a Sociology of Management*, London: Chapman & Hall, 1967.

[556] Vincenti, W.G., *What Engineers Know and How They Know It—Analytical Studies from Aeronautical History*, Baltimore, MD: Johns Hopkins University Press, 1990.

[557] Volk, H., Führen nach dem Verhaltensgitter (Leading According to the Behavior Grid), *Tech. Rundschau*, Vol. 51, 1984, and [310], 1985, pp. 71–73.

[558] von Fange, E., *Professional Creativity*, Englewood Cliffs, NJ: Prentice-Hall, 1959.

[559] Waldrop, M.M., *Complexity the Emerging Science at the Edge of Order and Chaos*, New York: Touchstone, 1993, pp. 312.

[560] Wales, C.E., Nardi, A.H. and Stager, R.A., *Professional Decision-Making*, Morgantown, WV: Center for Guided Design (West Virginia University), 1986.

[561] Wales, C., Nardi, A. and Stager, R., *Thinking Skills: Making a Choice*, Morgantown, WV: Center for Guided Design (West Virginia University), 1986.

[562] Wallace, P.J., *The Technique of Design*, London: Pitman, 1952.

[563] Wallas, G., *The Art of Thought*, London: Cape, 1926 (reprint 1931), pp. 79–96 (Reprinted in P.E. Vernon (ed.), *Creativity*, Harmondsworth: Penguin, 1970, pp. 91–97).

[564] Wankel, F., *Rotary Piston Machines*, London: Iliffe, 1965.

[565] Warfield, J.N., *Societal System: Planning, Policy and Complexity*, New York: Wiley, 1976.

[566] Warfield, J.N., *A Science of Generic Design*, Salinas, CA: Intersystems, 1989.

[567] Warnock, M., *An Intelligent Person's Guide to Ethics*, London: Duckworth, 2001.

[568] Watson, G.H., *The Benchmarking Workbook: Adapting Best Practices for Performance Improvement*, Cambridge, MA: Productivity Press, 1992.

[569] Watson, S.R. and Buede, D.M., *Decision Synthesis: The Principles and Practice of Decision Analysis*, Cambridge: University Press, UK, 1987.

[570] Watts, I., *The Improvement of the Mind*, London: Gale and Curtis, 1810.

[571] Wearne, S.H., *Principles of Engineering Organisation*, London: Edward Arnold, 1973.

[572] Wearne, S.H., A Review of Reports of Failures, *Proc. I Mech. E*, Vol. 193, 1979, pp. 125–136.

[573] Weber, C. and Vajna, S., A New Approach to Design Elements (Machine Elements), in [485], Vol. 3, 1997, pp. 685–690.

[574] Weber, C., Steinbach, M., Botta, C. and Deubel, T., Modelling of Product-Service Systems (PSS) Based on the PDD Approach, in *Proc. International Design Conference—Design 2004*, Dubrovnik, 2004, pp. 547–554.

[575] Weber, C., Lecture Presentation, sections SME00–SME06 and ME0102, 2005, http://www.cad.uni-saarland.de/index.html?rubrik=lehre/vorlesung&site=vorlesung.

[576] Weber, C., CPM/PDD—An Extended Theoretical Approach to Modelling Products and Product Develpoment Processes, in *Proc. PhD 2005*, November 7–9, 2005, Srni, Czech Republic, 2005, pp. 11–28.

[577] Weinbrenner, V., *Produktlogik als Hilfsmittel zum Automatisieren von Varianten- und Anpassungskonstruktionen* (Product Logic as Aid to Automation of Variant and Adaptation Design), München: Hanser, 1994.

[578] Weiss, M.P. and Gilboa, Y., More on Synthesis of Concepts as an Optimal Combination of Solution Principles, in *Proc. International Design Conference—Design 2004*, Dubrovnik, 2004, pp. 83–90.

[579] Wickelgren, W.A., *How to Solve Problems*, San Francisco, CA: Freeman, 1974.

[580] Wilson, C.C., Kennedy, M.E. and Trammell, C.J., *Superior Product Development*, Cambridge, MA, and Oxford, UK: Blackwell Publishers, 1996.

[581] Wilson, C.E., Sadler, J.P. and Michels, W.J., *Kinematics and Dynamics of Machinery*, New York: Harper & Row, 1983.

[582] Wimmer, W. and Züst, R., *Ecodesign Pilot*, Dordrecht: Kluwer, 2003.

[583] Woods, D.R., *Problem-Based Learning: How to Gain the Most from PBL*, Waterdown, ON: D.R. Woods, 1994.

[584] Woodson, W.E., Tillman, B. and Tillman, P., *Human Factors Design Handbook* (2nd edn.), New York: McGraw-Hill, 1992.

[585] Wörgerbauer, H., *Die Technik des Konstruierens*, München: Oldenbourg, 1943.

[586] Wundt, W., *Grundzüge der physiologischen Psychologie* (Basic Concepts of Physiological Psychology), Bd. 3, Leipzig, 1903.

[587] Yeats, D.E., *High-Performing Self-Managed Workteams: A Comparison of Theory to Practice*, Thousand Oaks, CA: Sage Publications, 1997.

[588] Yoshikawa, H., General Design Theory: Theory and Application, in *Conference on CAD/CAM Technology in Mechanical Engineering*, Cambridge, MA: MIT, pp. 370–376, 1981.

[589] Yoshikawa, H., General Design Theory and a CAD System, in T. Sata and E.A. Warman (eds.), *Man–Machine Communication in CAD/CAM, Proc. IFIP W.G. 5.2/5.3 Working Conference*, Amsterdam: North Holland, pp. 35–58, 1981.

[590] Yoshikawa, H., Scientific Approaches in Design Process Research, in [294], 1981, pp. 323–329.

[591] Yoshikawa, H., Designer's Designing Models, in [298], 1983, pp. 338–344.

[592] Yoshimura, M. and Papalambros, P.Y., Kansei Engineering in Concurrent Product Design: A Progress Review, in *Proc. TMCE 2004 Lausanne*, 2004, pp. 177–186, and on CD-ROM.

[593] Yoshioka, M. and Tomiyama, T., Towards a Reasoning Framework of Design as Synthesis, in *Proc. 1999 ASME Design Engineering Technical Conference*, paper DETC99/DTM-8743, 1999.

[594] Young, W.C. and Budnyas, R., *Roark's Formulas for Stress and Strain* (7th edn.), New York: McGraw-Hill, 2001, http://www.UTS.com/TheUltimateReference.

[595] Zangemeister, C., *Nutzwertanalyse in der Systemtechnik* (3rd edn.), München: Wittemann, 1973.

[596] Zaretsky E.V. (ed.), *STLE Life Factors for Rolling Bearings*, Park Ridge, IL: Society of Tribologists and Lubrication Engineers, 1992.

[597] Zwicky, F., *The Morphological Method of Analysis and Construction, Courant Anniversary Vol.*, New York: Wiley-Interscience, 1948.

[598] Zwicky, F., *Entdecken, Erfinden, Forschen im morphologischen Weltbild* (Discovering, Inventing, Researching in the Morphological World View), Zürich: Droemer Knaur, 1966.

Index